Thermodynamics, Microstructures and Plasticity

NATO Science Series

A Series presenting the results of scientific meetings supported under the NATO Science Programme.

The Series is published by IOS Press, Amsterdam, and Kluwer Academic Publishers in conjunction with the NATO Scientific Affairs Division

Sub-Series

I. **Life and Behavioural Sciences**	IOS Press
II. **Mathematics, Physics and Chemistry**	Kluwer Academic Publishers
III. **Computer and Systems Science**	IOS Press
IV. **Earth and Environmental Sciences**	Kluwer Academic Publishers
V. **Science and Technology Policy**	IOS Press

The NATO Science Series continues the series of books published formerly as the NATO ASI Series.

The NATO Science Programme offers support for collaboration in civil science between scientists of countries of the Euro-Atlantic Partnership Council. The types of scientific meeting generally supported are "Advanced Study Institutes" and "Advanced Research Workshops", although other types of meeting are supported from time to time. The NATO Science Series collects together the results of these meetings. The meetings are co-organized bij scientists from NATO countries and scientists from NATO's Partner countries – countries of the CIS and Central and Eastern Europe.

Advanced Study Institutes are high-level tutorial courses offering in-depth study of latest advances in a field.
Advanced Research Workshops are expert meetings aimed at critical assessment of a field, and identification of directions for future action.

As a consequence of the restructuring of the NATO Science Programme in 1999, the NATO Science Series has been re-organised and there are currently Five Sub-series as noted above. Please consult the following web sites for information on previous volumes published in the Series, as well as details of earlier Sub-series.

http://www.nato.int/science
http://www.wkap.nl
http://www.iospress.nl
http://www.wtv-books.de/nato-pco.htm

Series II: Mathematics, Physics and Chemistry – Vol. 108

Thermodynamics, Microstructures and Plasticity

edited by

Alphonse Finel
ONERA-LEM,
Châtillon-sous-Bagneux, France

Dominique Mazière
CEA-INSTN Saclay,
Gif-sur-Yvette, France

and

Muriel Veron
INPG-ENSEEG,
St Martin d'Hères, France

Kluwer Academic Publishers

Dordrecht / Boston / London

Published in cooperation with NATO Scientific Affairs Division

Proceedings of the NATO Advanced Study Institute on
Thermodynamics, Microstructures and Plasticity
Fréjus, France
September 2-13, 2002

A C.I.P. Catalogue record for this book is available from the Library of Congress.

ISBN 1-4020-1367-1 (HB)

Published by Kluwer Academic Publishers,
P.O. Box 17, 3300 AA Dordrecht, The Netherlands.

Sold and distributed in North, Central and South America
by Kluwer Academic Publishers,
101 Philip Drive, Norwell, MA 02061, U.S.A.

In all other countries, sold and distributed
by Kluwer Academic Publishers,
P.O. Box 322, 3300 AH Dordrecht, The Netherlands.

Printed on acid-free paper

TABLE OF CONTENTS

FOREWORD

Physical properties of materials rely on phenomena of various space and time scales ranging from nanometers to centimeters and from femtoseconds to seconds respectively. This immediately brings one face to face with the fundamental difficulty of establishing a connection between material behavior at the microscopic level, where understanding is to be sought, and macroscopic properties that need to be predicted. Among these, plastic deformation and microstructures evolution are probably the most challenging ones.

This book collects all the papers that have been presented as lectures and seminars during the NATO Advanced Study Institute entitled "Thermodynamics, Microstructures and Plasticity" held in Fréjus, France, September 2-13 2002. The aim of the ASI was precisely to assess the state of the art in both phase transformations and plasticity, with a particular emphasis on the interplay between the two fields. It was also our purpose to facilitate contacts, exchange of ideas and to initiate collaborations between the physicists belonging to the two communities and who have only scarce occasions to meet.

The papers provide a complete assessment of the various theoretical and numerical methods that are currently in use to investigate microstructural transformations and mechanical properties of inhomogeneous systems, from the atomic scale to the macroscopic one: kinetic mean field theories, Monte Carlo and molecular dynamics simulations, Ginzburg-Landau and phase field methods as applied to plasticity and microstructures transformations, discrete and stochastic dislocation dynamics, cluster dynamics. Overviews of major physical processes and properties were extensively presented: solidification, microstructural evolution in single and polycrystalline systems under internal and applied stresses, high temperature plasticity, recrystallisation, large plastic strain in multiphase systems, fatigue, fracture, diffusive transformations, fine grained materials.

In addition to lectures and seminars, various working groups on advanced or open topics and two poster sessions were organized. These informal sessions promoted very effective exchanges and discussions between the participants.

Our deepest appreciation goes to Michèle Cottin, who had in charge most of the administrative aspects of the organization, and who did not spare her time to insure the smooth running of the ASI and the preparation of the proceedings.

Finally, we are deeply grateful to the NATO Scientific Affairs Division, which provided most of the financial support for organizing the meeting, and to CEA, ONERA, INPG, IRSID, DGA for their additional and significant contributions.

Saclay, February 5th, 2003

A. Finel, D. Mazière, M. Véron.

I.   Introduction

INTRODUCTION TO MODELLING TECHNIQUES IN DIFFUSIVE PHASE TRANSFORMATIONS

Y.J.M. BRECHET
L.T.P.C.M., Institut National Polytechnique de Grenoble
Domaine Universitaire de Grenoble
Rue de la Piscine, BP75
38402 Saint Martind'Heres Cedex, France
email: yves.brechet@ltpcm.inpg.fr

Abstract:

An overview for the motivations for physically based model of microstructure evolution, and of the analytical tools available for the case of diffusive phase transformations is proposed. In the present paper, the classical solution of diffusion equations are recalled, both for parabolic growth and for self preserving shape propagating at constant velocity. Their application to some examples of interface mediated phase transformations are given. On case is discontinuous precipitation where a self organised pattern is selected via the cooperative motion of interfaces and interphases. The other case is the modelling of Bainite as a phase transformation kinetically controlled by the diffusion of carbon away from growing ferrite needles into parent austenite. In the companion paper by G.Inden, thermodynamic aspects and multicomponent systems are investigated.

3

A. Finel et al. (eds.), Thermodynamics, Microstructures and Plasticity, 3–23.
© 2003 *Kluwer Academic Publishers. Printed in the Netherlands.*

4

1. Introduction

The study of solid state phase transformations in metals and alloys is traditionnally divided in two main streams: the displacive transformations in which atoms move in a cooperative manner at a velocity approaching the one of sound waves, the diffusive transformations in which atoms move in a non cooperative manner, by diffusion processes. The emphasis laid upon various aspects of phase transition is different in the two cases. The distinction is certainly not as clear cut as it sounds (see for instance years of controversy concerning Bainite) but it remains useful at least to set limiting cases.

For the martensitic transformation, crystallography and back stresses generated during the transformation are a central issue, the thermodynamics of the problem enters mainly into the conditions for nucleation of the new phase, and since the propagation of the transformation is very rapid, the kinetics of invasion during continuous cooling is controlled by the possibility of repeated nucleation. The patterns emerging from these transformations reflect both the crystallographic constraints and the elastic interactions between different variants.

For the diffusive transformation, the central role in modelling is given to mass transport via diffusion processes. The scale of the microstructure results from the competition between available free energy (driving force) and interfacial energy. The kinetics results from diffusive transport, and limited mobility of the interfaces. Although crystallography and elastic stresses may play a role (see for instance the contribution of Khachaturyan in these proceedings [1]), they are not central to the main issue : transformation kinetics and morphology of the reaction products.

The present contribution is limited to diffusive phase transformations. We aim at presenting the different techniques classically used in the modelling of diffusive phase transformations in the solid state. This field presents striking (and still relatively unexplored) similarities with solidification (see for instance the paper by Karma in these proceedings [2]) with important differences : convection transport in the solid state is negligible, diffusion in the parent and product phase are within the same order of magnitude, diffusion in the bulk may be frozen whereas diffusion in the interface still operates, diffusion of different elements may differ by orders of magnitude...A systematic comparison between solid state reactions and solidification is still to be written.

The study of solid state transformations can be done for its own sake. It can also be attempted in relation with the consequences of the obtained microstructures on macroscopic properties. Since this school happily combines specialists of mechanical properties and of phase transformations, we have attempted in this paper to provide motivations for both communities.

The paper is structured as follows: Section 2 will outline the motivations for such an approach, and depending on the properties of interest, it will delimitated the level of precision required for the description of the microstructures and the limitation of purely phenomenological approach to phase transformation kinetics. Section 3 will propose a classification of the various problems encountered in solid state diffusive transformations: the well known precipitation kinetics from a supersaturated solution which will be treated elsewhere in this school will be only briefly recalled, and the emphasis will be laid upon the transformation mediated by a moving interface. Section 4 will propose an overview of the theoretical tools necessary for a continuum modelling of the kinetics of phase transformations. We will start by the evaluation of the available free energy, for which thermodynamic models are required. Then classical solutions of the diffusion equations, frequently used in modelling phase tranformations, will be recalled. The specificities of ternary systems , especially with respect to the interfacial concentrations , will be identified. Section 5 will illustrate on a serie of worked examples, application of these tools and Section 6 will conclude with open questions and possible input of new computer based modelling techniques.

2. Motivation for a physically based modelling

Metals and alloys have a remarquable (and sometimes annoying...) ability to change some of their properties (yield strength, formability, ductility, toughness...) simply via different thermomechanical treatments. This ability stems from the variety of microstructures which emerges from these treatments depending on the process parameters (Temperature, Cooling rate, Deformation and Deformation rate...). We focuss in this paper on the microstructures associated with new phases. Most of "useful" phase transitions in metallurgy are first order, meaning that they lead to coexisting phases. Most of transformation conditions are far from equilibrium, meaning that patterning involving important interface quantitites are a natural occurence. The most obvious quantity to be measured and modelled is the transformed volume fraction. It invariably exhibits a sigmoïdal shape as function of time as can naturally be expected from a process which can take place only at positions where it has not occured yet. Therefore a simple phenomenological description of this quantity is readily available [3]. Unfortunately it describes only the fraction transformed, and since its physical basis is non transparent, its generalisation outside the range of experimental parameters where the phenomenological constants have been measured is risky. The next two sections will emphasise these limitations and set the motivation for a more physically based approach of diffusive phase transformations.

2.1. THE NEED FOR A RELATION BETWEEN PROCESS PARAMETERS, ALLOY COMPOSITION AND PROPERTIES: MICROSTRUCTURAL FEATURES AS A NECESSARY INTERMEDIATE

The quantities the engineer is interested in are macroscopic properties (we will limit ourselves here to structural materials and therefore to mechanical properties). The lever he has access to are the composition and the process parameters. The relation between these two groups of observables is by no way straightforward. Mechanical metallurgy aims at using the microstructure resulting from processing to understand the macroscopic propeties. Physical metallurgy aims at predicting these microstructures from the process parameters. Both are required to have a science driven alloy design and optimisation [4]. Figure 1 shows a classical example of the dependance of mechanical properties (here the yield stress of a 7000 Aluminium alloy used in aeronautics) with respect to two process parameters: quenching rate and duration of precipitation heat treatment [5]. These alloys are first given a high temperature heat treatment above the solubility limit in order to obtain a solid solition of all the major elements. Then they are quenched in order to retain a metastable solid solution, and then a further heat treatment is performed in order to harden the alloy through a fine dispersion of precipitates. While the duration of the heat treatment increases, the hardness goes through a maximum although the volume fraction of precipitates remains roughly cosntante (as can be experimentally proven by TEM observations or SAXS). The only microstructural parameter which evolves significantly along the heat treatment is the size R of precipitates, which increases via a coarsening process. Theory of structural hardening [6] tells us that the yield stress resulting from dislocation pinning by precipitates is given by the following formulae, depending if the precipitates are sheared (below a critical radius) or bypassed (above a critical radius).

$$\Delta\sigma = K.(f.R)^{1/2}$$

$$\Delta\sigma = K\frac{f^{1/2}}{R}$$

The coarsening of precipitates with ageing therefore explains the non monotomic behaviour of the yield stress. Obviously a simple description of the transformed volume fraction f would not be able to do so.

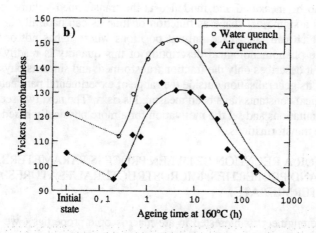

Figure 1: Hardening behaviour of a 7000 Aluminium alloy via precipitation [5]

The difference between water quench and air quench can also be explained by different scales of the microstructure: during air quench, the slow cooling rate leaves time for precipitation to occur at higher temperature, on heterogeneous nucleation sites (grain boundaries , other precipitates) . These precipitates are coarser and less efficient for structural hardening, and in addition they deprive the solid solution from elements which would otherwhise contribute to the fine precipitation during heat treatment. The total fraction of precipitates is the same as before, but their scale has a wide bimodal distribution, and this is responsible for the less efficient hardening effect: again focussing on the mere volume fraction would be insufficient to understand this large effect of the quenching rate.

Another example concerns the stability of grain structure: usually small grains are a positive feature in metallic alloys. But small grains store a substantial amount of surface energy which drives grain growth. A way of preventing this process is to pin grain boundaries by precipitates , just as one pins dislocations [7]. This effect, called "Zener Pinning" leads to a maximum grain size corresponding to the situation where capillary forces exactly balance pinning forces by precipitates. The derivation of this saturation grain size by Zener is a masterpiece of elegant simplicity. It leads to the following expression:

$$R_{max} \propto \frac{f}{r}$$

Again, both the volume fraction and the size of pinning centers matters. In addition, the elements entering the pinning precipitates are usually slow diffusers (precipitates have to be stable at high temperature) and therefore they retain often their segregation profiles following solidification, even during the further homogeneisation treatments. The distribution of precipitates is therefore spatially heterogeneous and this additionnal microstructural feature matters when the prediction of the stable ultimate grain size is of interest.

Mechanical metallurgy has investigated much more than the yield stress. Further hardening during deformation stems both from plastic incompatibilities between different phases and from dislocation storage. The fist contribution leads to a kinematic hardening and is responsible for the so called "composite effect"[8]. The standard (and efficient) tool to model this effect is the classic "Eshelby inclusion theory": size is irrelevant, only the shape and the volume fraction matters.

The second term (dislocation storage around bypassed precipitates) has been simply and elegantly approached by Ashby [9] with the concept of "geometry necessary dislocations". The density of stored dislocations is proportionnal to strain, and depends on both volume fraction and size of precipitates (and also in addition , on their shape)

$$\rho = \frac{\varepsilon r}{b}.(2\pi r)\frac{3f}{4\pi r^3} \propto \frac{f\varepsilon}{rb}$$

A last example is the modelling of the ductility of a ductile matrix containing hard particles where cavities responsible for ductile fracture nucleate. The model proposed by Embury and Brown [10] states that the strain leading to the coalescence of cavities decreases with increasing volume fraction, irrespectively of the particle size. By contrast, the strain for cavities to nucleate (either by particle fracture or particle decohesion) is smaller when the particles are larger [11]. Understanding ductility and formability requires therefore an understanding of both the volume fraction, shape, and spatial distribution of nucleation sites for cavities.

These examples , selected only on simple mechanical properties of metallic alloys, clearly identify situations for which a more detailed description of the microstructure than the mere volume fraction is required. Conversely, for people interested in mechanical properties, a proper heat treatment may allow to vary independantly size, shapes and volume fractions to test micromechanics models. Cast alloys are especially promising with this respect for studies in fracture mechanics.

2.2. THE LIMITS OF EMPIRICAL APPROACHES

Some empirical methods have been derived for modelling the transformation kinetics , aiming at predicting the volume fraction for a given time and a given temperature [3]. The input of these approaches are a nucleation rate, a growth rate,geometrical characteristics of the transformation, and hypothesis on the nature of nucleation sites. These approaches lead to transformation kinetics of the type Johnson Mehl Avrami Kolmogorov.

$$f(t) = 1 - \exp(-Kt^n)$$

A number of situations have been explored using these approaches , leading to different exponents. For instance, diffusionnal growth of spherical precipitates from a constant number of sites would give a 3/2 exponent, whereas the exponent would be 5/2 if a contant nucleation rate is assumed. Needle growth with a constant nucleation rate would give an exponent of 2. These approaches have two main limitations. From a given exponent , one cannot deduce a unique mechanism. Moreover the approach provides a description of the volume fraction with no information an the scale which results from the transformation. Other limitations dealing with possible evolution of the driving force while transformation occurs will be treated in the paper by Inden in these proceedings.

3. Classification of problems

3.1. PRECIPITATION AND INTERFACE MEDIATED TRANSFORMATIONS

Beside the classical displacive / diffusive transformation classification, a number of other classifications are useful for the selection of appropriate modelling tools for phase transformation. One is led to distinguish between continuous and interface controlled transformations. In continuous transformations [12], such as precipitation, the supersaturated solid solution evolves continuously in composition while transformation proceeds. In discontinuous transformations [13] , such as eutectoïd transformation, discontinuous precipitation or massive transformation, reaction takes place at a moving front (often initiated at preexisting grain boundaries) , and solid solution is progressively invaded by the reaction front.
 In precipitation reaction, it is usual to distinguish three stages, although the limits between these stages is somewhat fuzzy [12]. At the beginning of the decomposition, the driving force is sufficient to create a number of germs whose size is smaller when the driving force is larger. When this step proceeds, the supersaturation diminishes and

thus the driving force decreases so that only the existing germs can grow further. The number of precipitates remains constant and their radius scales as $t^{1/2}$. Such models present a remarquable efficiency, especially knowing their simplicity. The classic work by Aaronson et al. on allotriomorph ferrite [14] is shown in Figure 2. When such a simple model give such a good description, it seems pointless to search for more refined approaches.

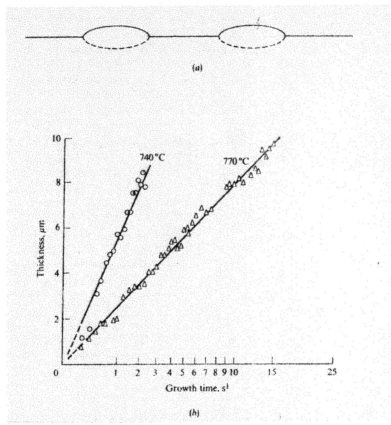

Figure 2: Diffusionnal growth of allotriomorph ferrite [14]

The growth of precipitates ultimately leads to a situation where the volume fraction of precipitates is very close to equilibirum (up to the capillary correction due to Gibbs Thomson effect). The only remaining driving force is the reduction in interface area, and only the larger precipitates can grow at the expense of the smaller ones: the system enters the coarsening regime during which the average radius scales as $t^{1/3}$ whereas the number of precipitates decreases [15].

These precipitation reactions span scales between nanometers to micrometers. They have been extensively studied by physicists [12,16,17], both using classical analytical tools (for nucleation kinetics, for growth kinetics, and for asymptotic coarsening regime), cluster dynamics (see the paper by Sigli in this volume [18]) , and Monte Carlo simulations (see the papers by Soissons [19], Pontikis [20] and Bellon [21]). No further details will be given here on these modelling tools. The current issues for precipitation is the validity of nucleation theory for real systems, and in particular the possible role of intermediate metastable phases, the extension of classical models to alloys with more than two elements, and the behaviour during non isothermal treatments possibly leading to precipitate reversion.

Interface mediated reactions [13] have been comparatively much less studied by the physicist community, in spite or because of a rich taxonomy and variety of phenomena. Among these transformations, the most classical example come from steel metallurgy whose richness is far from being fully understood. At high temperature, carbon steels are fcc (austenite) and they are bcc (ferrite) at low temperature. The tranformation from austenite to ferrite is associated with Carbon rejection, either in the parent austenite, or by decomposition of the product ferrite. Most of these reactions take place via an interface migration. The austenite to ferrite transition from the grain boundaries leads , depending on the driving force, to the grain boundary ferrite, or to the Wiedmanstatten ferrite which is the solid state equivalent of a dendritic growth. When carbon composition is sufficient, the transformation can lead to a lamellar product called perlite, where a set of parallel lamellae of ferrite and cementite follows the migration of an interface into parent austenite. At lower temperature, the morphology becomes needlelike, and the observed microstructure is called Bainite, with cementite precipitates either at the ferrite / austenite interface, or inside the ferrite itself. Another type of reaction called discontinuous precipitation occurs when decomposition of a supersaturated solution occurs in the wake of a moving grain boundary, leaving behind a set of lamellar precipitates separated by the same solid solution less supersaturated. The scales observed for these interface migration generated stuctures are usually of the order of the micrometer, and thus , at least for the growth regime, the usual modelling tools are analytical solutions of diffusion equations. Nucleation of these reactions is still very poorly understood. The final stages of the reaction occur either because of a progressive modification of the parent phase from diffusion from the product phase (soft impigement) , or simply because of the diminishing space to be invaded by the reaction (hard impigement). Two examples of such interface migration mediated reactions will be treated as examples in section 4.

3.2. CAPILLARY STABILITY OF THE MICROSTRUCTURE

In the previous examples, apart from the late stages of precipitation, the driving force for microstructural evolution is the available chemical free energy. As a rule of thumb, the higher is the driving force, the more interfaces the system can afford, and the finer will be the microstructure. The variety of shapes and self organised morphologies nature offers in solid state transformations is a consequence of the conditions which generates them which is often very far from thermodynamic equilibrium. As a consequence, when the reaction is completed, there is still a subtantial amount of interfaces to get rid of in order to reach thermodynamic equilibrium. This is the origin of the so called "capillary driven instabilities" of the microstructure. The most well known to physicists is the coarsening stage of precipitates [15], but again the zoology of observed phenomena is much richer and worth exploring.

Rod precipitates can be unstable through a mechanism known as Rayleigh instability [22]. As shown in figure 3,a pinching of a rod corresponds to an increase of one of the two radii of curvature together with the apparition of a negative curvature in the perpendicular direction. When the wavalength of the fluctuation is long enough, the net effect is to decrease the total curvature, resulting in a diffusion flow which destabilizes such a fluctuation. The theory of this phenomena was worked out by Nichols and Mullins for both bulk diffusion and interface diffusion. The "pinching rate" is exponential with time and the most unstable perturbation has a wavelength proportionnal to the rod radius. This results in the "spheroïdisation" of rod structures.

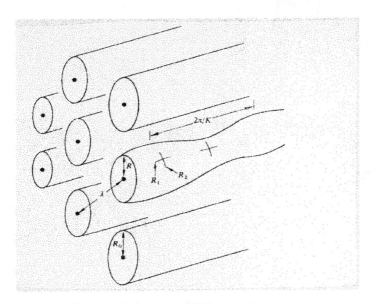

Figure 3: Spheroïdisation of a rod structure [22]

This is all fine and nice, but nature often chooses shorter path to reduce the total amount of surfaces: it takes advantage of existing defects such as rod branching (Fig.4) Rayleigh Mullins theory applied to infinite plates would predict no capillary instability. But plates also coarsen, and to do so, the instability is starting from the edges, via a Rayleigh instability!
Similar to discontinuous reactions , discontinuous coarsening [23] is another clever way nature finds to get rid of extra interfaces. It is even possible, supreme ingeniosity, to find systems in which several waves of discontinuous coarsening come one after the other (Figure 5) The simple theory of the problem has been worked out by Cahn et al. [24] under the unlikely assumption that capillary was the only driving force. Knowing that the lamellae between precipitates are still supersaturated, a chemical driving force has to be included in addition to the capillary term.

12

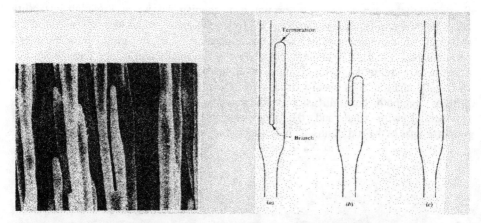

Figure 4: Defect initiated coarsening mechanisms of a rod structure [G5].

Figure 5: Discontinuous coasening reactions [23]

4. A "Tool Box" for Modelling Diffusive Phase Transformations

The variety of phenomena observed in solid state transformations is such that a number of tools have been developped. Precipitations has benefited from a very early stage from the development of atomistic simulations. More macroscopic techniques involving concentration waves on discrete grids have also been developped, leading to spectacular results , particularly for the investigation of elastic effects on the morphologies (see for instance Khatchaturyan in this volume [1]). We aim here at presenting the basic tools required for an analytic or a semianalytic model for diffusion controlled transformation kinetics.

4.1. THERMODYNAMICS : COMPUTING THE DRIVING FORCES

The driving forces for transformation are related to the supersaturation of the initial state. Phase diagrams are necessary to evaluate supersaturation, but computing these driving forces requires a more precise knowledge of the free energy curves as function of composition and temperature. A number of classical models can be used but very often the accurate estimation of the free energy curves, especially in multicomponent systems, requires the use of large thermodynamic databases such as THERMOCALC.

A detailed analysis of the content, the use and the reliability of these data is given in the companion paper by Inden [25].

In diffusion controlled transformations, the driving force is relevant either to model the nucleation step, or to model the growth regime when the mobility of the interface is considered as finite. For the solution of the diffusion problem, the question of concentrations at the interface becomes of central importance. Whereas for a binary alloy the standard hypothesis of local thermodynamic equilibirum defines uniquely the interface concentration, it is no longer the case for multicomponent systems when a range of possible operating conodes can be considered [25,4]. The situation has been studied in details for the case of ternary carbon steels Fe-C-X for which carbon diffusion vastly overcomes the diffusion of the other alloying elements [26]. Depending on the composition and the temperature, one is led to select either local equilibrium with partioning of X, or local equilibrium without partitioning of X (which would predict a spike in the X profile) , or paraequilibrium which is a constrained equilibrium assuming that the X diffusion profile is totally flat. A detailed discussion of these various interfacial conditions, and their limits is to be found in the companion paper by Inden [25].

4.2. KINETICS: SOLUTIONS OF CLASSICAL DIFFUSION PROBLEMS

In diffusion controlled phase transformation, an essential tool for modelling is the classical solutions of Fick's equation in various geometries [G3, 27]. We will successively investigate planar and spherical geometries leading to the so called "Parabolic solutions", and the needle like geometries leading to shapes propagating at constant velocities. Quasi planar fronts propagating at constant velocity require an equal composition on both sides of the reaction front and will be treated as an example in section 4. The fundamental equation to be solved is Fick's diffusion equation.

4.2.1. Parabolic solutions

The family of parabolic solutions relies on the classical solution of Fick's equation:

$$C = A + B erf(\frac{x}{\sqrt{4Dt}})$$

$$erf(u) = \frac{2}{\sqrt{\pi}} . \int_0^u \exp(-\lambda^2)d\lambda$$

In such solutions it is possible to impose a constant concentration at a position defined by

$$\xi = K\sqrt{4Dt}$$

K=0 corresponds to a static interface, whereas a non zero K corresponds to an interface propagating with a decreasing velocity. Depending on the boundary conditions, A, B and K can be determined , leading to a prediction of the diffusion field , and of the propagation rate. For instance, the case of precipitation of a β phase of composition C^{β} from a solid solution of composition C^{α} larger than the equilibrium concentration $C^{\alpha\beta}$, leads to the following results for the diffusion field and the parabolic constant K:

$$C^{\alpha}(x) = C^{\alpha\infty} + (C^{\alpha\beta} - C^{\alpha\infty})\left[\frac{1 - erf(x/\sqrt{4Dt})}{1 - erf(K)}\right]$$

$$K \approx \frac{(-1/\sqrt{\pi})(C^{\alpha\beta} - C^{\alpha\infty})}{(C^{\beta} - C^{\alpha\beta})}$$

This can be readily extended with the same method to situations where the composition of the precipitate is allowed to vary, where diffusion is allowed in the product phase, where an interface separates a one phase region from a two phase region...etc [27].

For a better grasp of the physical meaning of the method, it is worth going into detail through the derivation of the growth rate of a spherical precipitate from a supersaturated solid solution [G3]. In spherical coordinates, Ficks equation can be rewritten under the assumption of an invariant diffusion field:

$$\frac{d}{dr}(r^2.\frac{dc}{dr}) = 0$$

With appropriate boundary conditions at the interface and at infinite distance (i.e. assuming a diluted precipitation), one gets the expression for the diffusion field:

$$C(r) - C^{\alpha\infty} = (C^{\alpha\beta} - C^{\alpha\infty})\frac{R}{r}$$

The mass balance at the interface writes then:

$$4\pi R^2.dR.(C^{\beta} - C^{\alpha\beta}) = 4\pi R^2.Jdt$$

The expression of J can be then readily obtained from the expression of the diffusion field, from which the rate of growth of precipitate radius can be derived and by direct integration one finds the expression of the precipitate size as function of its initial size, of the diffusion coefficient, and of the supersaturation:

$$R^2(t) - R_0^2 = 2D.\frac{C^{\alpha\infty} - C^{\alpha\beta}}{C^{\beta} - C^{\alpha\beta}}.t$$

A similar solution exists for growing ellipsoïds.

4.2.2. Self preserving shapes

These solutions correspond to C(x-Vt). As can be readily seen, a planar solution is possible (with an exponential variation of the concentration field) only if the product phase has the same overall composition as the parent phase (as is the case with massive transformation, pearlitic reaction or discontinuous precipitation). But there is another family of solution propagating at constant velocity and preserving its shape : the cases where the surface of the reaction front is a quartic, and the most relevant situation in phase transformations, when it is a paraboloïd [28]. In ordre to prove this, the corresponding total differential equation for the function of x-Vt is rewritten in a system of confocal parabolic coordinates. In such a system, the parabolae can be lines of constant concentration . Solving the equation in this system of coordinates leads to an expression of the diffusion field, and the mass balance produces a relation between the dimensionless supersaturation S and the Peclet number P defined by:

$$S = \frac{C^{\alpha\beta} - C^{\alpha\infty}}{C^{\alpha\beta} - C^{\beta}}$$

$$P = \frac{V\rho}{2D}$$

This relation can be derived for all paraboloïds. The most frequently used are the cylindrical paraboloïd and the circular paraboloïd. For each of these geometries the relations between S and P are respectively:

$$S = \sqrt{\pi P}.\exp(P).\frac{2}{\sqrt{\pi}}\int_{P}^{\infty}\exp(-u^2)du$$

$$S = P.\exp(P).\int_{P}^{\infty}\frac{\exp(-u)}{u}du$$

These exact solutions do not survive to the introduction of capillary effects, which introduces higer order terms due to curvature, leading to similar mathematical difficulties to the ones found in the theory of dendritic solidification.

4.2.3. Conditions for analytical solutions

The above situations for which analytical solutions are available all correspond to simple and simply connected geometries, infinite or semi infinite systems, constant diffusion coefficients and constant interface conditions. The introduction of capillary terms in the interface conditions can be done within the constant field approximation for the growing sphere, or for the quasi planar solution provided one can neglect lateral diffusion flux. However all these conditions are highly idealised and in other situations one may have to solve numerically the diffusion equations with a moving boundary whose shape is itself a part of the problem. The so called "free boundary problems" [29], for which Stefan problem in solidification is a paradigm, are a numerical challenge. Classical methods such as Finite Element Methods, Finite Difference methods, Boundary Elements Methods have been applied sucesfully [30]. The trouble of "front tracking" is relieved using a diffuse description of the interface (the so called "phase field method" , offspring of the Cahn Hilliard model for diffuse interfaces). This method has been extensively used in solidification (see for instance Karma in this volume [2] but is still to be applied with its full strength to the variety of reactions observed in solid state phase transformations.

5. Examples of Applications

The following examples do not pretend to be exhaustive, they are to be understood as illustrations of the concepts described in the previous sections. We will insist on the modification required so that the academic situations described above can be adapted to realistic situations including industrial alloys.

5.1. APPLICATION OF DICTRA TO PHASE GROWTH

DICTRA is to date the most complete and comprehensive computer tool able to deal with solid state transformations combining both proper thermodynamic databases with diffusion data. It is not able yet to deal with all the subtleties of interface mediated phase transformations, but it is by far the most adapted "tool box" available on the market. As such it deserves a separate treatment and will be described in details in the companion paper by Inden in this volume [25].

5.2. EXAMPLE OF LAMELLAR GROWTH : DISCONTINUOUS PRECIPITATION

Discontinuous precipitation is an ubiquitous phenomena [13]. It is one of the tricks found by nature to come closer to thermodynamics while driving force is available and bulk diffusion is frozen due to a low reaction temperature. The system breaks symmetry, putting a grain boundary into motion: all the mass transport takes place through Grain Boundary diffusion, leaving behind, instead of the supersaturated solution which is the parent phase, a set of parallel lamellae of precipitates separated by interlamellae of a less supersaturated solid solution. This system has a unreasonable number of degrees of freedom: the spacing between lamellae in the product phase, the remaining supersaturation and the front velocity. Since the parent and the daughter phases have the same overall composition, a quasi planar front propagating at constant velocity is possible. Qualitatively, one can understand that the system will select a spacing and a velocity, through the competing effects of the driving force and the interface energy. A too large spacing between lamellae will lead to unpractical diffusion times and to a low velocity. A too small spacing will dissipate all the free energy into interface energy, and below a critical spacing the velocity should drop to zero. There is clearly an optimum spacing to give a reasonable propagation velocity. Similarly the driving force will be higher if the decomposition is more advanced. Since the reaction takes place only at the moving boundary, the slower will be the motion the more advanced will be the decomposition, but in turn the driving force will be higher and the velocity higher: again one can understand that the remaining supersaturation and the velocity are intimately related. In addition this system presents the peculiarity of a coexistence of two types of interfaces: a grain boundary between two solid solutions which are the same phase with different supersaturations, and an interface between two different phases : the precipitate and the solid solution. Along the grain boundary the driving force for diffusion is the concentration gradient. Along the interface, the driving force is the curvature gradient. These two different driving forces , together with the requirement of a constant velocity along a stable front, will select the spacing and the velocity without the need to introduce a heuristic principle such as maximum growth rate or maximum dissipation which is often used for perlitic growth for instance.
The typical geometry of a discontinuous precipitation front is shown in figure 6.

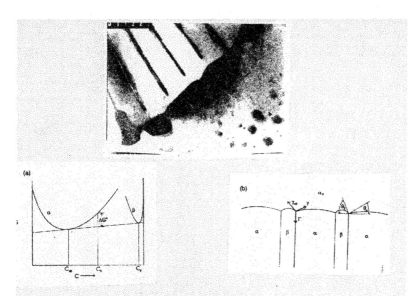

Figure 6: Discontinous precipitation front, Free energy diagram and schematics [31]

The analysis of interlamellae propagation requires the solution of a diffusion equation

$$\frac{\partial C}{\partial t} = D_{GB}.\frac{\partial^2 C}{\partial s^2} + \frac{CV}{K\delta}$$

including the incorporation of solute from the moving grain boundary of width δ. This solution was given by Cahn [32] and leads, for a steady state field, to a Cosh profile which is very accurately verified experimentally. This concentration profile allows to compute the local driving force and the local velocity of the grain boundary [33]. With a simple parabolic hypothesis for the free energy expression with concentration, one can derive the following relation [34]:

$$V = M.(\Delta G°.Q(a) - \frac{\gamma\Omega\sin(\theta_\alpha)}{S_\alpha}$$

where a is the so called "Cahn parameter" and Q is a function of a. M is a mobility coefficient. Set into dimensionless quantities, these equations, together with a global mass balance, provide a
first relation between spacing and velocity:

$$a = \frac{V.S_\alpha^2}{KD\delta}$$

$$b = \frac{Q(a).S_\alpha.\Delta G°}{\gamma\Omega\sin(\theta_\alpha)} = \frac{S_\alpha}{S_{\alpha 0}}$$

$$Q(a) = \frac{1}{2}.\left[1 + \frac{\tanh(\sqrt{a}/2)}{\sqrt{a}/2} - \tanh^2(\sqrt{a}/2)\right]$$

$$C* = \frac{\Delta G° KD\delta}{\left[M(\Omega\gamma\sin(\theta_\alpha))^2\right]}$$

$$b(b-1) = aQ(a).C*$$

The lamellae propagation is controlled by a driving force associated with curvature gradient. Following Mullins treatment for this class of problems [35], we obtain a set of equations whose numerical solution provide a second relation between spacing and velocities:

$$l = L/S_\beta, x = X/S_\beta$$

$$v = \frac{V(C_\beta - C_0)kTS_\beta^3}{KC_\beta D\delta\gamma\Omega}$$

$$\frac{\partial^3\theta}{\partial l^3} = -v.\cos(\theta)$$

$$\frac{\partial x}{\partial l} = \cos(\theta)$$

$$\theta(l=0) = 0, \theta(x=1) = -\theta_\beta, x(l=0) = 0$$

$$\frac{\partial\theta}{\partial l}(x=1) = -\frac{S_\beta}{R}$$

Assuming equilibrium at the triple junction allows to identify the local curvature and finally the dimensionless parameters representing velocity and spacing are related by two equations:

$$C * aQ(a) = b(b-1)$$

$$\tanh^2(\sqrt{a}) = A(1 - B\frac{\sqrt{a}.Q(a)}{b.\tanh(\sqrt{a})})$$

with the parameters A, B and C* some function of the supersaturation and of the thermodynamics of the system (for the detailed derivation see [34]). The graphical solution of this set of equations leads to the uniqueness of the velocity/spacing couple , when such a solution exists, which appears to be possible only for a range of supersaturations and values of the free energy second derivative with respect to concentration. The question of the stability of these uniquely selected solution, as well as the one of their adjustment when the temperature (and thus the supersaturation) is abruptly changes has not been treated up to now and seems an ideal problem for phase field models.

5.3. EXAMPLE OF NEEDLE GROWTH: BAINITIC TRANSFORMATION

Steel metallurgy is the school of phase transformations : all the possible phenomena can be observed in these family of materials. A long history of careful observations paved the way for precise taxonomy of microstructures. Austenite decomposition presents a variety of features still to be fully understood. In the jungle of steel metallurgy, the final microstructure depends drastically on the process conditions and especially on the cooling rate from austenite to ferrite. At low cooling rate, the transformation takes place by diffusion (Grain Boundary Ferrite, Pearlite, Widmanstatten Ferrite...). At high cooling rate , transformation takes place in a displacive manner (Martensite). At intermediate cooling rates, the transformation product is called Bainite: it consists of laths of ferrite with cementite precipitates either at the ferrite / austenite interface (upper Bainite) or within Ferrite (lower Bainite). The controversy concerning the mechanism controlling Bainite Kinetics has been raging for fifty years. The viewpoint presented here is controversial. The school of thought which assumes a displacive mechanism for Bainite growth (which put the emphasis on the nucleation kinetics) is detailed in a recent book by Badeshia [36]. A recent viewpoint set in Scripta Materialia [37] summarizes the state of the art on the Bainite controversy. We take here the opposite route, following Hillert and Purdy and assume that Bainite transformation (at least upper Bainite) is kinetically controlled by carbon diffusion from ferrite to austenite. The mechanisms for Fe to move from a fcc structure to a bcc structure may well be displacive, but we believe that it is not kinetically controlling the process. The aim of the present section is to illustrate the application of diffusive needle growth models in real systems, and to provide an example of comparison with experimental data [38]. The experimenal data are obtained using metallography and dilatometric measurements.

Macroscopic kinetics exhibit clearly a Johnson Mehl type of behaviour with an exponent close to 2, and the mesurement of lath extension clearly show a contant growth rate. This kinetic is compatible with a constant nucleation rate from the boundaries, and a constant growth of lath which can be described as paraboloïds. In order to deconvolute nucleation kinetics from growth kinetics, w have performed a two step heat treatment: the first step (high emperature) allows for the nucleation of Grain Boundary Ferrite, on which Bainite will grow without requiring a further nucleation step. Measuring the length of Bainite lath as function of time allows for different temperature, different compositions - the alloying elements are Ni, Mo and Si- and different carbon contents, to measure the transformation speed. Ni, Mo have barely any interaction with Carbon, whereas Si prevents Cementitte precipitation. With Silicon in sufficient quantity, one observes a "Bainite without carbides" (although the term Bainite is then inappropriate). The aim of the model here is to predict the transformation rate as a function of temperature and Carbon content with a minimal number of adjustable parameters [38]. The tool used is a modified version of Ivantsov solution proposed by Trivedi [39]. In order to apply this tool, three questions have to be solved: the interface conditions, the diffusion coefficient and the paraboloïd tip radius. From the position of the alloy representative point in the isotherm cut of the ternary phase diagram, it becomes obvious that local equilibrium with partitioning is inappropriate (with the diffusion of the alloying element allowed, the transformation rate would be 4 orders of magnitude slower than the observed one). We have chosen the hypothesis of paraequilibrium calculated using Thermocalc program [38]. In Trivedi approach, Carbon diffusion coefficient in austenite is assumed to be constant although it is known to vary substantially with Carbon composition. In order to account for this fact, we have taken an average value using the empirical formulation of Kaufmann. The problem of the tip radius entering the Peclet number is a difficult one: we have chosen the "maximum growth rate" criterion proposed by Zener to obtain an upper limit of the growth velocity. TEM observations show a much sharper tip , facetted, and it is likely that cristallography plays a key role in the selection of the tip radius. It remains that the comparison with experimental data should be done as a relative evolution of the growth

20

velocity with temperature and carbon content. The absolute value can be easily retrieved using the tip radius as an adjustable parameter.

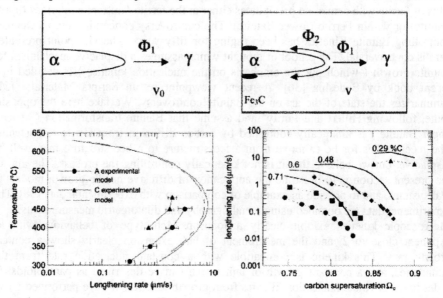

Figure 7: Bainite front velocity as function of temperature for alloys with different carbon contents, different Si content (A without Silicon, B with Silion). The schematics show the carbon diffusive flux away from the ferrite / austenite interface [38].

Figure 7 shows the results for different alloys , different carbon contents and different temperatures [38]. It can be seen that the trends can be accurately reproduced by the model both for the temperature dependence and for the Carbon content. The effect of Silicon is much more pronounced than a simple modification of the paraequilibrium conditions would predict. This has been accurately modelled by introducing cementite precipitates along the interface as extra carbon sinks which provide a carbon flux away from ferrite which has to be added to the diffusion field in ferrite. The effect of Ni and Mo is not well described by the model, indicating that a "solute drag effect" is to be expected (see the paper by Hutchinson in this volume for an introduction to the theory of Solute Drag). Back calculation of the nucleation rate from the overall JMAK kinetics using the measured growth velocities leads to a decreasing nucleation rate with decreasing temperature which rules out the possibility of a displacive nucleation.

6. Conclusions

Modelling of diffusive phase transformations is a demanding excercise. It requires access to good thermodynamic data (difficult, especially at lower temperatures where extrapolations of higher temperature measurements are suspicious), good diffusion data

(difficult, especially for cross terms, interface diffusion...), and good kinetic measurements (which can be obtained by a variety of techniques , metallography, hardness, dilatometry, TEM, X rays ...see Kostorz contribution in this volume). A number of modelling tools have been used over the years: the atomiscale (Monte Carlo simulations, cluster dynamics) are especially suited for precipitation kinetics , whereas the continuum level is still the major tool for modelling interface mediated reactions, with a range of analytical solutions or numerical methods for diffusion equations. Phase field methods now currently used in solidification can find in solid state transformations a number of possible applications.

It is the deep belief of the author that the field nowadays is ripe for a closer and fruitful interaction with other specialities in materials science. Culturally the most obvious is a systematic comparison with solidification phenomena. The recent study by Kurz on directionnal transformation from delta ferrite to austenite has shed a new and unexpected light on the everlasting problem of massive transformation by a clever analogy with the absolute stability criteria in solidification [40].

The persisting divorce between the communities of mechanical metallurgy and the community of phase transformations, which may have been necessary when these communities were constructing the concepts necessary to a modelling activity, is now to be overcome. The remarkable example given by the analysis of elastic effects in phase transformation has to be pursued [30]. Understanding phase transformations is a powerfull tool to prepare model materials to test predictions of micromechanics (ductile fracture of heterogeneous materials is a straighforward application of this strategy). Incorporating plasticity beside elasticity in phase transformation will probably enrich the field of thermodynamics of stressed solids. Transformation induced plasticity, TRIP steels [41], interplay betwen precipitation and recrystallisation [42] are open questions in which the contribution of both communities is urgently needed. It is the hope of the author that this school and hopefully the present paper may help into bringing together these two communities.

Acknowledgements

It is a pleasure to acknowledge here the collaboration over the years with Dr D.Duly, Dr L.Klinger, Dr D.Quidort and especially with my Mentor in the theory of phase transformations, Prof G.Purdy. Numerous discussions on phase transformations and Steel metallurgy with Prof P.Guyot, Prof J.Philibert, Prof D.Embury, Dr M.Durand, Dr M.Kandel, Dr P.Maugis and Dr T.Iung have greatly contributed to my interest in this field.

References

GENERAL REFERENCES
The field of phase transformations in metals and alloys has a long history, which has also led to a jargon which somewhat makes it less accessible to specialists of other fields. As a results, the general knowledge on precipitation for instance is much more developped than the one on phase transformations controlled by interface migration.

22

The general references listed here are, from the view point of the author, good overviews for a non specialist to enter the field.

G1. H.Aaronson "Lectures in the theory of phase transformations",TMS (1999), recently reedited and updated is a must. The articles by M.Hillert on "thermodynamics" (Chapter 1) and by R.Sekerka on "Moving boundary problems" (Chapter 5) are masterpieces.

G2. G.Kostorz "Phase transformations in Materials", Wiley VCH, (2002), is an updated version of the Treatise on Materials Science and Technology , volume 5 edited by R.Cahn, P.Haasen, E.Kramer. Of special relevance to the present topics are the review papers by R.Wagner et al. on "Homogeneous second phase precipitation" (Chapter 5) and G.Purdy et al. on "Transformations involving interfacial diffusion" (Chapter 7)

G3. J.Philibert, "Atom movements, Diffusion and Mass transport in solids", Editions de Physique (1991) provides an excellent textbook on diffusion and its applications

G4.M.Hillert "Phase diagrams and Phase Transformations", Cambridge University Press (1999) is the key reference on the application of thermodynamics to phase transformations.

G5. J.W.Martin, R.D.Doherty, B.Cantor "Stability of micostructures in metallic systems", Cambridge University Press (1997) is an encyclopedic visit of the world of microstructural evolution, and an invaluable source of references to original papers.

REFERENCES CITED IN THE TEXT

1. Kachaturyan, A. (2002) *(These proceedings)*
2. Karma, A. (2002) *(These proceedings)*
3. Christian , J.W. (1975) Theory of transformations in metals and alloys, Pergamon Press, Oxford
4. Shercliff, H. and Grong,O. (2002) , Physically based Process Modelling, Progress in Materials Science (under Press)
5. Deschamps, A, (1998) On the influence of quench and heating rates on the ageing response of Al-Zn-Mg-Cu , Mat.Sc.Eng, 251, 200
6. Gerold, V. (1979), Precipitation hardening in "F.R.N.Nabarro ed. *Dislocations in Solids*, North Holland Publishing Compagny, Vol 4 , 219,
7. Humphreys, J. and Heatherly, R. (1998), *Recrystallisation and related phenomena*, Pergamon Press
8. Brown, L.M. and Clarke, J., Acta Metallurgica, (1975) 23, 821
9. Ashby, M, (1971), in *Strengthening Methods in Crystals* , ed. A.Kelly, R.B.Nicholson, Applied Science Publishers
10. Brown, L.M. and Embury, J.D, (1964) Proc.ICSMA 3, Cambridge,1, 164, Institute of Metals
11. Goods S.H. and Brown,L.M., (1979) Acta.Met. 27, 1
12.Wagner, R., Kampmann, R., and Voorhees, P.W. (2002) Homogeneous Second Phase Precipitation, in G.Kostorz ed. *Phase transformations in Materials*, Wiley VCH, pp309-409
13.Purdy, G.R., and Brechet,Y, (2002) Transformations involving Interfacial diffusion, in G.Kostorz ed. , *Phase transformations in Materials*, Wiley VCH, pp481-519

14. Kinsman K.R. and Aaronson, H. (1967) Transformations and hardenability in steels, Cimax Molybdenum C, Ann Arbor, Michigan, p39

15. Lifschitz, I., and Slyosov, V.V., (1961) Phys.Chem.Solids, 19, 35

16. Binder,K. (2002) Statistical theories of phase transformations, in G.Kostorz ed. *Phase transformations in Materials*, Wiley VCH, pp 239-309

17. Langer,J., Bar-On and Miller,H.D.,(1975) Phys. Rev.A, 11, 1417

18. Sigli C. ,(2002) *(These proceedings)*

19. Soissons, F., (2002), *(These proceedings)*

20. Pontikis V. (2002) , *(These proceedings)*

21. Bellon P., (2002) *(These proceedings)*

22. Weatherly G., (1975) Stability of Eutectic structures at elevated temperatures, Treatises on Materials Technology, Vol 8

23. Pawlovski A., Zieba, P., (1991), *Phase transformations controlled by diffusions at moving boundaries*, Warsaw-Cracow

24. Livingston, J.D. and Cahn,J. (1974) Discontinuous coarsening, Acta Met, 22, 495

25. Inden G., (2002) *(These proceedings)*

26. Van Der Veen A., Delaey, (1996), Prog.Mat.Sc.40, 181

27. Sekerka R. and Wang, S-L, (1999) Moving Phase boundary problems, in H.Aaronson ed. , *Lectures in the theory of phase transformations*,TMS

28. Horvay G. and Cahn, J.W. (1961) Acta.Met, 9, 695-705

29. Crank,J. (1988) Free and moving boundary problems, Oxford Science Publications

30. Voorhees, P.W., (2002), *(These proceedings)*

31. Duly, D., Cheynet M.C., Brechet Y. (1994), Morphology and Chemical nanoanalysis of Discontinuous Precipitation in Mg-Al Alloys, Acta. Met.Mat., 42, 11, 3843-3854

32. Cahn J.W. (1959), On cellular reaction kinetics, Acta.Mat. 7, 18

33. Hillert M. (1982), Kinetics of Discontinuous Precipitation, Acta.Met, 30, 1689

34. Klinger L., Brechet,Y., and Purdy,G. (1997), On velocity and spacing selection in discontinuous precipitation, Acta.Mat. 45, 12, 5005-5013

35. Mullins, W.W., (1957), J.Appl.Phys. , 28, 333

36. Badeshia, H (2001) *Bainite*, Cambridge University Press

37. Hillert M. (ed.) (2002) A viewpoint Set on Bainitic transformation, Scripta.Mat

38. Quidort D., Brechet Y., (2001), Isothermal Growth of Bainite in 0.5% Carbon Steel, Acta.Mat., 4161-4170

39. Trivedi R., (1970) Metall.Trans., 1, 921

40. Kurz, W. (2002) Oral Presentation at the TMS symposium on massive transformation, Saint Louis

41. Jacques P. (2002) *(These proceedings)*

42. Hutchinson, C. , Zurhab, H., Brechet Y., Purdy, G., (2002) On the coupling between recrystallisation and precpitation , Acta.Mat. *(Under Press)*

II. Phase transformation and microstructures : mesoscopic methods

MEAN FIELD THEORIES AND GINZBURG-LANDAU METHODS

V. I. TOKAR

Institute of Magnetism, National Academy of Sciences
36-b Vernadsky str., 03142 Kiev-142, Ukraine

AND

PCMS-GEMM, UMR 7504 CNRS, 23, rue du Loess, F-67037
Strasbourg Cedex, France

1. Introduction

The *mean field* (MF) theory and the *Ginzburg-Landau* (G-L) method are
two related techniques widely used in condensed matter physics. Formally,
the MF approximation is a decoupling scheme which replaces the statistical
average of a product of microscopic variables by the product of independent
averages of each variable. The space- (and time- in the out of equilibrium
case) dependent average of an individual microscopic variable is called the
mean field function.

The G-L method [1] is based on the observation that the MF theory can
be formulated as a variational scheme in which equations of the MF type
can be derived as the Euler-Lagrange equations expressing the minimality
of the free energy (or Ginzburg-Landau) functional. The argument of the
G-L functional is traditionally called the *order parameter.*

Qualitatively the approximations underlying both approaches are the
same, the difference lying in their implementation. The MF theory starts
from a microscopic Hamiltonian. Therefore, with the use of the canonical
techniques the free energy of the system under consideration can be expressed as a functional of the mean field function, thus giving us both—the
order parameter (MF) and the G-L (the free energy) functional. So in this
case the G-L method is superfluous.

However, there are cases when the microscopic approach is hard to implement either because the microscopic Hamiltonian is not known or is too
complex (as in the case of the liquid-solid phase transition, for example)
that its MF treatment meets with severe difficulties [2]. In this and simi-

A. Finel et al. (eds.), Thermodynamics, Microstructures and Plasticity, 27–44.

lar cases the G-L approach is more efficient. In its framework the MF-type equations are derived by *postulating* the order parameter and the G-L functional on the basis of phenomenological considerations and the symmetry properties of the system studied [1, 2].

Thus, both approaches are very general and can be (and have been) applied to practically any problem in condensed matter physics. For concreteness, in the present paper the above techniques will be illustrated by their application in the theory of alloys. Because even in this relatively narrow field the applications of the MF/G-L methods are too numerous to be adequately covered in a short article, the presentation will be restricted to the very basic definitions and to a few techniques. The choice of the material was motivated by the desire that the present article could serve as a (very basic) technical introduction into the current literature on the *phase field method* [2] and its applications.

2. The Mean Field Theory

2.1. THE MEAN FIELD APPROXIMATION

In statistical mechanics physical quantities are calculated as averages over the fluctuations of microscopic variables. The mean field approximation (MFA) consists in the neglect of *correlations* between these fluctuations. Let $\phi(\mathbf{r})$ be a fluctuating microscopic variable or *field*. Denoting the statistical average by the angular brackets let us define the *mean field* $\bar{\phi}(\mathbf{r})$ as the average of $\phi(\mathbf{r})$: $\bar{\phi}(\mathbf{r}) = \langle \phi(\mathbf{r}) \rangle$. The (pair) correlation function of *fluctuations* [i. e., of deviations of $\phi(\mathbf{r})$ from $\bar{\phi}(\mathbf{r})$] is

$$G(\mathbf{r} - \mathbf{r}') = \langle [\phi(\mathbf{r}) - \bar{\phi}(\mathbf{r})][\phi(\mathbf{r}') - \bar{\phi}(\mathbf{r}')] \rangle_{|\mathbf{r}-\mathbf{r}'|\to\infty} \sim \exp[-|\mathbf{r} - \mathbf{r}'|/\xi(T)], \tag{1}$$

where $\xi(T)$ is the correlation length which is *defined* by this equality. Because in the MFA we neglect $G(\mathbf{r} - \mathbf{r}')$, from Eq. (1) follows that the approximation *can be* good when $\xi(T)$ is small, i. e., far from the *critical region* where $\xi(T)$ is large [3]. It should be stressed that this can take place at high temperatures above the critical temperature T_c as well as at low temperatures below T_c.

From a practical point of view the power of the MFA consists in the possibility to decouple the product of any number m of fields $\phi(\mathbf{r}_i)$ as

$$\langle \phi(\mathbf{r}_1) \dots \phi(\mathbf{r}_m) \rangle \overset{\mathrm{MF}}{=} \langle \phi(\mathbf{r}_1) \rangle \dots \langle \phi(\mathbf{r}_m) \rangle \equiv \bar{\phi}(\mathbf{r}_1) \dots \bar{\phi}(\mathbf{r}_m) \tag{2}$$

thus expressing all quantities through the values of the mean field $\bar{\phi}(\mathbf{r})$.

As is obvious from Eq. (1), however, the MFA may fail in the case of $\mathbf{r} \approx \mathbf{r}'$. To see what kind of problems one may encounter in this case, let us consider a model of an alloy with the fluctuating variables being the occupation numbers $n_i^K = 0, 1$ (i the lattice site index) of atoms of kind K. Obviously that because of the property $(n_i^K)^2 = n_i^K$ we have

$$\langle (n_i^K)^2 \rangle = \langle n_i^K \rangle = c_i^K, \tag{3}$$

where c_i^K is the concentration of the Kth component. On the other hand, substituting $\phi(\mathbf{r}) = n_i^K$ into Eq. (2) we arrive at

$$\langle (n_i^K)^2 \rangle \overset{MF}{=} \langle n_i^K \rangle \langle n_i^K \rangle = (c_i^K)^2. \tag{4}$$

Comparing Eqs. (3) and (4) we see that the MFA can be severely in error when applied to the fields at the same site because the concentration and its square can significantly differ. On the other hand, at low temperatures the alloys are usually ordered in which case $c_i^K \approx 1$ or $c_i^K \approx 0$. In both cases $(c_i^K)^2 \approx c_i^K$, the error being the smaller the better is the ordering. Thus we conclude that the MFA works best at low temperature. This is another cause for the popularity of the mean field methods. While at high temperatures there exist reliable and systematic approaches based on the high temperature expansions, at low temperatures such universal techniques are absent, the MFA being a good candidate for such technique.

2.2. BINARY ALLOY

Let us consider a model binary alloy with interatomic interactions restricted only to pair interactions. The configuration-dependent energy of such an alloy is

$$U = \frac{1}{2} \sum_{ij,KL} V_{ij}^{KL} n_i^K n_j^L, \qquad (K, L = A, B), \tag{5}$$

where V_{ij}^{KL} is the interatomic pair potential. Let us compute the internal energy in the mean field approximation:

$$E = \langle U \rangle \overset{MF}{=} \frac{1}{2} \sum_{ij,KL} V_{ij}^{KL} \langle n_i^K \rangle \langle n_j^L \rangle = \frac{1}{2} \sum_{ij,KL} V_{ij}^{KL} c_i^K c_j^L. \tag{6}$$

Noting that in the binary alloy $n_i^A = 1 - n_i^B$ and hence $c_i^A = 1 - c_i^B$, the above equation can be re-written as

$$E = -\frac{1}{2} \sum_{ij} V_{ij} c_i (1 - c_j) - (V^{AA} - V^{BB}) \sum_i c_i + V^{AA} N,$$

where $c_i \equiv c_i^B$, $V^{LL} = \sum_j V_{ij}^{LL}$, N the total number of atoms, and

$$V_{ij} = V_{ij}^{AA} + V_{ij}^{BB} - 2V_{ij}^{AB} \tag{7}$$

is the *ordering potential*. In the absence of spatial correlations (MFA) the entropy

$$S = -k_B \sum_i [c_i \ln c_i + (1 - c_i) \ln(1 - c_i)].$$

The Helmholtz free energy $F = E - TS$. The thermodynamic relation $\partial F/\partial c = \mu$ makes it convenient to replace F by $\bar{F} = F - \mu \sum_i c_i$

$$\bar{F} = -\frac{1}{2} \sum_{ij} V_{ij} c_i (1 - c_j) + k_B T \sum_i [c_i \ln c_i + (1 - c_i) \ln(1 - c_i)] - \mu \sum_i c_i \quad (8)$$

which satisfies

$$\partial \bar{F}/\partial c_i = 0 \quad (9)$$

and *at equilibrium* is at the *global* minimum.

2.2.1. *Alloy Decomposition*

Let us assume that the alloy is spatially homogeneous. Then $c_i = c = Const$ and $\bar{F}/N = \bar{f}(c) = f(c) - \mu c$, where

$$f(c) = -\frac{1}{2} V c(1 - c) + kT[c \ln c + (1 - c) \ln(1 - c)] \quad (10)$$

and $V = \sum_j V_{ij} = V(\mathbf{k} = 0)$ with $V(\mathbf{k})$ being the lattice Fourier transform of V_{ij}. The case of alloy decomposition corresponds to $V < 0$. At $\mu \neq 0$ there exists a unique global minimum of $\bar{f}$ at some c_α found from the equation

$$d\bar{f}(c)/dc\big|_{c=c_\alpha} = 0.$$

(see Fig. 2.2.1). In the symmetric case ($\mu = 0$) at low temperature there exist *two* equivalent minima at c_α and $c_\beta = 1 - c_\alpha$. These solutions appear for $T < T_c$ where the critical temperature $T_c = 0.25|V|$ is found from the condition

$$d^2 f(c)/dc^2\big|_{c=0.5} = 0. \quad (11)$$

Thus, the assumption of spatial homogeneity is *not* generally correct at low temperature. There are two solutions at equilibrium with each other and the alloys with composition c such that $c_\alpha < c < c_\beta$ simply do not exist at (and below) some temperature T_0 (see Fig. 2.2.1). The gap $\Delta c = c_\beta - c_\alpha$ in the allowed equilibrium alloy concentrations is called the *miscibility gap*.

2.2.2. *The Interphase Boundary*

One may ask what happens with the alloy quenched below T_c into the two-phase region (e. g., along the paths u or m shown in Fig. 2.2.1)? An obvious answer is that the alloy will strive to reach the equilibrium which means

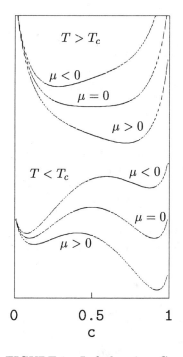

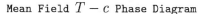

Mean Field $T - c$ Phase Diagram

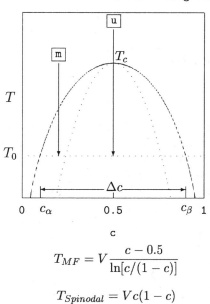

$$T_{MF} = V \frac{c - 0.5}{\ln[c/(1 - c)]}$$

$$T_{Spinodal} = V c(1 - c)$$

FIGURE 1. Left drawing: Concentration dependence of the mean field free energy $\bar{f}(c)$ of a decomposing binary alloy for different values of the chemical potential μ above and below the critical temperature T_c. Right drawing: Mean field phase diagram of the alloy (solid line) and the spinodal curve (curved dotted line). Δc—the miscibility gap corresponding to temperature T_0; u and m correspond to the unstable and metastable quenches, respectively (for further explanations see the text).

its decomposition into the mixture of the equilibrium phases α and β. The decomposition kinetics is a difficult problem dealt with in the *phase field* method [2]. We postpone the discussion of kinetics to section 4 and for the moment will restrict ourselves to the consideration of the asymptotic state of the system at the end of all kinetic processes, i. e., at the equilibrium. So we discard the homogeneity assumption and consider the symmetric ($\mu = 0$) mean field equation [see Eqs. (8) and (9)]:

$$\frac{\partial F}{\partial c_i} = -\frac{V}{2} + \sum_j V_{ij} c_j + kT \ln \frac{c_i}{1 - c_i} = 0$$

which is convenient to re-write in the form

$$\sum_j V_{ij} c_j - V c_i = -f'(c_i). \qquad (V = \sum_j V_{ij}) \qquad (12)$$

For concreteness we consider quasi-1D geometry, the BCC lattice and assume the interphase boundary in the [100] direction. Besides, for *simplicity* we consider the nearest neighbor ordering potential: $V = 8V_{NN}$.

$$\frac{V}{2}(c_{i+1} - 2c_i + c_{i-1}) = -f'(c_i), \qquad (13)$$

where now the index i numbers the successive planes in the [100] direction. On the left hand side of this equation we easily recognize the discrete second derivative. This is not coincidental and is connected to the known property of the short range pair interactions to have $\mathbf{k}^2$ as the leading quasimomentum dependence of their Fourier transform [3]. This can be used to approximate the discrete lattice equations similar to Eq. (13) by much more practical differential equations. Formally this is done as follows. The Fourier transformed Eq. (12) reads

$$\sum_j V_{ij} c_j - V c_i = \frac{1}{N} \sum_{\mathbf{k}} [V(\mathbf{k}) - V(0)] c(\mathbf{k})$$

$$\approx \frac{1}{N} \sum_{\mathbf{k},\gamma\delta} m_{\gamma\delta} k_\gamma k_\delta c(\mathbf{k}), \quad \text{where} \quad m_{\gamma\delta} = \frac{1}{2} \left[\frac{\partial^2 V(\mathbf{k})}{\partial k_\gamma \partial k_\delta} \right]_{\mathbf{k}=0}.$$

In the case of the *cubic* symmetry m_{ij} is diagonal and isotropic:

$$m_{ij} k_i k_j = m\mathbf{k}^2 \xrightarrow{FT} -m\nabla^2$$

In the 1D geometry Eq. (13) in the continuum approximation takes the form

$$m \frac{d^2 c}{dx^2} = f'(c), \qquad (14)$$

where $m = |V|/2$. In Fig. 2.2.2 the solutions for the IPB obtained with the use of the discrete Eq. (13) and its continuous analog Eq. (14) are compared. In the discrete case the impulses show the antiphase boundary profile (see subsection 2.3.1) which differ from the IPB only in its sign on every second plane. We see that despite the microscopic size of the IPB in only a few lattice constants the agreement of the discrete and continuous profiles is almost perfect. Because the differential equations are much more easy to use than the finite difference ones, in practical calculations only the former are usually being used. To make the formalism fully selfconsistent,

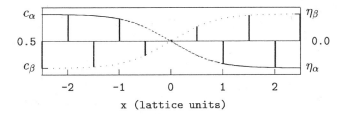

FIGURE 2. Interphase and antiphase boundaries calculated with the use of the continuous equation (14) (solid and dotted lines) and with the use of the discrete equation (13) (impulses). The values of the conserved (c) and nonconserved (η) order parameters can be read off from the left and right borders, respectively.

the continuous equations are derived by analogy with Eq. (9) through the minimization of a free energy functional

$$\bar{F}[c] = \int_{\mathcal{V}} \frac{d\mathbf{r}}{v} [\frac{m}{2}(\nabla c)^2 + f(c) - \mu c] \tag{15}$$

which can be formally derived from Eq. (8) with the use of the natural substitution

$$c_i \to \frac{1}{v} \int_{\mathbf{r} \, \epsilon \, v} c(\mathbf{r}) d\mathbf{r}, \tag{16}$$

where by v we denoted both the atomic cell associated with site i and its volume.

The shape of the concentration profile depends on temperature and *does not depend* on the average concentration. Its position may be found with the use of *the lever rule*.

2.3. ORDERING

In the next section we will show that in the MF/G-L approach all decomposing alloys belong to the same category (or universality class in the renormalization group language [3]). In contrast, ordering alloys exhibit much richer structural behavior. To begin with, let us consider the same model alloy with NN pair potential as in the previous subsection but this time with *positive* pairing potential $V_{NN} > 0$. At low temperature $T \simeq 0$ $\bar{F} \to E - \mu c N$. With μ, c, and N fixed, only the ordering energy E_0 is important [see Eq. (8)]

$$E = -\frac{1}{2} \sum_{ij} V_{ij} c_i (1 - c_j) = \frac{1}{2} \sum_{ij} V_{ij} c_i c_j - \frac{1}{2} V c N \equiv E_0 - \frac{1}{2} V c N. \quad (17)$$

Thus, for $V_{NN} > 0$ at low temperature the A and B atoms *"avoid"* each other. To be more specific, let us consider an equiatomic alloy ($c = 0.5$, $\mu = 0$) with a *bipartite* lattice (e. g., BCC) consisting of two equivalent interpenetrating sublattices: 1 and 2. The free energy (8) in this case is

$$F = -\frac{1}{2} \sum_{ij} V_{ij} c_i (1 - c_j) + kT \sum_i [c_i \ln c_i + (1 - c_i) \ln(1 - c_i)].$$

Now let us transform this equation by introducing two *order parameters* $\tilde{\eta}$ and η as
(i) Let $c_i = \frac{1}{2} + \tilde{\eta}_i$. (This is the so-called *pseudospin representation* widely used in the alloy theory [4].) Then

$$\begin{aligned}
F &= +\tfrac{1}{2} \sum_{ij} V_{ij} \tilde{\eta}_i \tilde{\eta}_j - \tfrac{1}{2} V \sum_i c_i \\
&+ kT \sum_i [(\tfrac{1}{2} + \tilde{\eta}_i) \ln(\tfrac{1}{2} + \tilde{\eta}_i) + (\tfrac{1}{2} - \tilde{\eta}_i) \ln(\tfrac{1}{2} - \tilde{\eta}_i)]
\end{aligned} \quad (18)$$

(ii) Now let $c_i = \frac{1}{2} + \eta_i$ on sublattice 1 and $c_i = \frac{1}{2} - \eta_i$ on sublattice 2

$$\begin{aligned}
F &= -\tfrac{1}{2} \sum_{ij} V_{ij} \eta_i \eta_j + \tfrac{1}{2} V \sum_i c_i \\
&+ kT \sum_i [(\tfrac{1}{2} + \eta_i) \ln(\tfrac{1}{2} + \eta_i) + (\tfrac{1}{2} - \eta_i) \ln(\tfrac{1}{2} - \eta_i)]
\end{aligned} \quad (19)$$

Because case (ii) with $V > 0$ reduces to case (i) with $V < 0$ considered earlier, on the basis of the results of subsection 2.2.1 above we immediately conclude:
There exists some critical temperature $T_c (= 0.25V)$ below which there exist two solutions with the same free energy. The solutions differ by the value of $\tilde{\eta}$ and η:
(i) $c_{\alpha,\beta} \to \tilde{\eta}_{\alpha,\beta}$: $\qquad \tilde{\eta}_{\alpha,\beta} = 0.5 - c_{\alpha,\beta} = \pm(0.5 - c_\alpha)$ because of the relation $c_\alpha = 1 - c_\beta$;
(ii) $\tilde{\eta}_{\alpha,\beta} \to \eta_{\alpha,\beta}$: $\eta_{\alpha,\beta} = (-1)^K \tilde{\eta}_{\alpha,\beta}$ (K the sublattice number).

2.3.1. *Antiphase boundary*

Let us consider the B2 structure (see Fig. 2.3.1). In this case the ordering can be described by the *concentration wave* [5] defined by the vector $\mathbf{k}_0 = \frac{2\pi}{a} [100]$. The transformation from η to $\tilde{\eta}$ with the use of this vector can be written as

$$\tilde{\eta}_{\mathbf{r}} = e^{i\mathbf{k}_0 \mathbf{r}} \eta_{\mathbf{r}}.$$

BCC lattice

Special points and stars
of the FCC lattice

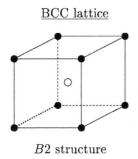

B2 structure

Star	Member vectors
$< 000 >$	$[000]$
$< 100 >$	$[100][010][001]$
$< 1\frac{1}{2}0 >$	$[1\frac{1}{2}0][10\frac{1}{2}]\ldots$
$< \frac{1}{2}\frac{1}{2}\frac{1}{2} >$	$[\frac{1}{2}\frac{1}{2}\frac{1}{2}][\bar{1}\frac{\bar{1}}{2}\frac{\bar{1}}{2}]\ldots$

FIGURE 3. Left: the B2 ordered structure. Right: special points and stars in the Brillouin zone of the FCC lattice the reciprocal to which is the BCC lattice (see the left drawing).

By analogy with the case of decomposition we can introduce the continuum representation and to write down the free energy for the *phase field* η as

$$F[0.5, \eta] = \int \frac{d\mathbf{r}}{v} [\frac{m}{2} (\nabla \eta)^2 + f(\eta)] + Const. \tag{20}$$

In the case of decomposing alloy ($V < 0$) the free energy for $c = 0.5$ would look the same in terms of $\tilde{\eta}$. So are the two models completely formally equivalent?

Let us consider the normalizations of the two parameters.

(i) $\quad \sum_i \tilde{\eta}_i = \sum_i (c_i - 0.5) = 0.$

(ii) $\quad \sum_i \eta_i = - \sum_{i \epsilon K=1}(c_i - 0.5) + \sum_{i \epsilon K=2}(c_i - 0.5)$
$= \sum_{i \epsilon K=2} c_i - \sum_{i \epsilon K=1} c_i \quad \epsilon \quad (-\infty, \infty)$ for $N \to \infty.$

Thus, in the conserved case the position of the interphase boundary is fixed by the alloy composition and the boundary is immobile. On the contrary, in the nonconserved case the antiphase boundary can freely move. In fact, the ground state corresponds to the alloy with no APB's because the field inhomogeneity raises the free energy. The same is true of the IPB, but the conservation of the order parameter requires that at least one boundary was present for the alloy compositions inside the miscibility gap.

Physically the two cases are completely different which is easily seen from the consideration of the extreme cases. While in the conserved case the two extremes (see Fig. 2.2.1) at $T = 0$ correspond to two different metals, in the nonconserved case it is the same equiatomic alloy with only the two sublattices being numbered differently.

3. The Ginzburg-Landau method

As was noted in the Introduction, the Ginzburg-Landau and the mean field techniques describe the same physics. From the formal point of view this is possible *only* if both methods are based on similar equations. This is indeed the case. The G-L functional for the free energy has essentially the same form as the continuous versions of the MF functionals Eqs. (15) and (20), except that the MF free energy density function $f(\eta)$ is replaced by the general Taylor series expansion with adjustable coefficients. The standard form of the G-L functional for the single component order parameter cases considered above [see Eqs. (15) and (20)] is

$$F_{GL}[\eta] = \int d\mathbf{r}[b(\nabla\eta)^2 - \mu\eta + a_2\eta^2 + a_3\eta^3 + a_4\eta^4 + \ldots], \qquad (21)$$

where the dots mean possible higher powers of the order parameter or possible higher-order derivatives. In contrast to the MF theory, in G-L approach only a minimal number of the coefficients a_2, a_4 ... sufficient to comply with the symmetry properties of the system under consideration is being kept. This is done in order to keep to a minimum the number of the fitting parameters because the values of these coefficients are chosen on the phenomenological grounds to fit some measurable quantities. In practice this often is more convenient because the MF theory is approximate and the values calculated within its framework are not accurate enough for practical needs. For example, the critical temperature of a second order phase transition T_c is calculated in the MFA with the accuracy rarely better than $\sim 30\%$. Instead, in the G-L approach the value of the coefficient a_2 is chosen as

$$a_2 = a_2'(T - T_c) \qquad (22)$$

so that the condition defining the second order phase transition temperature given by Eq. (11) is fulfilled automatically with T_c being equal to the experimentally measured transition temperature.

The power of the G-L approach in alloys theory lies mainly in its ability to classify the wealth of ordered structures by their symmetry properties. The group-theoretic approach of Landau and Lifshitz [1] can predict the most frequent and stable [5] ordered structures on the basis of the symmetry group G_0 of the disordered state. In its most abstract form the rules for the possible ordered structures are formulated in the language of irreducible representations of subgroups of G_0 [1, 4]. It is this abstract form of the Landau-Lifshitz theory which makes the approach equally applicable to virtually all types of phase transitions in condensed matter physics.

In the theory of alloys, however, it is possible to understand the Landau-Lifshitz theory in a simpler concentration wave formalism of Khachaturyan [5].

Special points and stars

At low temperature the ground state of an alloy is defined by its ordering potential. From the concentration wave representation of the alloy ordering energy (17)

$$E_o = \frac{1}{2} \sum_{\mathbf{k}} |c(\mathbf{k})|^2 V(\mathbf{k})$$

it is obvious that stable structures will correspond to the absolute minima of $V(\mathbf{k})$ at some points $\mathbf{k}_{0j}$. The necessary condition of a minimum is

$$\nabla_{\mathbf{k}} V(\mathbf{k})|_{\mathbf{k}=\mathbf{k}_{0j}} = 0. \qquad (23)$$

Definition: *Special* or *Lifshitz* points are defined as those which satisfy this condition *only on the symmetry grounds*, i. e., irrespective of the specific values of the potential $V(\mathbf{k})$. The symmetry operations unify these special points into the *special stars*. We note in passing that it can be shown that these stars correspond to the irreducible representations of the subgroups of group G_0 [5], in accordance with the Landau-Lifshitz theory.

The Fourier transform of the ordering potential in Eq. (23) is defined on the reciprocal lattice and in the case of short range interactions is a smooth function of $\mathbf{k}$. Its values are periodic in the reciprocal space, so $V(\mathbf{k})$ is fully characterized by its values inside the first Brillouin zone (BZ). In Fig. 2.3.1 the special points and stars of the FCC lattice are listed alongside with the BCC lattice which is reciprocal to the FCC lattice. It is easy to see that the stars $< 000 >$, $< 100 >$, and $< \frac{1}{2}\frac{1}{2}\frac{1}{2} >$ really correspond to such points of the BCC lattice in which Eq. (23) should be valid only because of the symmetry of these points. Indeed: the case of point $\Gamma = [000]$ (this and further notation is conventional) is obvious. This point is present in all structures and when the absolute minimum of $V(\mathbf{k})$ is at Γ it corresponds to the alloy decomposition. All other points lie at the BZ surface. The point $X = \frac{2\pi}{a}[100]$ lies in the middle of the cube side (its length is 2 in units of $2\pi/a$) and having chosen some arbitrary direction for $\nabla V(\mathbf{k})$ one can easily see that there is an equivalent direction obtained from the chosen one by a rotation and/or reflection. Similar reasoning holds for the point $L = \frac{2\pi}{a}[\frac{1}{2}\frac{1}{2}\frac{1}{2}]$ which lies in the middle of the line connecting the cube center and its corner. The case of the star $< 1\frac{1}{2}0 >$ is not so obvious but the consideration of the NN interaction

$$V_{FCC} = 4V[\cos(\frac{k_x a}{2})\cos(\frac{k_y a}{2}) + \cos(\frac{k_y a}{2})\cos(\frac{k_z a}{2}) + \cos(\frac{k_z a}{2})\cos(\frac{k_x a}{2})]$$

shows that for $W = \frac{2\pi}{a}[1\frac{1}{2}0]$ Eq. (23) indeed holds. More rigorous justification of this and similar special points requires either explicit construction of the BZ [4] or the group theoretic analysis [1].

In the theory of binary alloys the concentration waves in ordered structures are of the form

$$c_{\mathbf{r}} = c + \sum_{j} [\eta_j e^{i\mathbf{k}_{0j}\mathbf{r}} + (c.c.)], \tag{24}$$

where c is the average concentration and $(c.c.)$ means complex conjugate.

Now let us consider the alloy at a finite temperature. In this case we should augment the ordering energy by the entropic contribution as in Eq. (8). But because the function $f(c_i)$ is not known in the G-L approach, we replace it as in Eq. (21) above by the series expansion

$$F_L[c] = \sum_{\mathbf{r}} (-\mu c_{\mathbf{r}} + a_2 c_{\mathbf{r}}^2 + a_3 c_{\mathbf{r}}^3 + a_4 c_{\mathbf{r}}^4 + \ldots), \tag{25}$$

where the subscript L denotes the Landau functional which differs from the G-L functional by the absence of the gradient term [cf. Eq. (21)]. Substituting the concentration wave (24) into this equation we obtain many terms containing lattice sums of the type

$$\frac{1}{N} \sum_{\mathbf{r}} \eta_{j_1}^{(*)} \ldots \eta_{j_n}^{(*)} e^{(\pm \mathbf{k}_{0j_1} \ldots \pm \mathbf{k}_{0j_n})\mathbf{r}} = \eta_{j_1}^{(*)} \ldots \eta_{j_n}^{(*)} \delta(\pm \mathbf{k}_{0j_1} \ldots \pm \mathbf{k}_{0j_n} + \mathbf{K}),$$

where $\eta_j^{(*)}$ means the order parameter or its complex conjugate and $\mathbf{K}$ is the vector of the reciprocal lattice. Because of the δ-function, many terms of the expansion vanish. For example, there obviously cannot be linear terms, unless the special point is Γ with $\mathbf{k}_{0j} = 0$ which means the decomposition case only. In contrast, the quadratic term is always present because of the cancellation of exponents in the product $\eta_j \eta_j^*$. For the same reason a quartic term is also always present. The G-L functional usually includes only these two terms which are sufficient for the description of the second order phase transitions which require the free energy to be an even function of the order parameter. The third order terms occur quite rarely but the FCC lattice we are considering provides just such an exception in the case of ordering corresponding to the $< 100 >$ star. In this case the sum of three special vectors gives the vector [111] of the reciprocal BCC lattice (see Fig. 2.3.1). The "minimal" Landau functional in this case contains the third order term $\eta_1 \eta_2 \eta_3$ which means that the ordering corresponding to the star $< 100 >$ are of the first order. This is confirmed experimentally in the systems of the Cu-Au type (see Fig. 3.0.2) and illustrates the power of the Landau-Lifshitz approach in comparison with MF because the latter predicts the second order transition in this case. More detailed discussion of the FCC ordering can be found in Ref. [4].

FCC lattice

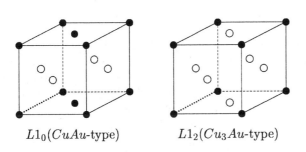

$L1_0(CuAu\text{-type})$ $L1_2(Cu_3Au\text{-type})$

FIGURE 4. Two ordered structures of an FCC binary alloy corresponding to the $< 100 >$ star of the special points.

The nature of the terms surviving in the above summation is most clearly understood in the framework of the original Landau-Lifshitz approach. The procedure of summation over all lattice sites effectively amounts to averaging over the symmetry group G_0 of the lattice which includes the point group and the translations. Thus, only the group invariants survive, as it should be because the free energy must be invariant with respect to all symmetry transformations. Thus, the result of the averaging can be predicted on the basis of the group theory according to which [1] the series expansion (25) should contain only the *invariant polynomials* of the order parameters $\eta_j^{(*)}$.

4. Kinetics

Besides the equilibrium thermodynamics considered in the preceding sections, the MF/L-G methods are widely used in nonequilibrium studies. In particular, they lie at the basis of the *phase field method* [2]. In this section we consider the derivation of the kinetic evolution equation for the case of the nonconserved order parameter. Our derivation follows the derivation of the *generalized diffusion equation* for conserved order parameter case due to Langer [6], adopted to the nonconserved case.

Microscopic atomic kinetics are rather complex [7, 8], so microscopic approach is difficult. Therefore, we will consider the kinetics in the framework of phenomenological G-L approach. This amounts to a transition from the atomic scale to a continuum description similar to the transition from the atomic hopping models to continuum Fick's laws in the theory of diffusion [7, 8].

4.1. COARSE-GRAINING

We already used the continuum description in the equilibrium case in previous sections. In the case of kinetics one has additionally take into account that the atomic exchanges take place between different sites, sometimes further than nearest neighbors [7, 8]. Therefore, the single-cell "average" like in Eq. (16) is inappropriate and a coarse-graining over several cells is necessary. Let us divide the N atoms into N/ν blocks containing ν atoms each. Statistical *variables* $\hat{\eta}_\alpha$ are defined as

$$\hat{\eta}_\alpha = \frac{1}{\nu} \sum_{i \epsilon \alpha(\mathbf{r})} \eta_i. \tag{26}$$

Now for any set of *numbers* $\{\eta\} = \eta_{\alpha_1}, \eta_{\alpha_2}, \dots, \eta_{\alpha_\nu}$ we can (in principle) compute the restricted average over η_i such that $\langle \hat{\eta}_\alpha \rangle = \eta_\alpha$:

$$\left\langle \exp\left(-\frac{H}{k_B T}\right) \right\rangle_{<\hat{\eta}_\alpha>=\eta_\alpha} = \exp\left(-\frac{F\{\eta\}}{k_B T}\right).$$

According to the Ginzburg-Landau approach [1], the equilibrium fluctuations of the order parameter are given by the probability distribution

$$\rho_{eq}\{\eta\} = \frac{1}{Z} \exp\left(-\frac{F\{\eta\}}{k_B T}\right). \tag{27}$$

Because in the coarse-grained variables the microscopic information about atomic movements is lost, their time evolution should be sought in the framework of some phenomenological approach similar in the spirit to the G-L method. Such an approach is considered in the next section.

4.2. THE MASTER EQUATION

In the absence of microscopic Hamiltonian, the evolution of a statistical system is usually described with the use of the *Master equation* (ME). The latter describes a stochastic Markovian process which means that the evolution is assumed to depend only on the current state of the system and does not contain any memory about its previous history. This is the essence of the ME approach.

The fundamental ME describes the evolution of a probability distribution $\rho(\{\eta\}, t)$ over the configurations $\{\eta\}$ as

$$\frac{\partial \rho}{\partial t} = \int \delta \eta' \left[P(\{\eta\}, \{\eta'\}) \rho\{\eta'\} - P(\{\eta'\}, \{\eta\}) \rho\{\eta\} \right], \tag{28}$$

where $P(\{\eta'\}, \{\eta\})$ is the transition rate (from η to η'). The meaning of Eq. (28) is very simple. It describes two processes: the first term corresponds to the growth of the probability to find the value η of the order parameter due to the transitions from other states (η') while the second term describes the loss of this probability due to the transitions from the state η into other states. The ME satisfies the requirement of the probability conservation

$$\int \delta\eta\rho\{\eta\} = 1 \qquad (29)$$

and *should satisfy* the fundamental condition of the detailed balance:

$$P(\eta \to \eta') \exp\left(-\frac{F\{\eta\}}{k_B T}\right) = P(\eta' \to \eta) \exp\left(-\frac{F\{\eta'\}}{k_B T}\right) \qquad (30)$$

which guarantees that the equilibrium distribution (27) is the end of all evolution:

$$\partial\rho_{eq}/\partial t = 0. \qquad (31)$$

By augmenting these two obvious requirements with some symmetry considerations and the requirement of simplicity in the spirit of the G-L approach it is possible to derive the equations describing the evolution of the order parameter. This is illustrated below for the case of the nonconserved order parameter.

4.3. NONCONSERVED DYNAMICS

Assuming *local* dynamics let us choose the local transition rate as

$$R(\eta' = \eta_\alpha + \epsilon, \eta = \eta_\alpha) = \bar{T}(\epsilon) \exp\left[\frac{F(\eta_\alpha) - F(\eta_\alpha + \epsilon)}{2k_B T}\right], \qquad (32)$$

where the arguments of F should be understood as in the left hand side, i. e., that the configuration η' is different from η only within one cell α where the value of the order parameter is changed by ϵ. It should be stressed that despite the awkward appearance, Eq. (32) is, presumably, the simplest choice compatible with conditions (29) and (30). This can be seen by substituting Eq. (32) into the ME (28):

$$\begin{aligned}
\frac{\partial\rho}{\partial t} &= \sum_\alpha \int_{-\infty}^{\infty} d\epsilon \bar{T}(\epsilon) \left\{\exp\left[\frac{F(\eta_\alpha + \epsilon) - F(\eta_\alpha)}{2k_B T}\right]\right. \\
&\quad \left. \cdot\rho(\eta_\alpha + \epsilon) - \exp\left[\frac{F(\eta_\alpha) - F(\eta_\alpha + \epsilon)}{2k_B T}\right]\rho(\eta_\alpha)\right\}.
\end{aligned} \qquad (33)$$

Thus, in view of Eq. (27) condition (30) is satisfied due to the exponential factors in Eq. (32) while condition (29) is fulfilled provided $\bar{T}$ is symmetric: $\bar{T}(\epsilon) = \bar{T}(-\epsilon)$. The transition intensity $\bar{T}(\epsilon)$ may be considered as describing random change of the order parameter due to thermal fluctuations. If the fluctuations around mean value of η referred to one atom have some unknown probability distribution with variance Γ then according to the *central limit theorem* the distribution of their mean (26) is Gaussian with the variance

$$\int_{-\infty}^{\infty} d\epsilon\, \epsilon^2 \bar{T}(\epsilon) \equiv \frac{\Gamma}{\nu}.$$

This makes $\bar{T}(\epsilon)$ symmetric and guarantees that the total transition rate do not depend on the artificial number ν (in view of $\sum_\alpha \to \nu v \int d\mathbf{r}$). Using this equality we expand the second multiplier in the integrand of Eq. (33) up to the second order in ϵ which amounts to taking the second derivative with respect to $\eta' = \eta + \epsilon$. This leads to the Fokker-Plank equation for the probability current

$$\frac{\partial \rho\{\eta\}}{\partial t} = -\sum_\alpha \frac{\partial J_\alpha}{\partial \eta_\alpha}, \tag{34}$$

where

$$J_\alpha = -\frac{\Gamma}{2\nu}\left(\frac{1}{k_B T}\frac{\partial F}{\partial \eta_\alpha}\rho + \frac{\partial \rho}{\partial \eta_\alpha}\right). \tag{35}$$

The continuity Eq. (34) guarantees the probability conservation while the probability current (35) turns to zero for the equilibrium distribution (27), as it should be.

4.4. THE ALLEN-CAHN EQUATION

Let us apply the above equations to the evolution of the average

$$\bar{\eta}_\alpha = \int \left(\prod_\beta d\eta_\beta\right)\rho\{\eta\}\eta_\alpha \equiv \int \delta\eta\, \rho\eta_\alpha:$$

$$\frac{\partial \bar{\eta}_\alpha}{\partial t} = \int \delta\eta \frac{\partial \rho}{\partial t}\eta_\alpha = \int \delta\eta\, J_\alpha = -\frac{\Gamma}{2k_B T\nu}\int \delta\eta \frac{\partial F}{\partial \eta_\alpha}\rho,$$

where we used the integration by parts together with Eqs. (34) and (35). In the continuum representation (using $\frac{\partial}{\partial \eta_\alpha} = \nu v \frac{\delta}{\delta\eta(\mathbf{r})}$) the last equation reads

$$\frac{\partial \bar{\eta}(\mathbf{r})}{\partial t} = -\frac{\Gamma v}{2k_B T}\left\langle \frac{\delta F}{\delta\eta(\mathbf{r})}\right\rangle.$$

Here we make the mean field approximation (2) by replacing the statistical average of a function by the function of the *mean field*:

$$\frac{\partial \bar{\eta}}{\partial t} = -\frac{\Gamma v}{2k_B T}\frac{\delta F\{\bar{\eta}\}}{\delta\bar{\eta}(\mathbf{r})}. \tag{36}$$

This equation turns into the *the Allen-Cahn equation* [9] with the conventional choice of $F[\eta]$ as in Eq. (20)

$$\frac{\partial \bar{\eta}}{\partial t} = -\frac{\Gamma}{2k_B T}\left(-m\nabla^2\bar{\eta} + \frac{df}{d\bar{\eta}}\right). \tag{37}$$

The equation corresponding to the conserved case is called the *Cahn-Hilliard equation* [10] and has a similar form:

$$\frac{\partial \bar{\eta}}{\partial t} = \frac{\Gamma}{2k_B T}\nabla^2\left(-m\nabla^2\bar{\eta} + \frac{df}{d\bar{\eta}}\right), \tag{38}$$

where the additional operator $\nabla \cdot \nabla$ guarantees the conservation of the order parameter in the course of the evolution. As we noted earlier, Eq. (38) was derived in the framework of the formalism used in this section by Langer [6].

5. Conclusion

In this introductory article only the most elementary topics pertinent to the broad field of the mean field and Ginzburg-Landau methods were considered. Much broader exposition of the subject can be found in the monographs of Landau and Lifshitz [1] (the group-theoretic approach), Ducastelle [4] (MF and G-L methods), Khachaturyan [5] (the concentration wave method and interphase boundaries). In connection with the phase field method [2] the following papers are considered to be the classics of the field: Ref. [9]—the Allen-Cahn equations and the sharp interface limit (for recent development of the latter see Refs. [11] and [12] and references therein). Refs. [10], [13], and [6] are devoted to the Cahn-Hilliard equation. In Refs. [14] and [15] the late stage coarsening or Oswald ripening corresponding to the metastable quench in Fig. 2.2.1 is considered. Finally, Ref. [16] is presumably the first review devoted to the theory of the solidification and dendrite growth. Recent literature on the subject is too numerous to be adequately reviewed here so we refer the reader to current publications and references therein.

6. Acknowledgments

The author expresses his gratitude to University Louis Pasteur de Strasbourg and IPCMS for their hospitality.

References

1. Landau, L.D. and Lifshitz, E.M. (1980) *Statistical Physics*, 3 Ed. Pergamon, Oxford.

44

2. Caginalp, G. and Fife, P. (1986) Phase-field methods for interfacial boundaries, *Phys. Rev. B* **33**, 7792–7794.
3. Ma, S.K. (1976) *Modern Theory of Critical Phenomena*. Benjamin, Reading, Mass.
4. Ducastelle, F. (1991) *Order and Phase Stability in Alloys*. North-Holland, Amsterdam.
5. Khachaturyan, A.G. (1983) *Theory of Structural Transformations in Solids*. Wiley, New York.
6. Langer, J.S. (1971) Theory of spinodal decomposition in alloys, *Ann. Phys.* **65**, 53–86.
7. Philibert, J. (1991) *Atome Movements*. Les Éditions de Physique, Les Ulis.
8. Allnatt, A.R. and Lidiard, A.B. (1993) *Atomic Transport in Solids*. Cambridge University Press, Cambridge.
9. Allen, S.M. and Cahn, J.W. (1979) A microscopic theory for antiphase boundary motion and its application to antiphase domain coarsening, *Acta Metallurgica* **27**, 1085–1095.
10. Cahn, J. W. and Hilliard, J. E. (1958) Free energy of a nonuniform system. I. Interfacial free energy, *J. Chem. Phys.* **28**, 258–267.
11. Grossmann, B. Guo, H., and Grant, M. (1991) Kinetic roughening of interfaces in driven systems, *Phys. Rev. A* **43**, 1727–1743.
12. Elder, K.R. Grant, M. Provatas, N. and Kosterlitz, J.M. (2001) Sharp interface limits of phase-field models, *Phys. Rev. E* **64**, 021604-1–18.
13. Cahn, J. W. and Hilliard, J. E. (1959) Free energy of a nonuniform system. III. Nucleation in a two-component incompressible fluid, *J. Chem. Phys.* **31**, 688–699.
14. Lifshitz, I.M. and Slyozov, V.V. (1961) The kinetics of precipitation from supersaturated solid solutions, *J. Phys. Chem. Solids* **19**, 35–50.
15. Wagner, C. (1961) Theorie der Alterung von Niederschlägen durch Umlösen. *Z. Elektrochem.* **65**, 581–591.
16. Langer, J.S. (1980) Instabilities and patter formation in crystal growth, *Rev. Mod. Phys.* **52**, 1–28.

MESOSCALE PHASE FIELD MODELING OF ENGINEERING MATERIALS : MICROSTRUCTURE AND MICROELASTICITY

YU U. WANG, YONGMEI M. JIN,
ARMEN G. KHACHATURYAN
Department of Ceramic and Materials Engineering,
Rutgers University
607 Taylor Road, Piscataway, New Jersey 08854, USA

Abstract

Recent development of the Phase Field Microelasticity theory has extended its applicability from diffusional and martensitic phase transformations to the dislocation dynamics, crack evolutions, and behavior of voids in elastically and structurally inhomogeneous solids under applied stress. The computational models based on this theory allow one to address problems of arbitrary microstructure evolution in complex systems. In particular, it enables one to investigate the structure-property relations of materials where different physical processes are simultaneously involved, such as phase transformations, dislocation plasticity, fracture, etc. It is the interplays between these distinct processes that determine the mechanical properties of engineering materials, where the long-range elastic interaction plays a key role. In this paper, the applications of the Phase Field Microelasticity theory to different nano- and meso-scale processes are discussed.

1. Introduction

The majority of engineering materials are multiphase coherent composites formed due to phase transformations. Their mechanical and other properties strongly depend on the shape, size, and mutual location of the product phase particles, i.e., on a group of factors that we collectively call microstructure. Processing of these materials is usually reduced to heat treatment schedules with or without external stress that lead to a microstructure providing optimal properties. Understanding the mechanisms controlling the microstructure-sensitive properties is important for the development of these materials. The most difficult problem here seems to be associated with the key role played by the transformation-induced strain in the formation and evolution of a stable or metastable microstructure as well as its long-range elastic interactions with crystal lattice defects, such as dislocations, microcracks, voids, etc. The misfitting strain is always generated by the phase transformations, both diffusional and diffusionless (martensitic). Because of a strong anisotropy and long-range nature of

47

A. Finel et al. (eds.), Thermodynamics, Microstructures and Plasticity, 47–64.

the strain field, the elastic energy depends not only on the product phase volume but also on its microstructure — the conventional volume-dependent "chemical" free energy does not depend on the microstructure. It is the interplays between multiple physical processes, phase transformations and crystal lattice defects, that determine the mechanical properties of material systems, such as strength and toughness.

The modeling of engineering materials is certainly far behind the modeling of macroscopic engineering structures. The computer-aided materials design is still a much more challenging problem due to the geometrical and physical complexity of a real material system. However, the advances in the materials theory and high-performance computers have been changing this situation drastically. In particular, the recent development of the Phase Field Microelasticity (PFM) theory provides a unified treatment of such apparently distinct physical processes as phase transformations, dislocation and crack dynamics, evolution of voids, etc. The realistic three-dimensional (3D) multi-physics modeling of engineering materials based on the PFM theory demonstrates that a computer-aided materials design via "virtual experiments" has become ever-closer than it seems.

The obvious and straightforward tool for multi-physics modeling could be atomic level simulation, such as first principle calculations and molecular dynamics. However, the requirement for simulating mesoscale phenomena is still far beyond the available computing power of state-of-the-art supercomputers. A viable alternative to the direct atomic level simulation is the macroscopic (continuum) approach based on the coarse-grained approximation. This is a continuum approximation, which assumes that a volume element of the inhomogeneous system consists of macroscopic number of atoms. The coarse-grained approximation drastically reduces the degrees of freedom required to describe the evolution of a system in the meso- and nano-scale. To characterize the system we need just several continuous density functions (Phase Fields) describing the mesoscale structure. For example, a concentration field can fully describe an arbitrary spatial pattern formed by concentration domains (precipitates) and its diffusional evolution. The concentration field alone is insufficient to characterize a system of multi-variant ordered precipitates. To describe such a system, we have to add the multi-component long-range order parameter field.

The PFM approach is an extended version of the Phase Field method that takes into account the configuration-dependent elastic strain energy generated by phase transformations and crystal lattice defects in the homogeneous elastic modulus approximation [1, 2]. The PFM theory is formulated in the spirit of Cahn-Hilliard [3] and Ginzburg-Landau [4] theories where the general symmetry rules related to the introduction of a multi-component long-range order parameter are employed. This approach has been developed for modeling of diffusional phase transformation (decomposition and ordering) [5-10] and martensitic transformation in elastically homogeneous single crystals [11-13] and polycrystals [14, 15].

Practically all phase transformations in solids are collective phenomena. They involve concomitant evolutions of two or more elastically interacting systems, viz., the system of evolving and stress-generating product phase particles and the system of spontaneously self-multiplying and self-organizing crystal lattice defects such as dislocations, cracks and voids.

A series of recent advances of the PFM theory significantly broadened its applicability, extending the PFM theory to dislocation dynamics [16, 17], crack evolutions [18], multi-void systems [18] and even more general elasticity problem of arbitrary elastically and structurally inhomogeneous solids [19] under applied stress. The latter development considerably extends the reach of the PFM theory to discontinuous systems and systems of different elastic moduli. Thanks to these advances, we have a real opportunity now to consider the evolution of several systems together as one common cooperative phenomenon.

In this paper, a unified treatment of apparently distinct physical processes via the PFM theory is presented. The emphasis is placed on the recent advances to cracks, voids, different modulus particles and dislocations. The relations and differences between specific models are discussed.

2. Phase Field Microelasticity Theory of Cracks and Voids

A body containing voids and/or cracks is no longer continuous or elastically homogeneous. Based on a variational principle established in a recent paper [18], the PFM theory provides a general approach to determine the 3D elastic field, displacement and elastic energy of an arbitrary macroscopically homogeneous pattern of multiple voids and/or cracks in an elastically anisotropic single crystal under heterogeneous internal stress and external load. Internal stress may be also caused by other crystal lattice defects such as dislocations and misfitting precipitates. In fact, the problem of voids and cracks under applied stress is reduced to the "equivalent" problem of the continuous elastically homogeneous body with mesoscopically heterogeneous misfit-generating stress-free strain. The stress-free strain minimizing the elastic energy functional in this equivalent problem is the exact solution of the elasticity problem for a discontinuous body with voids and/or cracks. In particular, it is the exact elasticity solution for a body with mixed mode cracks of arbitrary configuration.

Consider an elastically homogeneous body with the distributed misfit strain $\varepsilon_{ij}^{o}(\mathbf{r})$. We assume that $\varepsilon_{ij}^{o}(\mathbf{r})$ assumes non-zero values only within arbitrary-shaped domains and is zero outside the domains. These domains are chosen so that they exactly coincide with the voids and cracks. Domains characterizing cracks are thin platelets (slits) whose thickness is equal to the interplanar distance. The domains form a macroscopically homogeneous mesoscale pattern. The exact elastic energy of such a system is given by Khachaturyan-Shatalov theory [1, 2] as a functional of the misfit strain $\varepsilon_{ij}^{o}(\mathbf{r})$:

$$E^{el} = \frac{1}{2} \int_{V} C_{ijkl}^{o} \varepsilon_{ij}^{o}(\mathbf{r}) \varepsilon_{kl}^{o}(\mathbf{r}) \, d^{3}r - \bar{\varepsilon}_{ij} \int_{V} C_{ijkl}^{o} \varepsilon_{kl}^{o}(\mathbf{r}) \, d^{3}r + \frac{V}{2} C_{ijkl}^{o} \bar{\varepsilon}_{ij} \bar{\varepsilon}_{kl}$$
$$- \frac{1}{2} \int \frac{d^{3}k}{(2\pi)^{3}} n_{i} \tilde{\sigma}_{ij}^{o}(\mathbf{k}) \Omega_{jk}(\mathbf{n}) \tilde{\sigma}_{kl}^{o}(\mathbf{k})^{*} n_{l}, \tag{1}$$

where $\bar{\varepsilon}_{ij} = \frac{1}{V}\int_V \varepsilon_{ij}(\mathbf{r})d^3r$, V is the system volume, the integral $\dashint$ in the infinite reciprocal space is evaluated as a principal value excluding a volume $(2\pi)^3/V$ around the point $\mathbf{k} = 0$, $\mathbf{n} = \mathbf{k}/k$ is a unit directional vector in the reciprocal space, $\Omega_{ij}(\mathbf{n})$ is the Green function tensor inverse to the tensor $\Omega_{ij}^{-1}(\mathbf{n}) = C_{ikjl}^o n_k n_l$, C_{ijkl}^o is the elastic modulus, $\tilde{\sigma}_{ij}^o(\mathbf{k}) = C_{ijkl}^o\tilde{\varepsilon}_{kl}^o(\mathbf{k})$, the superscript asterisk indicates the complex conjugate, and $\tilde{\varepsilon}_{ij}^o(\mathbf{k})$ is the Fourier transform of the field $\varepsilon_{ij}^o(\mathbf{r})$, $\tilde{\varepsilon}_{ij}^o(\mathbf{k}) = \int_V \varepsilon_{ij}^o(\mathbf{r})e^{-i\mathbf{k}\cdot\mathbf{r}}d^3r$.

For such a system, we find the special misfit field $\varepsilon_{ij}^o(\mathbf{r}) = \varepsilon_{ij}^{oo}(\mathbf{r})$ confined within the domains that provides the minimum of the elastic energy functional (1). The minimum condition is the vanishing of the first variational derivative of the elastic energy (1) with respect to $\varepsilon_{ij}^o(\mathbf{r})$. This variational derivative is:

$$\frac{\delta E^{el}}{\delta \varepsilon_{ij}^o(\mathbf{r})} = -C_{ijkl}^o\left[\dashint\frac{d^3k}{(2\pi)^3}n_k\Omega_{lm}(\mathbf{n})\tilde{\sigma}_{mn}^o(\mathbf{k})n_n e^{i\mathbf{k}\cdot\mathbf{r}} - \varepsilon_{kl}^o(\mathbf{r}) + \bar{\varepsilon}_{kl}\right]. \qquad (2)$$

It can be shown that $\delta E^{el}/\delta\varepsilon_{ij}^o(\mathbf{r}) = -\sigma_{ij}(\mathbf{r})$ [18], where $\sigma_{ij}(\mathbf{r})$ is the stress caused by both the misfit strain $\varepsilon_{ij}^o(\mathbf{r})$ and the boundary constraint $\bar{\varepsilon}_{ij}$. Therefore, the minimum energy condition achieved at $\varepsilon_{ij}^o(\mathbf{r}) = \varepsilon_{ij}^{oo}(\mathbf{r})$, $\left[\delta E^{el}/\delta\varepsilon_{ij}^o(\mathbf{r})\right]_{\varepsilon_{ij}^o(\mathbf{r})=\varepsilon_{ij}^{oo}(\mathbf{r})} = 0$, implies that

$$\sigma_{ij}(\mathbf{r}_d) = 0, \qquad (3)$$

where $\mathbf{r}_d$ represents the points located within the domains, because the energy minimization is performed only at the points within the domains. This proves that the total stress within the domains vanishes when the elastic energy (1) reaches its minimum with respect to $\varepsilon_{ij}^o(\mathbf{r})$. It should be noted that although the total stress vanishes inside the domains at $\varepsilon_{ij}^o(\mathbf{r}) = \varepsilon_{ij}^{oo}(\mathbf{r})$, it does not vanish outside them. Since all domains in the continuous elastically homogeneous body with the misfit strain $\varepsilon_{ij}^{oo}(\mathbf{r})$ are stress-free, the material of the stress-free domains can be removed from the body, which is in the elastic equilibrium, without disturbing the elastic equilibrium in the system or affecting its total elastic energy. The removal of the domains leaves voids and/or cracks in the body, their shapes and locations coinciding with those of the domains.

The foregoing consideration proves the following variational principle:

If the misfit strain is located within domains of the equivalent homogeneous modulus body and minimizes its elastic energy, the equilibrium strain and elastic energy of this equivalent body is the equilibrium strain and elastic energy of the discontinuous body with voids and/or cracks.

The elastic energy minimizer $\varepsilon_{ij}^{oo}(\mathbf{r})$ could be found through a solution of the PFM kinetic equation at each point $\mathbf{r} = \mathbf{r}_d$, which is the Time-Dependent Ginzburg-Landau (TDGL) equation. In the case of static voids and/or cracks under applied stress, i.e., voids and cracks of fixed shape, the relaxing field $\varepsilon_{ij}^{o}(\mathbf{r})$ is contained only within the fixed-shaped domains (void volumes and/or crack slits). The TDGL equation in this case is:

$$\frac{\partial \varepsilon_{ij}^{o}(\mathbf{r}_d,t)}{\partial t} = -L_{ijkl}\frac{\delta E^{el}}{\delta \varepsilon_{kl}^{o}(\mathbf{r}_d,t)}, \tag{4}$$

where L_{ijkl} is the kinetic coefficient. The simplest choice for L_{ijkl} is $L_{ijkl} = L\delta_{ik}\delta_{jl}$. Using Eq. (2) into Eq. (4) gives the explicit form of the TDGL equation, which can be solved efficiently using the Fast Fourier Transform technique. Examples of numerical solution of Eq. (4) for specific crack/void systems and quantitative comparison with analytical solutions (if available) are shown in Figs. (1) and (2) and were presented in ref. [18].

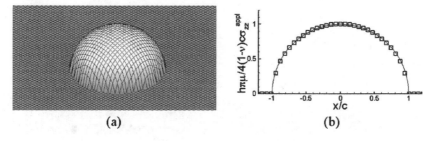

(a)   (b)

Figure 1. PFM computational results for a mode I circular (penny-shaped) crack in 3D elastically isotropic body ($\nu = 0.3$). (a) Ellipsoidal crack-opening profile determined by the PFM method for the case of linear elasticity; (b) a quantitative comparison of the computed (squares) and analytical [20] (solid line) crack-opening profiles along a cross-section through the crack center.

Fig. 1 shows the PFM solution for a penny-shaped crack in 3D elastically isotropic solid under mode I load. According to the analytical result of the linear elasticity theory [20], the crack-opening profile is an ellipsoid. The PFM solution reproduces the ellipsoidal shape of the crack-opening, as shown in Fig. 1(a). Fig. 1(b) shows a quantitative comparison of the computed result (squares) with the analytical solution (solid line) of the crack-opening along the cross-section through the crack center. They agree with each other.

Fig. 2 shows a 2D problem of a mode I crack in the vicinity of a cylindrical hole. Although this geometry shown in figure 2(a) is quite simple, this problem proves to be too complicated to be solved analytically. The PFM calculated stress fields are shown in Fig. 2(b)-(d). It illustrates the elastic interaction between the crack and the hole.

In the situation where the cracks in a multi-crack system can spontaneously nucleate, propagate and coalesce under applied stress, the effective misfit strain $\varepsilon_{ij}^o(\mathbf{r})$ is no longer constrained within fixed domains and is allowed to evolve driven by a reduction of the total system free energy. The field $\varepsilon_{ij}^o(\mathbf{r})$ can be considered as a non-conserved long-range order parameter. In this formalism, the field $\varepsilon_{ij}^o(\mathbf{r})$ describes the evolving cracks: regions where $\varepsilon_{ij}^o(\mathbf{r})$ is not zero are the laminar domains describing cracks, while regions where $\varepsilon_{ij}^o(\mathbf{r})$ vanishes are the solid material. This theory does not impose constraint on possible crack evolution path.

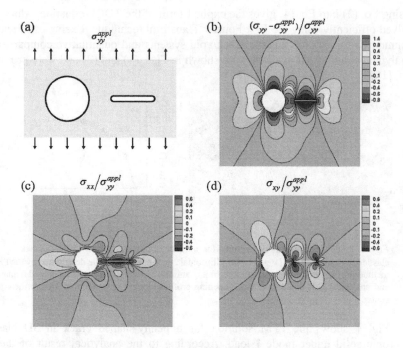

Figure 2. PFM calculated stress fields of a mode I crack in the vicinity of a cylindrical hole in 2D elastically isotropic body ($v = 0.3$) and their elastic interaction. (a) Schematic illustration of the system; the computed disturbed stress fields: (b) $\left(\sigma_{yy} - \sigma_{yy}^{appl}\right)/\sigma_{yy}^{appl}$; (c) $\sigma_{xx}/\sigma_{yy}^{appl}$; and (d) $\sigma_{xy}/\sigma_{yy}^{appl}$.

Consider a crystalline body whose crystallography is characterized by several cleavage planes described by their reciprocal lattice vectors. The reciprocal lattice vector $\mathbf{H}(\alpha)$ of the cleavage plane α can be presented as $\mathbf{H}(\alpha) = \mathbf{m}(\alpha)/d(\alpha)$, where $\mathbf{m}(\alpha)$ is the unit vector normal to cleavage plane α, and $d(\alpha)$ is the interplanar distance of this plane α. The index $\alpha = 1, 2, ..., p$ numbers all possible cleavage planes (for example, $p = 3$ for the body-centered cubic system where $\alpha = 1, 2, 3$ numbers three possible {100} cleavage planes). Under the load, the opposite surfaces of a planar crack in the cleavage plane α undergo relative displacement $\mathbf{h}(\alpha, \mathbf{r})$ associated with the crack opening. To equivalently describe a profile of the displacement discontinuity, $\mathbf{h}(\alpha, \mathbf{r})$, at each point $\mathbf{r}$ on the crack surface, we introduce an effective misfit strain distribution that takes a form of Invariant Plane Strain, $\varepsilon_{ij}^{o}(\alpha, \mathbf{r}) = h_i(\alpha, \mathbf{r}) H_j(\alpha)$. The vector $\mathbf{h}(\alpha, \mathbf{r})$ and thus the field $\varepsilon_{ij}^{o}(\alpha, \mathbf{r})$ describing the configuration of all cracks of α-th cleavage plane assume non-zero value only within laminar domains characterizing the crack volumes and vanish outside them. The total effective misfit strain generated by arbitrary crack ensemble in which all p cleavage planes are operative is

$$\varepsilon_{ij}^{o}(\mathbf{r}) = \sum_{\alpha=1}^{p} h_i(\alpha, \mathbf{r}) H_j(\alpha). \tag{5}$$

The elastic energy of the system is given by Eq. (1) with $\varepsilon_{ij}^{o}(\mathbf{r})$ given by Eq. (5).

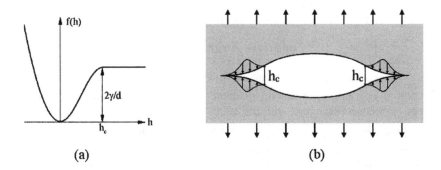

(a)   (b)

Figure 3. The crack tip cohesive force introduced through Orowan potential [21] shown in (a) is in spirit consistent to that introduced by Barenblatt [22] through an explicit distribution of the intense force acting in small crack tip zones illustrated in (b).

In order to describe the effect of cohesive forces resisting crack opening, we have to supplement the elastic energy (1) with the coarse-grained Landau-type "chemical" energy $f^{ch}(\mathbf{h}(\alpha))$ [18]. The energy, $f^{ch}(\mathbf{h}(\alpha))$, is defined as a cohesion energy required to provide a separation of two pieces of crystals with respect to the cleavage plane α by the distance $\mathbf{h}(\alpha)$. From the microscopic point of view, the energy $f^{ch}(\mathbf{h}(\alpha))$ is the atomistic energy required for a continuous breaking of atomic bonds across the cleavage plane and thus creating free surfaces during a process of crack formation. An accurate determination of the cohesion energy requires atomistic calculations. However, we can approximate it by the Orowan-like potential for mode I crack [21]. A general behavior of this function is illustrated in Fig. 3(a). It provides correct quadratic (harmonic) behavior at small crack-opening h corresponding to small strain. When the separation distance h is comparable with the interatomic interaction radius characterized by h_c, the deviation from the quadratic dependence caused by the anharmonic effects becomes significant: at large h, the effective elastic modulus of the material is significantly reduced until it vanishes. This effect is described as a smooth transition of the cohesion energy function from a quadratic form at small h to a plateau at large h. The value h_c is a critical separation distance above which all atomic bonds between separated planes are completely broken and further increase in h no longer affects the cohesion energy. The latter corresponds to a creation of new free surfaces. The value $f^{ch}(\mathbf{h}(\alpha))$ in this definition is a specific energy calculated per unit interplanar volume of the cleavage planes. Its saturation value achieved at the plateau is $2\gamma/d(\alpha)$ where γ is the surface energy coefficient. It is noteworthy that the crack tip cohesive force introduced through the Orowan potential [21] as shown in Fig. 3(a) is consistent with that introduced by Barenblatt [22] through an explicit distribution of intense force acting in small crack tip zones as schematically illustrated in Fig. 3(b).

For a multi-cleavage-plane system where all p cleavage planes are operative, the coarse-grained Landau energy can be formulated as:

$$E^{ch} = \sum_{\alpha=1}^{p} \int_{V} \left\{ f^{ch}(\mathbf{h}(\alpha,\mathbf{r})) + \beta(\alpha)[\mathbf{H}(\alpha) \times \nabla(\mathbf{h}(\alpha,\mathbf{r}) \cdot \mathbf{H}(\alpha))]^2 \right\} d^3r, \qquad (6)$$

where $\beta(\alpha)$ is a material constant for the cleavage plane α, and $\nabla = \partial/\partial\mathbf{r}$ is the differential operator. The second gradient term in Eq. (6) takes into account the energy correction associated with the effect of the crack surface curvature, since the first term, the cohesion energy $f^{ch}(\mathbf{h}(\alpha))$, only describes a "homogeneous" separation where both boundaries of the crack-opening are kept flat and parallel to the cleavage plane α. This specific form of the gradient term assumes that the crack tip energy does not depend on the direction of the crack front in the cleavage plane (directional isotropy). As usual in the Phase Field theories, it provides a short-range correlation between the values assumed by the displacement $\mathbf{h}(\alpha,\mathbf{r})$ at the neighboring points $\mathbf{r}$.

The total energy functional of the multi-crack system is a sum of the contributions from elastic energy (1) and "chemical" Landau energy (6):

$$F = E^{el} + E^{ch}. \qquad (7)$$

The fields $\mathbf{h}(\alpha,\mathbf{r})$ characterize the crack system configuration and crack openings. Using the relation (5) between the misfit strain $\varepsilon_{ij}^{o}(\mathbf{r})$ and the fields $\mathbf{h}(\alpha,\mathbf{r})$, we can formulate the total energy (7) as a functional of $\mathbf{h}(\alpha,\mathbf{r})$. The variational derivative, $\delta F/\delta h_i(\alpha,\mathbf{r},t)$, which is the driving force for the crack system development, can be also presented in terms of the fields $\mathbf{h}(\alpha,\mathbf{r})$. Assuming that the relaxation rate is proportional to the thermodynamic driving force gives the nonlinear TDGL stochastic equation in terms of $\mathbf{h}(\alpha,\mathbf{r})$:

$$\frac{\partial h_i(\alpha,\mathbf{r},t)}{\partial t} = -L_{ij}\frac{\delta F}{\delta h_j(\alpha,\mathbf{r},t)} + \xi_i(\alpha,\mathbf{r},t), \qquad (8)$$

where L_{ij} is the kinetic coefficient characterizing the crack propagation mobility, and $\xi_i(\alpha,\mathbf{r},t)$ is the Langevin Gaussian noise term reproducing the effect of thermal fluctuations. Eq. (8) is similar to Eq. (4). The difference is that in the former we use the energy functional (7) rather than solely the strain energy (1). With this modification, the temporal and spatial dependences of the crack-opening $\mathbf{h}(\alpha,\mathbf{r})$ and, thus, the crack system geometry and its evolution will be described by a solution of the TDGL kinetic equation (8).

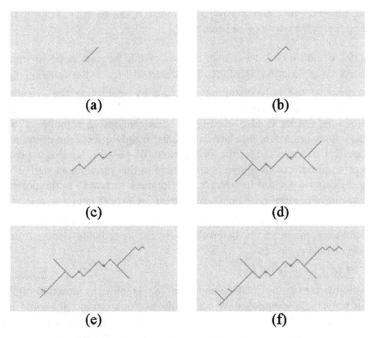

Figure 4. Under tensile stress in the vertical direction, an inclined crack seeks the most favorable cleavage planes to propagate while relieving as much energy as possible associated with the external loading device. This results in a complicated branching pattern of cracks.

Fig. 4 shows one example of the PFM solution of the TDGL equation (8). Two cleavage planes form equal angle with the vertically applied tensile stress. A pre-existing crack in one cleavage plane inclined to the tensile stress direction, as shown in Fig. 4(a), seeks the most favorable cleavage planes to propagate driven by the tensile stress. At the same time, it also tries to relieve as much energy as possible associated with the external loading device. This and the strain field generated by the crack system lead to a formation of complicated crack pattern with branches.

56

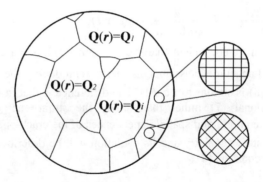

Figure 5. Schematic illustration of a polycrystal model. A rotation matrix field **Q(r)** can completely describe the polycrystalline structure [17].

To model a multi-crack system in a polycrystal, we just need to introduce a rotation matrix field function $\mathbf{Q(r)}$. As illustrated in Fig. 5, the function $\mathbf{Q(r)}$ can completely describe the polycrystalline structure by specifying the orientations of individual grains. If each grain is elastically isotropic, the equivalent polycrystalline system containing cracks is also elastically homogeneous: a rotation of the grain crystal lattices does not affect its elastic modulus; it only affects the crystallography-related properties of cracks, i.e., the orientations of possible cleavage planes. The variational principle discusses above is applicable to this problem as well. Therefore, the only modification required to address development of cracks in the polycrystal is a necessity to introduce the coordinate-dependent reciprocal lattice vector, $H_i(\alpha, \mathbf{r})$.

In the global coordinate system, the coordinate dependence of the reciprocal lattice vector of cleavage plane α is determined by the coordinate dependence of the rotation matrix of the grains as

$$H_i(\alpha, \mathbf{r}) = Q_{ij}(\mathbf{r}) H_j^o(\alpha). \tag{9}$$

With this modification, the TDGL equation (8) is able to describe multi-crack evolution in polycrystal composed of elastically isotropic grains. It should be noted that the limitation of elastic isotropy of grains could be lifted if we use the PFM theory of elastic inhomogeneity that will be present in the next section.

Fig. 6 shows an example of the PFM simulation of crack propagation in a 2D polycrystal. The polycrystalline structure is composed of randomly oriented elastically isotropic grains that are rods perpendicular to the plane of the figure. The grains of different orientation are shown by different gray scales. Two cleavage planes in each grain, (100) and (010), are assumed to be operative. In this specific simulation, the weakening of bonds at the grain boundaries was not taken into account. Under tensile stress applied in the vertical direction, the propagation of a pre-existing crack has been simulated by solving Eq. (8). The simulation results show that when the crack crosses the grain boundary and enters the neighboring grain, it is deflected from its original propagation path by automatically picking up the energetically favored cleavage plane in this grain. This results in the meandering cleavage path as shown in Fig. 6.

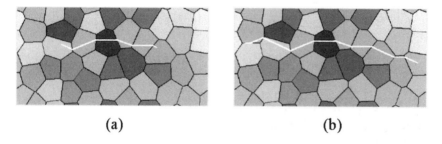

(a)   (b)

Figure 6. PFM simulation of intragranular crack propagation in a 2D polycrystal.

More PFM simulations of cracks in single crystals and material toughening due to the crack and second-phase obstacle interactions were presented in paper [18].

3. Phase Field Microelasticity Theory of Elastic and Structural Inhomogeneities

The variational principle proved in the previous section can be generalized to an elastically anisotropic solid of arbitrary inhomogeneous modulus that also contains arbitrary structural inhomogeneities. The elastic modulus of such a body is coordinate-dependent, $C_{ijkl}(\mathbf{r})$. The structural inhomogeneities are described by the fixed crystal lattice misfit strain, $\varepsilon_{ij}^*(\mathbf{r})$, arbitrarily distributed in the body. The misfit strain $\varepsilon_{ij}^*(\mathbf{r})$ can be generated by a fixed distribution of crystal lattice defects, such as coherent new phase inclusions, concentration heterogeneities, dislocations, *etc.* The system can be either a single crystal or polycrystal. This is the most general formulation of the problem. It is applicable to a majority of technologically important materials.

We can present the coordinate-dependent modulus $C_{ijkl}(\mathbf{r})$ as

$$C_{ijkl}(\mathbf{r}) = C_{ijkl}^o - \Delta C_{ijkl}(\mathbf{r}), \tag{10}$$

where $\Delta C_{ijkl}(\mathbf{r})$ is the modulus variation from the reference value C_{ijkl}^o. It can be shown that we can always find an "equivalent" elastically homogeneous system containing a "virtual" misfit strain $\varepsilon_{ij}^o(\mathbf{r})$ that is defined by [19]

$$C_{ijkl}^o \, \varepsilon_{kl}^o(\mathbf{r}) = C_{ijkl}^o \, \varepsilon_{kl}^*(\mathbf{r}) + \Delta C_{ijkl}(\mathbf{r})\left[\varepsilon_{kl}(\mathbf{r}) - \varepsilon_{kl}^*(\mathbf{r})\right], \tag{11}$$

where $\varepsilon_{kl}(\mathbf{r})$ is the equilibrium strain assumed by the original system. The equivalent system assumes both the same strain and the same stress as the original system does [19].

The elastic energy of the original elastically and structurally inhomogeneous system can be obtained through the elastic energy of the equivalent elastically homogeneous system with the virtual field $\varepsilon_{ij}^o(\mathbf{r})$, where the latter is given by Eq. (1). In fact, the elastic energy of the original system is [19]

$$E^{inhom} = \frac{1}{2}\int_V \left[C_{ijmn}^o \Delta S_{mnpq}(\mathbf{r}) C_{pqkl}^o - C_{ijkl}^o \right] \left[\varepsilon_{ij}^o(\mathbf{r}) - \varepsilon_{ij}^*(\mathbf{r}) \right] \left[\varepsilon_{kl}^o(\mathbf{r}) - \varepsilon_{kl}^*(\mathbf{r}) \right] d^3r$$

$$+ \frac{1}{2}\int_V C_{ijkl}^o \, \varepsilon_{ij}^o(\mathbf{r}) \, \varepsilon_{kl}^o(\mathbf{r}) \, d^3r - \bar{\varepsilon}_{ij} \int_V C_{ijkl}^o \, \varepsilon_{kl}^o(\mathbf{r}) \, d^3r + \frac{V}{2} C_{ijkl}^o \, \bar{\varepsilon}_{ij} \, \bar{\varepsilon}_{kl}$$

$$- \frac{1}{2}\int \frac{d^3k}{(2\pi)^3} n_i \, \tilde{\sigma}_{ij}^o(\mathbf{k}) \, \Omega_{jk}(\mathbf{n}) \, \tilde{\sigma}_{kl}^o(\mathbf{k})^* n_l. \tag{12}$$

It can be proved that this strain energy functional meets the following variational principle [19]:

The misfit strain minimizing the elastic energy functional (12) determines the equilibrium strain and stress of the elastically and structurally inhomogeneous body.

It is worth noting that the voids and cracks can be treated as special cases of elastic inhomogeneities, i.e., inhomogeneities of zero elastic modulus. As a matter of fact, assuming $\Delta C_{ijkl}(\mathbf{r}) = C_{ijkl}^o$ that results in $C_{ijkl}(\mathbf{r}) = 0$ reduces Eq. (12) to Eq. (1).

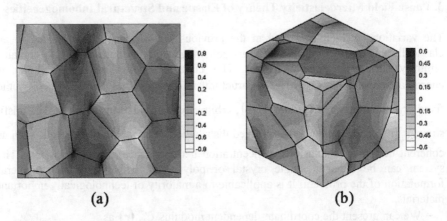

Figure 7. PFM determined equilibrium stress fields of (a) a 2D textured and (b) a 3D random polycrystal comprised of elastically anisotropic grains under uniaxial stress. The shades show the disturbed stress normalized by applied stress.

The energy minimizer $\varepsilon_{ij}^{oo}(\mathbf{r})$ could be found as a "steady state" solution of the TDGL equation for $\varepsilon_{ij}^o(\mathbf{r})$:

$$\frac{\partial \varepsilon_{ij}^o(\mathbf{r},t)}{\partial t} = -L_{ijkl} \frac{\delta E^{inhom}}{\delta \varepsilon_{kl}^o(\mathbf{r},t)}, \tag{13}$$

where L_{ijkl} is the kinetic coefficient, t is "time", and E^{inhom} is given by Eq. (12). The simplest choice for L_{ijkl} is $L_{ijkl} = L\delta_{ik}\delta_{jl}$. The equilibrium strain is given through the minimizer $\varepsilon_{ij}^{oo}(\mathbf{r})$ as

$$\varepsilon_{ij}(\mathbf{r}) = \bar{\varepsilon}_{ij} + \frac{1}{2}\int \frac{d^3k}{(2\pi)^3} \left[n_i \Omega_{jk}(\mathbf{n}) + n_j \Omega_{ik}(\mathbf{n}) \right] C_{klmn}^o \, \tilde{\varepsilon}_{mn}^{oo}(\mathbf{k}) \, n_l \, e^{i\mathbf{k}\cdot\mathbf{r}}. \tag{14}$$

A quantitative comparison of PFM calculations with analytical solutions for elastic inhomogeneous inclusions was presented in [19]. Fig. 7 shows examples of the equilibrium stress fields determined by the PFM method for a 2D textured and a 3D random polycrystal comprised of anisotropic (cubic symmetry) grains under the uniaxial stress.

4. Phase Field Microelasticity Theory of Dislocations

Dislocations can be treated as a special while much simpler case of cracks, where the variational principle is not required to determine the exact elastic field caused by multiple dislocations. In fact, the elasticity problem for dislocations can be solved directly by using Khachaturyan-Shatalov theory [1, 2]. This idea was employed in [23] to model the effect of static dislocations on decomposition. A model of dynamic dislocations was also developed in [24] employing the same idea, which is similar to the PFM dislocation model [16].

There is a profound analogy between cracks and dislocations. As a matter of fact, crack can be, in general, regarded as three self-organizing pileups of dislocations of distinct Burgers vectors under the action of stress. Let us consider the crack opening vector $h(\alpha, r)$. Under the load, the opposite surfaces of a planar crack in the cleavage plane α undergo relative displacement $h(\alpha, r)$. This results in the displacement discontinuity. If the vector $h(\alpha, r)$ is normal to the cleavage plane, the crack is of mode I (opening mode) and the displacement discontinuity can be regarded as a result of prismatic dislocation pileup. If the vector $h(\alpha, r)$ is parallel to the cleavage plane, the crack is of either mode II (sliding mode) or mode III (tearing mode) and the displacement discontinuity can be regarded as a result of either edge or screw dislocation pileup. In the case of an arbitrary direction of $h(\alpha, r)$, we deal with a mixed mode crack that is equivalent to three pileups of prismatic, edge and screw dislocations. It should be noted that the crack opening varies under varying stress, therefore the dislocation pileups also self-adjust to reproduce the same effect of the crack.

As in the case of cracks, in order to equivalently describe a Burgers vector distribution, $b(\alpha, r)$, associated with the slip system α at each point r in the body, we introduce an effective misfit strain distribution that takes a form of Invariant Plane Strain, $\varepsilon_{ij}^o(\alpha, r) = b_i(\alpha, r) H_j(\alpha)$, where $H(\alpha)$ is the reciprocal lattice vector of the slip plane corresponding to the slip system α [16]. The index $\alpha = 1, 2,..., p$ numbers all possible slip systems (for example, $p = 12$ for the face-centered cubic system where $\alpha = 1, 2,..., 12$ numbers 12 slip systems corresponding to 4 possible {111} slip planes with 3 possible <110> slip directions in each slip plane). The total effective misfit strain generated by arbitrary dislocation ensemble in which all p slip systems are operative is [16]

$$\varepsilon_{ij}^o(\mathbf{r}) = \sum_{\alpha=1}^{p} b_i(\alpha, \mathbf{r}) H_j(\alpha). \tag{15}$$

The exact elastic energy of the system is given by Eq. (1) with $\varepsilon_{ij}^o(\mathbf{r})$ given by Eq. (15).

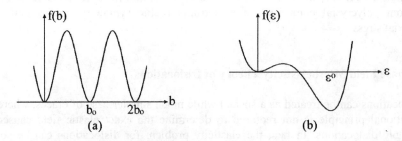

Figure 8. Schematic illustration of the general behavior of the coarse-grained Landau-type "chemical" energy function for (a) dislocation and (b) martensitic transformation.

In order to describe the effect of crystal lattice on dislocation motion, we have to supplement the elastic energy (1) with the coarse-grained Landau-type "chemical" energy $f^{ch}(\mathbf{b}(\alpha))$ [16], where $\mathbf{b}(\alpha)$ is the Burgers vector of the slip system α. The energy, $f^{ch}(\mathbf{b}(\alpha))$, is defined as a crystalline energy. A general behavior of this crystalline energy function is illustrated in Fig. 8(a). The difference between the neighboring minima is one elementary Burgers vector b_0. It is noteworthy that, in contrast to the Orowan potential of cracks (see Fig. 3(a)) where an energy plateau allows the crack surfaces to freely separate, minimize system energy and approach the elastic equilibrium, the crystalline energy of dislocations has infinitely degenerated global minima that are required to provide a discrete nature of a shear, i.e., the Burgers vector field $\mathbf{b}(\alpha,\mathbf{r})$ assumes a value that is integer times the elementary Burgers vector.

For a multi-slip system where all p slip modes are operative, the coarse-grained Landau energy can be formulated as:

$$E^{ch} = \sum_{\alpha=1}^{p} \int_{V} \left\{ f^{ch}(\mathbf{b}(\alpha,\mathbf{r})) + \sum_{\alpha'=1}^{p} \beta_{ijkl}(\alpha,\alpha')\nabla_i b_j(\alpha,\mathbf{r})\nabla_k b_l(\alpha',\mathbf{r}) \right\} d^3r, \qquad (16)$$

where $\beta_{ijkl}(\alpha,\alpha')$ is a material constant related to slip systems α and α'. The second term in Eq. (16), in fact a gradient term, takes into account the energy correction associated with the dislocation core and vanishes over slip planes as required for perfect dislocations. As usual in the Phase Field theories, it also provides a short-range correlation between the values assumed by the Burgers vector $\mathbf{b}(\alpha,\mathbf{r})$ at the neighboring points $\mathbf{r}$.

The total energy functional of the multi-dislocation system is a sum of the contributions from elastic energy (1) and "chemical" Landau energy (16), which is presented by the same equation (7) as for cracks. In analogy to cracks, the evolution of a multi-dislocation system is characterized by a TDGL equation similar to that in Eq. (8),

$$\frac{\partial b_i(\alpha, \mathbf{r}, t)}{\partial t} = -L_{ij} \frac{\delta F}{\delta b_j(\alpha, \mathbf{r}, t)} + \xi_i(\alpha, \mathbf{r}, t). \tag{17}$$

where L_{ij} is the kinetic coefficient characterizing the dislocation mobility. With the modification in the Landau energy, the temporal and spatial dependences of the Burgers vector $\mathbf{b}(\alpha, \mathbf{r})$ and, thus, the dislocation system configuration and its evolution will be described by a solution of the TDGL kinetic equation (17), which is structurally the same as that for cracks.

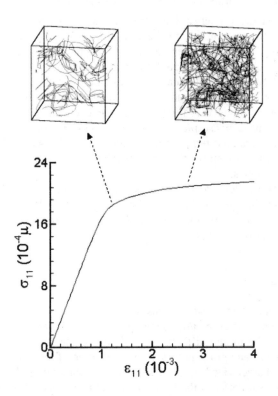

Figure 9. PFM simulated stress-strain curve for uniaxial loading during plastic deformation and the corresponding 3D dislocation structures.

A treatment for dislocation dynamics in a polycrystal can be also developed in the same way as that for cracks. Such a theory and model was presented in [17]. Fig. 9 shows one example of PFM simulation of self-multiplying and self-organizing dislocations in the single crystal plasticity. More PFM simulations of dislocation dynamics in single crystals and polycrystals were presented in [16, 17].

5. Phase Field Microelasticity Theory of Martensitic Transformation

In the PFM formalism, the dislocations and martensitic transformation can be treated in the same fashion. A dislocation is characterized by a pure shear strain (15), while martensitic transformation is characterized by a displacive transformation strain $\varepsilon_{ij}^o(\mathbf{r})$. The elastic energies of both systems are given by Eq. (1). The only difference between them lies in the specific form of the coarse-grained Landau energies. The Landau energy of the martensitic transformation is the chemical energy $f^{ch}(\varepsilon_{ij}(\alpha))$ [14], where $\varepsilon_{ij}(\alpha)$ is the transformation strain of martensite variant α. A general behavior of the chemical energy function for the martensitic transformation is illustrated in Fig. 8(b). In contrast to the crystalline energy of dislocations whose global minima are infinitely degenerated, the chemical energy of martensitic transformation has a finitely degenerated minima: the number of global minima with the same chemical free energy is equal to the number of martensite variants. While the driving force for cracks and dislocations is applied stress, the driving force for martensitic transformation is the difference in the chemical energies between the parent and the martensite phase plus the applied stress. For dislocations the energy contribution from the gradient term is proportional to the dislocation loop perimeter, while for the martensitic transformation this contribution is proportional to the surface area of martensite particles.

With this modification, a PFM model can be formulated for martensitic transformation in single crystals [11-13] and polycrystals [14, 15]. Fig. 10 shows one example of PFM simulation of low-symmetry martensitic transformation in the AuCd shape memory alloy polycrystal [14].

6. Concluding Remarks

The PFM theory and models of martensitic transformation, dislocation plasticity, crack evolution, voids and elastically inhomogeneous solids can be formulated in the similar formalisms. The formal difference is only in the analytical form of the coarse-grained Landau energy and gradient terms. The evolution of the microstructures in all these problems is characterized by conceptually similar theory and simulated by the similar method. The similarity in the PFM models of these distinct physical processes makes easy their integration into one unified model. A cost of this would be just an increase in the number of fields. The use of such a unified model would be a significant advance in addressing the scientific and engineering problems of the materials processing and, in particular, in realistic modeling of the microstructure and structure-property relations in the nano- and meso-scales. To illustrate a breadth of possible applications of the PFM approach, we just mention three examples of very different areas, where the theoretical characterization of a material system is impossible without utilizing the computationally effective model like PFM. They are the swelling of irradiated materials, the effect of the elastic modulus misfit on the morphology of coherent precipitates, and sintering under applied stress. These three areas are just several examples of the vast area of application of the PFM theory and computational approach.

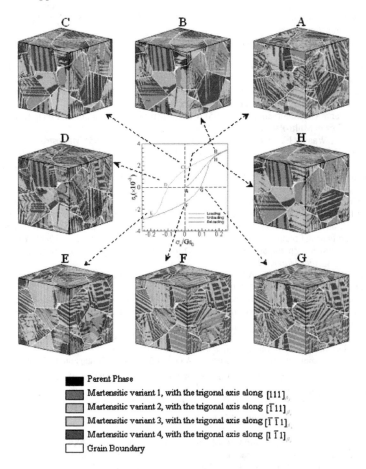

Figure 10. PFM simulation of the hysteresis loop and the corresponding 3D microstructures of ζ_2' martensite in AuCd polycrystalline alloys.

Acknowledgement

The authors gratefully acknowledge a support from National Science Foundation under grant DMR-9817235. The results were obtained using parallel code running on NPACI supercomputers.

64

References

1. Khachaturyan, A.G. and Shatalov, G.A. (1969) Theory of Macroscopic Periodicity for a Phase Transformation in the Solid State, *Zr. Eksp. Teor. Fiz.* **56**, 1032; (1969) *Sov. Phys. JETP* **29**, 557 (English Translation).
2. Khachaturyan, A.G. (1983) *Theory of Structural Transformations in Solids*, Wiley & Sons, New York.
3. Cahn, J.W. and Hilliard, J.E. (1958) Free Energy of a Nonuniform System. I. Interfacial Free Energy, *J. Chem. Phys.* **28**, 258.
4. Lifshitz, E.M. and Pitaevskii, L.P. (1980) *Statistical Physics*, 3rd edition, Part 1, *Landau and Lifshitz Course of Theoretical Physics*, Vol. 5, Pergmon Press, Oxford.
5. Wang, Y. and Khachaturyan, A.G. (1995) Microstructural Evolution during Precipitation of Ordered Intermetallics in Multi-Particle Coherent Systems, *Phil.Mag. A* **72**, 1161.
6. Wang, Y.Z., Wang, H.Y., Chen, L.Q. and Khachaturyan, A.G. (1995) Microstructural Development of Coherent Tetragonal precipitates in Mg partially Stabilized Zirconia, *J. Am. Ceramic Soc.* **78**, 657.
7. Wang, Y., Banerjee, D., Su, C.C. and Khachaturyan, A.G. (1998) Field Kinetic Model and Computer Simulation of Precipitation of L1$_2$ Ordered Intermetallics from FCC Solid Solution, *Acta Mater.* **46**, 2983.
8. Li, D.Y. and Chen, L.Q. (1998) Morphological Evolution of Coherent Multi-Variant Ti$_{11}$Ni$_{14}$ Precipitates in Ti-Ni Alloy under Applied Stress-Computer Simulation, *Acta Mater.* **46**, 639.
9. Poduri, R. and Chen, L.Q. (1997) Computer Simulation of the Kinetics of Order-Disorder and Phase Separation During Precipitation of Delta-Prime (Al$_3$Li) in Al-Li Alloys, *Acta Mater.* **45**, 245.
10. Rubin, G. and Khachaturyan, A.G. (1999) Three-Dimensional Model of Precipitation of Ordered Intermetallics, *Acta Mater.* **47**, 1995.
11. Wang, Y. and Khachaturyan, A.G. (1997) Three-Dimensional Field Model and Computer Modeling of Martensitic Transformations, *Acta Mater.* **45**, 759.
12. Artemev, A., Wang, Y. and Khachaturyan, A.G. (2000) Three-Dimensional Phase Field Model and Simulation of Martensitic Transformation in Multilayer Systems under Applied Stresses, *Acta Mater.* **48**, 2503.
13. Artemev, A., Jin, Y.M. and Khachaturyan, A.G. (2001) Three-Dimensional Phase Field Model of Proper Martensitic Transformation, *Acta Mater.* **49**, 1165.
14. Jin, Y.M., Artemev, A. and Khachaturyan, A.G. (2001) Three-Dimensional Phase Field Model of Low-Symmetry Martensitic Transformation in Polycrystal: Simulation of ζ_2' Martensite in AuCd Alloys, *Acta Mater.* **49**, 2309.
15. Artemev, A., Jin, Y.M. and Khachaturyan, A.G. (2002) Three-Dimensional Phase Field Model and Simulation of Cubic→Tetragonal Martensitic Transformation in Polycrystals, *Philos. Mag. A* **82**, 1249.
16. Wang, Y.U., Jin, Y.M., Cuitiño, A.M. and Khachaturyan, A.G. (2001) Nanoscale Phase Field Microelasticity Theory of Dislocations: Model and 3D Simulations, *Acta Mater.* **49**, 1847.
17. Jin, Y.M. and Khachaturyan, A.G. (2001) Phase Field Microelasticity Theory of Dislocation Dynamics in Polycrystal: Model and 3D Simulations, *Phil. Mag. Lett.* **81**, 607.
18. Wang, Y.U., Jin, Y.M. and Khachaturyan, A.G. (2002) Phase Field Microelasticity Theory and Simulation of Multiple Voids and Cracks in Single Crystals and Polycrystals under Applied Stress, *J. Appl. Phys.* **91**, 6435.
19. Wang, Y.U., Jin, Y.M. and Khachaturyan, A.G. (2002) Phase Field Microelasticity Theory and Modeling of Elastically and Structurally Inhomogeneous Solid, *J. Appl. Phys.* **92**, 1351.
20. Green, A.E. and Sneddon, I.N. (1950) The Distribution of Stress in the Neighbourhood of a Flat Elliptical Crack in an Elastic Solid, *Proc. Cambridge Phil. Soc.* **46**, 159.
21. Orowan, E. (1948) Fracture and Strength of Solids, *Reports on Progress in Physics* **12**, 185.
22. Barenblatt, G.I. (1962) The Mathematical Theory of Equilibrium Cracks in Brittle Fracture, *Advances in Applied Mechanics* **7**, 55.
23. Hu, S.Y. and Chen, L.Q. (2001) Solute Segregation and Coherent Nucleation and Growth near a Dislocation: a Phase-Field Model Integrating Defect and Phase Microstructures, *Acta Mater.* **49**, 463.
24. Finel, A. (2000) Phase field methods and dislocations (preprint).

PHASE-FIELD MODELS OF MICROSTRUCTURAL PATTERN FORMATION

ALAIN KARMA
*Department of Physics and
Center for Interdisciplinary Research on Complex Systems,
Northeastern University, Boston MA 02215*

1. Introduction

The last decade has witnessed an outburst of activity in phase-field modeling. This activity reflects to a large degree the success of diffuse interface models to reproduce complex interfacial patterns that have been historically very difficult (or virtually impossible) to simulate with sharp-interface models. The phase-field method is being applied to an increasing number of problems in different fields, but its main application to date has been to problems in materials science. A list of topics, which is by no means complete, includes alloy solidification (Boettinger *et al.*, 2002), various solid-state transformations involving the nucleation and growth of precipitates as well as structural transformations associated with symmetry changes (Chen and Wang, 1996), stress-driven interfacial instabilities (Muller and Grant, 1999; Kassner and Misbah, 2002), microstructural evolution in polycrystalline materials (Chen and Yang, 1994; Steinbach *et al.*, 1996; Kobayashi *et al.*, 1998; Lobkovsky and Warren, 2002; Gránásy *et al.*, 2003), surface evolution during epitaxial growth (Liu and Metiu, 1994; Karma and Plapp, 1998; Elder *et al.*, 2002), dislocation dynamics (Rodney and Finel, 2001; Wang *et al.*, 2001; Rodney *et al.*, 2003), and fracture (Aranson *et al.*, 2000; Karma *et al.*, 2001; Eastgate *et al.*, 2002; Wang *et al.*, 2002).

The goal of this article is to introduce the phase-field method at a level that is accessible to graduate students in science and engineering disciplines as well as researchers with no prior knowledge of this method. The next section summarizes the evolution of this method over the last two decades, with emphasis on the relationship between phase-field models and coarse-grained models of second-order phase transitions (Halperin and Hohenberg, 1974). Interestingly, these direct anscestors of phase-field models

A. Finel et al. (eds.), Thermodynamics, Microstructures and Plasticity, 65–89.
© 2003 Kluwer Academic Publishers. Printed in the Netherlands.

were developed to describe the large scale behavior of matter near critical points, where this behavior does not depend on microscopic details. In contrast, patterns formed by strongly first-order phase transitions such as solidification do depend on details of atomic interactions that dictate the energetics and kinetics of interfaces (see the accompanying article by Asta in this book (Asta, 2003)). In this latter context, order-parameter descriptions have proven useful both to simulate microstructural evolution on a mesoscale and to describe approximately microscopic equilibrium and nonequilibrium properties of a diffuse interface. In section 3, we describe in more details the application of the phase-field approach to model dendritic and eutectic solidification microstructures.

2. Overview

2.1. FROM CRITICAL POINTS TO FIRST-ORDER FREEZING

The phase-field approach was introduced originally with the idea to facilitate numerical simulations of dendritic solidification (Collins and Levine, 1985; Langer, 1986). In this problem, the main difficulty is to track the motion of a sharp boundary between two phases. This difficulty is ubiquitous in phase transformation problems and becomes exacerbated in three dimensions, with several phases or grain orientations present, or with singular changes of interface connectedness. Langer (Langer, 1986) and Collins and Levine (Collins and Levine, 1985) proposed that this difficulty could be circumvented for solidification by the introduction of a scalar field $\phi(\vec{r}, t)$ that takes on constant values in solid and liquid and varies smoothly across a thin, spatially diffuse, interface. This scalar field was termed "phase-field" because of its key role in distinguishing phases. Furthermore, these authors wrote down a set of coupled equations for the evolution of $\phi(\vec{r}, t)$ and the temperature field $T(\vec{r}, t)$ that reduce to the classical free-boundary model for the solidification of a pure melt in the "sharp-interface limit" where the width of the diffuse interface is vanishingly small.

Even though the idea to model dendritic evolution with a diffuse interface was novel at that time, the phase-field equations themselves were not. This point was emphasized by Langer in a proceeding paper of a conference held in the honor of Shang-Ken Ma (Langer, 1986). Halperin, Hohenberg and Ma had written down in the seventies a series of continuum models (denoted alphabetically by A, B, C, etc) to study the critical behavior of second order phase-transitions using a renormalization group approach (Halperin and Hohenberg, 1974). These continuum models, which are now part of the classical literature on critical phenomena, are based on a coarse-grained description of matter where order-parameters represent local volume averages of physical quantities. The total volume integral of a "conserved" order parameter (such as the density) does not vary in time and the opposite is true for a "non-conserved" order parameter such as the local magnetization (i.e. a ferromagnet can acquire a spontaneous total magnetization when quenched below the Curie temperature). Langer showed that the phase-field model of the solidification of a pure melt can be written in a form analogous to model C of critical phenomena with the non-conserved order parameter being the phase-field $\phi(\vec{x}, t)$ that distinguishes solid from liquid and the conserved order parameter being a linear combination of $\phi(\vec{x}, t)$ and $T(\vec{r}, t)$ that one can interpret as the energy density.

It is perhaps not entirely surprising that similar continuum equations can be used to model first- and second-order phase transitions. It is important to emphasize, however, that the justification for using these models is very different, and almost opposite, for these two cases. To illustrate this point, consider the free-energy functional

$$F = \int dV \left[\frac{\sigma}{2} |\vec{\nabla}\phi|^2 + a\phi^2 + b\phi^4 \right], \qquad (1)$$

where dV is a volume element of space and the integrand represents the free-energy density of the system. This type of expansion was introduced by Ginzburg and Landau in the study of superconductivity. In this context, ϕ represents the density of Cooper pairs (Ginzburg and Landau, 1950). Similar order-parameter descriptions have been applied to a wide range of phase transitions (Hohenberg and Halperin, 1977) and nonequilibrium pattern formation problems (Cross and Hohenberg, 1993). The form of the Ginzburg-Landau expansion of the free-energy is generally determined by the symmetry properties of the order-parameter(s) and the system. The simplest application of these models in the context of critical phenomena is to the Ising model near its Curie point, T_c, where $\phi(\vec{x}, t)$ is the local magnetization and $a \sim T - T_c$ in Eq. 1 (Ma, 1976). Expanding the free-energy density in the form of Eq. 1 is justified because the correlation length [1], or the width of the interface between up ($\phi > 0$) and down ($\phi < 0$) magnetic domains, diverges near a critical point. It therefore makes sense to only retain the lowest order gradient term in the expansion of the free-energy density near the Curie point (with $|\vec{\nabla}\phi|^2$ being the first term invariant under translation and rotation). In addition, since the magnetization is small near this point, it is reasonable to only include the first few polynomial terms that are even in ϕ, as imposed by the additional symmetry constraint that the free-energy must be be left unchanged by the operation $\phi \to -\phi$. At a more quantitative level, this expansion can be formally justified by a full blown

---

[1] The correlation length ξ measures the spatial rate of exponential decay of the two-point correlation function $\langle \phi(\vec{x})\phi(\vec{x}') \rangle \sim \exp(-|\vec{x} - \vec{x}'|/\xi)$.

renormalization group analysis (Ma, 1976). This analysis shows how thermal fluctuations influence the system behavior on larger and larger scales through a renormalization of the coefficients of the free-energy density, and why higher order derivatives and powers of ϕ are irrelevant. Namely, these "irrelevant" higher-order terms in the free-energy density do not affect the large scale critical behavior of the system (for example the value of the critical exponent β that characterizes how the average magnetization vanishes as $\bar\phi \sim (T_c - T)^\beta$ below the Curie point). Universal aspects of this behavior depend only on general properties such as spatial dimension, the symmetry of the order parameter, as well as conservation laws (Ma, 1976; Hohenberg and Halperin, 1977).

This universality is absent in strongly first-order phase transitions such as freezing where the interface width is microscopic in size and thermal fluctuations are small. Clearly, the energetic and kinetic properties of the solid-liquid interface, and hence the size, shapes, and growth rate of dendrites on a mesoscale, are ultimately controlled by details of inter-atomic forces (Asta, 2003). Therefore, to what extent is a coarse-grained continuum approach justified to model freezing and other phase transformations, and on what length and time scales?

2.2. MESOSCALE

The fact that the phase-field model is not fully realistic on a microscopic scale is not usually a major limitation for the purpose of simulating microstructural evolution. What is important is that this model reduces to the desired set of sharp-interface equations in some appropriate limit. First-order freezing can be modeled with the same free-energy functional defined by Eq. 1 with ϕ interpreted as some arbitrary measure of crystalline order and with the bulk free-energy density having the form of a double-well potential. A simple choice is $a = -h/2$ and $b = h/4$, where h is the height of the energy barrier between solid and liquid. The calculation detailed in Appendix A shows that this model describes a diffuse interface with a width $W \sim \sqrt{\sigma/h}$ and an excess free-energy $\gamma \sim Wh$. Furthermore, it can be shown to reproduce the classical Gibbs-Thomson condition

$$T_I = T_M \left(1 - \frac{\gamma}{L}\kappa\right), \qquad (2)$$

where T_I is the interface temperature, κ is the interface curvature, and L is the latent heat of melting per unit volume. Similarly, the kinetic law $V = \mu(T_M - T_I)$ for the crystal growth rate V of a planar interface can also be recovered from the phase-field model together with an expression for the kinetic coefficient μ that depends on the kinetic rate constant K_ϕ that governs the phase-field evolution. Anisotropic forms of γ and μ can also be incorporated into the model by letting σ and K_ϕ depend on the direction normal to the diffuse interface defined by $\vec{\nabla}\phi/|\vec{\nabla}\phi|$ (Zia and Wallace, 1985; Kobayashi, 1993 and 1994; Karma and Rappel, 1998).

Phase-field models have been used extensively to simulate dendrite growth with standard finite-difference schemes (Kobayashi, 1993 and 1994; Kupferman et al., 1994; Murray et al., 1995; Wheeler et al., 1993; Wang and Sekerka, 1996; Warren and Boettinger, 1995; Karma and Rappel, 1996; Karma and Rappel, 1998; Gránásy et al., 2003) as well as adaptive algorithms that solve efficiently the diffusion equation on a scale much larger than the dendrite tip (Braun and Murray, 1997; Provatas et al., 1998; Provatas et al., 1999; Plapp and Karma, 2000). The fact that the width (W) of the diffuse interface is typically much smaller than the size of the microstructure makes the phase-field equations extremely stiff. Therefore, carrying out mesoscale simulations on experimentally relevant length and time scales is still an ongoing challenge.

For the purpose of quantitative modeling, W should in principle be reduced until the simulation results become independent of W, meaning that they are "converged". The freedom to vary W may seem strange since the nanometer scale width of a real interface is fixed. However, the barrier height h of the double-well free-energy density and the time constant of the phase-field evolution can be varied simultaneously with W so as to reproduce the same magnitude of capillary and kinetic effects (e.g. by keeping the product Wh and hence $\gamma \sim Wh$ constant). This freedom can be exploited to carry out quantitative simulations with W small enough to accurately resolve the interface curvature, but large enough for mesoscale simulations to be computationally feasible, e.g. $W \sim 100$ nm for a 1 μm dendrite tip radius. This requires, however, to carry out a detailed asymptotic analysis of the thin-interface limit of the phase-field model that is valid for such a "mesoscopic interface width" (Karma and Rappel, 1996; Karma and Rappel, 1998; Almgren, 1999; Karma, 2001), as discussed further in Sec. 3.2.3 for a dilute binary alloy.

This analysis provides the guide to formulate the phase-field model to insure that spurious corrections to the sharp-interface model associated with a thick interface can be eliminated. In some applications, these corrections may be unimportant. For dendrite growth, however, the pattern evolution is highly sensitive to small modifications of the velocity-dependent form of the Gibbs-Thomson condition, which includes interface kinetics, or the heat/mass conservation (Stefan) condition. Therefore, experience has shown that all corrections need to be eliminated for this problem. While this has been achieved for the solidification of elementary (Karma and Rappel, 1996; Karma and Rappel, 1998; Bragard et al., 2002) and binary (Karma, 2001) liquids, it is not yet clear whether this is generally possible for more

70

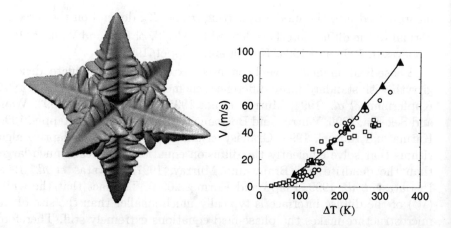

Figure 1. Phase-field simulations of the dendritic crystallization of pure liquid Ni (Bragard *et al.*, 2002) using anisotropic capillary and kinetic properties of the solid-liquid interface computed from atomistic simulations (Hoyt *et al.*, 1999; Hoyt *et al.*, 2001). Left: growth morphology for an undercooling $\Delta T = 87$ K. Right: comparison of dendrite velocity versus undercooling in simulations (filled triangles and solid line) and two sets of experiments by Lum et al. (Lum *et al.*, 1996) (open squares) and Willnecker et al. (Willnecker *et al.*, 1989) (open circles).

complicated cases. Achieving this goal more generally is crucially important given that mesoscale phase-field simulations are not likely to be feasible with a microscopic interface width any time soon (see Sec. 3.2.3).

Mesoscale simulations are also often limited by the lack of knowledge of interface properties that are difficult to obtain experimentally. Progress has been made recently to overcome this limitation by combining atomistic and phase-field simulations. In particular, molecular dynamics (MD) simulations have been used to compute anisotropic forms of γ (Hoyt *et al.*, 2001) and μ (Hoyt *et al.*, 1999) that were then used as input into mesoscale simulations of the crystallization of pure Ni for deep undercoolings (Bragard *et al.*, 2002). Results of this work are illustrated in Fig. 1.

Finally, thermal fluctuations usually have a negligible effect on microstructural evolution during first-order phase transformations. Dendritic solidification is an interesting counterexample where the amplification of thermal fluctuations on the sides of needle crystals suffices to give birth to sidebranches (Karma and Rappel, 1999). Essentially the same Langevin formalism used in continuum models of critical phenomena (Hohenberg and Halperin, 1977) can be used to incorporate thermal fluctuations in the phase-field model (Karma and Rappel, 1999; Pavlik and Sekerka, 1999). The phase-field approach has also been extended recently to model the effect of fluid flow on solidification patterns (Tonhardt and Amberg, 1998;

Anderson *et al.*, 2000; Beckerman *et al.*, 1999; Tong *et al.*, 2000 and 2001).

2.3. MICROSCOPIC SCALE

Order-parameter theories have also been used to model equilibrium (Ramakrishnan and Yussouff, 1979) and non-equilibrium (Shen and Oxtoby, 1996)[2] properties of the freezing transition on a microscopic scale. Density functional theory (DFT) uses a continuum approach where the free-energy is a functional of both the density and scalar order parameters ϕ_i that describe the crystallinity of the solid. Using an approximation where the density is represented as a sum of Gaussians centered around each atom of the crystal, Oxtoby and co-workers have shown that an infinite set of ϕ_i's can be reduced to a single order-parameter ϕ representing the amplitude of the density wave in the solid. Since this amplitude vanishes in the liquid, this order parameter varies smoothly across a diffuse interface, like ϕ in the phase-field approach. DFT has been used successfully to predict static thermodynamic properties of the freezing transition (Ramakrishnan and Yussouff, 1979) as well as the crystal growth rate (Mikheev and Chernov, 1991; Shen and Oxtoby, 1996) with the lattice structure as input. The DFT prediction of this rate for a Lennard-Jones liquid (Shen and Oxtoby, 1996) has been found to be in relatively good quantitatitve agreement with molecular dynamics simulations (Broughton *et al.*, 1982).

Despite having similar order-parameters, DFT and the phase-field approach differ in that the primary goal of the former is to predict energetic and kinetic properties of the interface while the goal of the latter is to model the effect of these properties on a mesoscale. It should be emphasized, however, that phase-field approaches have also been used successfully to model microscopic scale phenomena, even if they are more phenomenological than DFT on this scale. One example, which is analyzed in section 3 for the solidification of a dilute alloy, is solute trapping. This phenomenon occurs when the solutal diffusion length becomes comparable to the microscopic width of the diffuse interface (Ahmad *et al.*, 1998). The phase-field model predicts that the trapping velocity increases with the propensity of the alloy to partition solute in equilibrium, as observed experimentally (Aziz, 1996). In addition, a new phase-field strategy has been recently developed to model the evolution of polycrystalline materials with a conserved order parameter that is the local density, as in DFT, but that is not constrained to a fixed lattice (Elder *et al.*, 2002). This approach is able, for example, to model at the same time the evolution of a thin epitaxial growth layer and the microscopic formation of stress-induced defects within this layer. It is clear from the above examples that the phase-field approach, and more

---

[2]A list of earlier references can be found in this paper.

generally order-parameter models, are capable to describe phenomena on very different length and time scales.

3. Phase-field Models of Microstructural Pattern Formation

In this section, we discuss the phase-field approach to model dendritic and eutectic microstructures that form during the solidification of alloys (Wheeler *et al.*, 1992; Caginalp and Xie, 1993; Karma, 2001; Karma, 1994; Elder *et al.*, 1994; Wheeler *et al.*, 1996; Tiaden *et al.*, 1998; Kim *et al.*, 1999; Lo *et al.*, 2001; Boettinger *et al.*, 2002). We first write down a general form of the phase-field model for the isothermal solidification of a two-phase binary alloy and then explore in more details two specific cases: a dilute ideal solution and a eutectic alloy. The dilute alloy model (Karma, 2001b) is formulated such that the phase-field and solute profiles across the diffuse interface and the phase diagram parameters can be calculated analytically. This feature simplifies greatly the analysis of solute trapping at large growth rate. This analysis, which is similar to the one carried out by Ahmad *et al.* (Ahmad *et al.*, 1998), illustrates the use of the phase-field approach to model phenomena on the scale of the diffuse interface, in this case the departure from chemical equilibrium. The analysis of solute trapping also provides the basis to understand how to modify the phase-field model to quantitatively simulate microstructural evolution with a mesoscale interface width (Karma, 2001). The extension of the approach to model multicomponent alloys is briefly mentioned at the end (Plapp and Karma, 2002).

3.1. BASIC EQUATIONS

3.1.1. *Isothermal dynamics*
The starting point of the model is an expression for the total free-energy of the system that can be written in the form

$$F[\phi, c, T] = \int dV \left[\frac{\sigma}{2} |\vec{\nabla}\phi|^2 + \frac{\sigma_c}{2} |\vec{\nabla}c|^2 + f_{AB}(\phi, c, T) \right], \qquad (3)$$

where $f_{AB}(\phi, c, T)$ denotes the bulk free-energy density of a binary mixture of A and B atoms/molecules and c denotes the solute concentration defined as the mole fraction of B. One expects physically the system to relax to a global free-energy minimum. Dynamical equations that satisfy this constraint have the same form as the equations of model C of critical phenomena (Halperin and Hohenberg, 1974)

$$\frac{\partial \phi}{\partial t} = -K_\phi \frac{\delta F}{\delta \phi}, \qquad (4)$$

$$\frac{\partial c}{\partial t} = \vec{\nabla} \cdot \left(K_c \vec{\nabla} \frac{\delta F}{\delta c} \right), \qquad (5)$$

which couples a non-conserved and a conserved order parameter (here ϕ and c, respectively). $\delta F/\delta \phi$ denotes the functional derivative of F with respect to ϕ that is defined in Appendix A (and similarly for $\delta F/\delta c$). Eq. 5 is equivalent to the mass continuity relation with $\mu^c \equiv \delta F/\delta c$ identified as the chemical potential and $-K_c \vec{\nabla} \mu^c$ as the solute current density.

PROBLEM 1. Show that the above equations guarantee that F decreases monotonically in time (gradient dynamics).

SOLUTION. Let us differentiate both sides of Eq. 3 with respect to time and use Eqs. 4-5 to eliminate $\partial_t \phi$ and $\partial_t c$ inside the integrand on the right-hand-side. Furthermore, we use the fact that

$$\mu^c \vec{\nabla} \cdot \left(K_c \vec{\nabla} \mu^c \right) = \vec{\nabla} \cdot \left(\mu^c K_c \vec{\nabla} \mu^c \right) - K_c |\vec{\nabla} \mu^c|^2. \tag{6}$$

From the divergence theorem, the integral of the first term on the right-hand-side above gives zero if there is no solute flux through the boundaries of the system. We therefore obtain

$$\frac{dF}{dt} = - \int dV \left(K_\phi \left[\frac{\delta F}{\delta \phi} \right]^2 + K_c |\vec{\nabla} \mu^c|^2 \right) \leq 0. \tag{7}$$

It is important to emphasize that while the variational form of the dynamics is a nice way to motivate thermodynamically the form of the phase-field equations, this form need not be strictly enforced. In fact, non-variational formulations of the dynamics have been shown to offer more flexibility to obtain the desired thin-interface limit of the phase-field model in mesoscale simulations (Karma and Rappel, 1998; Karma, 2001). These formulations remain physically meaningfull because they produce essentially the same interface evolution as the known free-boundary problem of alloy solification. We return to this point below when when we discuss the difficulties of mesoscale simulations for a dilute alloy.

3.1.2. Equilibrium phase diagram

With little loss of generality, it is convenient to fix the value of ϕ to +1 in the solid and −1 in the liquid by requiring that

$$\left. \frac{\partial f_{AB}(\phi, c, T)}{\partial \phi} \right|_{\phi=\pm 1} = 0, \tag{8}$$

in which case $\partial_t \phi = 0$ away from the diffuse interface. With this choice, the Helmholtz free energies of the solid and liquid phases are defined by

$$f_s(c, T) = f_{AB}(+1, c, T), \tag{9}$$
$$f_l(c, T) = f_{AB}(-1, c, T). \tag{10}$$

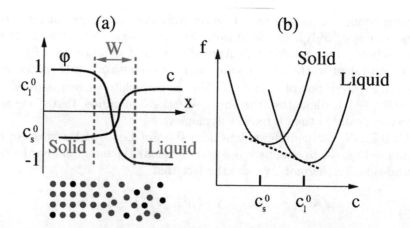

Figure 2. Schematic representation of (a) the phase-field and concentration profiles across the diffuse interface region of a two-phase binary alloy of A and B atoms; (b) the bulk solid and liquid free-energy densities (solid lines) and the common tangent (dashed line) that determines the equilibrium values of the bulk concentrations in the solid (c_s^0) and liquid (c_l^0).

A convenient choice for the free-energy density is

$$f_{AB}(\phi, c, T) = \frac{1 + g(\phi)}{2} \, f_s(c, T) + \frac{1 - g(\phi)}{2} \, f_l(c, T), \qquad (11)$$

where $g(\phi)$ is a function whose first derivative vanishes at $\phi = \pm 1$ and whose limits are $g(\pm 1) = \pm 1$. It follows from this choice that f_{AB} can reduce to an arbitrary form of the free-energy density of the solid (liquid) for $\phi = +1$ ($\phi = -1$). Hence, the phase-field model can be used to represent an arbitrary solid-liquid binary alloy phase-diagram. Physically, Eq. 11 implies that the thermodynamic properties of the diffuse interface region is an admixture of the properties of the bulk solid and liquid phases.

In equilibrium, $\partial_t \phi = \partial_t c = 0$, and Eqs. 4 and 5 reduce to

$$\frac{\delta F}{\delta c} = \mu_E^c, \qquad (12)$$

$$\frac{\delta F}{\delta \phi} = 0, \qquad (13)$$

where μ_E^c is the spatially uniform equilibrium value of the chemical potential. These two conditions uniquely determine the spatially varying stationary profiles of c and ϕ in the diffuse interface region. The c-profile varies between the equilibrium concentration $c_s^0(T)$ ($c_l^0(T)$) in the solid (liquid) phase, which define the equilibrium phase diagram of the alloy. These concentrations are determined by the standard common tangent construction that is shown schematically in Fig. 2. This construction is equivalent to requiring that the chemical potential and the grand potential (i.e. thermodynamic potential for a varying number of solute particles) be equal in the solid and liquid, or, respectively

$$\partial_c f_s(c,T)|_{c=c_s^0} = \partial_c f_l(c,T)|_{c=c_l^0} = \mu_E^c, \tag{14}$$

$$f_s(c_s^0,T) - \mu_E^c\, c_s^0 = f_l(c_l^0,T) - \mu_E^c\, c_l^0, \tag{15}$$

where ∂_c denotes partial derivative with respect to c. These conditions fix uniquely $c_s^0(T)$, $c_l^0(T)$, and μ_E^c.

PROBLEM 2. Derive Eqs. 14-15 starting from Eqs. 12-13.

SOLUTION. Let us start by writing out explicitly Eqs. 12-13 for the one-dimensional stationary interface depicted in Fig. 2

$$-\sigma_c \frac{d^2 c}{dx^2} + \partial_c f_{AB} = \mu_E^c, \tag{16}$$

$$-\sigma \frac{d^2 \phi}{dx^2} + \partial_\phi f_{AB} = 0. \tag{17}$$

The equality of chemical potentials (Eq. 14) follows directly from Eq. 16 and the fact that $dc/dx = 0$ in both phases in equilibrium. To obtain the equality of the grand potentials (Eq. 15), multiply both sides of Eq. 16 and Eq. 17 by dc/dx and $d\phi/dx$, respectively, and add the two resulting equations. All the terms in the sum of these two equations are simple to rewrite in the form of derivatives with respect to x

$$-\frac{d}{dx}\left(\frac{\sigma_c}{2}\left[\frac{dc}{dx}\right]^2 + \frac{\sigma}{2}\left[\frac{d\phi}{dx}\right]^2 \right) + \frac{df_{AB}}{dx} = \mu_E^c \frac{dc}{dx}. \tag{18}$$

Now integrate both sides of Eq. 18 from $x = -\infty$ (solid) to $x = +\infty$ (liquid). Using the fact that $dc/dx = d\phi/dx = 0$ in both phases and that f_{AB} reduces to f_s and f_l in solid and liquid, respectively, we obtain Eq. 15.

3.2. DILUTE LIMIT

3.2.1. Model

As a first application, we consider the dilute limit, $c \ll 1$. In this limit, $f_{AB}(\phi, c, T)$ can be written as the sum of the free-energy of the pure material, denoted here by $f(\phi, T)$, and the contribution due to solute addition. The latter is itself the sum of two terms. The first is the standard entropy of mixing $RTv_0^{-1}(c \ln c - c)$ for a dilute alloy that has the same form in both phases; R is the universal gas constant and v_0 is the molar volume assumed constant. The second is the change $\epsilon(\phi)c$ of the energy density due to solute addition. Following Eq. 11, we write

$$\epsilon(\phi) = \frac{1 + g(\phi)}{2}\,\epsilon_s + \frac{1 - g(\phi)}{2}\,\epsilon_l, \qquad (19)$$

which interpolates between the values ϵ_s and $\epsilon_l < \epsilon_s$ in the solid and liquid, respectively. Recall, that $g(\phi)$ has the limits $g(\pm 1) = \pm 1$. Adding together the pure and solute contributions, we obtain

$$f_{AB}(\phi, c, T) = f(\phi, T_M) + f_T(\phi)\Delta T + \frac{RT_M}{v_0}(c \ln c - c) + \bar{\epsilon}c + g(\phi)\frac{\Delta\epsilon}{2}\,c, \quad (20)$$

where we have defined $\bar{\epsilon} \equiv (\epsilon_s + \epsilon_l)/2$ and $\Delta\epsilon = \epsilon_s - \epsilon_l$. Moreover, we have expanded the pure part to first order in $\Delta T \equiv T - T_M$ by defining the function $f_T(\phi) \equiv \partial f(\phi, T)/\partial T|_{T=T_M}$, and we have replaced RT/v_0 by RT_M/v_0 since we can neglect terms $\sim \Delta Tc$ in the dilute limit.

Let us now determine the equilibrium properties of this model, and define by $c_0(x)$ and $\phi_0(x)$ the stationary concentration and phase field profiles, respectively, along a direction x perpendicular to a planar interface. We choose the standard form of the double-well potential

$$f(\phi, T_M) = h(-\phi^2/2 + \phi^4/4), \qquad (21)$$

where h measures the height of the energy barrier between local minima corresponding to the solid and liquid. We also choose $\sigma_c = 0$ in Eq. 3 to keep the calculation analytically tractable. Applying the first equilibrium condition (12), we obtain at once

$$\frac{RT_M}{v_0}\ln c + \bar{\epsilon} + g(\phi)\frac{\Delta\epsilon}{2} = \mu_E^c, \qquad (22)$$

from which we deduce the form of the equilibrium partition coefficient

$$k_E \equiv \frac{c_s^0}{c_l^0} = \exp\left(-\frac{v_0\Delta\epsilon}{RT_M}\right), \qquad (23)$$

and the stationary concentration profile

$$c_0(x) = c_l^0 \exp\left(b\left[1 + g(\phi_0(x))\right]\right), \qquad (24)$$

where we have defined the constant $b = \ln k_E/2$. Now, applying the second equilibrium condition (13), we obtain

$$\sigma\frac{d^2\phi_0}{dx^2} + h(\phi_0 - \phi_0^3) = f_T'(\phi_0)\Delta T + g'(\phi_0)\frac{\Delta\epsilon}{2}\,c_0, \qquad (25)$$

where the primes denote differentiation with respect to ϕ. At this point, we note that if we choose the function $f_T(\phi)$ of the form

$$f_T(\phi) = \frac{RT_M}{v_0 m} \exp\left(b\left[1 + g(\phi)\right]\right), \tag{26}$$

the right-hand-side of Eq. 25 vanishes provided that

$$T = T_M + mc_l^0, \tag{27}$$

such that the constant $m < 0$ in Eq. 26 is simply the liquidus slope of the dilute alloy phase diagram. Moreover, the vanishing of the right-hand-side of Eq. 25 implies that $\phi_0(x)$ is the same tangent hyperbolic profile as for the pure case analyzed in Appendix A. This phase-field profile also determines $c_0(x)$ via Eq. 24 for a given choice of function $g(\phi)$.

To complete the model, we need to specify the transport coefficient K_c. The appropriate choice in the dilute limit is $K_c(\phi, c) = v_0 D(\phi)c/(RT_M)$, which insures that Eq. 5 reduces to Fick's law of diffusion in the bulk phases. Namely, with this choice, we obtain that the solute flux $-K_c\vec{\nabla}\mu^c = -D_{s,l}\vec{\nabla}c$, where $D_{s,l} = D(\pm 1)$ is the diffusivity in solid ($\phi = +1$) or liquid ($\phi = -1$)[3]. In addition, it is useful to define

$$D(\phi) = D_l q(\phi) \tag{28}$$

where $q(\phi)$ is a dimensionless function that dictates how the solute diffusivity varies through the interface. For example,

$$q(\phi) = \frac{1 + \phi}{2} \frac{D_s}{D_l} + \frac{1 - \phi}{2}, \tag{29}$$

is the simplest choice that interpolates between the values of the solute diffusivity in the solid and liquid. Finally, the isothermal dynamics is governed by Eqs. 4 and 5, which now become

$$\tau \frac{\partial \phi}{\partial t} = W^2 \nabla^2 \phi + \phi - \phi^2 - \lambda g'(\phi)\left[c - \frac{\Delta T}{m} \exp\left(b\left[1 + g(\phi)\right]\right)\right], \tag{30}$$

$$\frac{\partial c}{\partial t} = D_l \vec{\nabla} \cdot \left(q(\phi) c \vec{\nabla}\left[\ln c - b\,g(\phi)\right]\right), \tag{31}$$

where we have defined $\tau = 1/(K_\phi h)$, $W = (\sigma/h)^{1/2}$, $b = (\ln k_E)/2 < 0$, $\lambda \equiv -bRT_M/(v_0 h) > 0$, and $\Delta T = T - T_M$.

[3]The solute flux is also often written as $-M_{s,l}c\vec{\nabla}\mu^c$ where $M_{s,l}$ is known as the atomic mobility. Note that $D_{s,l} \approx M_{s,l}RT_M/v_0$ in the dilute limit.

3.2.2. *Solute trapping*

Solute trapping is a direct consequence of the departure from chemical equilibrium at a moving interface (Aziz, 1996). Solute partitioning is established on the characteristic time $\sim W^2/D_l$ for solute atoms to diffuse across the solid-liquid interface. If this semi-liquid interface region crystallizes completely before partitioning is complete, the solid will trap excess solute resulting in a reduction of the peak concentration that is illustrated in Fig. 3. Since the time to crystallize a thickness W of interface is $\sim W/V$, trapping will occur when $W^2/D_l \sim W/V$, or $V \sim V_d$ where

$$V_d \equiv D_l/W$$

is the diffusive speed (Aziz, 1996). Note that, for $V \sim V_d$, the solutal diffusion length D_l/V is comparable to the interface thickness W. Ahmad *et al.* have shown that solute trapping is well reproduced by the phase-field model (Ahmad *et al.*, 1998). To analyze this phenomenon in the present dilute alloy model, we start by writing down Eq. 31 in a frame moving with the interface at velocity V in the positive x direction

$$-V\frac{dc}{dx} = D_l\frac{d}{dx}\left(q(\phi_0)\,c\frac{d}{dx}\left[\ln c + \frac{\ln 1/k_E}{2}\,g(\phi_0)\right]\right). \qquad (32)$$

The phase-field profile across the moving interface is close to the one for a stationary interface. We can therefore approximate ϕ by ϕ_0 in the above equation and treat solute trapping entirely within the context of Eq. 32 with the phase-field decoupled from the concentration field. Next, we integrate both sides of Eq. 32 with respect to x and fix the integration constant by mass conservation, which implies that the concentration in the solid must equal the liquid concentration far ahead of the interface, c_∞. Furthermore, we make the change of variable $y = x/W$ and focus on the large velocity regime $V \gg V_d$. Retaining only the first order term in an expansion where the small parameter is V_d/V, we obtain

$$c \approx c_\infty\left[1 + \frac{V_d\ln 1/k_E}{2V}\left(-q(\phi_0)\frac{dg(\phi_0)}{dy}\right)\right], \qquad (33)$$

where the dimensionless function $-q(\phi_0)dg(\phi_0)/dy$ is positive because both ϕ_0 and $g(\phi_0)$ decrease when crossing the interface from solid to liquid by increasing y. This function is maximum in the diffuse interface region in agreement with the expected formed of the solute profile at high velocity, which corresponds to the dashed line in Fig. 3.

This limiting expression for c shows that the relevant velocity for trapping is not V_d but $V_d\ln 1/k_E$ in agreement with the previous analysis by Ahmad *et al.* (Ahmad *et al.*, 1998). Consequently, a strongly partitioning

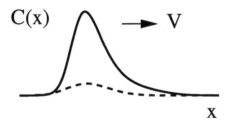

Figure 3. Schematic representation of the one-dimensional solute profile for two different velocities of the solid-liquid interface during the solidification of a dilute binary alloy. The higher velocity corresponding to the dashed line is close to the limit of complete trapping (massive transformation) examined here.

solute element (with a small equilibrium partition coefficient k_E) will have a larger trapping velocity than a weakly partitioning solute. The $\ln 1/k_E$ factor reflects the dependence of the time for solute atoms to diffuse across the interface on the gradient of chemical potential across the interface. A smaller k_E corresponds to a larger thermodynamic driving force and hence a shorter partitioning time and a larger trapping velocity.

Aziz and co-workers have found that a larger propensity to partition solute in equilibrium leads to a larger trapping velocity in several alloy systems (Aziz, 1996). This propensity is only approximately described by a $V_d \ln 1/k_E$ dependence of the trapping velocity (Ahmad *et al.*, 1998). The functions $q(\phi_0)$ and $g(\phi_0)$ in Eq. 33 imply that trapping can also be influenced by how the solute diffusivity and the non-entropic part of the free-energy density depend on the degree of crystalline order (ϕ_0) in the diffuse interface region. Therefore, it might be possible to gain further insight into the alloy-specificity of solute trapping by computing these functions that have so far been chosen arbitrarily in the phase-field model. This may be done by atomistic simulations (Asta, 2003) or using a more microscopic continuum approach like DFT (Shen and Oxtoby, 1996).

PROBLEM 3. Evaluate explicitly the solute profile in the limit of negligible solid-state diffusion, $q(\phi_0) \approx (1 - \phi_0)/2$, and for the simple cubic form $g(\phi_0) = 3(\phi_0 - \phi_0^3/3)/2$ that satisfies both $g'(\pm 1) = 0$ and $g(\pm 1) = \pm 1$. Where is the concentration peak located in the diffuse interface region?

SOLUTION. Using the chain rule $dg(\phi_0)/dy = (dg/d\phi_0)(d\phi_0/dy)$ and the fact that $d\phi_0/dy = -\left(1 - \phi_0^2\right)/\sqrt{2}$ for the tangent hyperbolic phase-field profile, $\phi_0(y) = -\tanh y/\sqrt{2}$, we can reduce Eq. 33 to

$$c \approx c_\infty \left[1 + \frac{3V_d \ln 1/k_E}{8\sqrt{2}V}(1 - \phi_0)^3(1 + \phi_0)^2 \right]. \tag{34}$$

Differentiating the above relation with respect to ϕ_0, we find that the peak of concentration occurs at $\phi_0 = -1/5$. Therefore this peak is slightly shifted towards the liquid with respect to the position ($\phi_0(y) = 0$) where the crystalline order is half-way between solid and liquid. Note that this answer is dependent on our choices for the functions q and g.

3.2.3. *Challenge of mesoscale simulations and anti-trapping*

The dendrite growth rate can vary from a few μm/sec to several m/sec depending on the solidification process (casting, welding, melt spinning, etc). In a simulation study based on the alloy phase-field model of Wheeler *et al.* (Wheeler *et al.*, 1992), Warren and Boettinger (Warren and Boettinger, 1995) demonstrated the power of the phase-field method to reproduce qualitatively many observed microstructural features of dendritic equiaxed grains. These authors also emphasized the difficulty of obtaining simulation results that are independent of interface thickness. This difficulty was present even though this study focused on large supersaturations, where the disparity between the microscopic capillary length and the mesoscale of the dendritic microstructure is greatly reduced.

Since the early use of phase-field models, it was hoped that adaptive meshing or multigrid algorithms could cope with the intrinsic stiffness of the phase-field equations. While such algorithms have been extremely helpful to extend simulations to a small velocity regime (Provatas *et al.*, 1998; Provatas *et al.*, 1999; Plapp and Karma, 2000), they do not fundamentally eliminate the stiffness introduced by a diffuse interface. For most solidification conditions (save perhaps very rapid rates), surmounting this difficulty requires to formulate the phase-field equations in such a way that they reduce to the correct set of sharp-interface equations for a mesoscale interface thickness. With rare exceptions[4], phase-field models based purely on thermodynamic considerations (Wheeler *et al.*, 1992) generally lack this crucial property. Thus, these models are of limited use for quantitative mesoscale simulations and need to be modified as discussed below.

To better understand the stiffness of the phase-field equations, let us set ourselves as a goal to simulate the isothermal solidification of a dilute alloy in a range of moderate growth rate where the interface can be assumed to be in local thermodynamic equilibrium. The appropriate free-boundary problem consists of the standard set of equations

$$\partial_t c = D_l \nabla^2 c, \tag{35}$$
$$c_l(1 - k_E)V_n = -D_l \partial_n c|_l, \tag{36}$$

---

[4]For a mesoscale interface width, the original phase-field model with equal thermal diffusivity in liquid and solid (Langer, 1986; Collins and Levine, 1985) reduces to the standard free-boundary problem of the solidification of a pure melt with a modified expression for the interface kinetic coefficient (Karma and Rappel, 1996; Karma and Rappel, 1998). However, this fortuitous property is lost when the thermal diffusivity is allowed to be unequal in the two phases (Almgren, 1999).

$$c_l/c_l^0 = 1 - (1 - k_E)d_0(1 - \alpha\cos 4\theta)\kappa, \tag{37}$$

which correspond to the diffusion equation in liquid, the condition of mass conservation at the interface where V_n is the normal interface velocity, and the Gibbs-Thomson relation where κ is the interface curvature and $d_0 = \gamma T_M/[L|m|(1-k)c_l^0]$ is the chemical capillary length; c_l^0 (c_s^0) is the concentration on the liquid (solid) side of a reference flat interface at the isothermal growth temperature. In addition, for simplicity, we have neglected solute diffusion in the solid.

The interface evolution is controlled by the surface tension anisotropy α and the dimensionless supersaturation

$$\Omega = \frac{c_l^0 - c_\infty}{c_l^0(1 - k_E)}, \tag{38}$$

where c_∞ is the initial alloy concentration. A useful scale to assess the stiffness of the phase-field equations is the dendrite tip radius

$$\rho = \frac{d_0}{\sigma(\alpha)p(\Omega)}, \tag{39}$$

where the dimensionless parameter

$$\sigma(\alpha) = \frac{2D_l d_0}{\rho^2 V} \tag{40}$$

is controlled by anisotropy, as predicted by solvability theory (Barbieri and Langer, 1989), and

$$p(\Omega) = \frac{\rho V}{2D_l} \tag{41}$$

is the Péclet number determined by the Ivantsov transport theory (Ivantsov, 1947). Note that $p = \rho/\ell$ where $\ell = 2D_l/V$ is the diffusion length that measures the scale over which c varies around the tip. For typical equiaxed grains in castings, p is in the range of 0.01 to 0.1. The adaptive meshing (Provatas et al., 1998; Provatas et al., 1999) and other hybrid (Plapp and Karma, 2000) algorithms developed recently for dendritic growth cope efficiently with the disparity between ρ and ℓ. They do not, however, eliminate the several orders of magnitude disparity between ρ and d_0 that follows from the smallness of p and σ in Eq. 39. The Ivantsov theory (Ivantsov, 1947) implies that $p \sim \Omega^2$ (or $p \sim -\Omega/\ln\Omega$) in 2-d (or 3-d) and solvability theory (Barbieri and Langer, 1989) implies that $\sigma \sim \alpha^{7/4}$, such that this disparity is exacerbated for a small supersaturation and/or a small anisotropy. Dendrite growth simulations with a mesoscale width ($0.1\rho \leq W \leq 0.3\rho$) typically take a few days on a fast workstation (Provatas et al., 1998;

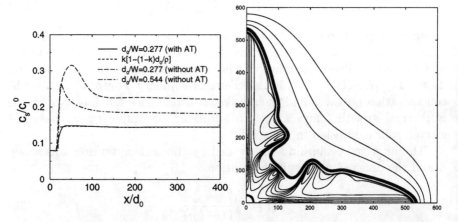

Figure 4. Results of phase-field simulations of the isothermal dendritic solidification of a dilute binary alloy with $k_E = 0.15$ and $\Omega = 0.55$. Simulations with the phase-field formulation of Sec. 3.2.1 lead to a supersaturated solid concentration inside the dendrite trunk. Here this concentration is scaled by the equilibrium concentration (c_l^0) on the liquid side of a flat interface and should be equal to k_E up to capillary corrections. In contrast, this concentration is well-described when this formulation is modified with an anti-trapping (AT) current that restores local chemical equilibrium at the solid-liquid interface. Right: Contour plots of concentration in the solid obtained with this new formulation: $c(x,y)/c_{liq}^0$ from 0.144 to 0.153 spaced by 0.001 and 0.2 to 0.8 spaced by 0.1. The simulation parameters are the same as Fig. 2 in Ref. (Karma, 2001).

Plapp and Karma, 2000). Simulations with a width even only ten times smaller would be $\approx 10^4$ (10^5) longer in 2-d (3-d)! Clearly, simulations with a microscopic interface width are unrealistic [5].

The consequences of this constraint generally needs to be analyzed with a detailed asymptotic analysis of the phase-field equations for a thin-interface, meaning $W \ll \rho$. This analysis has been carried out for pure melts (Karma and Rappel, 1996; Karma and Rappel, 1998; Almgren, 1999; McFadden *et al.*, 2000) and binary alloys (Karma, 2001). For the latter, this analysis reveals that the phase-field equations reduce to a modified free-boundary problem with a discontinuity of chemical potential at the

---

[5] The simulation time scales as $(\rho/W)^{d+2}$ with a volume factor of $(\rho/W)^d$ and the other factor of $(\rho/W)^2$ due to the stability constraint requiring a smaller time step for a smaller mesh size $\sim W$. This estimate is based on the use of an algorithm that makes the cost of solving the diffusion equation on a scale $\ell \gg \rho$ negligible in comparison to the cost of solving the phase-field equations on a scale ρ with standard finite differences (Plapp and Karma, 2000). Adaptive meshing and implicit time-stepping on a scale ρ could perhaps reduce this estimate, but not sufficiently to change this conclusion.

interface, and a modified mass conservation condition

$$c_l(1 - k_E)(1 - a_1 W \kappa)V_n = -D_l \, \partial_n c_l|_l - a_2 W D_l \, \partial_s^2 c_l, \qquad (42)$$

where a_1 and a_2 are coefficients of order unity that depend on the choice of functions in the phase-field model. The factor of $1 - a_1 W \kappa$ on the left-hand-side originates from the stretching of the arclength of the interface under the combined effects of curvature and motion. The term $\sim W D_l \, \partial_s^2 c_l$ on the right-hand-side corresponds to solute diffusion along this interface. Both of these terms have direct analogs for the solidification of a pure melt (Almgren, 1999). The discontinuity of chemical potential, in turn, is a direct manifestation of the departure from chemical equilibrium at the interface that we analyzed in Sec. 3.2.2, albeit here in the opposite small velocity regime, $V \ll V_d$. A similar correction was interpreted as heat trapping in the context of pure melts (McFadden et al., 2000). It is straightforward to estimate that these corrections are negligible at low velocity for a microscopic interface width. However, they are not for a mesoscopic width. For example, since $V_d \sim D_l/W$, a larger W will lead to an abnormally large amount of trapping. This is illustrated in Fig. 4 where dendrite growth was simulated using a phase-field formulation essentially identical to the one discussed in Sec. 3.2.1, which is based on the alloy phase-field model of (Wheeler et al., 1992). The solid dendrite trunk concentration is too elevated, and convergence to the limit of local equilibrium with decreasing interface thickness is difficult to achieve.

A simple way to remove the excess trapping caused by a thick interface is to add to the solute mass current an extra contribution

$$\vec{j}_{AT} \sim -c_l(1 - k_E)W \frac{\vec{\nabla}\phi}{|\vec{\nabla}\phi|} \frac{\partial \phi}{\partial t}, \qquad (43)$$

which produces a solute flux from solid to liquid along a direction normal to the interface ($\hat{n} = -\vec{\nabla}\phi/|\vec{\nabla}\phi|$). The magnitude of this anti-trapping (AT) current can be chosen so as to recover precisely local equilibrium at the interface. Furthermore, the addition of this current provides sufficient flexibility in the choice of functions in the phase-field model to eliminate the aforementioned corrections to the mass conservation condition (to obtain $a_1 = a_2 = 0$ in Eq. 42). In particular, eliminating surface diffusion requires a more elaborate choice of the diffusivity function $q(\phi)$ than the simple linear interpolation between D_s and D_l (Eq. 29). The details of this new phase-field formulation can be found in (Karma, 2001) and we only show here one result of simulation in Fig. 4.

It is important to emphasize that the phase-field model with the anti-trapping current no longer has the variational form of Eqs. 4-5. Enforcing such a form is important to describe phenomena like solute trapping that depend on having a thermodynamically motivated description of the system on the scale of the diffuse interface, and perhaps more generally when there is no clear separation between the transport scales and the interface width. Solute trapping falls in this category because the diffusion length that is usually mesoscopic becomes microscopic when $V \sim V_d$. When there is a clear separation of scale between the pattern and the interface width, variational and non-variational formulation have been shown to give identical results for the solidification of pure melts as long as they reduce to the same thin-interface limit (Karma and Rappel, 1998; Kim $et\ al.$, 1999). The same is true here for alloys.

Finally, the degree to which a mesoscopic interface thickness limits the accuracy of phase-field simulations depends sensitively on the growth rate but also on the pattern being modeled, with dendritic growth being generally more difficult to model than eutectic growth. The main reason is that the dendrite growth rate and size is determined by extremely small variations of concentration (or temperature) along the interface that are themselves controlled by the magnitude and anisotropy of surface tension and nonequilibrium effects at the interface. Hence, the departure of chemical equilibrium at the interface may be small in absolute term, but can nonetheless influence strongly the pattern evolution. In contrast, for eutectic growth (Sec. 3.3), departure from chemical equilibrium results mainly in a shift of the undercooling of the solidification front that does not influence as strongly this evolution (Karma, 1994; Plapp and Karma, 2002).

3.3. EUTECTIC ALLOYS

Phase-field models have been developed to model solidification microstructures that are composed of thermodynamically distinct solid phases, including eutectic (Karma, 1994; Elder $et\ al.$, 1994; Wheeler $et\ al.$, 1996) and peritectic (Tiaden $et\ al.$, 1998; Lo $et\ al.$, 2001) alloys. In the simplest case where the solid α and β phases have the same crystal structure, one can simply choose the bulk free-energy of the solid to be described by a single free-energy curve $f_s(c, T)$ with the form of a double well, with minima corresponding to the two solid phases, and the free-energy of the liquid by a curve $f_l(c, T)$ with a single minimum. The phase diagram, representing the stable or metastable equilibria between the three phases, is determined by the common tangents of these two free-energy curves. The phase field ϕ then simply distinguishes between solid and liquid as for a two-phase alloy. Thus, Eqs. 3, 4, and 5 are still applicable. In the solid, Eq. 5 becomes analogous in form to the well-known Cahn-Hilliard equation (Cahn and Hilliard, 1958), and σ_c determines the magnitude of the $\alpha - \beta$ interfacial energy. A simple polynomial form of $f_{AB}(\phi, c, T)$ for such a two-variable model

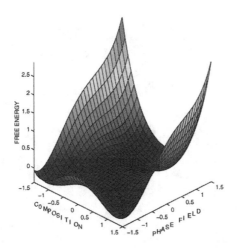

Figure 5. Plot of the equilibrium free-energy density of a simple two-field eutectic phase field model. $f_{AB}(\phi, \tilde{c}, T_E) = -\phi^2/2 + \phi^4/4 + \tilde{c}^4/4 + \phi\tilde{c}^2/2 - 0.14\,\phi$, where $\tilde{c}$ is a composition variable defined such that it vanishes in the liquid phase at the eutectic point (Karma, 1994). The bulk solid (liquid) free-energy curves correspond to sections of this free-energy surface at $\phi = -1.175$ ($\phi = 1.064$).

is plotted at the eutectic point in Fig. 5 to illustrate the three-minimum appearance of the free-energy landscape, where each minimum corresponds to one of the three phases. This form is restricted to model eutectic growth in a range of interface temperature reasonably close to the eutectic temperature (Karma, 1994). A more general form based on a regular solution model can also be written down that remains valid over the entire range of concentration of the binary phase diagram. In the more common case where the α and β phases have different crystal structures, it becomes necessary to describe each of these two phases, and the liquid, by three separate free-energy curves. This can be accomplished by introducing an additional phase field ψ to distinguish between α and β (Wheeler *et al.*, 1996).

PROBLEM 4. Construct the binary eutectic phase-diagram that corresponds to the free-energy curve defined in Fig. 5.

SOLUTION. Fig. 2 in (Karma, 1994).

As a last numerical example, we consider the case of a eutectic AB alloy with a dilute ternary impurity C (Plapp and Karma, 2002). The free-energy of this model is the sum f_{ABC} of the contribution of a binary eutectic, similar to f_{AB} in Fig. 5, and the contribution of a dilute ternary impurity analogous to the concentration dependent terms on the right-hand-side of Eq. 20 (with C as the dilute impurity). Fig. 6 shows a snapshot of a phase field simulation of such a model (Plapp and Karma, 2002). This simulation provides a striking example of the power of the phase field method to de-

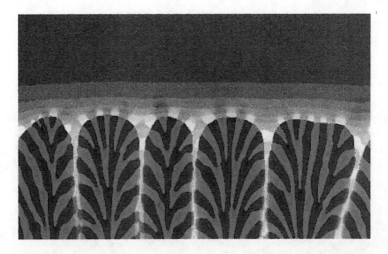

Figure 6. Snapshot of the late stage of two-phase cell formation subsequent to the morphological instability of a slightly perturbed eutectic interface in the presence of a dilute ternary impurity (Plapp and Karma, 2002).

scribe the evolution of complex microstructures. The resulting morphology is the classic two-phase cellular structure, or "eutectic colony", which has been widely observed experimentally.

4. Acknowledgments

I wish to acknowledges the U.S. Department of Energy and NASA for financial support as well as NERSC and NU-ASCC for computer time.

A. Excess free-energy of a diffuse interface

In this appendix, we review the calculus of variation that is generally used to derive the phase-field equations from a free-energy functional and we calculate as an example the excess free-energy γ of the solid-liquid interface for a pure substance. Consider the free-energy functional that describes the solid-liquid interface at the melting point

$$F[\phi, T] = \int dV \left[\frac{\sigma}{2} |\vec{\nabla}\phi|^2 + f(\phi, T_M) \right] \equiv \int dV F_v. \qquad (44)$$

The minimization of F is carried out by substituting $\phi(x) = \phi_0(x) + \delta\phi(x)$ into Eq. 44, where $\phi \equiv \phi_0(x)$ is the equilibrium phase-field profile and x is the direction normal to the interface. Let us first expand the integrand to linear order in $\delta\phi$, with the substitution $dV = A\,dx$, where A is the area of the interface, and $|\vec{\nabla}\phi|^2 = (d\phi/dx)^2 \equiv \phi_x^2$. After integrating once by parts, and using the fact that ϕ_x vanishes in bulk phases away from the interface, Eq. 44 can then be written in the form

$$\delta F = A \int dx \; \delta\phi(x) \left. \frac{\delta F}{\delta \phi} \right|_{\phi(x)=\phi_0(x)}, \tag{45}$$

where we have defined the functional derivative

$$\frac{\delta F}{\delta \phi} \equiv -\frac{d}{dx}\frac{\partial F_v}{\partial \phi_x} + \frac{\partial F_v}{\partial \phi}, \tag{46}$$

For $\phi = \phi_0(x)$ to minimize F, δF must vanish for an arbitrary smooth variation $\delta\phi(x)$. This requires $\delta F/\delta\phi = 0$ in the integrand, which is completely equivalent to the Euler-Lagrange equation for a point particle in mechanics (Landau and Lifshitz, 1976) (with the identification $x =$ time, $\phi =$ particle coordinate, $\sigma\phi_x^2/2 =$ kinetic energy, and $f(\phi) = -$ potential energy).

For $f(\phi, T_M)$ defined by Eq. 21, the equilibrium condition $\delta F/\delta\phi = 0$ together with Eq. 46 yields

$$W^2\frac{d^2\phi_0}{dx^2} + \phi_0 - \phi_0^3 = 0 \tag{47}$$

where we have defined the interface thickness $W = \sqrt{\sigma/h}$. This equation has the kink solution $\phi_0(x) = -\tanh[x/(\sqrt{2}W)]$. We can now calculate the excess free-energy of the solid-liquid interface, γ, which is the difference between the total free-energy of a two-phase and a single phase system occupying the same volume, divided by A, or

$$\gamma = \int_{-\infty}^{+\infty} dx \left[\frac{\sigma\phi_{0x}^2}{2} + f(\phi_0, T_M) - f_B \right], \tag{48}$$

where the bulk free-energy density $f_B \equiv f(\pm 1, T_M) = -h/4$. Both the gradient square term $\sigma\phi_{0x}^2/2$ and $f(\phi_0, T_M) - f_B$ yield equal contribution to this integral, such that Eq. 48 reduces to

$$\gamma = W h \int_{-\infty}^{+\infty} dy \; \phi_{0y}^2 = (2\sqrt{2}/3)Wh, \tag{49}$$

where we have defined the dimensionless variable $y = x/W$. Eq. 49 implies that crystalline anisotropy can be simply included by letting the interface thickness depend on the directional normal to the interface in the phase-field model, $\hat{n} = -\vec{\nabla}\phi/|\vec{\nabla}\phi|$, in which case $\gamma(\hat{n}) \sim W(\hat{n})$. This approach was first proposed by Zia and Wallace (Zia and Wallace, 1985) in the general context of first-order phase transitions and reintroduced independently by Kobayashi to simulate dendritic growth (Kobayashi, 1993 and 1994).

88

References

N.A. Ahmad, A.A. Wheeler, W. J. Boettinger, and G.B. McFadden, Phys. Rev. E **58**, 3436 (1998).

R.F. Almgren, SIAM J. Appl. Math **59**, 2086 (1999).

D.M. Anderson, G.B. McFadden, and A.A. Wheeler, Physica D **135**, 175 (2000).

I. S. Aranson, V. A. Kalatsky and V. M. Vinokur, Phys. Rev. Lett. **85**, 118 (2000).

M. Asta, article in this volume.

M. J. Aziz, Metall. Mater. Trans. A **27**, 671 (1996).

A. Barbieri and J. S. Langer, Phys. Rev. A**39**, 5314 (1989).

C. Beckermann, H.-J. Diepers, I. Steinbach, A. Karma, and X. Tong., J. Comp. Phys. **154**, 468 (1999).

W. J. Boettinger, J. A. Warren, C. Beckermann, and A. Karma, Ann. Rev. Mater. Res. **32**, 163 (2002).

J. Bragard, A. Karma, Y. H. Lee, and M. Plapp, Interface Science **10**, 121 (2002).

R.J. Braun and M.T. Murray, J. Cryst. Growth **174**, 41 (1997).

J. Q. Broughton, G. H. Gilmer and K. A. Jackson, Phys. Rev Lett., **49**, 1496 (1982).

G. Caginalp and X. Chen, in *On the Evolution of Phase Boundaries*, edited by M.E. Gurtin, G.B. McFadden, The IMA Volumes in Mathematics and Its Applications Vol. 43 (Springer-Verlag, New York, 1992) p. 1.

G. Caginalp and W. Xie, Phys. Rev. E**48**, 1897-1909 (1993).

J.W. Cahn and J.E. Hilliard, J. Chem. Phys. **28**, 258 (1958).

L.-Q. Chen and W. Yang, Phys. Rev. B **50**, 15752 (1994).

L.-Q. Chen and Y. Wang, JOM **48**, 12 (1996).

J. B. Collins and H. Levine, Phys. Rev. B **31**, 6119 (1985).

M. C. Cross and P. C. Hohenberg, Rev. Mod. Phys. **65**, 851 (1993).

L. O. Eastgate J. P. Sethna, M. Rauscher, T. Cretegny, C.-S. Chen, and C. R. Myers, Phys. Rev. E **65**, 036117 (2002).

K. R. Elder, F. Drolet, J. M. Kosterlitz, M. Grant, Phys. Rev. Lett. **72**, 677 (1994).

K. R. Elder, M. Katakowski, M. Haataja, and M. Grant, Phys. Rev. Lett. **24**, 245701 (2002).

V. L. Ginzburg and L. D. Landau, Soviet Phys. JETP **20**, 1064 (1950).

L. Gránásy, T. Pusztai, J. A. Warren, J. F. Douglas, T. Börzsönyi, and V. Ferreiro, Nature Materials **2**, 92 (2003).

B. I. Halperin, P. C. Hohenberg, and S-K. Ma, Phys. Rev. B**10**, 139 (1974).

P. C. Hohenberg and B. I. Halperin, Rev. Mod. Phys. **49**, 435 (1977).

J. J. Hoyt, B. Sadigh, M. Asta and S. M. Foiles, Acta Mater. **47**, 3181 (1999).

J. J. Hoyt, M. Asta and A. Karma, Phys. Rev. Lett. **86** , 5530-5533 (2001).

S-C. Huang and M. E. Glicksman, Acta Metall. **29**, 701 (1981).

G. P. Ivantsov, Dokl. Akad. Nauk SSSR **58**, 567 (1947).

A. Karma, Phys. Rev. E**49**, 2245 (1994).

A. Karma and W. J. Rappel, Phys. Rev. E**53**, R3017 (1996).

A. Karma and W.-J. Rappel, Phys. Rev. E**57**, 4323 (1998).

A. Karma and W.-J. Rappel, Phys. Rev. E **60**, 3614 (1999).

A. Karma and M. Plapp, Phys. Rev. Lett. **81**, 4444 (1998).

A. Karma, Phys. Rev. Lett. **87**, 115701 (2001).
  This article describes a quantitative phase-field model of the isothermal solidification of a dilute binary alloy that includes the anti-trapping current discussed in Sec. 3.2.3. This model can also be extended to model directional solidification with a frozen temperature approximation $T(z) = T_0 + G(z - V_p t)$. For this application, the thin-interface limit yields the expression for the interface kinetic coefficient $\beta = a_1 [\tau/(W\lambda) - a_2\{1 - (1 - k_E)(z - V_p t)/l_T\}W/D]$ where $l_T = |m|(1 - k_E)c_l^0/G$ is the thermal length. Note that it is still possible to make β vanish everywhere along the interface by simply letting τ depend on z in the phase-field model.

A. Karma, D. Kessler, and H. Levine, Phys. Rev. Lett. **87**, 045501 (2001).

A. Karma, "Phase Field Methods", in *Encyclopedia of Materials: Science and Technology*, edited by K. H. J. Buschow, R. W. Cahn, M. C. Flemings, B. Ilschner, E. J. Kramer, S. Mahajan, Volume 7, Elsevier, Oxford, pp. 6873-86 (2001b).

K. Kassner and C. Misbah, Europhys. Lett. **46**, 217 (1999).

S.-G. Kim, W. T. Kim, and T. Suzuki, Phys. Rev. E **60**, 7186 (1999).

Y-T. Kim, N. Provatas, N. Goldenfeld and J. A. Dantzig, Physical Review E **59**, 2549 (1999).

R. Kobayashi, Physica D**63**, 410 (1993); Experimental Math **3**, 60-81 (1994).

R. Kobayashi, J.A. Warren, and W.C. Carter, Physica D **119**, 415 (1998).

R. Kupferman, O. Shochet, and E. Ben-Jacob, Phys. Rev. E **50**, 1005-1008 (1994).

L.D. Landau and E.M. Lifshitz, *Mechanics*, (Third edition, Pergamon, New York, 1976).

J. S. Langer, in *Directions in Condensed Matter*, edited by G. Grinstein and G. Mazenko (World Scientific, Singapore, 1986), p. 164.

F. Liu and H. Metiu, Phys. Rev. E **49**, 2601 (1994).

T.-S. Lo, A. Karma, and M. Plapp, Phys. Rev. E **63**, 031504 (2001).

A. E. Lobkovsky and J. A. Warren, Phys. Rev. E **63**, 051605 (2001).

J. W. Lum, D. M. Matson and M. C. Flemings, Metall. Mater. Trans. B **27**, 865 (1996).

S-k. Ma, *Modern Theory of Critical Phenomena* (Frontiers in Physics Lecture Note Series, Benjamin/Cummings Publishing Co, 1976).

G. B. McFadden, A. A. Wheeler, R. J. Braun, S. R. Coriell, and R. F. Sekerka, Phys. Rev. E **48**, 2016 (1993).

G. B. McFadden, A. A. Wheeler, and D. M. Anderson, Physica D **154**, 144 (2000).

L. V. Mikheev and A. A. Chernov, J. Cryst. Growth, **112**, 591 (1991).

J. Muller and M. Grant, Phys. Rev. Lett. **82**, 1736 (1999).

B. T. Murray, A. A. Wheeler, and M. E. Glicksman, J. Cryst. Growth **47**, 386 (1995).

D. W. Oxtoby and P. R. Harrowell, J. Chem. Phys. **96**, 3834 (1992).

S.G. Pavlik and R.F. Sekerka, Physica A **268**, 283 (1999).

M. Plapp and A. Karma, Phys. Rev. Lett. **84**, 1740 (2000); J. Comp. Phys. **165**, 592 (2000).

M. Plapp and A. Karma, Phys. Rev. E **66**, 061608 (2002).

N. Provatas, N. Goldenfeld and J. A. Dantzig, Phys. Rev. Lett **80**, 3308 (1998).

N. Provatas, N. Goldenfeld and J. A. Dantzig, J. Comp. Phys. **148**, 265 (1999).

T. V. Ramakrishnan and M. Yussouff, Phys. Rev. B **19**, 2275 (1979).

D. Rodney and A. Finel, MRS Symposium Proc. **652**, Y4.9.1 (2001).

D. Rodney, Y. Le Bouar, and A. Finel, Acta Mater. **51**, 17 (2003).

Y. Shen and D. W. Oxtoby, J. Chem. Phys. **104**, 4233 (1996).

I. Steinbach, F. Pezzolla, B. Nestler, M. BeeBelber, R. Prieler, G.J. Schmitz, and J.L.L. Rezende, Physica D **94**, 135 (1996).

J. Tiaden, B. Nestler, H.J. Diepers, and I. Steinbach, Physica D **115**, 73 (1998).

X. Tong, C. Beckermann, and A. Karma, Phys. Rev. E **61**, R49 (2000); X. Tong, C. Beckermann, A. Karma, and Q. Li, Phys. Rev. E **63**, 061601 (2001).

R. Tonhardt and G. Amberg, J. Cryst. Growth **194**, 406 (1998).

S-L. Wang and R. F. Sekerka, Phys. Rev. E**53**, 3760 (1996).

Y. U. Wang, Y. M. Jin, A. M. Cuitino, and A. G. Khachaturyan, Acta Mater. **49**, 1847 (2001).

Y. U. Wang, Y. M. Jin, and A. G. Khachaturyan, J. Appl. Phys. **91**, 6435 (2002).

J. A. Warren and W. J. Boettinger, Acta Metall. Mater. A**43**, 689-703 (1995).

A. A. Wheeler, B. T. Murray, and R. Schaefer, Physica D **66**, 243 (1993).

A.A. Wheeler, W.J. Boettinger, and G.B. McFadden, Phys. Rev. A**45**, 7424 (1992).

A.A. Wheeler, G.B. McFadden, and W.J. Boettinger, Proc. Royal Soc. London A **452**, 495-525 (1996).

R. Willnecker, D. M. Herlach, and B. Feuerbacher, Phys. Rev. Lett. **62**, 2707 (1989).

R. K. P. Zia and D. J. Wallace, Phys. Rev. B **31**, 1624 (1985).

THE DYNAMICS OF INTERFACES IN ELASTICALLY STRESSED SOLIDS

K. THORNTON AND P.W. VOORHEES
Department of Materials Science and Engineering
Northwestern University
Evanston, IL 60208

1. Introduction

Describing the evolution of microstructure is central to predicting and controlling the properties of materials. There are many techniques that can be used to follow the evolution of microstructure, ranging from atomistic approaches such as the Monte Carlo and molecular dynamics methods to continuum phase field simulations. Here we will focus on so-called sharp interface descriptions of interface motion. In this approach the interface is envisioned as an infinitesimally thin region to which certain properties, such as an interfacial energy, are ascribed. This approach has its roots in Gibbs' thermodynamic treatment of interfaces [1]. While interfaces are clearly not infinitesimally thin, when the length-scale of interest is much larger than the thickness of the interface, this approach yields a very good approximation to the conditions governing the evolution of the interface. Furthermore, such a model has the distinct advantage of not requiring the detailed structure or length scale of the interfacial region to be tracked during a computation.

The sharp interface approach will be illustrated by focusing on the effects of elastic stress on interfacial motion. To accomplish this goal we shall first discuss the thermodynamic conditions governing interfacial equilibrium, since these are used to formulate a boundary condition that is central to the sharp interface description of interface motion. In particular we shall focus on the effects of elastic stress on interfacial equilibrium. The effects of elastic stress on interfacial motion will be illustrated by examining Ostwald ripening in coherent solids and the morphological stability of an interface in a stressed solid where the driving force for instability is the stress dependence of both the chemical potential and the interface mobility.

2. Thermodynamics

The thermodynamic equilibrium conditions in a stress-free system follow from an application of the first and second laws of thermodynamics. Assuming a binary alloy and two phases α and β, the equilibrium state of the system is given by a minimum in the total energy of the system subject to the constraints of constant entropy and total number of atoms of the two chemical components, 1 and 2. In the context of the sharp interface description the energy is written in terms of the energy of the two phases and energy of the interface, a plane of zero thickness. We assume for the moment that the interface is planar and thus,

$$\mathcal{E} = \int_\alpha E_v^\alpha \left(s, \rho_1, \rho_2 \right) dV + \int_\beta E_v^\beta \left(s, \rho_1, \rho_2 \right) dV + \int_a E_a \left(s^x, \Gamma_1, \Gamma_2 \right) da \qquad (1)$$

91

A. Finel et al. (eds.), Thermodynamics, Microstructures and Plasticity, 91–105.
© 2003 *Kluwer Academic Publishers. Printed in the Netherlands.*

where $\mathcal{E}$ is the total internal energy of the system, E_v is the internal energy per volume of the noted phase, s is the entropy, ρ_i is the density of atoms of component i, a is the interface, E_a is the internal energy of the interface, s^x is the excess entropy associated with the interface per area, Γ_i is the number of atoms per area of the interface, and a is the interfacial area. The interfacial equilibrium conditions obtained by minimizing $\mathcal{E}$ are,

$$
\begin{aligned}
\mathcal{T}^\alpha &= \mathcal{T}^\beta \\
P^\alpha &= P^\beta \\
\mu_1^\alpha &= \mu_1^\beta \\
\mu_2^\alpha &= \mu_2^\beta
\end{aligned}
\tag{2}
$$

where $\mathcal{T}$ is the temperature, P is the pressure, and μ_i is the chemical potential of component i. In reality the compositions of α and β are nonuniform in the interfacial region. Gibbs showed that this real system can be rigorously replaced by one in which each phase is uniform up to the interface and an interface with certain excess properties, regardless of the location of the interface. Thus the diffuseness of the interface has no effect on the resulting interfacial equilibrium conditions and the sharp interface description is exact.

If the interface between the two stress-free phases is curved, then the energy of the interface can, in principle, depend on the curvature of the interface and Eq. (1) is replaced by,

$$
\mathcal{E} = \int_\alpha E_v^\alpha \left(s, \rho_1, \rho_2 \right) dV + \int_\beta E_v^\beta \left(s, \rho_1, \rho_2 \right) dV + \int_a E_a \left(s^x, \Gamma_1, \Gamma_2, \kappa_1, \kappa_2 \right) da
\tag{3}
$$

where κ_1 and κ_2 are the principle curvatures at a point on the interface. In general the curvature dependence of the internal interfacial energy is nonzero, and thus to minimize the dependence of the equilibrium conditions on the curvature dependence of the internal interfacial energy Gibbs chose to place the interface at the point where $\partial E_a / \partial \kappa_1 + \partial E_a / \partial \kappa_2 = 0$. It is then possible to show that when the thickness of the interface is small compared to the radii of curvature of the interface, the derivatives of the interfacial energy with respect to curvature disappear from the equilibrium conditions and one is left with the familiar interfacial equilibrium conditions,

$$
\begin{aligned}
\mathcal{T}^\alpha &= \mathcal{T}^\beta \\
P^\alpha &= P^\beta + \sigma\kappa \\
\mu_1^\alpha &= \mu_1^\beta \\
\mu_2^\alpha &= \mu_2^\beta
\end{aligned}
\tag{4}
$$

where κ is the mean interfacial curvature, $\kappa = (\kappa_1 + \kappa_2)/2$ and σ is the grand canonical free energy of the interfacial energy. These conditions can be used to determine the compositions in each phase at an interface, since $\mu_i = \mu_i(C, P)$ where C is the composition of component 1, $C = \rho_1/(\rho_1 + \rho_2)$, see Ratke and Voorhees for a pedagogical treatment [2].

Gibbs did not develop the thermodynamics of stressed solids for solids that can undergo phase transformations as solid-state diffusion and crystalline defects, e.g. vacancies, were unknown at the time. It was not until recently that the equilibrium conditions in an stressed solid were derived, see the work of Herring [3], Larché and Cahn [4], and Leo and Sekerka [5]. There are a few insights that played a central role in the derivation of these conditions. First is the existence of a lattice in a crystalline solid. The importance of the lattice becomes clear when considering, for example, the morphological evolution of a pure solid in contact with a fluid. The crystalline lattice allows one to separate changes in shape of the crystal due to stress, those that occur with no change in the number of lattice points of the crystal, and changes due to melting or solidification, those that occur via addition or deletion of lattice points. The presence of the crystalline lattice also requires that the defects within the crystalline lattice be considered when developing the equilibrium conditions. Equilibrium conditions are obtained by again minimizing the total energy where, in contrast to that given for stress-free phases, we allow the energy of the bulk phases to depend on the deformation gradient of the solid, $\nabla \mathbf{u}$ where $\mathbf{u}$ is the displacement vector in the small displacement approximation. We will neglect the possible dependence of the internal interfacial energy on the interfacial deformation gradient. Thus, for a binary alloy,

$$\mathcal{E} = \int_\alpha E_v^\alpha \left(s, \nabla \mathbf{u}, \rho_1, \rho_2\right) dV + \int_\beta E_v^\beta \left(s, \nabla \mathbf{u}, \rho_1, \rho_2\right) dV + \int_a E_a \left(s^x, \Gamma_1, \Gamma_2\right) da \tag{5}$$

In minimizing the total energy we need to add constraints in addition to those used above of constant number of atoms of components 1 and 2 and entropy. If α or β is a crystal that within these phases the total number of lattice sites must remain fixed, since we are assuming that there are no climbable dislocations within the bulk. The total number of lattice sites of each phase can change, however, at the phase boundary. The ability to change the number of lattice sites requires that the structure of the interface be considered. If the interface is between two crystals and the lattice planes are continuous across the interface, the interface is coherent, then as the interface moves due a phase change the two phases may not separate or slide along the interface [4]. This introduces a constraint on possible variations in the displacement vector at the interface. For a solid-fluid interface, there is no such constraint. The result is that the equilibrium conditions at an interface separating two crystalline solids are different from those at an interface separating a fluid and crystalline solid.

We shall assume that the vacancy concentration is extremely small and that components 1 and 2 are substitutional atoms. For simplicity we take the lattice parameter of the solid and elastic constants to be independent of concentration. This yields the following interfacial equilibrium conditions [5]

$$
\begin{aligned}
\mathsf{T}^\alpha &= \mathsf{T}^\beta \\
\mu_1^\alpha - \mu_2^\alpha &= \mu_1^\beta - \mu_2^\beta \\
\omega^\alpha - \omega^\beta - [\![\mathsf{TE}]\!] &= -\sigma \kappa \\
\mathsf{T}^\alpha \mathbf{n} - \mathsf{T}^\beta \mathbf{n} &= 0
\end{aligned}
\tag{6}
$$

where ω is the grand canonical free energy per volume of the noted phase, T is the stress tensor, E is the total strain, $[\![a]\!]$ denotes the jump in a quantity at the interface: $a^\beta - a^\alpha$, and $\mathbf{n}$ is the normal to the interface. The last condition is the usual balance of forces at a coherent interface in the absence of interfacial stress. The grand canonical free energies are functions of the elastic energy density of the phases at the interface and the chemical potentials [4],

$$\omega = \rho \mu_2 + W_e \tag{7}$$

where ρ is the molar volume of the solid, and W_e is the elastic strain energy density. Using Eq. (7) in third of Eq. (6) yields,

$$\rho \left(\mu_2^\alpha - \mu_2^\beta\right) + [\![W_e]\!] - [\![\mathsf{TE}]\!] = -\sigma \kappa \tag{8}$$

This equation reflects the energy change on moving an atom from a phase with one elastic energy density to another via the presence of the term $[\![W_e]\!]$ and the energy required to keep the interface coherent as atoms are transferred from one phase to another via the term $[\![TE]\!]$. Since $\mu_i(C)$, Eq. (8) and the second of Eq. (6) form a system of two equations in the two unknown concentrations C^α and C^β at the interface. Assuming local equilibrium holds at the interface, these concentrations can then be used as boundary conditions for the diffusion equation that describes mass flow during a phase transformation.

For a single component solid with vacancies in equilibrium with a fluid, where the Lattice parameter is independent of the vacancy concentration, the equilibrium conditions are:

$$
\begin{aligned}
T^\alpha &= T^f \\
\mu_1^\alpha - \mu_v^\alpha &= \mu_1^f \\
\omega^\alpha + P &= -\sigma\kappa \\
\mathsf{T}^\alpha \mathbf{n} - P\mathbf{n} &= \mathbf{0}
\end{aligned}
\tag{9}
$$

where P is the pressure in the fluid at the interface and μ_v is the chemical potential of a vacancy. For a solid with vacancies [6],

$$
\omega^\alpha = \rho\mu_v + W_e
\tag{10}
$$

Using Eq. (8) in Eq. (10) yields an expression for the chemical potential of vacancies at the interface,

$$
\rho\mu_v = -W_e - \sigma\kappa
\tag{11}
$$

Since there is no elastic energy stored in the fluid, the elastic energy density of the fluid does not appear in Eq. (11). In addition, due to the lack of a coherency constraint, there is no term related to the jump in the inner product of the stress and strain as is the case in Eq. (8). This chemical potential can now be used to determine the evolution of surface in contact with the gas where the atoms move by surface diffusion.

3. Applications

Using the thermodynamics discussed above, the concentrations or chemical potential at an interface can be specified. A solution to the relevant diffusion equation determines that mass fluxes in the system and thus the rate of motion of the boundary. The advantage of the sharp interface formulation, as will become clear from the following examples, is that the small scale diffuseness of the interface does not have to be computed along with the much longer length scale diffusion fields. Moreover, in some cases it is possible to map the diffusion field in the bulk phases to the interfaces themselves, thus reducing the effective dimensionality of the system by 1. The disadvantages of the sharp interface approach are that the resulting equations can be difficult to solve in a numerically efficient manner and that coalescence or splitting of domains is not allowed. As we will see below it is possible to develop methods that address the former issue, the later is an inherent limitation of the method.

3.1. COARSENING IN ELASTICALLY STRESSED SOLIDS

Ostwald ripening or coarsening, in which larger second-phase particles grow at the expense of smaller particles, occurs in a vast range of two-phase mixtures, from liquid-liquid to solid-solid mixtures. Recent theoretical and experimental work has indicated that the coarsening process in systems with elastic stress can be qualitatively and quantitatively different from more classical interfacial-energy-driven coarsening processes. The elastic stress can come from many sources, such as the difference between the lattice parameters of the particle and the matrix, or an applied stress. As virtually all precipitates have a different lattice parameter from that of the matrix, or a misfit, elastic stress is generic to the Ostwald ripening in two-phase coherent solids. We shall focus on such systems.

In two-phase solids, the dynamics of ripening are driven by the reduction of the sum of the interfacial energy and the elastic energy. The elastic stress can have a large effect on the coarsening process in coherent solids because the total elastic energy of the system can easily be of the same order as the total interfacial energy. This can be shown by examining the magnitude of a dimensionless parameter, which is a measure of the relative importance of elastic and interfacial energies in the system [7]. This parameter is given by

$$L = \epsilon^2 l C_{44}/\sigma, \tag{12}$$

where ϵ is the particle-matrix misfit, l is a characteristic length of a particle, C_{44} is an elastic constant. For a system of multiple particles, $\langle L \rangle = \epsilon^2 \langle l \rangle C_{44}/\sigma$, where $\langle L \rangle$ is the average L and $\langle l \rangle$ is the average characteristic length associated with the particles in the system. L is a ratio of a characteristic elastic energy, $\epsilon^2 C_{44} l^3$, to a characteristic energy due to the presence of interfaces, σl^2 (in three dimensions). In many systems, even those with small misfits, it is likely for particles to attain an average size during coarsening to yield $\langle L \rangle$ of order 1 or larger. For example, in a Ni-Al alloy, $L = 5$ corresponds approximately to a particle size of $0.09 \mu m$, which is well within the range of technological interest. Since the elastic and interfacial energies are then of the same order of magnitude, the coarsening process must proceed by a decrease in the *sum* of the elastic and interfacial energies in the system. Thus, the dynamics of the coarsening process may not be similar to those given by an interfacial-energy-driven coarsening process.

This qualitatively different coarsening process in the presence of elastic stress, compared to that in the absence of stress, is illustrated by many theoretical and experimental investigations. The classic work of Ardell and Nicholson showed clearly that particles change their morphology from spheres to cuboids to plate-like shapes and align along the elastically soft directions of the crystal as they increase in size [8], or equivalently, $\langle L \rangle$. Theoretical investigations have shown, for example, that elastic stress can give rise to large-scale particle migration through the matrix [9, 10, 11] and inverse coarsening where small particles *grow* at the expense of large particles [10, 12, 13, 14].

In simulating Ostwald ripening in elastically anisotropic solids, it is absolutely necessary that the particle morphology be unconstrained, even if the particle shapes are not of direct concern. For example, two elastically and diffusionally interacting particles were found to be unstable with respect to coarsening; a small change in the area of one particle will always lead to the smaller particle shrinking [13]. In contrast, when the morphology of the particles was fixed to be a circle, the two particles could be stable with respect to coarsening. When the morphology of the particles are not constrained, the extra degrees of freedom associated with changes in morphology allow the particle to access evolutionary pathways that decrease the energy of the system and that are not open when the particle is constrained to a simple shape.

3.1.1. *Formulation*

We consider a system with a misfitting particle phase (β phase) and a matrix phase (α phase) with identical elastic constants, coherent interfaces, and isotropic interfacial energy. The misfit strain, ϵ, is taken to be purely dilatational. The elastic constants are that of pure Ni. The elastic constants and lattice parameters of α and β are taken to be independent of composition. We approximate an infinitely large system by a computational cell of unit size that is repeated periodically to fill all space, thus eliminating edge effects. The calculations are performed in two dimensions.

One of the central equations governing the coarsening process is found by expanding the chemical potentials appearing in Eq. (6) and Eq. (7) to first order in the difference in concentration between a flat interface in a stress-free system, C_∞^α, and the concentration in the matrix at an interface in the system with stress and curvature, C^α, and solving the resulting linear system of equations yields the Gibbs-Thompson equation,

$$C^\alpha = C_\infty^\alpha + B \left([\![W_e]\!] - [\![TE]\!] + \sigma\kappa \right) \tag{13}$$

where $B = 1/[\rho G_m''(C_\infty^\beta - C_\infty^\alpha)]$ and G_m'' is the second derivative of the molar Gibbs free energy of the α phase with respect to composition evaluated at C_∞^α. This sets the concentration at each location on the particle-matrix interface. It is clear that the concentration at a point on the interface is a function of both stress and curvature. Thus, both stress and interfacial energy will affect the dynamics of the interface.

Using the dimensionless variables given in [9], assuming that the temporal variation of the concentration in the matrix is fast on the time scale of the interfacial motion, the dimensionless concentration c in the matrix satisfies the steady-state diffusion equation

$$\nabla^2 c = 0. \tag{14}$$

The concentration at the particle-matrix interface in dimensionless form is,

$$c(\mathbf{x}) = \kappa + L\left\{ W_e - [\![TE]\!] \right\}, \tag{15}$$

The parameter L appears in the dimensionless form of the Gibbs-Thompson equation. As was mentioned above and as is clear from Eq. (15), the magnitude of L sets the relative importance of stress and interfacial energy. Eq. (14) is solved with the boundary condition, Eq. (15), and a global mass conservation condition. The evolution of the interface in time is computed using the interfacial mass balance,

$$V = \frac{\partial c}{\partial n}, \tag{16}$$

where V is the dimensionless normal velocity at a point on the interface and n is the coordinate along the direction normal to the interface.

The stress and strain fields at the interfaces can be written via an integral over the interfaces [15]. The stress and strain that appear in Eq. (15) are then calculated using

$$u_{j,k}(\mathbf{x}) = c_{ilmm} \int_s g_{ij,k}(\mathbf{x}, \mathbf{x}') n_l' ds' \tag{17}$$

where u_j is the j-th component of the dimensionless displacement vector, comma denotes differentiation with respect to the noted index, c_{ijkl} is the dimensionless elastic constant tensor, g_{ij} is the dimensionless elastic Green's function tensor, $\mathbf{x}$ and $\mathbf{x}'(s')$ are points on the interface, s' is the arc length, and the integral is taken over the interface, s. The stress, strain and elastic energy density are then determined using the derivatives of the displacement and the constitutive relations of linear elasticity.

The solution to Laplace's equation in the bulk phases is determined using boundary integrals [16]. In this approach the two-dimensional diffusion field is mapped onto the interfaces. Therefore, only the particle-matrix interfaces need to be discretized and one solves one dimensional singular integral equations to determine the two-dimensional diffusion field. We avoid the stiffness associated with integrating the equations in time via a small-scale decomposition method [17]. Finally, the

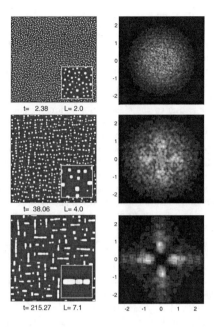

Figure 1. The microstructures (left) and the corresponding scattering functions (right) at $\langle L \rangle$ = 2.0, 4.0, and 7.1, from the top to the bottom, using $\langle L_0 \rangle$ = 1.5. The size of the boxes for the microstructures is the computational domain. The axes in scattering function plots are $k_x \langle r \rangle$ and $k_y \langle r \rangle$. A small section of the microstructure is magnified to show the detailed morphology of a few particles.

fast multipole method is employed to evaluate the diffusion and elastic fields [18, 19]. As a result, the computational cost scales linearly with the number of mesh points and 2-4×10^6 mesh points can be employed routinely in the calculations. Since only the interfaces are descretized we obtain very good resolution of the particle morphologies. Details of numerical methods can be found in [18].

3.1.2. *Results*

When only one particle exists in the system, a particle evolves towards an equilibrium shape that depends on the parameter, L. The equilibrium shapes have been computed by Thompson, Su, and Voorhees [7] for the same parameters used in our simulations. They found that for $L = 0$ the particles are circular. As L increases but remains less than 5.6, the equilibrium shape of particles are four-fold symmetric square-like shapes with rounded corners and relatively flat sides. For $L >$ 5.6, these four-fold symmetric shapes are energy maxima and two-fold symmetric, rectangular-like shapes are energy-minimizing equilibrium shapes. The diagram depicting the equilibrium shapes as a function of L is a classical supercritical bifurcation diagram with the bifurcation point at 5.6. Thus, the particle morphology can change continuously from four- to two-fold symmetric shapes with increasing L.

Fig. 1 shows the spatial arrangement of the particles and its scattering function at three different times when $\langle L \rangle = 2.0$, 4.0, and 7.1. The runs are started at an initial $\langle L \rangle$ or, $\langle L_0 \rangle$, of 1.5 with the area fraction, $\phi = 10\%$. We find a general agreement between the microstructures calculated in our simulations and those in experiments. During coarsening, the particles lose circular symmetry and become four-fold symmetric square-like shapes, even at a low $\langle L \rangle$ of 2. In addition, we find particle alignment along the vertical and the horizontal directions that correspond to the elastically soft $\langle 100 \rangle$ directions. The shape changes of the particles at the early stage are mainly due to the elastic energy generated by the particle itself, rather than by elastic interactions, and the particle alignment in the later stages is due to the elastic interactions or configurational forces generated by other particles [11]. An exception to this is when two particles or more are relatively close or at high $\langle L \rangle$, and the interaction causes the shapes to be distorted significantly.

Although the effects of the elastic-self energy are obvious from the non-circular particle morphology, the alignment of particles is not as clear at $\langle L \rangle = 2$. This is likely to be because the elastic interaction energy is typically a small fraction of the total elastic energy at the low particle area fraction of 10%. However, early signs of alignment are evident where chains establish themselves by chance due to fluctuations in the initial configuration. Elastic interaction helps to stabilize such conformations, and they survive for a long period of time at the expense of particles not aligned with others.

By $\langle L \rangle = 4.0$, most particles are aligned with several others, making the microstructure consist of more ordered congregations of chains. Particles in the middle of chains are stabilized by the elastic interaction energy, and they may grow even if they are smaller than the surrounding particles. A chain may grow as a whole at the expense of isolated particles or smaller chains.

In the last snapshot taken at $\langle L \rangle = 7.1$, the average L is well beyond the four-fold to two-fold shape bifurcation point of 5.6, and therefore the majority of the particles are clearly elongated. However, it is seen that the shape is not solely a function of the value of L for the individual particles. For example, we find large, almost square-like particles for which $L > 5.6$ (several of them are seen near the upper right corner), as well as much smaller particles that are elongated significantly. This indicates that the local arrangement of particles strongly influences the particle shapes. An arrangement of particles that has a four-fold symmetry with respect to a particle is likely to cause the particle to take on a four-fold symmetry if elastic interactions are important, while a two-fold arrangement is likely to encourage elongation.

The particles do not coalesce in our calculations as a result of the short-range repulsion due to the elastic stress. However, as $\langle L \rangle$ increases, the separation between the interfaces, particularly those interfaces of particles that are aligned, decreases. Thus, at sufficiently large $\langle L \rangle$, this separation can be less than a few lattice parameters, indicating that coalescence may be possible. As a result, we stop the calculation before $\langle L \rangle$ becomes so large that coalescence, which we do not allow in our simulations, becomes an important factor.

The corresponding scattering functions show the changes in the shapes and the arrangement of the particles, in agreement with the evolution seen in the spatial arrangement. At $\langle L \rangle = 2.0$, the scattering function is similar to that of circular particles distributed randomly. The characteristic circular symmetry of the scattering function seen in Fig. 1 indicates that the scattering function is not very sensitive to the four-fold symmetric shapes of the particles, at least when viewed using intensity plots. By $\langle L \rangle = 4.0$, the scattering function shows the effects of the alignment of the particles. The originally radially symmetric scattering function now shows a well developed four-fold-symmetric structure, with high intensities concentrated at the dominant wave vectors at magnitude of $\simeq 1/\langle r \rangle$ along the $\langle 100 \rangle$. The four-fold symmetry of the scattering intensity indicates that there are as many particles elongated or aligned in the [100] direction as in the [010] direction, while there are fewer particles aligned along the $\langle 110 \rangle$. The trend continues at $\langle L \rangle = 7.1$, where the

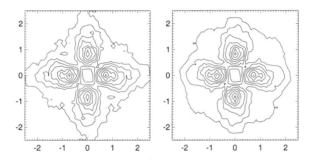

Figure 2. Scattering intensity as a function of the scattering vector. The plot on the left shows the scattering function for the microstructure found at $\langle L \rangle = 7.1$. On the right, the scattering function is calculated using particles at the same location as on the left, but assuming circular shapes. The axes are $k_x \langle r \rangle$ (horizontal) and $k_y \langle r \rangle$ (vertical).

four-fold symmetry becomes even more defined.

Although the scattering function is much more sensitive to the spatial distribution of particles than to the shapes of the individual particles, the effect of changing Fourier shape factors (the scattering functions of the individual particles) is not negligible. When the particle shapes are examined directly, the change in the shape of the particles is clearly seen between $\langle L \rangle = 4.0$ and 7.1. However, it is difficult to determine if this change is responsible for the change of the scattering function. To examine this issue, we calculate the scattering function of a microstructure identical to that for $\langle L \rangle = 7.1$, except that the shapes of the particles are replaced by circles. The comparison is shown in Fig. 2. As expected, the scattering function at small k, which is sensitive to correlation at long distances, remains relatively unchanged by altering the morphology of the particles. The scattering function at large k, which is sensitive to the particle shapes, is more isotropic when the particles are circular.

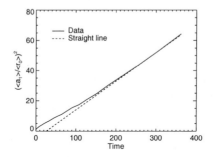

Figure 3. $(\langle a_1 \rangle / \langle r_o \rangle)^{1/2}$ is plotted against time. The straight line is not a best fit, but given as a guide to the eye.

We can also examine the coarsening kinetics of the two-phase mixture. One measure of particle size is the average particle size, defined here as $\langle r \rangle = (\langle A \rangle / \pi)^{1/2}$, where A is the area. Such a

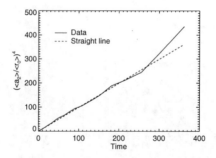

Figure 4. $(\langle a_2 \rangle / \langle r_o \rangle)^{1/4}$ is plotted against time. The straight line is not a best fit, but given as a guide to the eye.

measure is independent of particle shape. In other studies, the particle size has been measured using the side length. The closest measure we have is the size along the axis along which a particle is elongated (the major axis), a_1, and the size along the axis perpendicular to that axis (the minor axis), a_2. These quantities have been studied by Finel in his numerical study using a Monte Carlo method [23]. The side lengths, a_1 and a_2, provide measures of the particles size and shape. It has been suggested in [23] that $\langle a_1 \rangle$ follows a power law of $t^{1/2}$, while $\langle a_2 \rangle$ follows $t^{1/4}$. To examine our results from this viewpoint, we have plotted $(\langle a_1 \rangle / \langle r_0 \rangle)^2$ and $(\langle a_2 \rangle / \langle r_0 \rangle)^4$ against time in Figs. 3 and 4, respectively, where $\langle r_o \rangle$ is the initial average particle size. Since these power laws apply when particles are elongated, a poor fit at early time is expected. For $\langle a_1 \rangle$, the data are not inconsistent with a power law with exponent of $1/2$, and in fact, a good fit is obtained for $t > 200$. For $\langle a_2 \rangle$, we find that the fit is good at $t < 270$, or separately at $t > 260$ (there are many data points in the range $t > 260$). However, due the limited change in length scale over which there is a good fit between the power law and the data we cannot say conclusively that the simulations agree with Finel's results.

3.2. THE STABILITY OF AN ELASTICALLY STRESSED CRYSTAL:
THE ROLE OF INTERFACIAL MOBILITY

A planar interface separating an elastically stressed solid and a fluid is unstable to infinitesimally small morphological fluctuations. This instability, identified by Asaro and Tiller [24] and Grinfeld [25] (ATG), and others [26, 27, 28], results from the dependence of the chemical potential on the elastic energy density. Consider a solid under uniaxial stress in contact with a fluid. Although the stress in the unperturbed or basic state is uniform, this is not the case when the interface is nonplanar. When the surface is perturbed the stress is concentrated at the troughs, in analogy with the stress concentration of a crack, and relieved at the peaks. Because the chemical potential is linearly proportional to the elastic energy density, see Eq. (11), the chemical potential of an atom is higher at the trough than at the peak and atoms thus move from the trough to the peak tending to make the peaks higher and the troughs deeper. Opposing this, of course, is surface energy that would tend to move atoms in the opposite direction and lead to a planar surface. The stress concentration in the trough appears in both compression and tension, and thus the AT instability is independent of the sign of the applied stress.

Given the basic mechanism, the development of the elastic energy driven, or AT, instability must also depend on the mode of mass transport. Thus, the elastic energy driven instability has been examined in the context of bulk diffusion [24], surface diffusion [26, 27, 28] and evaporation-condensation [26, 29]. The evaporation-condensation model is essentially a completely kinetically controlled interface in which the rate-limiting mass transport process is atom motion across an interface between a solid and another solid, a vapor, or even a liquid. In this case the velocity of the interface is linearly proportional to the difference between the chemical potential at the surface and that in the parent phase. The type of mass transport process does not alter the critical wavenumber of the instability, it alters only the dependence of the growth rate of the instability on wavenumber.

All of this work, however, assumes that the relevant transport coefficient, the surface diffusion coefficient for surface-diffusion-limited interface motion, or the interface mobility for interface-reaction-limited interface motion, is independent of elastic stress. Atomistic simulations have shown that the surface diffusion coefficient can be linearly proportional to the elastic stress on the surface. More importantly, recent experiments by Barvosa-Carter et al. (BC) have shown that a stress-dependent interfacial mobility alone can generate an interfacial instability when the transformation is interface-limited [30]. The mechanism for the kinetic instability follows from the observation that the change in the mobility with stress in certain cases can be linearly proportional to the stress. Consider a basic state consisting of a moving planar interface under uniform uniaxial stress. Assume that the stress compressive, i.e. *negative*, and that the mobility increases with increasingly positive stress. Then at the troughs, where the stress is the more negative than that of the planar state, the mobility is smaller than that of the planar interface. Conversely at the peaks, where the stress is larger than that of the basic state, the mobility is larger than that of the planar interface. Thus the troughs grow slower and peaks grow faster than the moving planar interface. In this case the stress-dependent mobility is destabilizing.

Moreover, experiments have shown that the amorphous to crystalline (a-c) transformation in Si is an ideal system in which to explore this kinetic instability. The stress dependence of the interfacial mobility has been measured [31]. The mobility, M, was found to depend upon stress, T, as

$$M\left(\sigma_{ij}\right) = \bar{M} \exp\left(V_{ij}^{*} T_{ij}/k\mathcal{T}\right) \tag{18}$$

where V_{ij}^{*} is the activation strain [31], $\bar{M}$ is the mobility of the interface in the absence of stress, k is Boltzmann's constant and T is temperature. For the Si(001) a-c interface, $V_{11}^{*} = V_{22}^{*} = 0.14\Omega$, and $V_{33}^{*} = -0.35\Omega$, where Ω is the atomic volume, and the off-diagonal elements are zero. The transformation is completely interface reaction-limited, and other relevant materials parameters have been measured. It is possible to carefully control the morphology of the initial interface using lithography. Finally, because, $V_{11}^{*} > 0$ a compressive uniaxial stress $T_{11} < 0$, should result in an interfacial instability. Such an instability was observed by BC.

We illustrate the application of the sharp interface model to determine the linear stability of a planar interface where the motion of the interface is interface reaction-limited. Unlike in the ripening problem discussed above it is not necessary to determine the diffusion field in the matrix. Such a problem is thus particularly ideal for a sharp interface description as the mass flow is entirely local to a point on the interface. We allow for a stress dependent interfacial mobility and examine its effect on the stability of the interface. Using the materials and processing parameters of the BC experiments, we make a comparison between theory and experiment with no adjustable parameters. However, we assume that the amorphous phase into which the crystal grows is stress-free. We also assume M to be independent of crystallographic orientation. We follow closely the work of Voorhees and Aziz [32].

3.2.1. *Theory*

The crystal grows into the amorphous phase at a constant velocity v in the positive x_3-direction where x_3 is parallel with the [001] direction. All quantities except stress and activation strain are assumed to be isotropic.

The motion of the a-c interface is kinetically limited. The equation for the motion of the boundary can be found by considering first an interface in equilibrium with the fluid with nonzero stress and curvature. Substituting Eq. (11) into the second of Eq. (9) gives,

$$\rho\left(\mu_1^\alpha - \mu_1^f\right) + W_e + \sigma\kappa = 0 \tag{19}$$

where α is the crystal and f is the amorphous phase. The chemical driving force for interfacial motion in a stress-free crystal, $\Delta G = (\mu_1^\alpha - \mu_1^f)\rho$. Thus,

$$\Delta G + W_e + \sigma\kappa = 0 \tag{20}$$

If the LHS of Eq. (20) is zero then the system is at equilibrium. If this is not the case then there is driving force acting on the interface. For kinetically limited motion of an interface in the small driving force limit, the velocity is linearly proportional to the deviation in the LHS of Eq. (20) from zero. In the limit of small driving forces the velocity of the interface, v is given by [30, 31]

$$v = -M\left(\mathsf{T}\right)\left[\Delta G + W_e + \gamma\kappa\right] \tag{21}$$

where $M\left(\mathsf{T}\right)$ is the stress-dependent interface mobility. The chemical driving force for crystallization ΔG is the Gibbs free energy change ($\Delta G < 0$) per unit volume of growing phase upon crystallization of stress-free material at a planar interface. Since the transformation is isothermal and occurs in a pure material, the only field equation that must be solved is that for the elastic stress.

In the unperturbed, or basic state, the interface is planar and moving at a constant velocity into the amorphous phase. The applied stress in the crystal is biaxial and of magnitude $T_{11}^0 = T_{22}^0 = T_a$. Assuming that the pressure in the amorphous phase is zero, the third of Eq. (9), implies $T_{i3} = 0$ throughout.

To examine the linear stability of the moving planar interface located at $x_3 = 0$, we perturb the height of the interface in the x_3-direction, $h = h\left(x_1, x_2\right)$ and all other quantities in normal modes,

$$\begin{pmatrix} h \\ v \\ u_i \\ M \\ W_e \\ \kappa \end{pmatrix} = \begin{pmatrix} 0 \\ v^0 \\ u_i^0 \\ M^0 \\ W_e^0 \\ 0 \end{pmatrix} + \delta \begin{pmatrix} \hat{h} \\ v^1 \\ u_i^1 \\ M^1 \\ W_e^1 \\ \kappa^1 \end{pmatrix} \Phi \tag{22}$$

where $\Phi = \exp\left(i\mathbf{q}\cdot\mathbf{x} + rt\right)$, $\mathbf{q}$ and $\mathbf{x}$ are the wavevector and position vector, respectively, in the plane of the interface, $\delta \ll 1$ and t is time. If $r > 0$ the perturbation grows and if $r < 0$ the perturbation decays. If r is imaginary then an oscillatory instability is possible.

Using Eq. (22) in Eq. (21) we obtain in the basic state, $\delta = 0$,

$$v^0 = -M^0\left(\Delta G + W_e^0\right) \tag{23}$$

The perturbed velocity is

$$v^1 = -\left(\Delta G + W_e^0\right)M^1 - M^0 W_e^1 - M^0\sigma\kappa^1 \tag{24}$$

We now need to find $M^0, M^1, W_e^0, W_e^1, v^1$ and κ^1.

The elastic strain energy density in the basic state follows from the stresses given above,

$$W_e^0 = T_a^2/Y \qquad (25)$$

where $Y = E/(1 - \nu)$. The strain energy density in the perturbed state is determined by solving the conditions for elastic equilibrium with the boundary condition given by the fourth of Eq. (9) [33],

$$W_e^1 = 2\hat{h}(1 + \nu)qT_a^2/Y \qquad (26)$$

where $q = |\mathbf{q}|$.

The mobility of the interface in the basic and perturbed states, M^0 and M^1 are [32]

$$M^0 = \bar{M}\left(1 + \frac{2V_{11}^* T_a}{k\mathcal{T}}\right) \qquad (27)$$

and

$$M^1 = -\bar{M}2\hat{h}(1 + \nu)V_{11}^* T_a q/k\mathcal{T} \qquad (28)$$

Thus, if $V_{11}^* > 0$, a compressive stress decreases the mobility of the planar interface. For compressive stresses the perturbed mobility M^1 is in phase with the interface shape. Therefore the mobility is the highest at the peaks and the lowest at the troughs.

The remaining terms in Eq. (24) are functions only of the shape of the interface. Using the definition of the curvature in cartesian coordinates, the form of the perturbed interface, and by taking the derivative with respect to time we obtain, $\kappa^1 = \hat{h}q^2$ and $v^1 = \hat{h}r$.

Using Eq. (25)-Eq. (28) in Eq. (24) yields the dispersion relation,

$$
\begin{aligned}
\frac{r}{\bar{M}} &= \left[(1 + \nu)\,\alpha\Delta G_v T_a + 2\beta T_a^2 + 3\alpha\beta T_a^3\right]q \\
&\quad - (1 + \alpha T_a)\,\sigma q^2
\end{aligned} \qquad (29)
$$

where $\alpha = 2V_{11}^*/k\mathcal{T}$ and $\beta = (1 + \nu)/Y$. The dispersion relation is linear in r and thus instability cannot be oscillatory.

The wavenumber at which the growth rate of the instability is zero, q_c, is found by setting $r = 0$.

$$q_c = \frac{(1 + \nu)\,\alpha\Delta G_v^0 T_a + 2\beta T_a^2 + 3\alpha\beta T_a^3}{(1 + \alpha T_a)\,\sigma}. \qquad (30)$$

For $q < q_c$, $r > 0$; for $q > q_c$, $r < 0$. The wavenumber with the maximum growth rate, q_m, is given by $q_m = q_c/2$. Thus q_c determines the length-scale of the instability.

3.2.2. Discussion

The dispersion relation is quadratic in the wavenumber, q, with the terms related to the elastic stress being linear in q and the interfacial energy related terms going as q^2. Interfacial energy is always stabilizing, because by assumption $T_a V_{11}^*/k\mathcal{T} \ll 1$. Thus any instability is driven solely by the elastic stress. Of the terms that depend upon the applied stress, the term that is quadratic in the stress is the classical elastic-energy-driven ATG instability. In this reaction-limited case the AT instability is linear in q and not cubic as it is in the usual case where the instability develops by surface diffusion. The term that is linear in the applied stress can be either stabilizing or destabilizing depending on the sign of the product αT_a. If $\alpha T_a < 0$, as in the a-c Si BC experiments, the stress dependence of the mobility itself can drive an interfacial instability. The term that is cubic in the applied stress is the AT instability modified by the stress dependent mobility.

Because $\alpha T_a < 0$ under compressive stresses for the Si(001) a-c interface, both the stress dependent mobility and the elastic energy promote instability. As shown in Eq. (29), both effects have the same scaling with the wavenumber and thus it can be difficult to determine which phenomenon is responsible for an experimentally observed interfacial instability. However, the elastic energy driven instability scales with stress in a different manner than the mobility driven instability. It is clear that for a sufficiently small applied stress, the stress-dependent mobility will dominate. When $\alpha T_a < 0$, this will occur when

$$\frac{(1+\nu)\,\alpha\Delta G_v}{2\beta T_a + 3\alpha\beta T_a^2} \gg 1. \tag{31}$$

Using the values for the a-c Si interface under the conditions employed in the BC experiments we find that LHS of the inequality is approximately 12, and thus the instability of the interface is predicted to be due to the kinetic instability.

Conversely, if $\alpha T_a > 0$, then the stress-dependent mobility is stabilizing. The critical stress, T_a^c below which the stress-dependent mobility will stabilize the AT instability is found by setting the terms linear in q in Eq. (29) to zero,

$$T_a^c = -1/\left(3\alpha\right) + \left[1 - 3\alpha^2\left(1+\nu\right)\Delta G_v/\beta\right]^{1/2}/\left(3\alpha\right) \tag{32}$$

For the a-c Si system, this yields a stress of 9.6 GPa. All tensile stresses below this value result in an interface that is stable against the AT instability. Preliminary experimental results [34] confirm this prediction. Thus in the a-c Si system very large (perhaps impossibly large) tensile stresses are required to overcome the stabilization due to the stress-dependent mobility.

Using $T_a = -0.5GPA$ [30], $Y = 100GPA$, $\nu = 0.25$, $T = 520C$, $\bar{M} = 4.52 \times 10^{-20} m^2/Js$ [35], $\Delta G_v = -6.27 \times 10^8 J/m^3$ [36] and $\sigma = 0.4 J/m^2$ [37, 38] the length scale of the interfacial instability, provided by the critical wavelength, which follows from Eq. (30), is 8.5 nm. Thus perturbations with wavelengths in excess of 8.5 nm will amplify and those with wavelengths less than 8.5 nm will decay. The critical wavenumber of the AT instability, q_c^{AT} is,

$$q_c^{AT} = 2T_a^2\beta/\sigma \tag{33}$$

For the BC experiment, this yields $\lambda_c^{AT} = 2000nm$. It is thus clear that the length scale of the kinetically driven instability can be quite different from that driven solely by elastic strain energy. The wavelength of the kinetically driven instability with the maximum growth rate is 17 nm. Unfortunately, the BC experiments were unable to determine q_c because, for reasons of experimental practicality, they started not with a planar interface but rather with an interface with a large initial sinusoidal perturbation with a wavelength of 400 nm. They found that a perturbation of this wavelength is unstable, and thus is consistent with this prediction. The experimentally observed amplification rate of such a perturbation is $6 \times 10^{-5}/sec$ whereas the predicted amplification rate is $1.6 \times 10^{-4}/sec$. This is very good agreement considering the uncertainties in the materials parameters.

4. Acknowledgments

We are grateful for the financial support of the National Science Foundation, Grant DMR-9707073.

References

1. J.W. Gibbs, *The Scientific Papers of J.Willar Gibbs Volume 1: Thermodynamics* Dover, New York 1961.
2. L. Ratke and P.W. Voorhees, *Growth and Coarsening: Ripening in Materials Processing*, Springer Berlin, 2002.
3. C. Herring, *The Physics of Powder Metallurgy*, Ed. R.E. Kingston, McGraw-Hill, New York, 1951, p. 143.
4. F.C. Larché and J.W. Cahn, *Acta. Metall.* **33**, 331 (1985).
5. P.H. Leo and R.F. Sekerka, *Acta Metall.* **37**, 3139 (1989).
6. B.J. Spencer, P.W. Voorhees and S.H. Davis, *J. Appl. Phys.*, **73**,4955(1993).
7. M.E. Thompson, C.S. Su and P.W. Voorhees, *Acta Metall.*, Mater. **42**, 2107 (1994).
8. A.J. Ardell and R.B. Nicholson, *Acta Metall.*, **14**, 1295 (1996).
9. P.W. Voorhees and W.C. Johnson, *Phys. Rev. Lett.*, **61**, 2225 (1988).
10. Y. Wang, L.Q. Chen, and A.G. Khachaturyan, *Acta Metall. Mater.*, **40** 2979 (1993)
11. C.H. Su and P.W. Voorhees, *Acta Mater.*, **44**, 2001 (1996).
12. W.C. Johnson and J.W. Cahn, *Acta Metall.*, **32**, 1925 (1984).
13. C.H. Su and P.W. Voorhees, *Acta Mater.*, **44**, 1987 (1996).
14. H.J. Jou, P.H. Leo and J.S. Lowengrub, *J. Comp. Phys.*, **131**, 109 (1997).
15. T. Mura, *Micromechanics of Defects in Solids* (Kluwer Academic Publishers, Dordrecht, The Netherlands, 1987).
16. A. Greenbaum, L. Greengard and G.B. McFadden, *J. Comput. Phys.*, **105**, 267 (1993)
17. H.-J Hou, J.S. Lowengrub, and M.J. Shelly, *J. Comput. Phys.* **114**, 312 (1994).
18. N. Akaiwa, K. Thornton and P.W. Voorhees,*J. Comput. Phys.*, **173**, 61 (2001).
19. L. Greengard and V. Rokhlin, *J. Comput. Phys.*, **73**, 325 (1987).
20. I.M. Lifshitz and V.V. Slyozov, *J. Phys. Chem. Solids*, **19**, 35 (1961).
21. C. Wagner, *Z. Elektrochemie*, **65**, 581 (1961).
22. K. Thornton, Norio Akaiwa, and P.W. Voorhees, *Phys. Rev. Lett.* **86**, 1259 (2001).
23. A. Finel, *Phase Transformation and Evolution in Materials*, Ed. P.E.A. Turchi and A. Gonis, The Minerals, Metals and Material Society, Warrendale, PA (2000), p. 371.
24. R.J. Asaro and W.A. Tiller, *Metall. Trans.*, **3**, 1789 (1972).
25. M.A. Grinfeld, *Dok. Akad. Nauk SSSR*, **290**, 1358 (1986).
26. D.J. Srolovitz, *Acta Metall.*, **37**, 621 (1989).
27. J. Gao, *Int. J. Solids Structures*, **28**, 703 (1991).
28. B.J. Spencer, P.W. Voorhees and S.H. Davis, *Phys. Rev. Lett.*, **67**, 3696 (1991).
29. K.-S. Kim, J.A. Hurtado and H. Tan, *Phys. Rev. Lett.*, **83**, 3872 (1999).
30. W. Baravosa-Carter and M.J. Aziz, *Phys. Rev. Lett.*, **81**, 1445 (1998).
31. M.J. Aziz, P.C. Sabin, and G.-Q. Lu,*Phys. Rev. B*, **44**, 9812 (1991).
32. P.W. Voorhees and M.J. Azis, *Interfaces for the 21st Century* ed. Marc. K. Smith et. al, Imperial College Press London, 2002 pg. 167.
33. J.E. Guyer and P.W. Voorhees, *Phys. Rev. B*, **54**, 10710 (1996).
34. J.F. Sage, W. Barvosa-Carter and M.J. Aziz (unpublished).
35. G.L. Olson and J.A. Roth, *Mater. Sci. Rep.*, **3**, 1 (1988).
36. E. P. Donovan, F. Spaepen, D. Turnbull, J.M. Poate, and D.C. Jacobson, *J. Appl. Phys.*, **57**, 1795 (1998).
37. N. Bernstein, M.J. Aziz, and E. Kaxiras, *Physical Review B*, **58**, 4579 (1998).
38. C.M. Yang, Ph.D. thesis, California Institute of Technology, 1997.

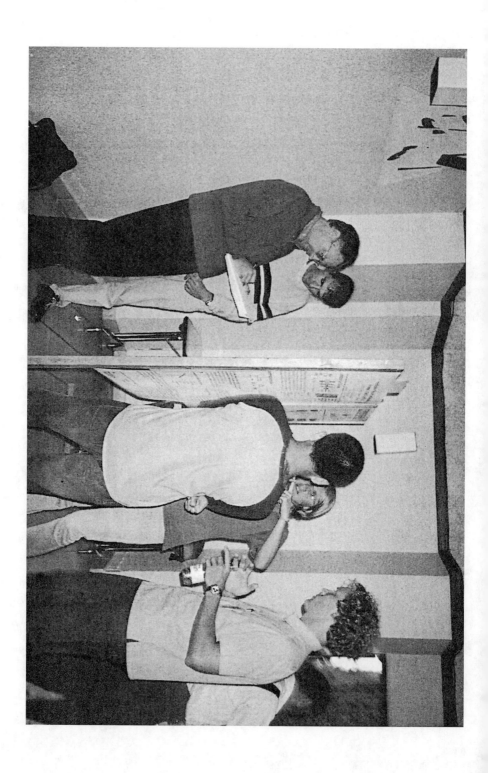

CLUSTER DYNAMICS

P. GUYOT[1], L. LAE[1], C. SIGLI[2]
[1]: *LTPCM/ENSEEG, UMR 5614, Domaine Universitaire, B.P. 75, 38 402 Saint Martin d'Hères Cedex, France*
[2]:*Pechiney Centre de Recherches de Voreppe, B.P. 27, 38341 Voreppe, Cedex, France.*

Abstract: First-order transformations are first introduced. A distinction is then developed between heterophase and homophase statistical fluctuations. The basic elements of the classical theory of heterophase nucleation are the thermodynamic driving force and the kinetics master equation. The nucleation, growth and coarsening stages emerge naturally from this theory. In the following section the kinetic ingredients of the classical theory, i.e. the monomer condensation and evaporation rates are examined in detail. A numerical solution of the master equation is given for dilute binary alloys. Cluster size distribution, nucleation current, growth and coarsening are analyzed and discussed. A generalized cluster dynamics framework, involving more complex inter-cluster exchange mechanisms is then considered.

1 Introduction

1.1 FIRST ORDER "PHASE TRANSFORMATION"

This lecture deals with the dynamics of phase transformations which occur when a one-phase homogeneous material in thermal equilibrium state is quenched through a first order phase transition in a two-phase region. The quenched one-phase system is no longer in equilibrium and evolves toward a new thermodynamic equilibrium state that consists of two coexisting phases.

Such a transformation is found in different materials like fluids, glasses, polymer blends, or metallic solid solutions. The situation is schematized in Figure 1 for a binary mixture of two atomic components A and B forming a high temperature disordered solid solution, and having homo-atomic attractive interactions. Figure 1a) is the schematic phase diagram of such a solid solution that displays a miscibility gap (full line) in the temperature-concentration representation. After a rapid temperature drop from T_i, in the one-phase region, to T below the gap, the solid solution of composition x_0 separates in A rich and B rich solid solutions. This lecture addresses the understanding of the kinetics rate and path towards the two-phase equilibrium between the solid solutions of composition x_{eq} and x_β. The composition x_{eq} and x_β are determined by the common tangent to the curve of Figure 1b). This curve represents, for a given temperature, the free enthalpy of mixing of the element B in a homogeneous solid

A. Finel et al. (eds.), Thermodynamics, Microstructures and Plasticity, 107–121.

solution of element A. We will see later that we need the knowledge of this curve which determines the driving force of the transformation. The dotted line represents the spinodal curve. More complex cases may occur when an ordering process is involved. The parent phase and the new phase have generally different crystal lattices and structures as well as different chemical compositions, as for instance in a first order ordering.

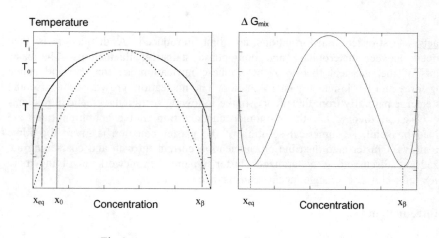

Fig.1.a   Fig. 1.b

Figure 1   a) miscibility gap and spinodal line for a binary solid solution
b) mixing free enthalpy at temperature T

1.2   HOMOPHASE AND HETEROPHASE FLUCTUATIONS

Two different types of mechanisms may initiate the early beginning of such phase transformation. They correspond to different types of statistical fluctuations and are described by different theoretical models. Homophase and heterophase fluctuations are represented in Figure 2. "Homophase" concentration fluctuations have small amplitude and smooth "interfaces" and can be considered as packets of concentration waves. "Heterophase" fluctuations, on the other hand, have a sharp interface (width of the order of the inter-atomic distance), and their amplitude varies between the concentrations of the coexisting phases.

Theoretical concepts have been developed for each fluctuation type. Delocalized homophase fluctuations are at the basis of **"phase field"** approaches, where the dynamical evolution of fluctuations is driven by a free energy functional density including a concentration gradient effect [1-4]. Heterophase fluctuations are on the other hand at the basis of the **"classical nucleation theory"** [5-10] and of a later **"cluster dynamics theory"** extension developed by Binder and others [11-13]. The heterophase fluctuations are nothing but droplets or clusters, and the phase separation kinetics consists in the time evolution of their nucleation, growth and coarsening. Historically, there has been a long debate on the meaning of spinodal decomposition and on its experimental observation as well. This aspect is not covered in this lecture.

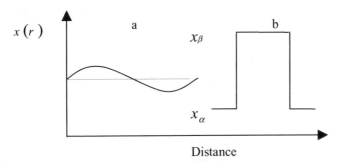

Figure 2. Homophase (a) and heterophase (b) concentration fluctuations

2   Classical Theory of Nucleation and Growth

2.1   CLUSTER THERMODYNAMICS

In the conventional theory of nucleation and growth the new phases are treated as localized heterophase fluctuations, droplets or clusters. These clusters are cut in the new β equilibrium phase and embedded in the matrix parent phase α. Each cluster is defined by its size, shape and homogeneous chemical composition. In the capillary model, the free enthalpy of formation of a ζ_n atoms or molecules cluster is given by:

$$\Delta G_n = \zeta_n \Omega \Delta G_v + A \zeta_n^{2/3} \gamma \qquad (1)$$

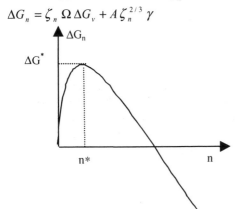

Figure 3. Typical free enthalpy change associated with cluster formation

The first term is the bulk driving force (chemical energy + elastic strain energy if the average atomic volume Ω in β is different from the average atomic volume in the matrix parent phase), γ is the matrix-β interface specific free energy, assumed further to be independent of n. A is a cluster shape factor. ΔG_v is positive when the temperature stands above the equilibrium temperature T_0 (see Figure 1). When crossing downward from T_0, it becomes negative and balances the positive surface term for a critical size ζ_n^*, according to the well known scheme of Figure 3.

The critical size of a spherical cluster and the enthalpy barrier are:

$$R^* = \frac{2\gamma}{|\Delta G_v|} \text{ or } \zeta_n^* = \frac{32\pi}{3} \frac{1}{\Omega} \left(\frac{\gamma}{|\Delta G_v|} \right)^3 \qquad (2.a)$$

$$\Delta G^* = \frac{16\pi}{3} \frac{\gamma^3}{\Delta G_v^2} \qquad (2.b)$$

For an assembly of non interacting clusters the Gibbs free energy of formation is given by:

$$\Delta G = \sum_n \Delta G_n \, C_n - T \, S_{mix} \tag{3}$$

where C_n is the atomic fraction of the n atom clusters and S_{mix} is the clusters configuration entropy of mixing. For a dilute system the equilibrium cluster distribution which minimizes (3) is [5,14]:

$$\overline{C}_n = \Theta \exp\left(-\frac{\Delta G_n}{kT}\right) \tag{4}$$

where Θ is a normalization constant obtained from the solute content balance. Such an equilibrium distribution is physically achieved only above the equilibrium temperature (T_0 in Fig. 1.a) and corresponds to what is usually termed as local order in solid solution. The cluster nucleation current is then controlled by the rate at which clusters pass the critical size. In the stationary state, it can be written in the form:

$$J_s \propto \exp\left(-\frac{\Delta G^*}{kT}\right) \tag{5}$$

Equation (5) results from an analysis of the master equation which controls the time evolution of the cluster distribution $C(n,t)$.

2.2   KINETICS: MASTER EQUATION

The time evolution of a cluster distribution must follow the general differential equation:

$$\frac{\partial C_n}{\partial t} = \sum_{n' \neq n} C_{n'} \, W_{n' \to n} - C_n \, W_{n \to n'} \tag{6}$$

where $W_{n \to n'}$ is the probability per unit time that a cluster of size n transforms to n' by catching a $(n'-n)$ cluster. This master equation is the basis of the cluster dynamics treatment that will be further detailed. Within the classical theory the exchange between clusters is limited to monomers. In that case, equation (6) only covers exchanges between the three classes: $n-1$, n and $n+1$ and can be written:

$$\frac{\partial C_n}{\partial t} = C_{n-1} W_{n-1 \to n} - C_n \left(W_{n \to n-1} + W_{n \to n+1} \right) + C_{n+1} W_{n+1 \to n} \tag{7}$$

or in terms of cluster flux J:
$$\frac{\partial C_n}{\partial t} = J_{n-1 \to n} - J_{n \to n+1} \tag{8}$$

with
$$J_{n \to n+1} = C_n W_{n \to n+1} - C_{n+1} W_{n+1 \to n} \tag{9}$$

The rate probabilities $W_{n \to n\pm1}$ will be evaluated later. In equilibrium conditions, all fluxes $J_{n \to n+1}$ must vanish; probabilities $W_{n \to n+1}$ are then related by:

$$\frac{W_{n+1 \to n}}{W_{n \to n+1}} = \frac{\overline{C}_n}{\overline{C}_{n+1}} \tag{10}$$

where $\overline{C}_n$ is given by equation (4). In classical nucleation theory, this is supposed to hold even in the two-phase region (this assumption is often referred to as "constraint equilibrium"). Substituting (10) in (9) yields:

$$J_{n\to n+1} = W_{n\to n+1}\,\overline{C_n}\left(\frac{C_n}{\overline{C_n}} - \frac{C_{n+1}}{\overline{C_n}}\right)$$

(11.a)

Considering n as a continuous variable and developing (11.a) to first order

$$J_{n\to n+1} = -W_{n\to n+1}\,\overline{C_n}\,\frac{\partial}{\partial n}\left[\frac{C_n}{\overline{C_n}}\right]$$

(11.b)

and combining it with the continuity equation for C_n (see Equation 8), i.e.

$$\frac{\partial C_n}{\partial t} = -\frac{\partial J_{n\to n+1}}{\partial n}$$

(12.a)

one obtains a Fokker-Planck type equation:

$$\frac{\partial C_n}{\partial t} = -\frac{\partial}{\partial n}\left[W_{n\to n+1}\,\overline{C_n}\,\frac{\partial}{\partial n}\left(\frac{C_n}{\overline{C_n}}\right)\right]$$

(12.b)

This describes the cluster diffusion along the n axis in a cluster gradient and a free enthalpy gradient as well. Examples of the numerical resolution of the master equation are given later for a model Fe-Cu alloy.

2.3   PRINCIPAL RESULTS OF THE CLASSICAL THEORY

2.3.1   The Steady-State Current

When a steady state is achieved, the continuity equation (12.a) indicates that all currents $J_{n\to n+1}$ are independent of cluster size. The value of this stationary current is obtained by integration of (11.b). It can be shown that this current is defined by [15]:

$$J_s = Z\,W_{n^*\to n^*+1}\,\frac{\Theta}{\Omega}\exp\left(-\frac{\Delta G^*}{kT}\right)$$

(13)

where the Zeldovich parameter Z is given by [7]:

$$Z = \left[-\frac{1}{2\pi kT}\left|\frac{\partial^2 \Delta G_n}{\partial n^2}\right|_{n^*}\right]^{1/2} = \frac{1}{\zeta_{n^*}}\left[\frac{\Delta G^*}{3\pi kT}\right]^{1/2}$$

(14)

Z is of the order of 10^{-2} and corresponds to the probability that clusters have to back jump to sub-critical sizes.

2.3.2   Incubation Time, Transient Current

The incubation time is the time necessary for the stationary nucleation current to be established. That is approximately the time needed for the probability of back-dissolution of the clusters to become negligible; it can be shown to be [7,16]:

$$\tau = \frac{\zeta_{n^*}^2}{4W_{n^*\to n^*+1}} \quad \text{or} \quad \tau = \frac{1}{4W_{n^*\to n^*+1}\,Z^2} = \frac{\zeta_{n^*}^2}{4W_{n^*\to n^*+1}}\left(\frac{3\pi kT}{\Delta G^*}\right)$$

(15)

and the transient nucleation current can be written:

$$J(t) = J_s\exp\left(-\frac{\tau}{t}\right)$$

(16)

2.3.3   Nucleation Slowing-Down, Growth and Coarsening

The nucleation rate decreases with time together with the supersaturation; the solute consumption decreases the transformation driving force, increases the critical free energy barrier ΔG^*. Consequently the nucleation current, via the equation (13), slows down dramatically with time.

Secondly, the formation of over-critical clusters introduces a solute depletion around each cluster and cluster growth occurs by diffusion through the concentration gradient. This is the well known stage of growth by long-range diffusion. It must be reminded that this growth stage can alternatively be controlled by the rate at which the solute atoms stick to the cluster-matrix interface. This point will be further developed in § 3.1.

The deterministic cluster growth rate by long-range diffusion is obtained by solving Fick 's laws and writing the solute conservation at the particle matrix interface. For a spherical particle of radius R, the growth rate is, with certain approximations, [17]:

$$\frac{dR}{dt} = \left(\frac{x_\infty - x_R}{x_\beta - x_R}\right)\frac{D}{R} \tag{17}$$

where x_∞ is the solute concentration at an infinite distance of the particle interface, x_R is the concentration at the interface and D is the diffusion coefficient of the solute. The integration of (17) gives an initial parabolic growth law. Again, the growth rate decreases with time due to the decrease of $(x_\infty - x_R)$. Ultimately, when the average matrix solute content reaches approximately the equilibrium value given by the phase diagram, the average particle size continues to grow. This phenomenon is due to the Gibbs-Thomson effect: the solute concentration in equilibrium with a curved interface is higher than with a flat interface (R=∞). In a binary dilute alloy, this equilibrium concentration can be written [18]:

$$x_R = x_{eq,R=\infty}\, \exp\!\left(\frac{2\gamma\Omega}{kT\,x_\beta\,R}\right) \tag{18}$$

The Gibbs-Thomson effect is at the origin of a coarsening of the average particle size, through a multi-particle diffusion process. The asymptotic law of this coarsening, when it occurs by long-range diffusion, has been established by Lifshitz-Slyozov-Wagner; for spherical clusters the average precipitate volume grows linearly with time [18]:

$$\overline{R}^3(t) = \overline{R}_0^3 + \frac{8\gamma\Omega D x_{eq}\left(1 - x_{eq}\right)}{9kT\left(x_\beta - x_{eq}\right)^2}t \tag{19}$$

In parallel, the supersaturation, expressed in solute concentration, follows the law:

$$x - x_{eq} = \left[\frac{9\gamma^2\Omega^2 x_{eq}^2}{(kT)^2 D}\right]^{1/3}\left(\frac{1}{t}\right)^{1/3} \tag{20}$$

and a scaled cluster distribution function is predicted:

$$C(\rho) = \frac{4}{9}\rho^2\left(\frac{3}{3+\rho}\right)^{7/3}\left(\frac{3/2}{3/2-\rho}\right)^{11/3}\exp\!\left(-\frac{\rho}{3/2-\rho}\right) \text{ with } \rho = R/\overline{R} \tag{21}$$

2.3.4   Numerical model (Kampman-Wagner, [19])

To quantify the precipitation kinetics, Kampman and Wagner have applied the previous models of nucleation, growth and coarsening concomitantly rather than successively. The approach is similar to what we address in the next section for the numerical simulation of the cluster dynamics equations. They use an algorithm where a cluster size distribution $C(R,t)$ is divided in a finite number of classes. The total cluster atomic fraction is given by:

$$C = \sum_{1}^{j_0} C_j \tag{22}$$

and the average particle radius is:

$$\overline{R} = \sum_{1}^{j_0} C_j \frac{(R_j + R_{j+1})}{2} \tag{23}$$

The continuous time evolution of $C(R,t)$ is calculated for small time steps related to sufficiently small radius and supersaturation changes. There are three physical ingredients: a stationary nucleation current, given by equation (13), a growth rate which includes the Gibbs-Thomson effect at the precipitate interface, equation (17) and (18), and of course, the conservation of the solute.

3   Cluster dynamics

As already pointed out, the classical nucleation theory results from the simplest assumption which can be made at the microscopic level for the exchange between clusters, i.e. through monomer diffusion. In this section we consider this simple assumption again. But instead of solving separately the nucleation step as in § 2.3.4, and then applying the growth and coarsening of the supercritical clusters, we solve directly the set of cluster kinetics equations. We show that all the characteristic features merge naturally by solving numerically the kinetic equations. We then give an insight of a generalized cluster dynamics theory where more complex cluster reactions may occur. This theory has been described by Binder and co-workers [11-13]; it gives a more accurate description of systems characterized by large supersaturations.

3.1   CLUSTER DYNAMICS: MONOMER EXCHANGES

3.1.1   Evaluation of the Absorption and Emission Rates

The classical theory uses the simplified form (7), where the reactions between clusters involve only monomers, i.e. evaporation $n \rightarrow (n-1)+1$, and condensation $n+1 \rightarrow (n+1)$.

Calling $W_{n \rightarrow n+1} = \beta_n$ the condensation or absorption rate of the n size cluster, and $W_{n \rightarrow n-1} = \alpha_n$ its evaporation or emission rate, equation (7) can be written for n values larger than unity:

$$\frac{\partial C_n}{\partial t} = \alpha_{n+1} C_{n+1} + \beta_{n-1} C_{n-1} - (\beta_n + \alpha_n) C_n \tag{24}$$

For $n = 1$, the following equation holds true:

$$\frac{\partial C_1}{\partial t} = \sum_2^\infty \alpha_n C_n - \sum_2^\infty \beta_n C_n - 2\beta_1 C_1 + \alpha_2 C_2 \tag{25}$$

The next step consists in specifying the coefficients β_n and α_n. As already mentioned, the absorption rate is either controlled by the last jump probability for a monomer to impinge on a cluster (sticking), or by long-range diffusion in the matrix when a solute depletion exists around large enough clusters. Accordingly, two regimes are expected: the rate of the first one varies with the cluster surface area, i.e. like $n^{2/3}$, whereas the second one varies with size, i.e. $n^{1/3}$.

We present here an expression of β_n proposed by Waite [20] in a theoretical treatment of the kinetics of diffusion-limited reactions. Before the sticking reaction (last jump) the monomers must diffuse; the two mechanisms (sticking and long-range diffusion) are in series. If we assume that the n cluster is immobile, and that only the reacting monomers move by diffusion, the following expression for β_n is obtained:

$$\beta_n = 4\pi R_n^2 \frac{D}{x_\beta}\left(\frac{x_1}{\Omega}\right)\frac{\omega}{R_n\omega+1}\frac{1}{\zeta_{n+1}-\zeta_n} \tag{26}$$

where D is the solute diffusion coefficient and R_n is the spherical cluster radius. ω characterizes the last atom jump:

$$\omega = \frac{av\exp\left(\dfrac{-\Delta E}{k.T}\right)}{D} \approx \frac{\exp\left(\dfrac{-\Delta E}{kT}\right)}{a\exp\left(-\dfrac{Q}{kT}\right)} \tag{27}$$

a is the monomer jump distance, v is its jump frequency, Q is the activation energy of the bulk monomer diffusion and ΔE is the energy required for the condensation to occur.

If $\Delta E \ll Q$, or $R_n\omega \gg 1$, (26) reduces to the long-range diffusion term:

$$\beta_n = 4\pi R_n \frac{D}{x_\beta}\left(\frac{x_1}{\Omega}\right)\frac{1}{\zeta_{n+1}-\zeta_n} \tag{28}$$

If $\Delta E \gg Q$, or $R_n\omega \ll 1$, the last jump controls the absorption and (26) reduces to the interface term:

$$\beta_n = 4\pi R_n^2 \frac{D}{x_\beta}\frac{x_1}{\Omega}\frac{1}{a}\exp\left(-\frac{\Delta E-Q}{kT}\right)\frac{1}{\zeta_{n+1}-\zeta_n} \tag{29}$$

ΔE can be approximated by the free enthalpy barrier that a solute atom must overcome to transform a cluster of size n in size $n+1$. ΔE can then be written:

$$\Delta E = Q + \left(\frac{\Delta G_{n+1}-\Delta G_n}{2}\right) = Q + \frac{1}{2}\frac{\partial \Delta G_n}{\partial n} \tag{30}$$

Substituting ΔE in (26) and (27) gives β_n; α_n is deduced using (10; 4), leading for a spherical cluster, to:

$$\beta_n = 4\pi R_n \frac{D}{x_\beta}\left(\frac{x_1}{\Omega}\right)\frac{1}{1+\dfrac{a}{R_n}\exp\left(\dfrac{\Delta G_{n+1}-\Delta G_n}{2kT}\right)}\frac{1}{\zeta_{n+1}-\zeta_n} \tag{31.a}$$

$$\alpha_{n+1} = 4\pi R_n \frac{D}{x_\beta}\left(\frac{x_1}{\Omega}\right) \frac{\exp\left(\dfrac{\Delta G_{n+1} - \Delta G_n}{kT}\right)}{1 + \dfrac{a}{R_n}\exp\left(\dfrac{\Delta G_{n+1} - \Delta G_n}{2kT}\right)} \frac{1}{\zeta_{n+1} - \zeta_n} \tag{31.b}$$

For the special case of a **sufficiently dilute solid solution**, the following expression hold true:

$$\Delta G_{n+1} - \Delta G_n = -kT\left(\zeta_{n+1} - \zeta_n\right)x_\beta \ln\left(\frac{x_1}{x_{eq}}\right) \tag{32}$$

where ζ_n is the atom number in the nth cluster. In general, ζ_n can be chosen equal to

$$\zeta_n = \frac{n}{x_\beta} \tag{33}$$

In that case, equation (32) becomes:
$$\Delta G_{n+1} - \Delta G_n = -kT \ln\left(\frac{x_1}{x_{eq}}\right) \tag{34}$$

$$\beta_n = 4\pi R_n D\left(\frac{x_1}{\Omega}\right) \frac{1}{1 + \dfrac{a}{R_n}\left(\dfrac{x_1}{x_{eq}}\right)^{-1/2} \exp\left(\dfrac{\left((\zeta_n+1)^{2/3} - \zeta_n^{2/3}\right)\gamma A}{2kT}\right)} \tag{35.a}$$

$$\alpha_{n+1} = 4\pi R_n D\left(\frac{x_{eq}}{\Omega}\right) \frac{\exp\left(\dfrac{\left((n+1)^{2/3} - n^{2/3}\right)\gamma A}{kT}\right)}{1 + \dfrac{a}{R_n}\left(\dfrac{x_1}{x_{eq}}\right)^{-1/2} \exp\left(\dfrac{\left((n+1)^{2/3} - n^{2/3}\right)\gamma A}{2kT}\right)} \tag{35.b}$$

Mathon [26], in her simulation of the Cu precipitation in Fe has used α_n and β_n coefficients corresponding to those only limited by long-range diffusion, i.e. suppressing the second term in the denominator of equations (35.a-b). In that particular case, the evaporation rates depend only on the size n.

3.1.2   *Numerical Method for the Resolution of the Kinetics Equations*

Equation (24) may be re-written having time shown explicitly.
In an explicit formalism, we may write:

$$C_n^{t+dt} - C_n^t = \alpha_{n+1} \, dt \, C_{n+1}^t + \beta_{n-1} \, dt \, C_{n-1}^t - (\beta_n + \alpha_n) dt \, C_n^t \tag{36}$$

We will however favor an implicit formalism; it is numerically more stable and allows larger time steps; equation (24) is then written:

$$C_n^{t+dt} - C_n^t = \alpha_{n+1} \, dt \, C_{n+1}^{t+dt} + \beta_{n-1} \, dt \, C_{n-1}^{t+dt} - (\beta_n + \alpha_n) dt \, C_n^{t+dt} \tag{37.a}$$

or
$$a_n C_{n-1}^{t+dt} + b_n C_n^{t+dt} + c_n C_{n+1}^{t+dt} = C_n^t \tag{37.b}$$

with
$$a_n = -\beta_{n-1} \, dt \; ; \; b_n = \{1 + (\beta_n + \alpha_n) dt\} \, ; c_n = -\alpha_{n+1} \, dt \tag{38}$$

For practical reasons, we will truncate the cluster ensemble and only consider clusters with an index smaller or equal to N. The value of N must be sufficiently large so that at any step of the calculation C_N remains negligible.

Knowing $\{C_n^t\}_{n=1,N}$, $\{C_n^{t+dt}\}_{n=1,N}$ is deduced by inverting a tri-diagonal matrix (except the first row). This type of system is efficiently solved by first evaluating X_n, Y_n such that:

$$C_n^{t+dt} = X_n + Y_n C_{n-1}^{t+dt} \tag{39}$$

It can be shown that: $X_N = \dfrac{C_N^t}{b_N}$, $Y_N = \dfrac{-a_N}{b_N}$;... $X_n = \dfrac{C_n^t - b_n X_{n+1}}{b_n + C_n Y_{n+1}}$, $Y_n = \dfrac{-a_n}{b_n + C_n Y_{n+1}}$ ...;

$$X_2 = \frac{C_2^t - b_2 X_3}{b_2 + C_2 Y_3}, Y_2 = \frac{-a_2}{b_2 + C_2 Y_3} \tag{40}$$

Now we can express all C_n^{t+dt} in terms of C_1^{t+dt} with the generic equation:

$$C_n^{t+dt} = F_n + G_n C_1^{t+dt} \tag{41}$$

with $F_1=0$, $G_1=1$...; $F_n = X_n + Y_n F_{n-1}$, $G_n = Y_n G_{n-1}$;... $\tag{42}$

The monomer atomic fraction at time (t+dt) can now be evaluated by using the solute content balance equation:

$$\sum_{n=1}^{N} C_n^{t+dt} \zeta_n x_\beta = x_0 \tag{43}$$

or (using equation (33)): $\left(\sum_{n=1}^{N} n F_n \right) + \left(\sum_{n=1}^{N} n G_n \right) C_1^{t+dt} = x_0 \tag{44}$

which can be re-written: $C_1^{t+dt} = \dfrac{x_0 - \left(\displaystyle\sum_{n=1}^{N} n F_n \right)}{\left(\displaystyle\sum_{n=1}^{N} n G_n \right)} \tag{45}$

The cluster atom fractions are then deduced from equation (41). We can then move on to the next time increment and iterate the whole process.

3.2   RESULTS

The results obtained with the monomer exchange cluster model are presented in this paragraph. The absorption and emission coefficients are calculated using equations (35.a-b) . The following values of the parameters have been used; they are representative of pure Cu precipitation in Fe:

$\Omega=1,18.10^{-29}$ m^3 $\gamma=0,38$ J.m^{-2} T=773 K   D=1,89.10^{-20} m^2.s^{-1}

$x_{eq}=0,0034$ $x_\beta=1$ $x_0/x_{eq}=20$

The results are given in Figures 4-7. Figures 4 shows the time evolution of the cluster distribution $C(R)$ at different times for an ideally quenched solid solution (initially, all solute atoms are in the monomer state). At short times, the curve only consists in a monotically decreasing cluster distribution. At intermediate times, a minimum and a maximum develop in the distribution. It turns out that the minimum corresponds to the initial value of the critical radius whereas the right end part of the distribution corresponds to the formation of overcritical clusters, i.e." the precipitates". With increasing times, this maximum shifts toward larger R and its magnitude decreases. Simultaneously, the left part of the distribution below the minimum converges toward an equilibrium ($\Delta G_v=0$) distribution given by equation (4).

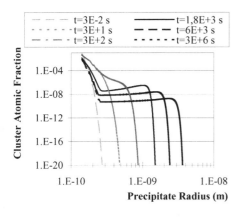

Figure 4. Cluster size distribution at different times.

Figure 5. Time evolution of atomic fraction of monomers and of solid solution concentration.

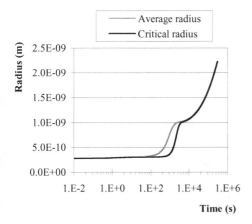

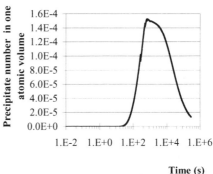

Figure 6. Time evolution of precipitate radii.

Figure 7. Time evolution of precipitate density.

In order to analyze the occurrence of the different mechanisms involved in the precipitation kinetics, we present in Figure 5-7, the time evolution of: the atomic fraction of solute monomers, the average and critical cluster radii and the precipitate density. For times shorter than approximately 40 seconds, which corresponds to the **nucleation incubation period**, only a quasi equilibrium distribution of small clusters is formed in the solid solution whose concentration stays constant (Fig. 5). For times larger than 40 seconds, **nucleation - growth** starts and the supersaturation quickly decreases. The rapid growth of the first nucleated clusters allows the average cluster radius to be larger than the critical size (Fig. 6). At approximately 2000 seconds, the precipitate density reaches its maximum (Fig. 7), the critical and average radii become nearly identical (Fig. 6) and the supersaturation decreases from 20 to 3 (Fig.5).

The variation of $\overline{R}^3$ versus time is shown in Fig. 8. A linear relation is followed for long enough times; the slope is $3.24 \ 10^{-32} \ m^3.s^{-1}$ which may be compared with the slope predicted by the Lifshitz-Slyozov-Wagner theory which is equal to $3.46 10^{-32}$ $m^3.s^{-1}$. Finally the scaled distribution function is shown in Figure 9 for an elapsed time of $3\ 10^5$ s. As can be seen, the prediction is very close to the Lifshitz-Slyozov-Wagner (LSW) one.

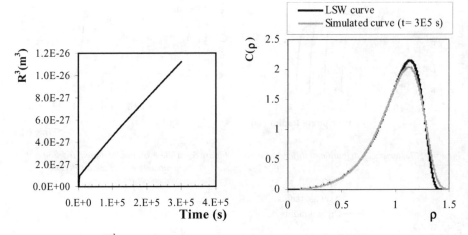

Figure 8.   Variation of $\overline{R}^3$ as a function of time for a   Figure 9.   Simulated and LSW asymptotic scaled
supersaturation of 20.   distribution function; for LSW theory,
                                                                                                                 $C(\rho)$ is given by equation (21).

3.3   GENERALIZED CLUSTER DYNAMICS

Binder et al. [11-13] have considered the possibility for the clusters to react according to more complete schemes. Besides the evaporation and condensation reactions previously considered involving monomers previously considered, other reactions between n and n' clusters are envisaged, like those shown in Figure 10 where clusters of B atoms are sketched on a square lattice. The first part of the Figure 10.a shows the cluster contours, one atom belonging to a cluster when it has at least one other B atom among their nearest neighbours. Figure 10.b shows successively several cluster reactions: a) evaporation/condensation of monomers, b) coarsening/splitting and c) cluster diffusion. In the latter, the number of B atoms in the cluster is unchanged, but its center of gravity is shifted by the atomic rearrangement on the surface; a cluster diffusion provides a mechanism for clusters to get in contact and coagulate.

$$\frac{\partial C_n(t)}{\partial t} = \sum_{n'=1}^{\infty} S_{n+n',n'}\, C_{n+n'}(t) - \frac{1}{2}\sum_{n'=1}^{n-1} S_{n,n'}\, C_n(t)$$

$$+ \frac{1}{2}\sum_{n'=1}^{n-1} A_{n-n',n'}\, C_{n'}(t) C_{n-n'}(t) - \sum_{n'=1}^{\infty} A_{n,n'}\, C_n(t) C_{n'}(t)$$

(46)

In (46), $A_{n,n'}$ holds for the rate coefficients of the association of two clusters n and n' to give a $(n+n')$ cluster, $S_{n,n'}$ holds for the reverse splitting reaction rate. Note that the condensation and evaporation coefficients used previously are included in the above terms for $n'=1$.

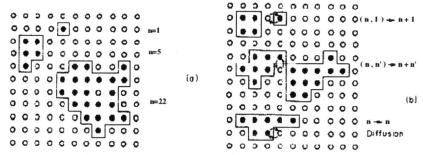

Figure 10. Scheme of clusters of B atoms on a square lattice; a) clusters of various sizes. b)cluster reactions, from top to bottom: condensation-evaporation, splitting-coagulation and diffusion. (from[13]).

The determination of the coefficients $S_{n,n'}$ and $A_{n,n'}$ is not trivial. As before, for evaporation and condensation, they must however satisfy a detailed balance at equilibrium:

$$S_{n+n',n'} \, \overline{C_{n+n'}} = A_{n,n'} \, \overline{C_{n'}} \, \overline{C_n} \qquad (47)$$

which allows the splitting coefficients to be deduced from the coagulation coefficients. The numerical resolution of (46) leads qualitatively to similar general features as those obtained in our n'=1 classical case (§3.1 and 3.2). The cluster distribution $C(n,t)$ converges rapidly at small size towards an equilibrium distribution (the fluctuating solid solution). At increasing time a minimum and a maximum appear at large cluster sizes, with an amplitude decrease and shift towards larger sizes (the growing clusters).

The average size follows an asymptotic law $n \approx t^a$, with $a \approx 0.5$ instead of 1 for a LSW regime. The difference is attributed to cluster coagulation and diffusion, not taken into account in the classical nucleation theory nor in LSW, which are only based on a condensation-evaporation regime. It is expected that an increase of the cluster number density via a higher initial super-saturation would favor the coagulation participation. This has been shown by Monte Carlo simulations with an Ising model.

4 Ising model and Monte-Carlo Method

Cluster dynamics treats clusters as small objects having well defined bulk and interface free energies. The extrapolation down to the nanometric scale, as required for the understanding of the nucleation stage, is therefore somewhat limited. An analysis at the atomic scale is then required. Such an approach of clustering is given, for example, by an Ising model where atoms A and B and vacancies are positioned on the lattice sites and interact with pair interactions. Once a probability law for the interchange of atoms on neighbouring lattice sites has been chosen, the time evolution of the may be simulated by Monte Carlo techniques.

The 2- and 3-dimensional kinetic Ising model has given rise to a number of studies, starting in 1976 with the pioneer work of Lebovitz and co-workers [21] in the field of first order phase transitions. The kinetic Ising model is treated further in this School. It will be shown that the characteristic features which emerge from the classical cluster dynamics description, where only condensation and evaporation of monomers are considered, are recovered in the Ising model simulations in the case of the decomposition of supersaturated solid solution [23,24]: the four classical stages, incubation, nucleation, growth and coarsening are well observed. But the Ising model shows also, for certain atomic migration parameters, that small clusters coagulation may occur as described by Binder (in Fig. 7). This coagulation may accelerate the nucleation stage at high supersaturation. The link between the microscopic Ising model and the mesoscopic cluster dynamics approach is therefore enlightening.

5 Conclusions

Cluster dynamics, and in its simplest form the classical nucleation theory, appears to be a powerful tool to describe precipitation kinetics. The knowledge of two types of parameters is required: the thermodynamic and the kinetic parameters. The thermodynamic parameters are the driving force and the cluster interfacial energy. The kinetic parameters are the ones involved in the atomic exchanges between clusters; they allow the system to reach its new equilibrium state after quench in the two-phase region. The precipitation path, obtained by numerical integration of the kinetic master equation, follows the usual steps of nucleation, growth and coarsening. Condensation and evaporation of monomers are the only inter-cluster reactions which have been taken into account. Other mechanisms can be introduced in the master equation like cluster coagulation and cluster diffusion. The difficulty is to develop realistic kinetic rate coefficients. In this regard, it is very helpful to rely on Monte Carlo simulations and study, at the atomic scale, the whole path of the cluster kinetics.

Elastic strain effects due to atomic size differences can be taken into account in cluster dynamics. They decrease the driving force and therefore slow-down the kinetics. However, the clusters are then coupled at long-range by the elastic strains; this coupling cannot be handled with cluster dynamics. Cluster dynamics is also not a good tool to study other effects such as: cluster shape and faceting, vacancy trapping at the interface, "vacancy pump" invoked in the past to explain not well understood kinetics aspects of the Guinier-Preston precipitation in aluminum alloys, and interconnected or bi-percolated two-phase structure in the case of highly supersaturated solid solutions. Phase fields methods (see for instance [24,25]) and Monte Carlo simulations are more appropriate.

Acknowledgements:

This work has been performed in the framework of the CPR (Contrat de Programme de Recherche) "PRECIPITATION" which has the following members: CNRS, CEA, INPG, Université de Rouen, INSA de Lyon, Université Aix-Marseille 3, ONERA, USINOR and PECHINEY

6 Bibliography

[1]   J.W. Cahn ; Trans. Met. Soc. AIME, 242, 166 (1966)

[2]   J.S. Langer, Bar-on and H. Miller; Phys. Rev.A 11, 1417, (1975)

[3]   M. Allen, J.W. Cahn, Acta Metall, 24, 425 (1976)

[4]   L.Q. Chen, A.G. Katchaturyan, Acta Metall. Mater. 39, 2533 (1991)

[5]   J. Frenkel, J. Phys. URSS, 1, 315 (1939)

[6]   R. Becker, W. Döring, Ann. Physik, 24, 719 (1935)

[7]   J.B. Zeldovich, Acta Physicachimica URSS, 18, 1 (1933)

[8]   M. Volmer, A. Weber, Z. phys. Chem., 119 (1926)

[9]   K.F. Kelton, A.L. Greer, C.V. Thompson, J. Chem. Phys. 79, 6261 (1983)

[10] for reviews: K.C. Russel in Phase Transformations, ASM, Chapman et Hall, 219 (1970). G. Martin in Solid state phase transformations in metals and alloys, Ecole d'été d'Aussois, Les éditions de physique, 336, (1978). H.I. Aaronson, J.K. Lee, in Lectures on the Theory of Phase Transformations, 2ond edition, TMS, 165 (1999). R. Wagner, R. Kampmann, in Materials science and technology, 5, Phase transformations in materials, ed. P. Haasen, 213 (1991)

[11] K. Binder, D. Stauffer, Adv. Phys. 25, 343 (1976)

[12] K. Binder, Rep. Prog. Phys., 50, 783 (1987)

[13] P. Mirold, K. Binder, Acta Met., 25, 1435 (1977)

[14] J.W. Christian, the theory of Transformations in Metals and Alloys, Pergamon Press, 383 (1965)

[15] K.C. Russel, Adv. Colloid Interface Sci.,13, 205 (1980)

[16] J. Feder, K.C. Russel, J. Lothe, G.M. Pound, Adv. Phys.,15, 111 (1966)

[17] H.B. Aaron, D. Fainstain, G.R. Kotler, J. Appl. Phys., 41, 4404 (1970)

[18] J.W. Martin, R.D. Doherty, Stability and Microstructures in Metallic Systems, 2nd ed., Cambridge Univ. Press, 256 (1997)

[19] R. Kampmann, R. Wagner, Decomposition of Alloys: the early stages. 2nd Acta Scripta Met. Conf., Pergamon, 91 (1983); R. Kampman, Th. Eberl, M. Haese, R. Wagner, Physt. Stat. Sol. (b) 172, 295 (1992)

[20] T.R. Waite, J. of Chem. Phys.,28, 103 (1958)

[21] J.L. Lebovitz, M.H. Kalos, Scripta Metall. ,10, 9 (1976); J.L. Lebovitz, J. Marro, J. Kalos Acta Metall; 30, 297 (1982)

[22] F. Soisson and G. Martin, Phys. Rev.B, 62, 1 (2000)

[23] M. Athènes, P. Bellon, G. Martin, Acta Mater. 48, 2675 (2000)

[24] A.A. Wheeler, W.J Boettinger, G.B. McFadden, Phys. Rev. A, 45, 7424 (1992)

[25] Phase Transformations and Evolution in Materials, ed. E.A.Turchi and A. Gonis, TMS, Warrendale (2000)

[26] M.H. Mathon, Thèse, Université de Paris-Sud, U.F.R. Scientifique d'Orsay (1995)

SENSITIVITY OF TEXTURE DEVELOPMENT DURING GRAIN GROWTH TO ANISOTROPY OF GRAIN BOUNDARY PROPERTIES

Anthony D. Rollett
Department of Materials Science & Engineering
Carnegie Mellon University
Pittsburgh, PA 15213

1. Introduction

Microstructure is a term that includes many different aspects of the structure of materials. In crystalline materials it is accepted as the set of defects contained within a body of material, as distinct from the lattice itself. Subsets of the set of defects include point defects (which are not generally treated as microstructural elements), line defects or dislocations and planar defects such as grain boundaries and interphase boundaries. Setting aside vacancies, essentially all defects are non-equilibrium features. Therefore thermal activation and migration mechanisms will lead to evolution of the microstructure to lower the free energy of the material. For the particular example of single phase polycrystals, coarsening of the grain structure (grain growth) decreases the total energy associated with the grain boundaries via curvature-driven boundary migration. In metals such as aluminum and iron that exhibit well developed dislocation cell structures from plastic deformation, recrystallization can be understood as a continuous coarsening process (grain growth) even when the heterogeneity of the process makes it appear to be a discontinuous process when characterized by optical microscopy. As discussed briefly below, the structure and properties of grain boundaries vary widely as a function of the misorientation between the grains on either side of a boundary as well as the orientation of the boundary normal in relation to the crystal lattices (i.e. the boundary inclination). The properties of boundaries occuring in a material are thus highly dependent on the texture (crystallographic preferred orentation) of the material. The migration rate of a grain boundary is proportional to the (anisotropic) mobility and the driving force which in turn is proportional to the (anisotropic) grain boundary energy and the curvature. Grain growth can therefore lead to changes in texture because of the influence of energy and mobility on boundary migration rates.

As a practical example,. consider the marked changes in texture that occur during the annealing of *fcc* metals have been studied over a period of many years. The cube texture component, <100>//RD and <001>//ND, is technologically important in a wide range of applications from aluminum beverage can stock, to nickel foils used as substrates for high temperature superconductors. A common, albeit oversimplified view of the origins of this

A. Finel et al. (eds.), Thermodynamics, Microstructures and Plasticity, 123–133.
© 2003 *Kluwer Academic Publishers. Printed in the Netherlands.*

annealing texture component is that a combination of oriented nucleation (ON) and oriented growth (OG) is responsible. The former, ON, emphasizes the inheritance from the deformation structure in the sense that certain small subsets of the deformation texture are favorably situated for producing new grains. OG on the other hand emphasizes the anisotropy of the properties of the grain boundaries. There is evidence, reviewed below that boundaries in the vicinity of a 38° rotation about a <111> misorientation axis have high mobility. Motivated by these considerations, this paper examines the dependence of texture development on assumptions about grain boundary properties. As expected texture evolution is highly sensitive to assumptions about the constitutive behavior.

2. Boundary Structure

A discussion of properties must take account of the structure of the grain boundaries [1]. Although much remains to be investigated in this area, the general features are known. In what follows, only the five macroscopic degrees of freedom for grain boundary structure will be discussed. In most cases it is convenient to parameterize the boundaries in terms of three variables that describe misorientation, and two more variables that describe inclination. Given that the properties of boundaries are often dependent on the common crystallographic axis between the two lattices, an axis-angle description is adequate for most purposes. For low angle boundaries, or if a low-index rotation axis is chosen, the angle itself is the most useful single parameter. Specific parameterizations include Euler angles, quaternions and Rodrigues vectors, of which the latter are most commonly used. For boundaries where the lattices are related by a rotation that brings a fraction of the lattice sites into (exact) coincidence, the properties adopt special values because of the high symmetry of the boundary. These specific geometrical lattice relationships are called Coincident Site Lattice boundaries or CSLs; they are parameterized by the reciprocal ratio of the coincidence fraction, or Σ value. By convention, low angle boundaries are deemed to have $\Sigma=1$. Twin related grains, i.e. 60°<111> have $\Sigma=3$, for example. Useful as the CSL concept has been, the simplicity of microstructural analysis has resulted in its over use. Clearly the properties of boundaries will, to first order, depend on the local atom arrangement. Not all CSL structures will possess special properties. The properties illustrated in this paper are therefore based on a combination of experimental and theoretical knowledge.

2.1 Grain Boundary Properties

The general picture for grain boundary properties is as follows [2, 3, 4]. The excess free energy of grain boundaries is uniform to first order except in the vicinity of special boundary structures. Each special structure, such as the $\Sigma3$ coherent twin, has a central singular minimum surrounded by a well that can be characterized by either the classic logarithmic Read-Shockley function or, more simply, by a *sine* function. By analogy, the energy function resembles a mesa-top landscape with sinkholes of varying depths and widths, including a large well for low-angle boundaries. Grain boundary mobility, by contrast, exhibits a plateau at moderate values for high angle boundaries. Maxima with varying heights and widths are centered on certain special boundary types which are not coincident with the minima in the energy function. The mobility maxima can be described by Gaussians ($exp\{-x^2\}$). For low angle boundaries, there is a sharp transition from low mobilities at misorientations less than 10° to the moderate values of high angle grain boundaries with misorientations greater than 15°. This transition can be described by exponential functions such as $exp\{x^n\}$; $n\sim9$.

The grain boundaries properties used in the simulations described in this paper are derived from two sources. One source is the atomistic simulation [5, 6] of curvature-driven migration of high angle grain boundaries using molecular dynamics (MD) allied with experimental results for low angle grain boundaries. MD simulations of migration in pure aluminum show sharp maxima in mobility allied with equally narrow wells in the energy function. Based on these results, the energy function is shown in Fig. 1a as a solid line with a Read-Shockley dependence at low angles and several sharp minima for special misorientations based on rotations about <111>; the mobility function is shown in Fig. 1b again as a solid line with sharp maxima for certain <111> misorientations. The second source is experimental results on boundary migration under stored energy driving forces (recrystallization). Based on these results, the mobility function is shown in Fig. 1b as a dot-dash line showing a single broad maximum based on a 38°<111> misorientation; in the absence of information on energies, the energy function, shown as a dot-dash line in Fig. 1a, is assumed to be flat except at low angles where a Read-Shockley variation is also assumed. A large ratio of peak to plateau values of 100:1 was adopted in order to emphasize the effects of anisotropic boundary properties. This ratio was varied in one set of simulations, however.

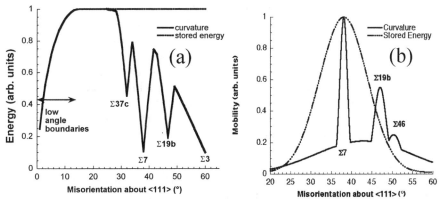

Figure 1. Plot of grain boundary properties as a function of misorientation about a <111> axis: (a) energy, showing Read-Shockley variation at low angles and minima at several CSL types; (b) mobility, showing local maxima around three different CSL types. In the latter plot, the transition from very low mobilities for low angle boundaries has been omitted.

3. Simulation Method

The Monte Carlo method for simulating grain growth has been described previously so only the relevant details are given here [7, 8, 9]. Briefly, the microstructure to be modeled is discretized, and the orientation number at each grid point is associated with a 3-parameter crystallographic orientation. Disorientations and their properties between all possible juxtapositions of orientations, i.e. all possible boundary types, are stored in a lookup table at the beginning of the simulation. Each voxel i in a simple cubic lattice of

$N=100x100x100=10^6$ elements was assigned an orientation number $S_i \leq 500$. Grain correspond to continguous regions with the same value S_i. The total system energy i given by

$$E = \sum_{j}^{N} \sum_{i}^{n} J(S_i, S_j)(1 - \delta_{S_i S_j}$$

(1

where the inner sum is taken over the n first and second nearest neighbors of element i, δ i the Kronecker delta function, and J is the energy of a unit of boundary between element of indices S_i and S_j. To evolve the structure, an element and a new index were chosen a random. The element was reoriented to the new index with probability

$$P(S_i, S_j, \Delta E, T) = \begin{cases} \dfrac{J(S_i, S_j)}{J_{max}} \dfrac{M(S_i, S_j)}{M_{max}} & \Delta E \leq 0 \\ \dfrac{J(S_i, S_j)}{J_{max}} \dfrac{M(S_i, S_j)}{M_{max}} \exp(-\Delta E / kT) & \Delta E > 0 \end{cases}$$

(2

where ΔE is the energy change for the reorientation, M is the boundary mobility betwee elements of indices S_i and S_j, J_{max} and M_{max} are the maximum boundary energy an mobility respectively, k is the Boltzmann constant, and T is the lattice temperature. Afte each reorientation attempt, the time is incremented by $1/N$ Monte Carlo Steps (MCS), and new reorientation is selected. An efficient algorithm that avoids the wasted effort c computing unsuccessful reorientation attempts was utilized in these simulations [9]. finite temperature, $T=0.9$, was used in order to minimize lattice anisotropy and to ensu that small differences in boundary energy did not lead to irreversible motion of boundar segments. As discussed previously [10], this basic algorithm was modified in order to b able to use a wide range of grain boundary energies by scaling the Boltzmann facto according to the grain boundary energy.

3.1 Volume Fraction Estimation

Volume fractions, v_f of each texture component were calculated using a simple voxe count approach. For each voxel, the misorientation angle was calculated between it orientation and every member of the list of texture components of interest. That voxel wa then added to the count for the component with the smallest misorientation, provided th; the angle was less than a cutoff value of 15°. An orientation that lies within the cutoff c multiple components was assigned to the component with which it has the lowe: misorientation angle. The following equation illustrates the procedure for the cub component, {001}<100>.

$$v_f(g_{cube}) = \sum_{i=1,N} w_i V_i \Big/ \sum_{i=1,N} V_i; \quad w_i = \begin{cases} 0, if \left| g_{cube} g_i^{-1} \right| > 15° \\ 1, if \left| g_{cube} g_i^{-1} \right| \leq 15° \end{cases}$$

(3

3.2 Initial Microstructures and Textures

The initial grain structure with approximately 10,000 grains was taken from normal grain growth. Texture was superimposed on these grain morphologies by assigning individual orientations from a list generated by either simulation of rolling (plane strain compression) in *fcc* metals or from a set of randomly chosen orientations. The texture simulations were performed with *LApp* which is a polycrystal plasticity code with {111}<110> slip geometry [11]. In order to study the growth (or lack of it) of the cube texture component, an additional set of orientations were generated with a Gaussian distribution about the cube component, {001}<100>. Twenty orientations from this list were inserted into the set of 500 to give an initial volume fraction of cube of ~6%.

4. Results

4.1 Influence of the Mobility Function.

Grain growth was simulated with an initial deformation texture with a cube component added as described above. The simulations were typically run to 10^9 Monte Carlo Steps (MCS) at which point the number of grains had dropped to ~10. This is equivalent to three decades of decrease in grain number or a one decade increase in grain size. The energy function was as shown in Fig. 1a. Both a sharp, multiply peaked mobility function and a mobility function with a single broad peak was used, Fig. 1b. Previous simulations in 2D systems [10] indicated that the sharply peaked mobility function (with its associated energy function) resulted in consistent growth of the cube component whereas the single broad peak function did not result in growth of the cube component. The same work indicated that variations in the energy function had little effect on the outcome. By contrast, in 3D simulations, both functions led to consistent growth of the cube component to the eventual exclusion of all other components.

4.2 Orientation Spread of the Cube Grains.

The effect of varying the width of the cube-oriented grain was investigated. Figure 2 shows that the orientations must be tightly clustered around the symmetric position in order for cube to dominate the final texture. The texture component that dominates at the larger spreads is the normal direction (ND)-rotated cube, {001}<110>.

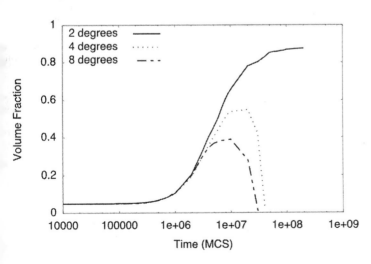

Fig. 2. Plot of the cube volume fraction against time for three different spreads in orientation of the cube component. As the dispersion increases, the cube fraction becomes less likely to dominate at long times.

128

4.3 Influence of the Energy Function.
Simulations were performed with a simplified energy function that included the Read-Shockley variation for low-angle boundaries but no other features. Anisotropy in the energy function leads to variations in the dihedral angles at triple junctions which must affect grain boundary curvature. In general this might be expected to affect the progress of grain growth as indicated by Upmanyu et al. [12]. In these simulations, however, there was no discernable change in the texture development from changing the energy function in contrast to the sensitivity observed for the mobility function. As pointed out by Upmanyu et al., however, sensitivity of evolution to the boundary energy function is most obvious for the case of 2D orientations which have only one degree of freedom. When orientation is fully 3D, high densities of low energy boundaries only occur for highly textured materials, which is not the case here.

4.4 Evolution of Volume Fraction of Principal Texture Components.
In addition to the cube component itself, other texture components are of interest such as the principal components of the rolling texture (copper, {112}<111>, S, {412}<121>, and brass, {110}<112>) and the Goss annealing component, {110}<001>. Figure 3 shows that variation in fraction with time. The copper, brass and Dillamore components consistently lose area. The S component gains volume fraction at short times but then decreases. The Goss component initially gains volume but then gives way to the cube component.

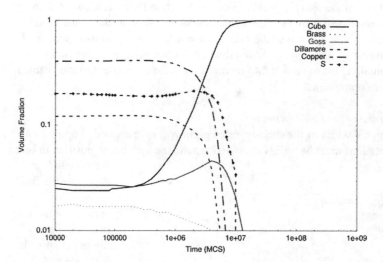

Fig. 3. Plot of various important texture components for a simulation with a sharply peaked mobility function and a narrow spread cube component. The rolling texture components decrease at different rates with the S component being the last to disappear. The Goss component increases at first and then is eliminated in favor of the cube component.

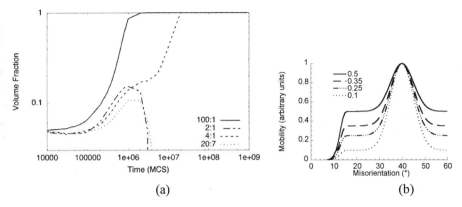

(a)   (b)

Fig. 4. (a) Variation in cube fraction with time as a function of the peak:plateau ratio. (b) Plot of the mobility function for four different peak:plateau ratios.

4.5 Effect of Peak-to-Plateau Ratio.

An as yet undetermined feature of the mobility function is the ratio between the mobility of boundaries at the peak around the 40°<111> type and general high angle grain boundaries. This ratio is called the peak:plateau ratio and figure 4b illustrates the shape of the mobility function for various values of the ratio. Simulations with all other parameters held constant, fig. 4a, show that the effect of the ratio on texture evolution is marked. For large enough values of the ratio (more peaked), the cube component is dominant. If the mobility function is relatively flat then the cube component loses its advantage.

4.6 Initial Volume Fraction of Cube Component

The amount of cube component present initially affects the outcome, as might be expected. Figure 5 shows that as the initial cube fraction is decreased, so the the likelihood of the cube dominating the final texture decreases. One note of caution here is that the result may

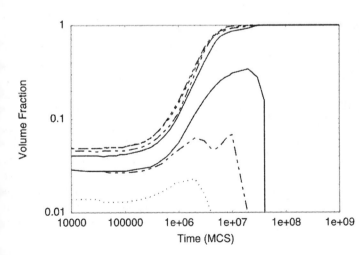

well be sensitive to the statistical details. Varying the total number of orientations (q-value) for example, may change the amount of cube required for eventual dominance.

Fig. 5. Variation in cube fraction with time as a function of the initial volume fraction of the cube component.

4.7 Texture Evolution

The changes in texture that occur during grain growth are dramatic. As mentioned above, the initial texture is that of a strong *fcc* rolling texture with a small (6%) volume fraction of near-cube grains added in, fig. 6a. The rolling texture is apparent as the strong beta fiber that extends throughout the sections and the initial volume fraction of cube-oriented grains is also apparent. After grain growth in the case where no cube component is inserted, components grow that bear no special relationship to the deformation texture, fig. 6b. The growing components correspond to the high intensities in the $\phi_2 = 10°$ and $30°$ sections. Also some of the original rolling texture components persist as is evident from the intensity in the sections at $\phi_2 = 75°$ and $80°$. If, on the other hand, the mobility function is sufficiently anisotropic and some cube oriented grains are present at the outset, then cube grows to dominate the microstructure, fig. 6c.

5. Discussion

Abnormal grain growth theory suggests that a given grain will grow abnormally if the mobility of its perimeter is higher than that of the average mobility in the matrix. Higher mobility ratios lead to a greater potential for abnormal growth. Lower grain boundary energies on the periphery also promote abnormal growth. Abnormal growth in this context means that the abnormal grain eventually attains a size that is a multiple of the current average grain size of the matrix. Consequently, it is interesting to consider the mobility and energy of boundaries between a component of interest and its surroundings, and the average mobility and energy of boundaries in the material as a whole.

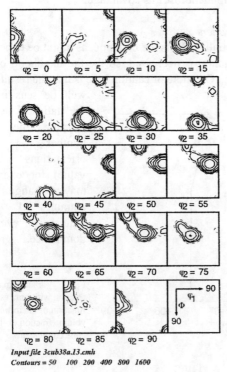

$\phi_2 = 0$ $\phi_2 = 5$ $\phi_2 = 10$ $\phi_2 = 15$

$\phi_2 = 20$ $\phi_2 = 25$ $\phi_2 = 30$ $\phi_2 = 35$

$\phi_2 = 40$ $\phi_2 = 45$ $\phi_2 = 50$ $\phi_2 = 55$

$\phi_2 = 60$ $\phi_2 = 65$ $\phi_2 = 70$ $\phi_2 = 75$

$\phi_2 = 80$ $\phi_2 = 85$ $\phi_2 = 90$

Input file 3cub38a.13.cmh
Contours = 50 100 200 400 800 1600

Fig. 6.. (a) Sections through (Bunge) Euler angle space showing the initial texture showing the combination of *fcc* rolling texture with a small volume fraction (6%) of near-cube grains.

Thus by evaluating $<M(S_{cube}, S_{non-cube})>$, $<\gamma(S_{cube}, S_{non-cube})>$ $<M>$, $<\gamma>$, it is possible to evaluate the potential for rapid growth or shrinkage of certain components. Clearly individual grains will have local environments that vary from these averages, but the simulations average the behavior to a certain extent. This approach has been verified for the case of sub-grain growth in which an occasional grain acquires a large misorientation around most of its periphery and consequently grows or shrinks abnormally.

The results for mobility and energy for the sharply peaked mobility function with the same combination of ~6% cube in an *fcc* rolling texture are shown in figure 7. Histograms of $<M(S_i,S_j)>$ and $<\gamma(S_i,S_j)>$ where the average is taken over the second orientation, S_j, and then the results were binned to show a distribution of mobilities and energies. $<M(S_{cube},S_{non-cube})>$ and $<\gamma(S_{cube},S_{non-cube})>$ are shown as arrows on the histograms. The results indicate that the cube grains have a mobility that is higher than the average and an energy that is slightly lower than the average. Note also that the mobility distribution is clustered around $M=0.03$ in a system in which the maximum mobility is $M=1$ because the mobility is generally low except close to the high mobility boundary types. The energies, on the other hand, are closer to one as expected for a function whose value is one except for special boundary types.

The histograms for the broadly peaked mobility function are shown in figure 8. In this case, the mobilities are generally higher as expected for this function but the cube orientations have an average mobility that is similar to the median value in the matrix, which suggests that abnormal growth is less likely than in the case of the sharply peaked function. The distribution of energies is broader than in the previous case and the cube orientations have somewhat lower energies than the matrix average.

Future work will focus on applying the theory of abnormal grain growth to individual grains (in comparison to the coarsening behavior of the matrix) in order to explore whether the growth or shrinkage of any given orientation is predictable in terms of its environment and the combined energy and mobility functions.

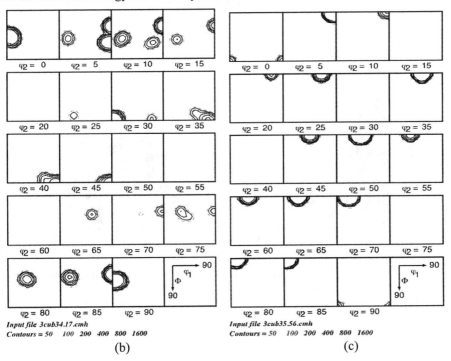

Fig. 6, *contd.*. Sections through (Bunge) Euler angle space showing the texture after (b) 10^7 MCS for grain growth with a shallow mobility function (small peak:plateau ratio) where components other than the cube become dominant; (c) after (a) 5.10^6 MCS for grain growth with a peaked mobility function (large peak:plateau ratio) where the cube component does become dominant.

132

6. Summary

Previous work on two dimensional grain growth has been extended to three dimensional coarsening. In general, a sufficiently anisotropic mobility function is sufficient to allow the cube component to dominate the final texture when it is present as a 'seed' within an *fcc* rolling texture. Only when the mobility function becomes very flat does the cube lose its apparent growth advantage. The eventual dominance of the cube depends on many factors such as the initial volume fraction and the spread or dispersion in the set of orientations used to represent the cube component. The fact that such a drastic change in texture can occur after coarsening of the grain size by only a factor of ten gives some indication of the applicability to the early stages of recrystallization. Coarsening of the recovered subgrain network in high stacking fault energy metals is known to lead to nucleation of recrystallization. Abnormal grain growth during this process has long been thought to be the key factor [13]. The results discussed here appear to support the idea that significant changes in texture during annealing may be the result of the early stages of recovery before macroscopic recrystallization has taken place.

7. Acknowledgements

This work was supported primarily by the MRSEC program of the National Science Foundation under Award Number DMR-0079996. Support of the Computational Materials Sciences Network, a program of the Office of Science, US Department of Energy, is also acknowledged.

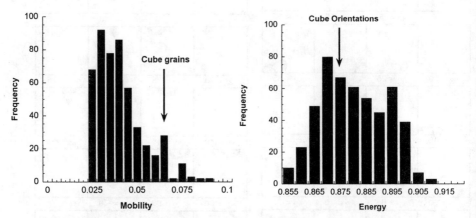

Figure 7. Histograms of the distribution of (a) mobilities and (b) energies for grain boundaries with a sharply peaked mobility function and an energy function with several low energy boundary types. The average mobility and energy for cube grains against all others is noted in each diagram.

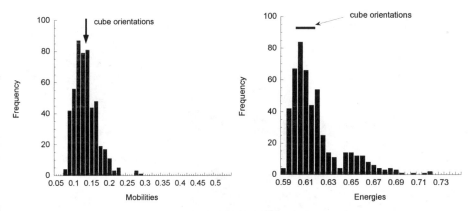

Figure 8. Histograms of the distribution of (a) mobilities and (b) energies for grain boundaries with a broadly peaked mobility function and an energy function with several low energy boundary types. The average mobility and energy for cube grains against all others is noted in each diagram.

8. References

1. Sutton, A.P. and Balluffi, R.W. (1995), *Interfaces in Crystalline Materials*, Clarendon Press, Oxford, UK.
2. Goux, C. (1974), *Can. metall. quar.* **13**, 9-31.
3. Wolf, D. (1989), *Scripta metall.* **23**, 1913-1918.
4. Otsuki, A. (1996), *Mater. Sci. Forum* **207-209**, 413-416.
5. Upmanyu, M., Srolovitz, D. and Smith, R. (1998), *Int. Sci.* **6**, 41-58.
6. Srolovitz, D., Upmanyu, M., Shvindlerman, L. and Gottstein, G. (1999), *Acta mater.* **47**, 3901-3914.
7. Anderson, M.P., Srolovitz, D.J., Grest, G.S. and Sahni, P.S. (1984), *Acta metall.* **32**, 783-791.
8. Srolovitz, D.J., Anderson, M.P., Sahni, P.S. and Grest, G.S. (1984), *Acta metall.* **32**, 793.
9. Hassold, G.N. and Holm, E.A. (1993), *Comp. in Phys.* **7**, 97-107.
10. Rollett, A.D. (2002), *Mater. Sci. Forum* **1**, 251-256.
11. Kocks, U.F., Tomé, C. and Wenk, H.-R. (1998), Eds., *Texture and Anisotropy* Cambridge University Press.
12. Upmanyu, M., *et al.* (2002), *Int. Sci.*, in press.
13. Rollett, A.D. and Holm, E.A. (1996), in *Proc. Rex-96*, Monterey, CA, McNelley, T.R., Eds., TMS, Warrendale, PA, pp. 31-42.

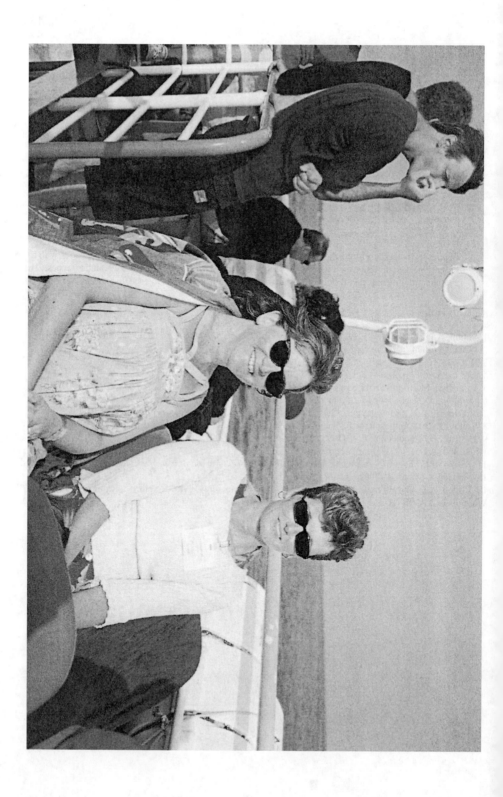

COMPUTER MODELLING OF DIFFUSION CONTROLLED TRANSFORMATIONS

G. INDEN
Max-Planck-Institut für Eisenforschung GmbH
Max-Planck-Str. 1
D-40237 Düsseldorf
Germany

1.   Introduction

About 35 years ago computational thermodynamics started with the pioneering work of Larry Kaufman [1], evolving to what is known today as the CALPHAD approach [2]. Meanwhile, a variety of thermodynamic databases and software has been developed covering large parts of the wide field of materials and phases [3]-[7]. The main purpose of these developments was and still is the calculation and prediction of multi-component phase diagrams. With this background of thermodynamics now available, it is quite natural to start the next step from equilibrium towards transformations. The merit of computer simulations of phase transformations is the possibility of treating complex systems, i.e. multi-component systems and complex heat treatments or long term service conditions. It is virtually impossible to guess the kinetic path of such complex systems. It is only by performing the simulation that the quantitative aspects can really be appreciated, often leading to unexpected results.

The thermodynamic properties represent a major ingredient of the treatment of kinetics of phase transformations. They define driving forces for reactions and, in the case of diffusion controlled transformations, contribute strongly to kinetic parameters like diffusivities. The new applications of thermodynamic data also reveal shortcomings of current thermodynamic descriptions, despite the fact that they yield excellent phase diagram data. A large variety of phases is treated as stoichiometric compounds. This description does not allow to treat diffusion within such phases.

Steels are very complex multi-component systems. Although the amount of some of the elements added to steels may seem to be negligibly small, this small quantity may still have a strong influence on the kinetics of phase transitions and thus on the formation of the microstrucure. Furthermore, the kinetic parameters of alloying elements vary by orders of magnitude in different phases, e.g. in ferrite (α) compared to austenite (γ). A similar ratio exists between the mobilities of substitutional (e.g. Mn, Cr, Ni,...) and interstitial elements (C, N). As a result of these differences, various reaction regimes may appear as function of time and heat treatment temperature before equilibrium is attained. It is also possible that, because of a fast kinetics, metastable phases appear at intermediate stages consuming a large part of the available driving force. The time to reach equilibrium will then be drastically increased, often far

A. Finel et al. (eds.), Thermodynamics, Microstructures and Plasticity, 135–153.

136

beyond industrial time scales. The consequences will then appear by a degradation of e.g. creep properties during long term service of such material.

Steels thus represent excellent candidates for illustrating the problems encountered in materials development and their solutions. A variety of examples out of the field of steels will be presented illustrating the capability and the limitations of presently available software like the program DICTRA. The software DICTRA [8] is the result of more than a decade of development emerging from a co-operation between the Royal Institute of Technology/ Stockholm (Hillert/Ågren et al.) and the Max-Planck-Institute for Iron Research/Düsseldorf (Inden et al.) started by a 5 years project in 1988. For details the reader is referred to [10]-[15].

2.   CALPHAD contribution to steel development

CALPHAD is a method of calculating phase equilibria [2]. This technique has been widely used in the development of materials including steels. For the purpose of illustration the development of martensitic/ferritic super heat-resistant steels for application in 650°C Ultra Super Critical (USC) Power Plants shall be considered. In order to achieve a high creep strength, these steels should contain finely distributed precipitates ($M_{23}C_6$, MX, Laves phase) in a tempered martensitic microstructure. The alloys are complex multicomponent alloys based on 9-12% Cr steels.

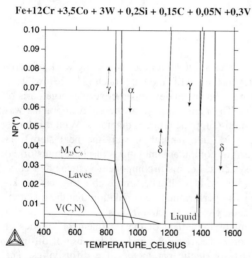

Fe+12Cr +3,5Co + 3W + 0,2Si + 0,15C + 0,05N +0,3V

Figure 1: Calculated phase fractions as function of temperature (NP=number of moles of phase). At the austenitization temperature of 1070°C there is only γ (austenite) present plus some V-carbo-nitride. At the annealing temperature of 780°C $M_{23}C_6$ and a small amount of Laves phase is predicted.

There is a variety of questions to be answered by thermodynamic calculations: (a) what is the amount of phases present at a given temperature (e.g. annealing temperature of about 780°C), (b) how to reach a purely martensitic microstructure after quenching from austenitization temperature (e.g.1070°C), i.e. no δ-ferrite after quenching, (c) no precipitation in the liquid state (such precipitates would coarsen too much and thus have no effect on the creep resistance). Two typical outputs of it shall be recollected in order to illustrate the merits and limitations: phase fractions and Schaeffler diagrams.

2.1   PHASE FRACTIONS

Figure 1 shows the calculated phase fraction of the various phases present at a given temperature. This diagram confirms that for this alloy no δ-ferrite is expected at the austenitization temperature. At the tempering temperature the three phases $M_{23}C_6$, MX and Laves are predicted to be present. For the chosen composition the precipitation of V(C,N) starts just below the solidus temperature.

The question still remains what time it takes to precipitate these phases, how much of each phase is precipitated during tempering, and what is the evolution during service at 650°C.

2.2   SCHAEFFLER DIAGRAMS

In this type of alloy development it is important to know the effect of alloying elements on the stability of the various phases.

It is common practice to classify the elements according to their stabilizing effect on various phases, e.g. ferrite or austenite stabilizer. The stabilizing effect of an element is determined by its effect on the phase boundaries. If adding the element leads to a widening of the ferrite (austenite) phase field, the element is called a ferrite (austenite) stabilizer. The effect is scaled relative to reference elements, Cr for ferrite stabilizers, and Ni for austenite stabilizers.

So-called "Cr equivalents" and "Ni equivalents" are introduced as axes such that, instead of having many component axes, multi-component property diagrams are projected into a plane with these two axes. Schaeffler introduced diagrams characterizing the effect of alloying elements on the microstructure of steel welds. The diagram displays the range of compositions at which given microstructural elements are present after cooling to room temperature. This type of diagrams is very attractive in view of its simplicity. The approach has also been proposed for ferritic steels in order to determine limits for the occurrence of δ-ferrite after an austenitization treatment. Ferritic steels transform fully into martensite during cooling from the austenization temperature. δ-ferrite is thus present only if it existed before cooling at the austenitization temperature. Therefore, in this particular case, a Schaeffler type diagram can be deduced from equilibrium calculations. Two questions shall now be answered by thermodynamic calculations: (a) is it justified to assume the same equivalence factor for both ferrite and austenite boundaries; (b) are the calculated values of the equivalence factor in agreement with empirically fixed values like those in Figure 2. The present calculations were performed with the software THERMOCALC [21] using the solid solution database SSOL [9].

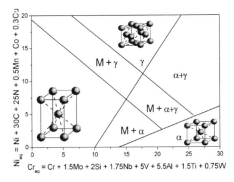

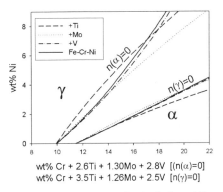

Figure 2: Schaeffler diagram characterizing microstructures in ferritic 9-12%Cr steels after austenitisation, according to [20]. On the axes Cr and Ni equivalents are plotted. These equivalents are defined such that the boundaries of an Fe-X-Y system coincide with the ternary Fe-Cr-Ni system.

Figure 3: Schaeffler diagram for the ferrite stabilizers Ti, Mo, V. The zero phase fraction lines for α and γ were calculated for the addition of Ti, Mo, V to Fe-Cr-Ni. The scaling factors (Cr equivalents) were determined such that the lines coincide as much as possible with those of Fe-Cr-Ni.

138

If no other elements than Cr and Ni are considered the boundaries (α+γ) in Figure 2 correspond to the phase boundaries α/γ and γ/α of the ternary Fe-Cr-Ni system at a certain temperature. The temperature 1150°C was selected such that the intercepts of the calculated boundaries α/γ and γ/α of the Fe-Cr-Ni system come closest to the intercepts in Figure 2. It turns out that for the selected ferrite stabilizers the Cr equivalence factors turn out almost the same for both boundaries. They differ, however, slightly from the values given in Figure 2. Other elements like Al cannot be scaled to fit the Fe-Ni-Cr boundaries in the composition range presented here [22], see Figure 6.

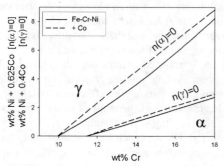

Figure 4: Schaeffler diagram for the austenite stabilizer Co. The zero phase fraction lines for α and γ were calculated for the addition of Co to Fe-Cr-Ni. The scaling factors (Ni equivalents) were determined such that the lines coincide as much as possible with those of Fe-Cr-Ni.

Figure 5: Schaeffler diagram for the austenite stabilizer C. The zero phase fraction lines for α and γ were calculated for the addition of C to Fe-Cr-Ni. The scaling factors (Ni equivalents) were determined such that the lines coincide as much as possible with those of Fe-Cr-Ni.

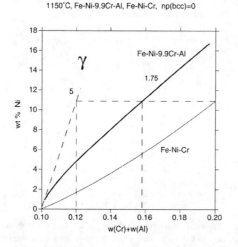

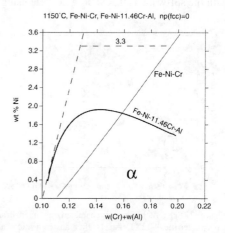

Figure 6: Calculated effect of Al on the phase boundaries ferrite and austenite compared to the effect of Cr, T=1150°C. Depending on the concentration range a factor of 5 or 1.75 applies to the γ/(γ+α)–boundary, while for the α/(α+γ) boundary a factor 3.3 can only be used for low concentrations. On the abscissa the sum of weight fractions Cr+Al is plotted.

The situation becomes much less satisfactory when austenite stabilizers are considered. There is no way to get Ni-equivalents for C close to each other for the two boundaries, see Figure 5. The Ni-equivalents are 18 and 50, respectively, and have to be compared with 30 proposed in Figure 2. The effect of Co is shown in Figure 4. Co turns out to be a ferrite stabilizer at this temperature and Cr composition range. This is recognized by a Ni equivalent factor less than one. Similar results hold for Mn.

Although the mentioned simplified approaches may have been used with some success as a guide line, the limitations become transparent with thermodynamic calculations. It appears that even on a qualitative level there are serious deficiencies, the major deficiency being the lack of any predictive capability.

3. Numerical treatment of diffusion controlled transformations

There are several fundamental concepts forming the basis of the treatment with the software DICTRA [8]:

1. The phase interface is treated as a sharp interface. Its movement is controlled by the mass balance obtained from the fluxes of the diffusing elements across the interface [10].

2. Local equilibrium is assumed at moving interfaces. In a binary system this equilibrium determines the compositions on either side of the boundary in terms of the tie-line of the two-phase equilibrium. In multi-component systems the operating tie-line varies with time. It is determined by the condition that the mass balance of every diffusing element gives the same interface velocity. The local equilibrium may be subject to constraints such as a finite interface velocity. The local equilibria are calculated using THERMOCALC as a subroutine [11].

3. Diffusion is treated considering true thermodynamic driving forces, i.e. chemical potential gradients. This is essential in multi-component systems where diffusion may take place against the composition gradient.

4. The thermodynamic driving forces imply that the kinetic parameters are the mobilities rather than diffusivities. A kinetic database has been established covering a large number of elements and phases [16]-[19].

5. Transformation and diffusion are treated in geometries with only one space coordinate (linear, ellipsoidal etc.). Space may be subdivided into "cells" with one or more different phases [13]. Within each phase the space coordinate is discretized by grid points. Boundary conditions are imposed to define the interaction between the cells [11].

Alloy compositions will be given in wt%. In some cases it may be more convenient to use mole fractions x_i. In steels with interstially dissolved carbon the substitutional elements are on one sublattice, the interstitial element on a second sublattice. It is useful to consider so-called "u-fractions" u_i which refer the number of i-atoms to the total number of substitutional ele-

ments S: $u_i = \dfrac{x_i}{\sum\limits_{j \neq C} x_j}$.

140

4. Boundary conditions

Only sharp interfaces will be considered, diffuse interfaces being treated in other reports on the phase field method. For sharp interfaces it is generally assumed that the interface does not present any hindrance for diffusion. Consequently, no driving force is needed to cross the interface. Diffusion will be treated in the volume fixed frame of reference defined by $\sum_{k \in S} J_k = 0$ where S denotes substitutional elements.

Considering a moving α/γ interface the flux balance is given by

$$\frac{v^{\alpha}}{V_{m,S}^{\alpha}} u_k^{\alpha} - \frac{v^{\gamma}}{V_{m,S}^{\gamma}} u_k^{\gamma} = J_k^{\alpha} - J_k^{\gamma} \qquad k = 1, 2, \ldots n \qquad (1)$$

Making use of the two relations $\sum_{k \in S} u_k = 1$ and $\sum_{k \in S} J_k = 0$ we can eliminate one velocity $\frac{v^{\alpha}}{V_{m,S}^{\alpha}} = \frac{v^{\gamma}}{V_{m,S}^{\gamma}} \underset{def}{=} v$ and one flux equation. The flux balance thus reduces to

$$v \cdot \left(u_k^{\alpha} - u_k^{\gamma} \right) = J_k^{\alpha} - J_k^{\gamma} \qquad k = 1, 2, \ldots (n-1) \qquad (2)$$

The fluxes are given by

$$J_k = -M_k \frac{u_k}{V_{m,S}} \frac{\partial \mu_k}{\partial z} \qquad (3)$$

4.1 LOCAL DIFFUSIONAL EQUILIBRIUM (LDE)

In the case of interdiffusion, i.e. diffusion of substitutional elements on a common sublattice, the condition $\sum_{k \in S} J_k = 0$ reduces the number of independent fluxes by one and the diffusion potential may be written $\Phi_k = \mu_k - \mu_n$, where n is one arbitrarily selected element. In the case of an interstitial element the corresponding potential is $\Phi_k = \mu_k$ since the interstitial lattice is empty and the atoms can diffuse without interchanges with other atoms.

Diffusional equilibrium at an α/γ interface means $\Phi_k^{\alpha} = \Phi_k^{\gamma}$ k=1,2,..(n-1). We may thus fix one of the Φ_k and determine v and the (n-2) remaining Φ_k by means of the (n-1) equations (2). It is to be emphasized that the solution will generally not lead to $\mu_k^{\alpha} = \mu_k^{\gamma}$, the condition of (full) local equilibrium. This is illustrated for a binary system A-B in 7. The LDE condition is met with a parallel tangent construction. For a selected value of $\Phi_B^{\gamma} = \mu_B^{\gamma} - \mu_A^{\gamma}$ corresponding to a composition $x^{\gamma/\alpha} > x_0$ at the interface, the corresponding composition $x^{\alpha/\gamma}$ in α at the interface is obtained by the parallel tangent. The driving force for diffusion is related to the difference $\Phi_k^{\gamma/\alpha} - \Phi_{k0}^{\gamma}$. It becomes evident from Figure 7 that this driving

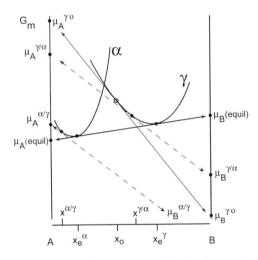

Figure 7: Schematic illustration of the boundary condition at moving interfaces. x_0 is the composition of the γ matrix, $x^{\alpha/\gamma}$ and $x^{\gamma/\alpha}$ are the compositions in α and γ at the interface, respectively. x_e^{α} and x_e^{γ} are the compositions of coexisting α and γ at equilibrium.

force increases the more $x^{\gamma/\alpha}$ approaches x_e^{γ}. Beyond x_e^{γ} the driving force for diffusion increases further, but the Gibbs energy release per reaction step decreases. Therefore, in general, the maximum of Gibbs energy release rate is obtained for $x^{\gamma/\alpha} = x_e^{\gamma}$. In that case the parallel tangents coincide forming the common tangent of the (full) local equilibrium, i.e. the chemical potentials on either side are equal. The LDE condition becomes the local equilibrium condition LE.

Instead of fixing one of the Φ_k, it is also possible to impose a velocity v of the interface and determine the $(n-1)$ diffusion potentials Φ_k by means of the $(n-1)$ equations (2). This situation is encountered in reality when a particle, precipitated at a grain boundary, grows asymmetrically, i.e. only into one of the adjacent grains. The interface with the highest coherency is generally least mobile or even pinned, i.e $v=0$. The resulting conditions are schematically drawn in Figure 8. The situation is particularly interesting in the Fe-C system since C diffu-

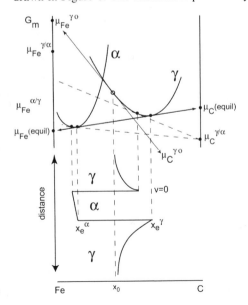

Fe x_0 C

Figure 8:

(a)   Local conditions at α/γ phase boundaries in a binary system, e.g. Fe-C, for two situations:

I.   mobile interface:

$$LE \Leftrightarrow \mu_C^{\gamma/\alpha} = \mu_C^{\alpha/\gamma} = \mu_C(equil)$$

II.   immobile interface:

$$LDE \Leftrightarrow \mu_C^{\gamma/\alpha} = \mu_C^{\alpha/\gamma} \neq \mu_C(equil) .$$

(b) α particle growing under local equilibrium at the mobile interface and under diffusional equilibrium at the immobile interface, $v=0$. At the pinned interface the flux balance must be zero. At the moving boundary the composition is given by the equilibrium tie-line in the phase diagram. The resulting velocity v is controlled by the mass balance.

sivity is about 100 times faster in α than it is in γ. The chemical potential (and thus composition) gradient through α leads to a short-circuit diffusion such that the mobile interface moves twice as fast as it did if both interfaces were moving under LE conditions. The resulting growth rate is thus practically the same in both cases.

4.2 LOCAL EQUILIBRIUM (LE)

LE at an α/γ interface means $\mu_k^\alpha = \mu_k^\gamma$ k=1,2,...n. The degree of freedom in this 2-phase equilibrium is n-2. Thus, the n-2 free chemical potentials and v are determined by the (n-1) equations (2). In a binary system the degree of freedom is zero. LE is identical to the global equilibrium and the compositions at either side of the boundary are given by the corresponding tie-line. The situation in higher order systems is different and may be illustrated with a ternary system as shown in Figure 9. The system Fe-Mn-C was selected with C being a fast diffuser. Given an alloy composition in the two-phase field α+γ, the operating tie-line is defined by the same velocity of the interface obtained from the flux balance with respect to every diffusing species. To meet this condition, C as a fast diffuser needs only a small driving compared to Mn. The first operating tie-line is thus very close to the intersection of the iso-activity line of the austenite matrix with the phase boundary. The reaction proceeds with this boundary condition until the driving force for C-diffusion has decreased to an extent that Mn diffusion becomes competitive. Then the operating tie-line shifts towards the global equilibrium tie-line.

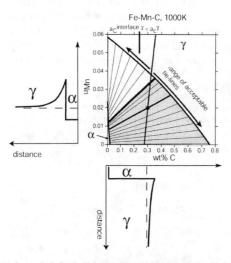

Figure 9: Boundary condition at a moving interface in a ternary system: ferrite (α) precipitation from austenite (γ) in Fe-Mn-C. The operating tie-line is defined by the equality of interface velocities obtained from the flux balance of both C and Mn. In the case of fast diffusing C the first operating tie-line is close to the intersection of the isoactivity line of austenite with the phase boundary. During the reaction this interface moves towards the global equilibrium tie-line.

4.3 LOCAL EQUILIBRIUM WITH NEGLIGIBLE PARTITIONING (LENP)

In the case of systems with fast diffusing elements like C, N, H the domain of operating tie-lines may be subdivided into a range where local equilibrium is possible only with partitioning of the substitutional elements (LE), and a range of local equilibrium with negligible or no partitioning (LENP).

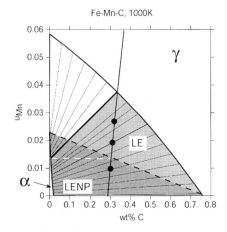

Figure 10: Composition domains of different reaction regimes: local equilibrium with partitioning (LE) of substitutional elements and local equilibrium with negligible partitioning (LENP). The broken line separating the two regimes is obtained by the intersection of the isoactivity line with the line of constant Mn composition.

At compositions within the LENP domain, the tie-line corresponding to a constant substitutional element composition is a possible operational tie-line since the C-activity in γ at the interface is higher than in the bulk. This provides the driving force required for transporting C released from α away from the interface, thus allowing ferrite to grow. In the domain LE the opposite holds: the C-activity in γ at the interface is lower than in the bulk. The driving force for C diffusion now leads to a flux of C from γ to α. A ferrite particle, instead of growing, would thus have to transform back to austenite. This thermodynamic condition for a fast reaction is sharp. This limit is shown in Figure 10 by the black broken line. It is obtained by the intersection of the line of constant Mn content (white broken line) with the line of constant activity in γ defined by the equilibrium with constant Mn ferrite. These basic concepts were introduced long ago and discussed in [23]-[27]. For an overview see [28]. With the new computational instruments the kinetics of these phase transformations can be calculated quantitatively for any given time-temperature profile and composition.

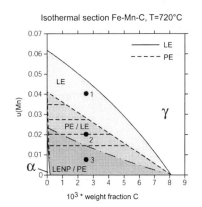

Figure 11: Isothermal section of the Fe-Mn-C system with phase boundaries according to local equilibrium and para-equilibrium. Alloy 1 can transform only according to LE. Alloy 2 may transform either under LE or PE. Alloy 3 may transform under LENP or PE.

4.4 PARA-EQUILIBRIUM

The concept of para-equilibrium (PE) was introduced by Hultgren in 1947 [29]. The basic idea of PE is to consider all substitutional elements as immobile and to treat a multicomponent system Fe-X-C as a pseudo-binary system M-C where all substitutional elements Fe and X are replaced by one average component M. The thermodynamic properties of M have to be constructed such that the Gibbs energy function of the pseudo-binary system M-C is identical to the Gibbs energy function within the corresponding vertical section of the system Fe-X-C. An example of this procedure has recently been outlined in [30]. Figure 11 shows the isothermal section of the Fe-Mn-C system at 720°C with the equilibrium phase boundaries, with the limit of LENP and with the PE phase diagram. The tie-lines of the PE phase diagram are, of course, all parallel to the C-axis.

5. Ternary Systems

5.1   LE AND LENP IN FE-SI-C

The precipitation of ferrite from austenite in a ternary Fe-1.15 %Si-0.5 %C alloy at T=749°C shall be considered. Figure 12 shows the isothermal section. It is seen that the tie-line obtained from the intersection of the line of constant number of Si atoms with the ferrite phase boundary fulfills the criterion for the fast reaction according to LENP.

Experiments performed at this temperature and analysed metallographically by serial sectioning and three-dimensional analysis are shown in Figure 13. It was found that nucleation took place at grain vertices and the density of nuclei was determined as 2200 ± 400 / mm^3. This density fixes the average cell size (e.g. spherical cell of about 48μm radius) to be used in the calculation. This size controls the time at which soft-impingement of carbon starts and reduces the driving force for C diffusion, leading to a switch from LENP to LE. The results of DICTRA calculations are shown in Figure 13 for linear and spherical geometry. The experiments are perfectly reproduced by the calculation for spherical geometry. Both calculations clearly show the existence of two reaction regimes. It shall be emphasised that the excellent agreement is obtained without any fitting. At the end of the fast regime the plateau defines an almost constant volume fraction for very long times, far beyond industrial time scales.

The evolution of composition during the transformation is shown in Figure 14 and Figure 15

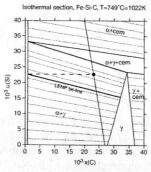

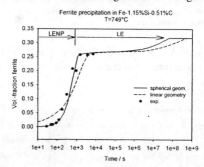

Figure 12: Isothermal section of the Fe-Si-C system at 749°C. The point defines the composition Fe-1.15%Si-0.51%C. The solid line through the point represents $a_C^\gamma = c^{st}$, the dotted line represents constant Si-content.

Figure 13: Growth of ferrite in Fe-1.15%Si-0.51%C at 749°C. The experimental data were determined by serial sectioning. The calculations show that the plateau value defines the end of the LENP regime and not yet the equilibrium state. The experiments follow closely the calculations for spherical geometry.

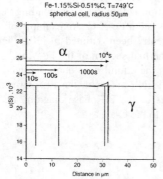

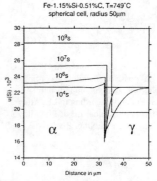

Figure 14 Growth of ferrite under LENP during the first 10^{4}s. Ferrite inherits the Si-content of austenite. Growth requires only C diffusion.

Figure 15 Growth of ferrite under LE at later stages. Growth requires diffusion of both Si and C.

in form of composition profiles for various time steps. Figure 14 corresponds to the time interval of the fast reaction. The position of the interface is recognised by the spike of Si composition. During this period the boundary condition is constant. In the slow regime, Figure 15, the spike broadens and develops a Si profile. During this period the boundary condition varies with time, as seen by the change in Si composition at the interface.

5.2   LE AND LENP IN FE-MN-C

The limit of transition from LE to LENP is a sharp boundary faced in heat treatments. As an example the alloy Fe-1Mn-0.25C will be considered. Figure 16 and Figure 17 show the isothermal sections at the temperatures 1000K and 1050K. The LENP tie-line shows that at 1000K the fast reaction is possible, while at only 50K higher temperature only the LE reaction is possible. Figure 18 and Figure 19 show the composition profiles for various time steps obtained from the calculation. The differences of volume fraction formed are tremendous: at 1000K, 63% of ferrite is formed within 1000s, while at 1050K only 0.2% is formed within 10^6s. This shows the overwhelming possibilities of microstructure control offered by appropriate heat treatments. There may be ways of selecting the temperature in order to avoid the precipitation of an unwanted phase during industrial heat treatments.

Fast and slow reaction regimes are also obtained in the process of dissolution of phases. This has been treated in [14].

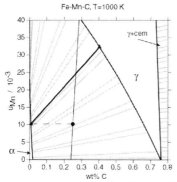

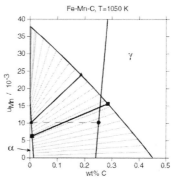

Figure 16:   Calculated isothermal section of the Fe-Mn-C system at 1000K. The point corresponds to the composition Fe-1%Mn-0.25%C. The precipitation of ferrite from austenite under LENP condition is possible

Figure 17: Calculated isothermal section of the Fe-Mn-C system at 1050K. The point corresponds to the composition Fe-1%Mn-0.25%C. The precipitation of ferrite from austenite under LENP condition is not possible

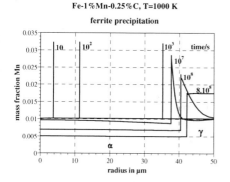

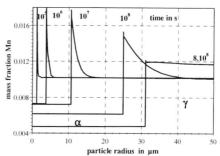

Figure 18: Fast reaction according to LENP. Calculated composition profiles of Mn at various time steps. Most of the transformation occurs during the first 1000s.

Figure 19: Slow reaction according to LE. Calculated composition profile of Mn at various time steps. Only a small volume fraction is formed even after 10^{5s}.

6. Quaternary Systems

6.1 EFEECT OF TRACES ON THE GROWTH OF GRAIN BOUNDARY CEMENTITE

Si is an element that has little solubility in cementite Fe_3C. In the current thermodynamic databases cementite is treated as a phase with no solubility for Si. Lateral growth of grain boundary cementite was studied experimentally by Ando and Krauss [31] on a (nominally) ternary Fe-2.26%Cr-1.06%C at 738°C.

Figure 20 shows the isothermal section of the Fe-Cr-C system with the ternary alloy composition converted into u-fraction and mole-fraction. The starting tie-line corresponds to a fast LENP reaction. The experimental data are shown in Figure 21. In order to comply with the experimental situation in [31] a spherical cell with diameter 70μm was chosen, cementite growing from the outer shell of the spherical grain towards the centre. The results for short times are shown in Figure 21. The calculation shows that for for 0%Si the time to reach a

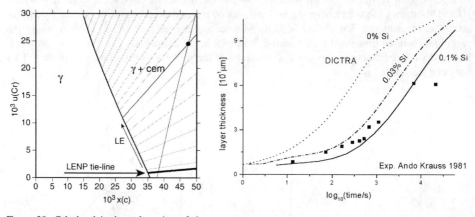

Figure 20: Calculated isothermal section of the Fe-Cr-C phase diagram at 1011K. The point corresponds to the composition Fe-2.26%Cr-1.06%C, expressed in u- and mole-fractions.

Figure 21: Lateral growth of grain boundary cementite in Fe-2.26%Cr-1.06%C at T=738°C. The alloys contained a trace of 0.03%Si.

given layer thickness is more than an order of magnitude larger than the experiments. It is only after taking into account that the alloys contained a trace of 0.03%Si that the calculations come close to the experimental data. The rejection of the little amount of Si into the matrix slows drastically down the rate of reaction. Taking three times more Si, 0.1%Si, makes only little difference. This demonstrates the effectiveness that traces may have on the reaction rate.

6.2 SIMULTANEOUS GROWTH OF $M_{23}C_6$ AND LAVES PHASE: TWO CELL CALCULATION

The simultaneous growth of $M_{23}C_6$ and Laves phase in a ferritic steel with composition Fe-12Cr-3W-0.15C will now be treated. It is assumed that the two particles grow in individual spherical cells of ferritic matrix with 5µm radius. The two cells are coupled together by the condition of equal chemical potentials of all diffusing components i, as schematically shown in Figure 22 .The value of this surface potential is determined by the overall mass balance of

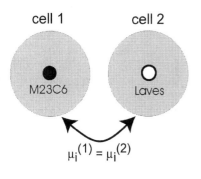

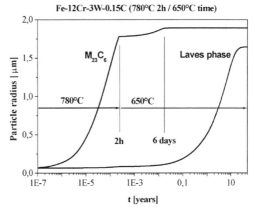

Fe-12Cr-3W-0.15C (780°C 2h / 650°C time)

Figure 22: Simultaneous growth of $M_{23}C_6$ and Laves phase in two cells of ferrite. The cells are coupled by the equilibrium condition of equal chemical potentials

Figure 23: Simulation of the growth of $M_{23}C_6$ and Laves phase in two cells of ferrite. The heat treatment consits of two steps: 2h annealing at 780°C, followed by 650°C for a given time.

the whole system. The calculation is performed for a two step heat treatment of ferrite at 780°C for 2h, followed by a long term annealing at 650°C corresponding to a service condition. It is seen that within the first 2 hours $M_{23}C_6$ grows fast, while the Laves phase particle is smaller by more than a factor of 10. The precipitation of $M_{23}C_6$ has not yet come to completion after 2h. The change in temperature to 650°C decreases the growth rate of $M_{23}C_6$ considerably. After about 6 days the precipitation of $M_{23}C_6$ is complete. The precipitation of Laves phase, on the contrary, takes more than 10 years to reach equilibrium.

The carbide $M_{23}C_6$ is treated as phase with no diffusion inside. Therefore, the carbide grows at every time step with a composition given by the boundary condition and this profile does not level out. The boundary condition varies considerably with time as shown in Figure 24 and Figure 26. The Laves phase is treated as a two sublattice phase $(Fe,Cr)_2W$, stoichiometric with respect to W, see Figure 27, but the content of Cr is not fixed. Therefore the boundary condition also varies with time, see Figure 25. Changing temperature after 2h from 780°C to 650°C and continuing the heat temperature for longer times at 650°C leads to the results in Figure 28-Figure 31. This last heat treatment shall simulate the behaviour at long term service conditions.

It has to be mentioned that this first calculation cannot yet reproduce correctly the real situation. The Laves phase nucleates preferentially at the surface of $M_{23}C_6$ rather than homogeneously in the bulk. $M_{23}C_6$ nucleates preferentially at grain boundaries and sub-grain boundaries. Therefore, growth is not only controlled by volume diffusion, but also by boundary diffusion. It is also to be expected that the Laves phase may grow much faster when in contact with an $M_{23}C_6$ particle by taking up Cr and W from the carbide in addition to volume diffusion.

148

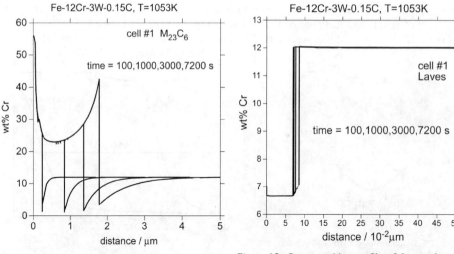

Figure 24: Cr composition profile of $M_{23}C_6$ at various timesteps of annealing at 780°C.

Figure 25: Cr composition profile of Laves phase at various timesteps of annealing at 780°C.

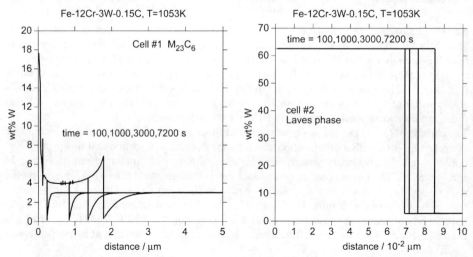

Figure 26: W composition profile of $M_{23}C_6$ at various timesteps of annealing at 780°C.

Figure 27: W composition profile of Laves phase at various timesteps of annealing at 780°C.

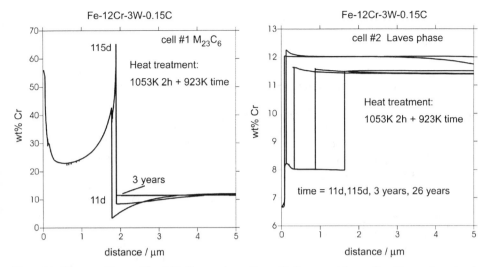

Figure 28: Cr composition profile of $M_{23}C_6$ at various timesteps of further annealing at 650°C.

Figure 29: Cr composition profile of Laves phase at various timesteps of further annealing at 650°C.

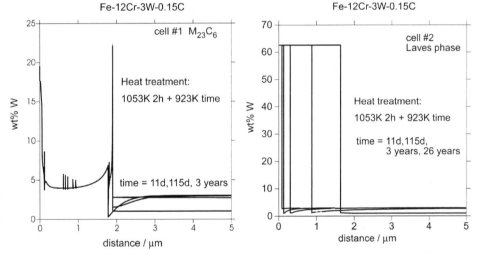

Figure 30: W composition profile of $M_{23}C_6$ at various timesteps of further annealing at 650°C.

Figure 31: W composition profile of Laves phase at various timesteps of further annealing at 650°C.

7. Continuous cooling

The phenomena discussed so far under isothermal conditions are also encountered during continuous cooling or heating. In a cooling experiment an alloy enters first into the LE region of a two phase field, then into the PE region well before it enters eventually into the LENP region. By continuous cooling it is thus possible to observe these regimes within one specimen.

Such experiments were performed with two alloys, a ternary Fe-1.5Mn-0.09C and a quaternary Fe-1.37Mn-0.42Si-0.18C with a cooling rate of 1°C/s from the austenite region [32]. The cooling experiments were interrupted at given temperatures to determine the microstructures obtained. The fraction of ferrite was determined from metallographic analysis. The experimental results are shown in Figure 1. Calculations were performed for the same conditions using a cell size in accordance with the austenite grain size. The calculated temperatures of penetration into the regions LE and PE are marked by arrows. The volume fraction vs. temperature curves were calculated according to LE and LENP. The volume fraction precipitated under LE is negligibly small. The onset of LENP reaction is characterised by a sudden increase in volume fraction.

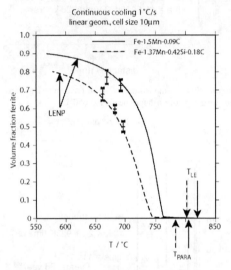

The experimental data were determined at temperatures below the LENP starting point. They are perfectly located on the calculated LENP branches. Since the growth rates of PE and LENP reactions are very similar, the PE reaction, if occurring, should show a similar behaviour, except that the start should occur at considerably higher temperatures leading to a shift of the corresponding volume fraction curves to higher temperatures. The results thus indicate that in these alloys there is no indication of any PE reaction.

Figure 32: Volume fraction of ferrite determined experimentally after interrupted continuous cooling from austenite. Cooling rate 1K/s. Experiments courtesy IRSID [32]. Calculations: linear geometry, cell size 10μm.

There are several arguments which may be put forward to explain this finding. At first, we must recall that PE assumes the substitutional elements to be almost immobile compared to the fast diffusing interstitials. Strictly speaking, they must be immobile, otherwise they could not be subsumed under an average component M. The question remains at what temperature this criterion is fulfilled. On the other hand, these elements have to cross the interface. At least this process is controlled by the diffusion potential. Assuming for simplicity that the transfer mobility of the S elements is the same, we would thus require that the diffusion potential at either side of the moving boundary must be the same. This requirement cannot be fulfilled within an isopleth corresponding to PE.This is illustrated in Figure 33 for the ternary Fe-Mn-C system. Considering a compostion of γ according to point 2 we can draw the

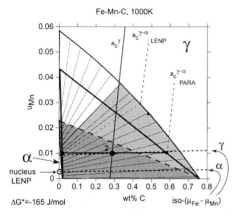

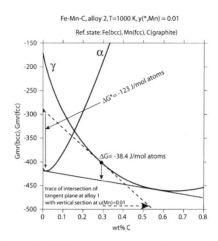

Figure 33: Isothermal section of the Fe-Mn-C system. The boundaries for global equilibrium and para-equilibrium are shown. Alloy 2 falls within the domain where LENP or PE are possible alternatives for the boundary condition at the moving interface.

Figure 34: Gibbs energy contours within the section u_{Mn}=.01 corresponding to PE. Determination of the driving force for nucleation according to PE.

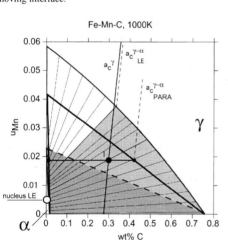

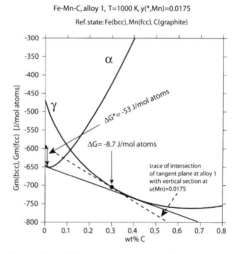

Figure 35: Isothermal section of the Fe-Mn-C system. The boundaries for global equilibrium and para-equilibrium are shown. Alloy 1 falls within the domain where LE or PE are possible alternatives for the boundary condition at the moving interface.

Figure 36: Gibbs energy contours within the section u_{Mn}=.01 corresponding to PE. Determination of the driving force for nucleation according to PE.

$iso - \Phi_{Mn}^{\gamma}$ contour passing through point 2. The contour is practically a line in this instance. The iso-potential contour for α with the same value as γ is clearly displaced from the section point defining the nucleus of α precipitating from γ. It is very clear that in PE the equality of diffusion potential cannot be fulfilled.

We may also evaluate the driving forces for nucleation. Considering the nucleus defined by LENP we obtain the value for ΔG^* given in Figure 33. Figure 34 shows the Gibbs energy contours with the section $u_{Mn}=0.01$ corresponding to PE. The driving force for nucleation under PE condition given in Figure 34 is clearly smaller than for LENP in Figure 33. It should be reminded that ΔG^* enters with the power of two in the probability for nucleation. We may thus conclude that for compositions within the domain where, in principle, LENP and PE may occur there is no argument in favour of PE.

Let us now turn to alloy 1 falling into the domain where LE or PE are possible alternatives. The situation for this case is illustrated in Figure 35 and Figure 36. From Figure 35 it follows that the nucleus for LE condition is located at a composition far away from the section with constant $u_{Mn}\cong0.0175$ corresponding to PE. Figure 36 shows the Gibbs energy contours within the section at $u_{Mn}=0.0175$ and the evaluation of the driving force for nucleation. It turns out that at this composition the driving force for nucleation is almost three times less for PE that it is for LE. This may be one reason why PE does not start at the corresponding transition temperatures as shown in Figure 32.

There are even more aspects to be investigated carefully: the dissipation of Gibbs energy during the transformation. In order for a spontaneous transition to occur there must be a Gibbs energy gain. It is thus necessary to evaluate carefully the balance between driving force available and dissipation, here the dissipation by diffusion within the parent γ-phase.

Summing up all these aspects it appears very difficult to meet situations where PE may occur. It is thus not surprising that in the cases studied in Figure 32 the PE reaction was not observed.

8. Conclusion

The results presented show that the numerical calculation of diffusion controlled phase transformations has reached a high level of accuracy and agreement with experimental data. It has become a valuable tool for research and alloy development. The limitations are given by the quality of the thermodynamic descriptions which have previously been determined only for the purpose of phase diagram calculations, discarding any aspect relevant for diffusion. The mobilities have to be defined in accordance with the thermodynamic description. Consequently, they suffer from the same limitations.

9. Acknowledgement

Part of the results presented was obtained in collaboration with P. Cugy (IRSID/France) and A. Schneider (MPI-Düsseldorf). This is gratefully acknowledged. Thanks are also due to M. Kandel (IRSID) for making available unpublished experimental results in Figure 32.

References

[1]   L. Kaufman and H. Bernstein (1970), *Computer Calculation of Phase Diagrams*, Academic Press, New York
[2]   N. Saunders and A.P. Miodownik (1998), CALPHAD, Pergamon Materials Series Vol. 1, R.W. Cahn Ed., Elsevier Science, Oxford
[3]   G. Eriksson and K.Hack (1990), "ChemSage - A Computer Program for the Calculation of Complex Chemical Equilibria", *Met.Trans.B*, **21B**, 1013-1023

[4]   R.H. Davies, A.T. Dinsdale, T.G. Chart, T.I. Barry, M. Rand (1990), "Application of MTDATA to the model-
        ling of multicomponent equilibria", *High Temp. Science* **26**, 251-262
[5]   R.H. Davies, A.T. Dinsdale, J.A. Gisby, S.M. Hodson, R.G.J. Ball (1994), "Thermodynamic Modelling using
        MTDATA: A Description showing applications involving oxides, alloys and aqueous solution", Proc. Conf.
        ASM/TMS Fall Meeting: Applications of thermodynamics in the synthesis and processing of materials, Rose-
        mont, IL, USA, 2-6 Oct 1994
[6]   S.-L. Chen, S. Daniel, F. Zhang, Y. A. Chang, W. A. Oates, R. Schmid-Fetzer (2001),
        "On the Calculation of Multicomponent Stable Phase Diagrams", *J. Phase Equilibria* **22**, 373-378
[7]   PANDAT, Software for Multicomponent Phase Diagram Calculation, Computherm LLC, 437 S. Yellowstone
        Drive, Suite 217, Madison WI 53719, USA.
[8]   DICTRA (Diffusion Controlled Transformations), software developed within the project COSMOS (Com-
        puter Supported Modelling of Steel Transformations), funded by Volkswagen-Stiftung and Land Nord-Rhein-
        Westfalen (1988-1993)
[9]   THERMO-CALC, DICTRA, SSOL (Solid Solution Data Base) (1994), Mobility Database, Foundation of
        Computational Thermodynamics, Royal Institute of Technology, Stockholm/Sweden
[10]   S. Crusius, G. Inden, U. Knoop, L. Höglund and J. Ågren (1992) On the numerical treatment of moving
        boundary problems, *Z. Metallkunde* **83**, 673-678
[11]   S. Crusius, L. Höglund, U. Knoop, G. Inden and J. Ågren (1992) On the growth of ferrite allotriomorphs in
        Fe-C, *Z. Metallkunde* **83**, 729-738
[12]   G. Inden and P. Neumann (1996) Simulation of diffusion controlled phase transformations in steels, *steel re-
        search* **67**, 401-407
[13]   P. Franke and G. Inden (1997) Diffusion Controlled Transformations in Multi-Particle Systems, *Z. Metall-
        kunde* **88**, 917-924
[14]   G. Inden (1997) Cinétique de transformation de phases dans des systèmes polyconstitués - aspects thermody-
        namiques, *Entropie* **202/203**, 6-14
[15]   A. Borgenstamm, A. Engström, L. Höglund and J. Ågren (2000) DICTRA, a Tool for Simulation of Diffu-
        sional Transformations in Alloys, *J. Phase Equilibria* **21**, 269-280
[16]   J.-O. Andersson and J. Ågren (1992) Models for numerical treatment of multicomponent diffusion in simple
        phases, *J. Appl. Phys.* **72**, 1350-1355
[17]   B. Jönsson (1994) Ferromagnetic Ordering and Diffusion of Carbon and Nitrogen in bcc Cr-Fe-Ni Alloys, *Z.
        Metallkunde* **85**, 498-509
[18]   A. Engström and J. Ågren (1996) Assessment of Diffusional Mobilities in Face-centered Cubic Ni-Cr-Al Al-
        loys, *Z. Metallkunde* **87**, 92-97
[19]   P. Franke and G. Inden (1997) An assessment of the Si mobility and the application to phase transformations
        in Si steels, *Z. Metallkunde* **88**, 795-799
[20]   H. Schneider (1960) Investment Casting of High-hot-strength 12%Chrome Steel, *Foundry Trade Journal* **108**,
        562-563
[21]   B. Sundman, B. Jansson and J.-O. Andersson (1985) The THERMO-CALC databank system, *CALPHAD* **9**,
        153-190
[22]   G. Inden (2002) Computational Thermodynamics and Simulation of Phase Transformations, in *CALPHAD
        and Alloy Thermodynamics*, P.E. Turchi, A. Gonis and R.D. Shull (Eds), The Minerals, Metals and Materials
        Soc (TMS), Warrendale/PA, 107-130
[23]   A. Hultgren (1920), A Metallographic Study of Tungsten Steels, John Wiley, New York NY
[24]   M. Hillert (1953) Paraequilibrium, Internal Report, Swedish Institute of Metals Res., Stockholm
[25]   J.S. Kirkaldy (1962) Theory of Diffusional Growth in Solid-Solid Transformations, in *Decomposition of Aus-
        tenite by Diffusionla Processes*, V.F. Zackay and H.I. Aaaronson (Eds), Interscience Publ., New York, 39-130
[26]   A.A. Popov and M.S. Miller (1959) On the kinetics of ferrite formation in the decarburization of carbon and
        alloy steels, *Physics Metals Metallography* **7**, 36
[27]   L.S. Darken (1961) Role of Chemistry in Metallurgical Research, *Trans AIME* **221**, 654-671
[28]   M. Hillert (1998) *Phase Equilibria and Phase Transformations – Their Thermodynamic Basis*, Cambridge
        University Press
[29]   A. Hultgren (1947) Isothermal Transformation of Austenite, *Trans ASM* **36**, 915-989
[30]   G. Ghosh and G.B. Olson (2001) Simulation of Para-equilibrium Growth in Multicomponent Systems, *Metall.
        Mater. Trans.* **32A**, 455
[31]   T. Ando and G. Krauss (1981) The isothermal thickening of cementite allotriomorphs in a 1.5Cr-1C steel,
        Acta Metall. **29**, 351
[32]   P. Cugy and M. Kandel, IRSID/France, unpublished work, private communication

SOLUTE DRAG :

A review of the 'Force' and 'Dissipation' approaches to the effect of solute on grain and interphase boundary motion.

C. R. HUTCHINSON AND Y. BRECHET
Laboratoire de Thermodynamique et Physico-Chimie Métallurgiques
Domaine Universitaire, St. Martin D'Hères, FRANCE.

1. Introduction

It is now well known that segregated impurity atoms can drastically reduce the mobility of grain boundaries in pure metals. Lucke and Detert [1] developed the first quantitative treatment of this effect to explain their observations that the recrystallisation rate of high purity Al could be reduced by many orders of magnitude by the addition of 0.01% Mn or Fe. They attributed the effect to an interaction between the solute atoms in solution and the moving grain boundaries. The phenomena is now considered to be a general effect and is usually referred to simply as the 'solute-drag effect'. A characteristic feature of this effect is that it's magnitude depends on the boundary velocity. It is worthwhile at the outset to distinguish this effect from the Zener pinning of boundaries by particles, the latter being a threshold force usually considered independent of boundary velocity.

The solute-drag effect has received considerable attention both because of it's scientific interest and because of the central role of grain boundary motion in recrystallisation and grain growth processes during the thermo-mechanical processing of many industrially important alloys. Since the initial treatment of Lucke and Detert, two quantitative treatments of the solute-drag effect on grain boundaries have come to dominate the literature. These are the treatments of Cahn [2] (the force approach, 1962) and Hillert [3,4] (the dissipation approach, 1969, 1976). More recently, an analogous solute effect on interphase boundaries has been proposed [5]. Cahn published his solute drag paper before an interaction between solute atoms and interphase boundaries was suggested and he never discussed the application of 'the force approach' to phase transformations. Purdy and Brechet [6] were the first to apply this approach to phase transformations and this has since been repeated with various modifications, e.g. [7]. The possibility of an interaction between solute atoms and a migrating interphase boundary was well known to Hillert when he published his 'dissipation approach' and he discussed applications to both grain and interphase boundaries. In particular, he provided a critical comparison of the force and dissipation approaches and identified the conditions

A. Finel et al. (eds.), Thermodynamics, Microstructures and Plasticity, 155–164.
© 2003 *Kluwer Academic Publishers. Printed in the Netherlands.*

under which they would be expected to give the same result.

A review and comparison of these two approaches to the 'solute-drag effect' forms the basis of this paper. In sections 2 and 3, we discuss and compare the theories of solute drag applied to grain and interphase boundary motion, respectively. In section 4 we discuss the factors limiting a quantitative comparison between theory and experiment and highlight one of the important criticisms of the current treatments of solute-drag. In Sections 5 we conclude with a brief discussion of possible future directions for research in the field of 'solute-drag'.

2. Solute Drag Models for Grain Boundaries

Any quantitative description of the interaction between solute atoms and a moving grain boundary must make some assumptions about the nature of grain boundary motion and of the interaction with the solute. The treatments of Cahn and Hillert both assume a continuum model (as opposed to a discrete or atomistic model) for the grain boundary and assume that it remains planar during motion. As far as the solute atoms are concerned, the grain boundary is represented by a simplified local interaction energy of the type shown in Fig. 1. Lucke and Detert [1] originally postulated that this potential variation arises from the strain energy differences between the misfitting solute atoms and the matrix. Cahn used a wedge shaped well (Fig. 1) to describe the interaction energy and Hillert has used both the square [3] and the truncated wedge [4] for the energy profile of the boundary.

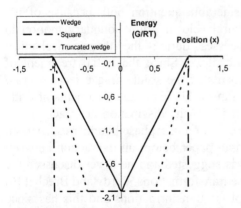

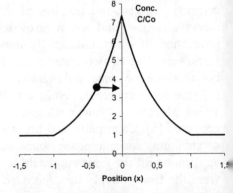

Figure 1. Examples of interaction energy profiles that have been used to represent the grain boundary.

Figure 2. Equilibrium solute profile across a stationary grain boundary with a wedge shaped, attractive interaction energy (Fig.1)

The treatments of Cahn and Hillert can be illustrated in two steps. The first, which is common to both approaches, involves the evaluation of the solute concentration profile across a moving grain boundary. The behaviour of the solute under the extremes of grain boundary motion is self-evident. When the grain boundary moves with a velocity that is very slow with respect to the diffusion of the solute in the vicinity of the grain boundary, then the solute concentration profile will be close to the equilibrium solute profile for a stationary boundary (Eq. 1, Fig. 2).

$$C(x) = Co \exp\left[-\frac{{}^{o}G(x)}{kT}\right], \qquad \text{where } Co \text{ is the bulk alloy content.} \qquad \text{(Eq. 1)}$$

If the grain boundary velocity is very high with respect to the diffusion of the solute, then the concentration profile would be expected to approach the uniform bulk alloy composition Co, through the boundary. However, at intermediate velocities, much more interesting concentration profiles are possible. The profile through the boundary can be found by solving Fick's law for diffusion, assuming descriptions of the interaction energy ${}^{o}G(x)$, and the solute diffusivity $D(x)$, across the boundary in addition to a description of the solute thermodynamics. Apart from the sign of interaction between the solute and the grain boundary (attractive or repulsive), very little is quantitatively known about any of these parameters. For example, it is not clear that diffusion across and along a grain boundary are the same.

2.1. SOLUTE CONCENTRATION PROFILE

Both Cahn and Hillert provided solutions to the diffusion equation for the solute profile across a moving grain boundary under steady-state conditions. Cahn's analytical solution for arbitrary interaction energy ${}^{o}G(x)$, and diffusivity $D(x)$ is shown below (Eq. 2). The only assumption made in the derivation of Eq. 2 is that of ideal (dilute) solution thermodynamics for the solute.

$$C(x) = CoV \cdot \exp\left[-\frac{{}^{o}G(x)}{kT} - V\int_{xo}^{x}\frac{d\eta}{D(\eta)}\right]$$

$$\times \int_{-\infty}^{x}\exp\left[\frac{{}^{o}G(\xi)}{kT} + V\int_{xo}^{\xi}\frac{d\eta}{D(\eta)}\right]\frac{d\xi}{D(\xi)}$$

(Eq. 2)

where $C(x)$ is the solute concentration at any point, Co is the initial bulk alloy composition and V is the velocity with which the boundary is moving. Xo is an arbitrary value of the co-ordinate x which is required for integration but does not enter the final result.

In light of the uncertainty in the descriptions of ${}^{o}G(x)$, $D(x)$ and the solution thermodynamics, Cahn chose the simplest descriptions that capture the physics of his proposed treatment to illustrate the resulting concentration profiles. He assumed a wedge shaped interaction energy for the grain boundary (Fig. 1), a constant diffusivity for the solute and ideal (dilute) solution thermodynamics in the vicinity of the grain boundary. To illustrate this, we have evaluated the solute concentration profile across a moving grain boundary, under the same assumptions (using Eq. 2) for a range of dimensionless velocities (Vw/D), where w is the half-width of the interface, assuming an attractive interaction between the solute and the boundary (Fig. 3). The solute profile can be readily evaluated numerically for alternative choices of $^oG(x)$, $D(x)$ and solute thermodynamics using the solution to the diffusion equation provided by Hillert [4].

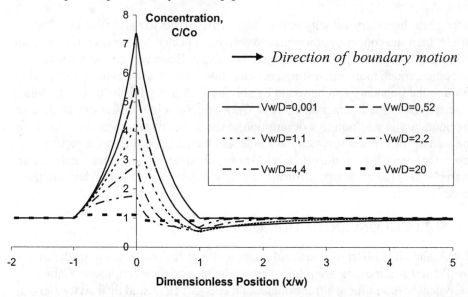

Figure 3. Solute profile through a grain boundary moving with velocity V, for which there is an attractive interaction with the solute. A wedge shaped interaction $G(x)$, a constant diffusivity D, and ideal (dilute) solution thermodynamics are assumed.

It is clear from Fig. 3 that the solute profiles are no longer symmetric about the moving boundary, as was the case for the stationary boundary (Fig. 2), and that the solute concentration significantly ahead of the interface is perturbed by the motion of the interface.

2.2. INTERPRETATION & QUANTIFICATION OF THE SOLUTE EFFECT

2.2.1   J. W. Cahn: The Force Approach

Cahn [2] and Lucke and Stuwe [8] argued that an impurity atom such as that shown in Fig. 2, will be attracted to the centre of the boundary with a force $d(^oG)/dx$. The solute atom exerts an equal and opposite force on the boundary and the net force can be found by summing the forces from each of the atoms over the entire boundary. Clearly, for a stationary boundary with the symmetric equilibrium solute concentration profile (Fig. 2), the number of atoms pulling the boundary to the left and the number pulling it to the right are equal and the net force sums to zero. For the asymmetric profiles in Fig. 3, the magnitude of the force pulling the boundary to the left is greater than that pulling it to the right and a net drag force results. Lucke and Stuwe [8] summed the forces in the following way:

$$P = -\int_{-\infty}^{+\infty} \frac{C}{V_m} \frac{d(^oG)}{dx} dx \qquad \text{(Eq. 3)}$$

where P is the net drag pressure, V_m is the molar volume and C is the atomic fraction of solute.

Cahn subtracted Co from C and used Eq. 4 for the evaluation of the net force. This is allowed because the interaction energy ^oG, is the same on both sides of the grain boundary far from the interface, i.e. $\int_{-\infty}^{+\infty} Co \cdot \left(\frac{d(^oG)}{dx} \right) dx = 0$.

$$P = -\int_{-\infty}^{+\infty} \frac{(C - Co)}{V_m} \frac{d(^oG)}{dx} dx \qquad \text{(Eq. 4)}$$

It can be seen from these equations that a contribution to the drag pressure arises only from those regions where a gradient in the interaction energy exists, i.e. where $d(^oG)/dx$ is non-zero. The integration is made only over the width of the interaction energy well. This is the case even though the concentration profile well outside this region can be affected by the motion of the boundary (Fig. 3).

2.2.2   M. Hillert: The Dissipation Approach

Hillert's [3,4] interpretation of the retarding effect of the solute differs from that of Cahn. Hillert argues that the retarding effect of the solute on the moving boundary corresponds to some work done by the boundary. He identifies this work as a dissipation of Gibbs free energy (ΔG^{diss}) due to the irreversible nature of diffusion. Hillert considers the total chemical potential rather than only the energy and derives the following expression for the dissipation of Gibbs energy.

$$\Delta G^{diss} = PV_m = -\int_{-\infty}^{+\infty} (C - Co)\frac{d\mu}{dx} dx, \quad \mu \text{ is the chemical potential} \qquad \text{(Eq. 5)}$$

Hillert [9] points out that for a stationary grain boundary with the equilibrium concentration profile, it is obvious that there is no force acting on the solute atoms since no gradients in chemical potential exist at equilibrium. According to the dissipation approach, a contribution to the retarding effect of the solute arises wherever there is a deviation of the concentration from the equilibrium value. This includes the region in front of the boundary where $d(^oG)/dx$ is zero, but where the concentration is removed from the equilibrium value, Co (Fig. 3), giving non-zero values for $d(\mu_A - \mu_B)/dx$. This is apparently different to the approach of Cahn where the integration occurs only over the region where $d(^oG)/dx$ is non-zero.

2.3 COMPARISON OF THE FORCE AND DISSIPATION APPROACHES

Cahn chose to describe the thermodynamics of the alloy system as ideal (dilute) when deriving his expression for the solute concentration profile (Eq. 2). Hillert [4] has shown that when the dependency of the chemical potential on position and composition come through separate terms, then the dissipation approach of Hillert and the force approach of Cahn give identical results. Hillert illustrated this by considering the dilute solution used by Cahn.

For a dilute solution of B atoms in A

$$\mu_A(x) = ^oG_A(x) + RT \cdot \ln[C_A(x)] \qquad \text{Raoult's Law} \qquad \text{(Eq. 6)}$$

$$\mu_B(x) = ^oG_B(x) + RT \cdot \ln[C_B(x)] \qquad \text{Henry's Law} \qquad \text{(Eq. 7)}$$

$$\frac{d(\mu_B - \mu_A)}{dx} = \frac{d(^oG_B - ^oG_A)}{dx} + \frac{RT}{C_A C_B} \frac{dC_B}{dx} \qquad \text{(Eq. 8)}$$

Substitution into Hillert's expression for the free energy dissipation (Eq. 5) yields:

$$P = -\int_{-\infty}^{+\infty} \frac{(C_B - Co)}{V_m} \cdot \frac{d\mu}{dx} dx = -\int_{-\infty}^{+\infty} \frac{(C_B - Co)}{V_m} \cdot \frac{d(^oG_B - ^oG_A)}{dx} dx$$

$$-\int_{-\infty}^{+\infty} \frac{RT}{V_m} \frac{(C_B - Co)}{C_A C_B} \cdot \frac{dC_B}{dx} dx \qquad \text{(Eq. 9)}$$

The second term on the right hand side vanishes when the composition far from the interface is the same on both sides, as is the case for grain boundaries. Hillert points out that this will always be the case when the dependence of the chemical potential on position and composition come through different terms. When the second term in Eq. 9 vanishes, the dissipation expression reduces exactly to the expression used by Cahn to evaluate the solute drag (Eq. 4). Under these conditions, the treatments, although arrived at by different reasoning, give identical results.

The drag has been evaluated as a function of velocity under these simplified conditions and is shown below (Fig. 4). The important characteristics are that the drag increases from zero for a stationary grain boundary, goes through a maximum and then decreases with increasing boundary velocity. This simply reflects the symmetry of the solute concentration profile when $V \to 0$ (equilibrium) and when $V \to \infty$ (flat profile).

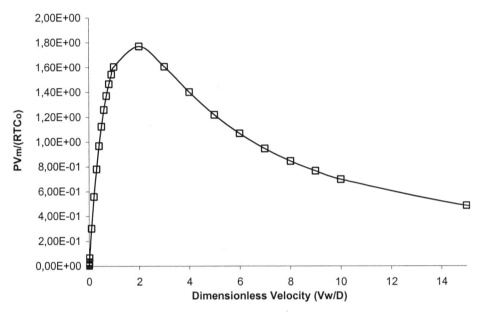

Figure 4. Plot of the Drag Force as a function of dimensionless velocity for a moving grain boundary assuming a wedge shaped interaction profile $^oG(x)$, a constant solute diffusivity D and dilute solution thermodynamics.

3. Solute Drag Models for Interphase Boundaries

Hillert considered both grain and interphase boundaries when evaluating the retarding effect of solute using 'the dissipation approach'. The only restriction was that the solute concentration (but not chemical potential) must be the same on both sides of the boundary removed from the interface. Cahn never discussed the application of 'the force approach' to interphase boundaries. It was Purdy and Brechet [6] who first extended Cahn's treatment in this direction when they applied it to the growth of proeutectoid ferrite from austenite in Fe-C-X systems. They used Eq. 4 to sum the contribution to the drag from each of the solute atoms, even though the inclusion of Co in Eq. 4 (derived for grain boundaries) was made under the assumption that the interaction energy $^oG(x)$, between the solute and the matrix is the same in the grains on both sides of the interface. This is not the case for interphase boundaries and it would appear at first sight that Eq. 4 is not applicable to interphase boundaries. Instead, we may expect that Eq. 3, presented by Lucke and Stuwe [8], would be a better choice for the evaluation of the solute drag effect, via the force approach. However, Hillert [4,10] demonstrated that it is Eq. 4 and not Eq. 3 that is applicable to phase interfaces and has recently shown that even though 'the force approach' and 'the dissipation approach' integrate over different volumes, under conditions where the dependence of the solute chemical potential on position and composition enter through different terms (as in the ideal or dilute case), the two treatments are identical. This is demonstrated below (Fig. 5) where we have calculated the solute drag (Eq. 4) and the dissipation of Gibbs free energy (Eq. 5) over a moving interphase boundary. Clearly, the two treatments give identical results under these conditions.

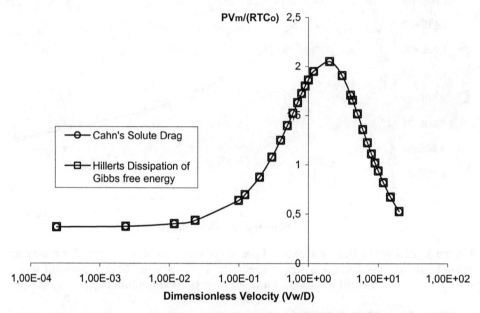

Figure 5. Cahn's solute drag and Hillert's dissipation of Gibbs free energy as a function of interphase boundary velocity. A wedge shaped interaction profile $^{o}G(x)$, a constant diffusivity D, and dilute solution solute thermodynamics were assumed.

4. Comparison of Model Predictions with Experimental Observations of Boundary Motion.

Quantitative comparisons between model predictions and the experimental observations of boundary motion are limited in two ways. The first is the uncertainty surrounding the choice of input parameters ($^{o}G(x)$, $D(x)$ and solution thermodynamics). There is currently little guidance available for suitable choices of these parameters and as a result the simplest descriptions are usually chosen. Hillert [4] has shown that these choices can have a strong influence on the calculated solute drag. The second limitation stems from a lack of suitable experimental data. Ideally, boundary migration data would be obtained from very high purity materials with only one impurity addition but because of the technical difficulties associated with alloy purification this is rarely the case. Most experimental alloys contain varying amounts of several impurities and the resulting solute drag effect is a convolution of each of the additions. It is because of these limitations in comparing theory with experiment that it is difficult to critically evaluate the theoretical treatments. However, there is one criticism that deserves particular attention. This is the question of the validity of approximating the atomistic structure of the grain boundary by simple continuous functions ($^oG(x)$ and $D(x)$) that migrate in a planar manner. The continuum solute-drag models have so far been qualitatively compared with some limited experimental data, e.g. Cu in Al and Fe in Au (see [11] for a discussion) and they do appear to give acceptable qualitative descriptions of the effect of temperature and solute content on boundary motion. Considering the uncertainty in the input parameters this suggests that, at least in a qualitative sense, this approximation is not too bad. If and when better guidance on the choice of ^oG, D and solute thermodynamics and suitable experimental boundary migration data becomes available, this approximation will require re-visiting. In the case of a critical quantitative comparison, important questions concerning the atomic structure of the migrating boundary that will need to be considered include; What is the atomic mechanism of grain boundary migration and it's dependence on driving force (velocity)?; What is the effect of the boundary atomic structure on the solute interaction?; What is the subsequent effect of segregated solute on the boundary structure and migration mechanisms?; How do these effects vary with boundary orientation? etc.

5. Future Directions for Research

Research efforts in this field are probably now best directed towards facilitating a critical quantitative comparison between theory and experiment. This would include work concerning both the choice of input parameters and experimental boundary migration experiments in high purity materials.

The control of impurity content during alloy preparation is now becoming very good and this is particulary evident in the microelectronics industry. It may now be possible to conduct grain boundary migration experiments over a wide boundary velocity range (using recrystallisation and grain growth experiments) in ultra high purity materials with carefully controlled impurity additions. Furthermore, the technological advances necessary to experimentally measure the solute concentrations at boundaries are now also available using three-dimensional atom probe field ion microscopy (3D-APFIM) and some analytical transmission electron microscopy techniques, e.g. [12,13]. These measurements as a function of boundary velocity over a wide velocity range in suitable materials would greatly aid the quantitative comparison between theory and experiment.

A comparison also requires the identification of appropriate input parameters, $^oG(x)$, $D(x)$ and the solute thermodynamics in the vicinity of the boundary. Useful efforts in this direction might include the use of atomistic calculation techniques such as molecular dynamics. In particular, guidance on the continuous forms of $^oG(x)$, $D(x)$ and of the solute thermodynamics that are likely to best reflect the real

164

boundary would contribute greatly to 'solute-drag' theory. Such calculations may also provide some insight into the expected role of the atomic boundary structure on boundary motion and solute segregation.

Acknowledgements

CRH greatly acknowledges the financial support of IRSID.

References

1. Lucke, K. and Detert, K., (1957) A quantitative theory of grain boundary motion and recrystallisation in metals in the presence of impurities, *Acta Metall.*, **5**, 628-637.
2. Cahn, J. W., (1962) The impurity-drag effect in grain boundary motion, *Acta Metall.*, **10**, 789-798.
3. Hillert, M., (1969) The role of interfaces in phase transformations, <u>Monograph and Report Series</u>, No. 33, 231-247, Inst. of Metals.
4. Hillert, M. and Sundman, B., (1976) A treatment of the solute drag on moving grain boundaries and phase interfaces in binary alloys, *Acta Metall.*, **24**, 731-743.
5. Kinsman, K. R and Aaronson, H. I., (1967) Transformations and hardenability in steels, Climax Molybdenum Corp., Ann Arbor, MI.
6. Purdy, R. R and Brechet, Y., (1995) A solute drag treatment of the effects of alloying elements on the rate of the proeutectoid ferrite transformation in steels, *Acta metall. mater.*, **43**, 3763-3774.
7. Enomoto, M. (1999) Influence of solute drag on the growth of proeutectoid ferrite in Fe-C-Mn alloys, *Acta mater*, **47**, 3533-3540.
8. Lucke, K and Stuwe, H., (1963) Recovery and Recrystallisation, p. 131, New York, Interscience.
9. Hillert, M., (1979) Grain boundary mobility in solid solution alloys, *Metal Science*, March-April, 118-124.
10. Hillert, M., Odqvist, J. and Agren, J., (2001) Comparison between solute drag and dissipation of Gibbs energy by diffusion, *Scripta mater.*, **45**, 221-227.
11. Martin, J. W., Doherty, R. D. and Cantor, B., (1997) *Stability of Microstructure in Metallic Systems,* pp. 196-202, Cambridge Solid State Science, Cambridge.
12. Maruyama, N., Smith, G. D. W. and Cerezo, A., (2002) Interaction of the solute niobium or molybdenum with grain boundaries in α-iron, submitted to Materials Science and Engineering A.
13. Fletcher, H. A., Garratt-Reed, A. J., Aaronson, H. I., Purdy, G. R., Reynolds, W. T. and Smith, G. D. W., (2001) A STEM method for investigating alloying element accumulation at austenite-ferrite boundaries in an Fe-C-Mo alloy, *Scripta Mater.*, **45**, 561-567.

ON THE JERKY NATURE OF MARTENSITIC TRANSFORMATION

G.ANANTHAKRISHNA (`garani@mrc.iisc.ernet.in`)
Materials Research Centre, Centre for condensed Matter Theory, Indian Institute of Science, Bangalore-560012, India.

S.SREEKALA
Materials Research Centre, Indian Institute of Science, Bangalore-560012, India.

RAJEEV AHLUWALIA
Jawaharlal Nehru Centre for Advanced Scientific Research, Jakkur, Bangalore.
Materials Research Centre, Indian Institute of Sciences, Bangalore-560012, India.
Presently at Theoretical Division, Los Almos National Lab, Los Almos, New Mexico 87545, USA.

Abstract. We devise a two dimensional model of martensitic transformation for the observed jerky nature of the transformation reflected in acoustic emission (AE) experiments, in particular the power law statistics of the AE signals. Our model includes inertial effects, dissipation, long-range interaction between the transformed domains apart from a local sixth order free energy. An inhomogeneous stress field is used to describe the effect of lattice defects which serve as nucleation centers. Using a Lagrangian formalism, we drive an equation of motion for a strain order parameter. Both single-site nucleation and multi-site nucleation are studied for a single quench situation as well as thermal cycling situation. The latter shows the existence of hysteresis in the transformation. More importantly, the rate of energy dissipated occurs in the form of bursts with power law behaviour for the amplitudes and their durations. The morphological features are similar to those in real systems.

Keywords: martensitic transformation, hysteresis, Acoustic emission, powerlaw

1. Introduction

Martensitic transformation is considered as an atypical first order transformation as it exhibits some features of second order transitions. For instance the precursor effect [1] and power law distribution of AE signals, both are signatures of critical fluctuations in a second order transition. The power law distributions of AE signals was demonstrated in a set of careful AE experiments on Cu-Al-Zn singles crystals subjected to a broad range of cooling and heating rates [2]. There has been attempts to simulate power law behaviour of avalanches occuring in first order transitions using disorder based Ising models [3] where the power law behaviour of avalanches arises in the presence of a critical amount of disorder. Since quenched-in-disorder (defects) plays an essential role in the nucleation process of martensitic transformation, it appears that these kinds of models [3] may be relevant to martensitic transformation.

167

A. Finel et al. (eds.), Thermodynamics, Microstructures and Plasticity, 167–178.
© 2003 *Kluwer Academic Publishers. Printed in the Netherlands.*

However, by subjecting the system to repeated hysteresis cycles, these authors ascertained that the system evolves towards the critical state independent of the initial treatment of the alloy suggesting a dynamical evolution of the system towards such a state. According to these authors, the interpretation is that in real martensites, although there are quenched-in defects, the disorder that is responsible for the power law is actually generated during the transformation itself.

Thus, one has to look for an alternate explanation. This has been provided recently by a model which is [4] based on the concept of self-organized criticality (SOC) introduced by Bak *et al.* [5]. The essential idea of the model is to incorporate the typical features of systems evolving to a SOC state, namely, slow driving, threshold dynamics, approriate relaxational mode and lack of tuning of any relevant parameter. Power law statistics can arise in a large class of slowly driven spatially extended dissipative systems and hence they are ubiquitous. Such systems naturally evolve towards a marginally stable state [5]. Since the introduction of this concept, there are several reports of physical systems exhibiting SOC features, for example, earthquakes [6] acoustic emission from volcanic rocks [7], stress drops during the Portevin Le-Chatelier effect [8, 9, 10] and biological evolution [11], to name a few. The purpose of the present paper is to present the results of extensive numerical simulations on this two dimensional model describing square-to-rectangle martensitic transformation [4]. As we shall see, number of features of martensitic transformation are typical of SOC systems.

The martensitic transformations are first order, solid-solid, diffusionless structural phase transformations. Starting from a higher symmetry parent phase, nucleation of thin plate like product domains with internal twinned structure is seen. (Nucleation of the product phase is athermal usually occurring at defects such dislocations [12, 13, 14, 15]). The difference between the unit cell structure of the parent and product phases leads to long-range strain fields which block the transformation leaving the system in a two phase metastable state. Given a certain quench, the amount of transformed phase is fixed and further undercooling is required for additional growth. (For the same reason, the transformation occurs in a broad range of temperatures.) This also implies that thermal flucatuations play little role in the transformation kinetics. The physical mechanism attributed to hysteresis is the relaxation of the system through a sequence of metastable states accompanied by dissipation [16, 4]. Part of the increase in the driving force arising from the decrease in temperature goes in surmounting the barrier and creating new interfaces, and the rest of the energy is dissipated in the form of AE signals due to the rapid motion of the interface [4].

2. The Model

We briefly recall the model for 2D square-to-rectangle transition. The free energy is a function of all the three components of strain defined by the bulk dialation strain $\epsilon_1 = (\eta_{xx} + \eta_{yy})/\sqrt{2}$, the deviatoric strain $\epsilon_2 = (\eta_{xx} - \eta_{yy})/\sqrt{2}$, and the shear strain $\epsilon_3 = \eta_{xy} = \eta_{yx}$. Here, $\eta_{ij} = \frac{1}{2}(\frac{\partial u_i}{\partial x_j} + \frac{\partial u_j}{\partial x_i})$ refers to the components of the strain tensor and u_i's are the components of the displacement field in the direction i ($i = x, y$). In our model, we use only the deviatoric strain $e_2 = \epsilon(\vec{r})$ as the principal order parameter since volume changes are usually small [12, 13, 14]. The effect of other components of the strain, i.e., the bulk and shear strain is accounted phenomenologically by considering a long-range interaction between the deviatoric strains.

The scaled free-energy functional is given by $F\{\epsilon(\vec{r})\} = F_L\{\epsilon(\vec{r})\} + F_{lr}\{\epsilon(\vec{r})\}$, where F_L is the local free energy and F_{lr} is an effective long-range term that describes the interaction between the transformed domains. The local free energy is given by

$$F_L = \int d\vec{r}\left[f_l(\epsilon(\vec{r})) + \frac{\alpha}{2}(\vec{\nabla}\epsilon(\vec{r}))^2 - \sigma(\vec{r})\epsilon(\vec{r})\right], \qquad (1)$$

where both α and σ are in scaled form. In Eqn. (1), $f_l(\epsilon(\vec{r}))$ is taken as the usual Landau polynomial for a first order transition given by $f_l(\epsilon(\vec{r})) = \frac{\tau}{2}\epsilon(\vec{r})^2 - \epsilon(\vec{r})^4 + \frac{1}{2}\epsilon(\vec{r})^6$, where $\tau = (T - T_c)/(T_0 - T_c)$ is the scaled temperature. T_0 is the first-order transition temperature at which the free energy for the product and parent phases are equal, and T_c is the temperature below which there are only two degenerate global minimia $\epsilon = \pm\epsilon_m$. It is known that nucleation of the martensite domains occurs at localized defect sites [15]. This is mimicked by an inhomogeneous stress field, $\sigma(\vec{r})$ [17]. This term modifies the free-energy f_l and can locally render the austenitic phase unstable leading to the nucleation of the product phase. Long-range interactions between the transformed domains (F_{lr}) are expected to arise due to coupling with the other components of the strain order parameter [1, 18]. This is included in the model in a phenomenological way by considering the symmetry allowed long-range interaction between the order parameter [19]. Following this, in the Fourier space, it is given by

$$F_{lr}\{\epsilon\} = \int d\vec{k}B(\vec{k}/k)\{\epsilon^2(\vec{r})\}_k\{\epsilon^2(\vec{r})\}_{k^*}, \qquad (2)$$

where $\{\epsilon^2(\vec{r})\}_k$ and $\{\epsilon^2(\vec{r})\}_{k*}$ are the Fourier transform of $\epsilon^2(\vec{r})$ and its complex conjugate respectively. For square-to-rectangle transformation, the favorable directions of growth of martensite domains are [11] and [1$\bar{1}$] with lowest free energy barriers in these directions, and highest in the [10] and [01] directions. A simple choice of $B(\vec{k})$ having these features is $B(\vec{k}/k) = -\frac{1}{2}\beta\theta(k - \Lambda)\hat{k}_x^2\hat{k}_y^2$, where $\hat{k}_x$ and $\hat{k}_y$ are the unit vectors in the x and y directions, and β is the strength of the interaction. The step function $\theta(k - \Lambda)$ has been introduced to impose a cutoff on the range of the interaction.

Acoustic emission during the transformation implies that inertial effects are important [20]. This is included through the kinetic energy of the displacement fields $\mathcal{T}$ defined by

$$\mathcal{T} = \int d\vec{r}\rho\left[\left(\frac{\partial u_x(\vec{r},t)}{\partial t}\right)^2 + \left(\frac{\partial u_y(\vec{r},t)}{\partial t}\right)^2\right], \tag{3}$$

where ρ is the mass density. Experiments show that the AE activity depends on the acceleration of the microplates. Since, the parent-product interface moves in the parent phase, it is associated with dissipation which we represent by the Rayleigh dissipative functional Further, since deviatoric strains are the dominant order parameter fields, we represent the dissipative functional entirely in terms of the deviatoric strains [20]. Thus,

$$R = \frac{1}{2}\gamma \int d\vec{r}\left(\frac{\partial}{\partial t}\epsilon(\vec{r},t)\right)^2. \tag{4}$$

This is consistent with the fact that shear and bulk strains are known to equilibrate rapidly and hence do not contribute. The equations of motion for the displacement fields are calculated using the Lagrangian $L = \mathcal{T} - F$, where F is the total free energy through

$$\frac{d}{dt}\left(\frac{\delta L}{\delta \dot{u}_i}\right) - \frac{\delta L}{\delta u_i} = -\frac{\delta R}{\delta \dot{u}_i}, \quad i = x, y. \tag{5}$$

After scaling out ρ and α, the equation of motion for the deviatoric strain $\epsilon(\vec{r}, t)$ is given by

$$\frac{\partial^2}{\partial t^2}\epsilon(\vec{r},t) = \nabla^2\left[\frac{\delta F}{\delta\epsilon(\vec{r},t)} + \gamma\frac{\partial}{\partial t}\epsilon(\vec{r},t)\right], \tag{6}$$

where, β and γ now refer to the scaled parameters. The structure of Eqn. (6) is similar to that derived in [20] for 1-d except for the long-range term. There are three adjustable parameters in the model, namely, the scaled temperature τ, the strength of the long-range interaction β and that of dissipation γ.

3. Numerical Simulations

We now describe the results of our numerical simulation on the morphological features. We discretize Eqn. (6) on a $N \times N$ grid using the Euler's scheme with periodic boundary conditions. The mesh size of the grid is $\Delta x = 1$ and the smallest time step $\Delta t = 0.002$. Most results reported here correspond to $N = 128$, 256 or 512. A psuedo-spectral technique is employed to compute the long-range term [4]. In all simulations reported in the paper, the cutoff Λ in the long-range interaction is chosen to be 0.2. An inhomogeneous stress field $\sigma(\vec{r})$ is appropriately chosen to describe the defect configuration (see below). We consider nucleation at a single defect site and at several defect sites.

3.1. SINGLE-DEFECT AND MULTI-DEFECT NUCLEATION

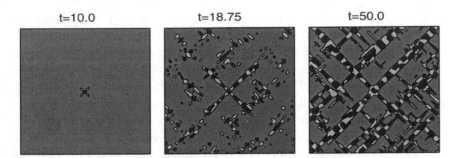

Figure 1. Snapshots of the morphology for a single defect situation at selected intervals of time, N=128

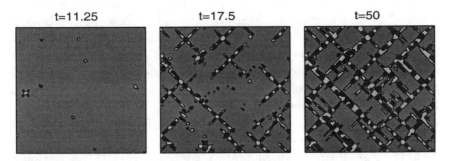

Figure 2. Snapshots of the morphology for multi-defect nucleation at selected intervals of time, N=128

The multi-defect nucleation case is mimicked by assuming a random distribution of defects, described by the stress field

$$\sigma(\vec{r}) = \sum_{j}^{j_{max}} \sigma_0(\vec{r_j}) exp\left(\frac{-|\vec{r} - \vec{r_j}|^2}{\zeta^2}\right),\qquad(7)$$

where $\vec{r_j}$ represents the coordinates of randomly chosen defect sites and j_{max} the total number of defect sites. Clearly, $j_{max} = 1$ represents the single-defect case. For the multi-defect case, we choose $j_{max} = 163$ (nearly one percent of the total number of sites and N $=128$) and consider $\sigma_0(\vec{r_j})$ to be uniformly distributed in the interval $[-0.3, 0.3]$.

First we consider a single isotropic defect with it's core located at the centre. At $t = 0$, we start with $\epsilon(\vec{r}, 0)$ distributed in the interval $[-0.005, 0.005]$ representing the austenite phase and simultaneously turn on the stress field $\sigma(\vec{r})$ as we quench the system to $\tau = -2.0$. Although, a single spherically symmetric defect is a rather artificial system, it is useful to clarify the physics of nucleation and growth in martensitic systems. With a value of $\sigma_0 = 0.3$, the system becomes locally unstable at the core $\vec{r}_0$. The parameters chosen for the simulations are $\beta = 50$, $\gamma = 4$ and $\zeta = 1$. Figure 1 shows the nucleation and growth of the martensite domains from the defect core at various instants of time. (Grey regions represent the austenite phase $\epsilon = 0$, black and white regions represent the two variants characterized by $\epsilon = \pm\epsilon_m$.) In a short time $t \sim 10$, we find the emergence of a small nucleus of the positive having $\epsilon = \pm\epsilon_m$. In addition, we also see the emergence of domains of the other variant ($\epsilon = -\epsilon_m$) adjacent to the nucleus in the [11] and [1$\bar{1}$] directions. The structure further develops into twinned arrays, propagating along [11] and [1$\bar{1}$] directions. Observation of the snap shots reveals an interesting feature, namely, the creation of a nuclei ($\epsilon = \pm\epsilon_m$) located at a finite distance from the propagating arrays ($t = 18.5$). This can be attributed to the accumulation of long-range stress fields at these sites. These new nuclei give birth to secondary fronts that also propagate along [11] and [1$\bar{1}$]. These observations in our simulation are in accordance with a similar collective nucleation behaviour seen in experiments [15] and in simulations [19]. The propagation of these new secondary fronts continues till they 'collide' with pre-existing martensite domains and stop ($t = 25$). The evolution practically ceases around $t = 50$.

Now consider the multi-defect nucleation. In Fig. 2, we show the time evolution of the system at specifically choosen instants of time. As in the case of single-site nucleation, around $t = 11.25$, nucleation of the product phase is seen to occur at several sites. Soon these nuclei grow into twinned lenticular shape. Several additional nuclei emerge

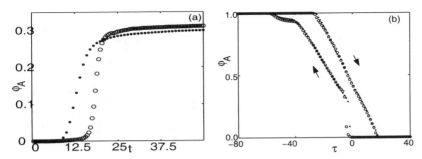

Figure 3. (a) Growth of area fraction for a single quench situation: (○) multisite, (●) signle site, N=128. (b) Hysteresis cycle under thermal cycling: (○) multisite, (●) single defect case.

at finite distances from these original domains. These new nucleation sites most often coincide with the pre-existing defect sites. However, occasionally, nucleation does occur at sites where there are no defects, due to stress accumulation arising from long-range term, as in the case of single-site nucleation. There is a rapid growth of the domains forming an irregular criss-crossed pattern of martenstic domains. We find that there is very little growth beyond $t = 20$, and by $t = 50$ the growth practically stops. The adaptive nature of the nucleation observed in the single-defect nucleation case is evident in this case also. A comparison of the final morphology with the corresponding single-defect nucleation shows that they are similar, implying that it is independent of the original defect (stress-field) configuration. We also see thin needle-like structures emerging from the larger domains. The final configuration has domains of several sizes.

In Fig. 3a, we have shown the evolution of the area fraction ϕ of the martensitic regions for both the single (●) and multi-defect cases (○) for a single defect quench situation. As can be seen, the growth of ϕ starts earlier for the single-defect case and takes longer time to saturate compared to the multi-defect case for identical conditions. This can be attributed to the nucleation occuring at several sites in the latter case. But the saturation value for the multi-defect case is only marginally higher than the single defect case. This might suggest that quenched-in defects play a role only in the intial stages of the growth.

4. Thermal Cycling and Hysteresis

One key feature of the martensitic transformation is hysteresis under thermal cycling conditions. We have performed 'continuous cooling' and 'heating' simulations where we change τ at a constant rate: the

174

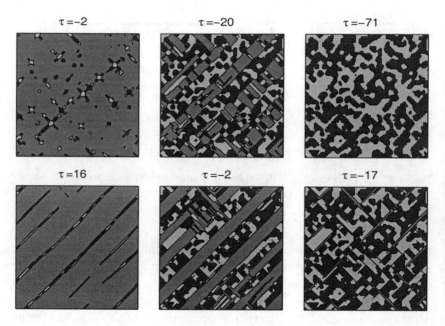

$\tau = -2$ $\tau = -20$ $\tau = -71$

$\tau = 16$ $\tau = -2$ $\tau = -17$

Figure 4. Snap shots of the morphology (clock wise) during thermal cycling.

interval $\tau = 40$ to -80 is cooled (or heated from -80 to 40) in 1000 time steps. We have monitored both the morphology and the area fraction of the transformed phase for single-defect nucleation as well as multi-defect nucleation cases. We use the same initial conditions for both these situations as that used for a single quench situation. The initial condition for the reverse transformation is the final configuration of the cooling run.

Figure 3b shows the variation of the area fraction ϕ with τ for the heating and the cooling runs for the single defect case ($\bullet$) and multi-defect case ($\circ$) for $N = 252$. In the cooling run, for the multi-defect case, the transformation starts around $\tau_{ms} \sim -2.0$ showing a rapid increase in ϕ ($\sim 30\%$). Thereafter, there is a nearly linear increase up to 90% at which the growth rate tapers off. The system gets fully transformed at $\tau_{mf} \sim -60$. In the heating run, the reverse transformation does not start till $\tau_{as} \sim -22.0$ and thereafter, ϕ decreases almost in a linear way till the transformation is nearly complete around $\tau_{af} \sim 18$. As can be seen from the figure, the difference between the hysteresis cycles corresponding to the multi-site defect, single-defect nucleation is small except that the initial increase in ϕ is faster for the single-site case.

We have followed the morphological evolution of the martensitic domains both for the single and multi-defect cases. However, we shall only discuss the more realistic multi-defect case. Figure. 4, shows snap shots of the morphology at specifically chosen values of τ. In the snapshot corresponding to $\tau = -2$, one can see the nucleation of martensitic domains at multiple locations. As the system is further 'undercooled' to lower values of τ, not only does these twins propagate in the [11] and [1$\bar{1}$] directions, the thickness of the martensite domains also increases reaching a saturation value around $\tau = -59$. The near final configuration of the cooling run is shown for $\tau = -71$. One can notice that the domain walls are curved under the action of mutual interaction of the various domains.During heating some regions of the austenite phase reappear as can be seen from the morphology corresponding to $\tau = -17$. As τ is further increased, we find that the martensite phase gradually disappears. This is shown in the snapshots corresponding to $\tau = -2$ and 16. Further, one can see that the overall morphology in the final stages of the heating run is significantly different from the initial stages of the cooling run. Equivalently, in our model, there is no long term memory effects, through there is short term memory. Finally, by $\tau = 18.0$, martensite phase completely disappears.

5. Power Law Statistics of Energy Release During Thermal Cycling

From Fig. 3b, it appears that changes in the area fraction ϕ are smooth on the scale shown in the figure. In reality, the changes in ϕ are actually jerky. Jerky nature of the transformation is better understood in terms of the rate of energy dissipated $R(t) = -dE/dt$. We have calculated $R(t)$ during the heating and cooling runs. Figure 5a, shows $R(t)$ as a function of t with the inset showing the enlarged section of the peak. The figure clearly shows that the rate of energy release occurs in bursts consistent with acoustic emission studies [2] during thermal cycling. Similar spiky behaviour is seen for the heating run also.

Since experiments demonstrate that the AE signals show power law statistics, we have investigated the distributions of the amplitudes of the AE signals and their durations. Denoting the amplitude of $R(t)$ by R_A, we find that the distribution of R_A has a tendency to approach a power law $D(R_A) \sim A^{-\alpha_R}$ with an exponent α_R. Figure 5b shows a log-log plot of $D(R_A)$ as a function of R_A, for both the single-site case ($\bullet$) and the multi-defect case ($\circ$). It is clear that both these cases exhibit the same exponent value $\alpha_R \sim 2.5$ over one and half orders of R_A. Similarly, we have also plotted the distribution $D(\Delta t)$ of the durations Δt of energy bursts for both the single and multi-defect cases. We find that $D(\Delta t) \sim \Delta t^{-\beta_R}$ with an exponent value $\beta_R \approx 3.2$, although, the scaling regime is not as impressive as for R_A. (Even in models of SOC,

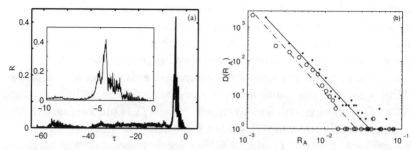

Figure 5. (a)Plot of energy dissipated during cooling run. (b) Plot of the distribution of R_A.

the scaling regime for the durations of the events is much smaller than that for the amplitudes[5].) We have also calculated the conditional average $< R_A >_c$ for a given value of Δt. This is expected to obey a power law given by $< R_A >_c \sim \Delta t^{x_R}$. The value we get is about $x \approx 1.36$ for both these cases. Using these values, we find that the scaling relation $\alpha = x(\beta - 1) + 1$ is satisfied quite well. It must be mentioned that in experiments, one actually measures the exponents corresponding to the amplitude of the AE signals, ie., $R_A \sim A_{AE}^2$ where A_{AE} is the amplitude of the AE signal. Using the relation between the two joint probability distributions $D(R_A, \Delta t) \propto D(A_{AE}, \Delta t)/A_{AE}$, one easily finds that $\alpha_R = (\alpha_{AE} + 1)/2$, with the other two exponent values remaining unchanged. Using the experimental values [2] of $\alpha_{AE} \approx 3.8$, ($\tau_{AE} \sim 3.8$ and $x_{AE} \sim 1$), we see that the calculated value of $\alpha_R \approx 2.4$. Considering the fact that our model is two dimensional, we see that the agreement of the exponent values is reasonable. We have also carried out a similar analysis on $R(t)$ for the cooling run. Even, though, the changes in $R(t)$ occurs in bursts, we find the scatter is considerably more than for the cooling run.

6. Summary and Discussions

In summary, we have presented the results of a comprehensive study of the dynamics of strain driven martensitic transformations within the framework of a two dimensional square-to-rectangle transition that explains some generic features of martensitic systems. One dominant feature that emerges from the model is that the elastic energy stored is released in the form of bursts. Second, the model provides a proper basis for explaining the power law statistics of the AE signals observed in experiments. Further, it is interesting to note that the morphological patterns bear good similarity with real micrographs. Patterns studied in the context of single-defect and multi-defect nucleation show that the eventual morphology is nearly independent of the the original defect configuration. Experiments of Vives *et al* appear to provide an indirect evidence [2]. In addition, the model shows hysteresis under thermal cycling. Even though, the model explains several generic features stated above, we stress that our model is not material specific.

Several feature of the model deserve comments in the lager context of SOC dynamics. The origin of the power law statistics in the model can be traced to the fact that we have included important ingredients of SOC dynamics, namely, the threshold dynamics, dissipation, the generation of large number of metastable states during the transformation, and a relaxation mechanism for the stored energy. The relaxation of the stored energy occurs at the scale set by the speed of sound. Compared to this the driving force increases with temperature at a slow pace which is characteristic to all SOC systems. Another important feature is the creation of large number metastable states during cooling or heating runs. This is direct consequence of an interplay between the local free energy (free energy barrier) and the long-range interaction between the transformed domains as can be seen from the following reasoning. The latter modifies the local free energy at specific points of space. We note that the value of this term at any location is the result of the superposition of the contributions arising from the spatial distribution of the already transformed domains which in turn leads to a complex terrain of local barriers (metastable states). It must be noted that these local thresholds are *self generated* (transformation induced). At a given time, these local thresholds must be overcome by the increase in the driving force arising from the slow cooling (or heating). We note that once a local barrier is overcome, part of the driving force goes in creating a new twin and the rest is dissipated in the form of burst of energy due to the advancing one or more interfaces. The fact that long-range interaction is at the root of creating the local thresholds is *further supported by the fact that we find a power law distributions even in the single-site nucleation case.* (See • in Fig. 5b.) The presence of defect sites only triggers the initial nucleation process. To the best of our knowledge this is the first model which explains the power law distribution of the energy bursts in martentic transformation.

References

1. Kartha, S., Krumhansl, J.A., Sethna, J.P., Wickham, L.K.: 1995, 'Disorder-Driven Pretransitional Tweed Pattern in Martensitic Transformations', *Phys. Rev. B* **Vol 52**, pp.803–822

178

2. Vives, E., Ortín, J., Mañosa, L., Rãfols, I., Pérez-Magrané, R., Planes, A.:1994, 'Distributions of Avalanches in Martensitic Transformations', *Phys. Rev. Lett* **Vol 72**, *pp.1694–1697*

3. Sethna, J.P., Dahmen, K., Krumhansl, J.A., Roberts, B.W., Shore, J.d.:1993, 'Hysteresis and Hierarchies: Dynamics of Disorder-Driven First-Order Phase Transformations', *Phys. Rev. Lett* **Vol 70**, pp.3347–3350

4. Ahluwalia, R., Ananthakrishna, G.:2001, 'Power-Law Statistics for Avalanches in Martensitic Transformation', *Phys. Rev. Lett.* **Vol 86**, pp.4076–4079

5. Bak, P., Tang, C., Wiesenfeld, K.:1988, 'Self-Organised Criticality', *Phys. Rev. A* **Vol 38**, pp.364–374

6. Carlson, J.M., Langer, J.S.:1989, 'Mechanical Model of an Earthquake Fault', *Phys. Rev. A* **Vol 40**, pp.6470–6484

7. Diodati, P., Marchesoni, F., Piazza, S.:1991, 'Acoustic Emission from Volcanic Rocks: An Example of Self-Organised Criticality', *Phys. Rev. Lett.* **Vol 67**, pp.2239–2243

8. Ananthakrishna, G., Noronha, S.J., Fressengeas, C., Kubin, L.P.:1999, 'Crossover from Chaotic to Self-Organized Critical Dynamics in Jerky Flow of Single Crystals', *Phys. Rev. E* **Vol 60**, pp.5455–5462

9. Bharathi, M.S., Ananthakrishna, G.:2002, 'Chaotic and Power Law States in the Portevin-Le Chatleier Effect', In print *Euro Phys. Lett.*

10. Bharathi, M.S., Lebyodkin, M., Ananthakrishna, G., Fressengeas, C., Kubin, L.P.: 2001, 'Multifractal Bursts in the Spatio-Temporal Dynamics of Jerky Flow', *Phys. Rev. Lett* **Vol 87**, pp.165508-1–165508-4

11. Bak, P.:1996, 'How Nature Works', *Springer- Verlag*, New york.

12. Christian, J.W., Olson, G.B., Cohen, M.:1995, 'Classification of Displacive Transformations: What is a Martensitic Transformation?', *J. de Physique C* **Vol 8** pp.3–10

13. Roitburd, A.L., Kurdjumov, G.V.:1979, 'The Nature of Martensitic Transformations', *Materials Science and Engineering* **Vol 39**, pp.141–167

14. Kachaturyan, A.G.:1983, *Theory of Structural Transformations in Solids* (Wiley, New York).

15. Ferraglio, P.L., Mukherjee, K.:1974, 'The Dynamics of Nucleation and Growth of a Thermoelastic Martensite in a Splat Quenched Au-47.5 at.% Cd Alloy', *Acta Metall.*, **Vol 22**, pp. 835–845

16. Vives, E., Ràfols, I., Mañosa, L.,Ortín, J., Planes, A.:1995, 'Statistics of Avalanches in Martensitic Transformations. I. Acoustic Emission Experiments', *Phys. Rev.B* **52**, pp.12644–12650

17. Cao, W., Krumhansl, J.A., Gooding, R.J.:1990, 'Defect-Induced Heterogeneous Transformations and Thermal Growth in Athermal Martensite', *Phys. Rev. B* **41**, pp.11319–11327

18. Shenoy, S.R., Lookman, T., Saxena, A., Bishop, A.R.:1999, 'Martensitic Textures: Multiscale Consequences of Elastic Compatiblity', *Phys. Rev. B* **Vol 60**, pp.R12537–R12541

19. Wang, Y., and Kachaturyan, A.G.:1997, 'Three-Dimensional Field Model and Computer Modeling of Martensitic Transformation', *Acta Mater.*, **Vol 45**, pp. 759–773).

20. Bales, G.S., Gooding, R.J.:1991, 'Interfacial Dynamics at a First-Order Phase Transition Involving Strain: Dynamical Twin Formation', *Phys. Rev. Lett.* **Vol 67**, pp.3412–3415

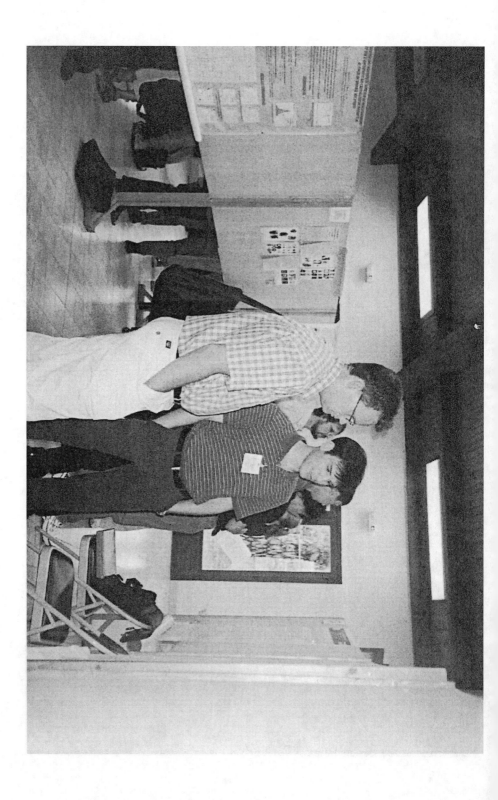

III. Microstructures and plasticity

THE DEVELOPMENT OF ULTRA HIGH STRENGTH MATERIALS BY MICROSTRUCTURAL REFINEMENT

J.D. EMBURY
Department of Materials Science and Engineering
McMaster University
Hamilton, Ontario, CANADA

1. Introduction

In the past five decades the basic understanding of the methods by which materials can be strengthened has increased to the point where simple models can be used to analyse the mechanical response of a variety of alloy systems. A number of excellent review articles[1,2] and textbooks[3,4] provide very comprehensive coverage of the topic.

However the increase in yield strength is only one aspect of the overall mechanical response of materials. It is important to consider a variety of other aspects such as the competition with fracture events, the rate of work hardening related to the stability of plastic flow, the stability of the microstructures used to develop high strength materials, and the path dependence of plastic flow including cyclic response. These aspects are important in the task of relating materials development to design and utilization of materials but although the basic understanding of a number of these topics is available[5,6] a comprehensive view remains to be developed.

An alternative viewpoint is to consider the theoretical strength of materials estimated from the lattice forces[4] and to compare these with the maximum strength level attained by a variety of mechanisms. This approach is illustrated in Table 1. The data indicates that strength levels close to the theoretical strength level can be attained by a variety of routes which produce ultra fine scale microstructures.

The purpose of this brief review article will be first to consider some basic aspects of strengthening mechanisms encompassing a range of features from microscopic to mesoscopic and macroscopic. In later sections these concepts will be related to the behaviour of practical engineering materials.

The final portion of the review will raise a variety of questions relating both to aspects of strengthening mechanisms and the development of future materials including those that combine both structural and functional properties.

A. Finel et al. (eds.), Thermodynamics, Microstructures and Plasticity, 183–203.

Table 1

Strengthening Mechanism	Material	Maximum Observed Strength	Theoretical Strength E/30
Precipitation Hardening	Al-Zn-Mg-Ag	750 MPa	2.3 GPa
Work Hardening	Cu deformed at 4.2K to large stress	1.7 GPa	6.3 GPa
Amorphous Solids	$Fe_{80}B_{20}$ metallic glass	3.6 GPa	5.6 GPa
Drawn Composites	Drawn pearlitic wire	5.6 GPa	7 GPa

2. Review of Strengthening Mechanisms

An initial if somewhat simplified view of the development of models of strengthening mechanisms can be considered as follows. The strengthening contribution can be considered following the seminal papers of Foreman and Makin[7] and Kocks[8] in terms of a dislocation moving through a dilute array of point obstacles (Figure 1). Three essential features of estimating the strength due to the obstacles are:

i) Considering the geometry of how flexible dislocations contact the obstacles

ii) The strength of individual obstacles is related to the angle through which the dislocation is bent in order to overcome the obstacle.

iii) The appropriate additivity laws if more than one set of obstacles is present.

In a very elegant treatment Brown and Ham[9] have shown that if the strength of the obstacle is represented by the angle of φ through which the dislocation is bent to overcome the obstacle and the spacing of obstacles L is written as $n^{-1/2}$ where n is the number of obstacles per unit area the critical strength overcome the obstacles can be written as

$$\tau_c = \frac{0.8\,Gb}{L} \cos\frac{\phi}{2} \qquad \phi \leq 100° \qquad (1)$$

and

$$\tau_c = \frac{Gb}{L}\left(\cos\frac{\phi}{2}\right)^{3/2} \qquad \phi \geq 100° \qquad (2)$$

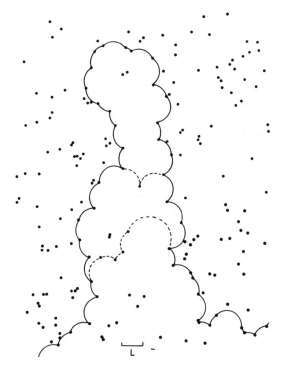

Figure 1   Schematic diagram of a dislocation passing through an array of point
obstacles (from ref. 9).

where G is the shear modulus and b the Burgers vector of the glide dislocation.

Clearly detailed consideration must be given to the nature of the dislocation particle interaction and this may reflect the surface energy, the ordering energy and the elastic properties of the particles. In general[7], the effects of the details of particle distribution on the resulting flow stress are small. However, there are important effects both of particle shape and orientation[10] They have been explored to a limited extent but certainly deserve further experimental and theoretical consideration.

The important distinction to be made is between the shearing of particles during plastic flow and the Orowan process in which loops are accumulated around particles or some from of local plastic location around the particles is enforced.

In most engineering materials a number of strengthening mechanisms such as solution hardening, precipitation hardening and strain hardening act in parallel and indeed the relative contributions can vary with strain or time. Thus the consideration of the appropriate additivity laws for a set of obstacles of strength φ_1 and number density n, together with other obstacles of strength φ_2 and number density n_2 is very important and the treatments by Foreman and Makin[7] and Koppenaal and Kuhlmann-Wilsdorf[11] are of great value.

Linear additivity of the individual component τ_1 and τ_2 such that

$$\tau = \tau_1 + \tau_2 \tag{3}$$

is found to be a poor approximation except where a few strong obstacles are introduced amongst many weak obstacles. In general a better approximate solution is

$$\tau^2 = \tau_1^2 + \tau_2^2 \tag{4}$$

A useful general consideration of superposition laws in strain hardened materials has been given by Kocks[12]. If we consider the contribution to the flow stress of the dislocations accumulated by strain hardening to be expressed in term of the density of dislocations ρ we can ignore details of the dislocations distribution and write

$$\Delta\tau = \alpha \, Gb\rho^{1/2} \tag{5}$$

where α is a constant of order 0.5 and G is the shear modulus and b the Burgers vector. The important question is whether the dislocations provide a high work hardening rate as in a single phase material or whether in the presence of other obstacles the effective hardening rate is low leading to some form of shear localization or instability.

The features considered so far in the review have been microscopic in character and described in terms of obstacles to dislocation motion. In addition we need to consider more mesoscopic or macroscopic aspects of the plasticity of heterogeneous materials. In many cases descriptions are possible at both the microscopic and mesoscopic level. An example of this is in the case of polycrystals where the conventional Hall-Petch scaling law can be expressed as

$$\sigma = \sigma_0 + k \, d^{-1/2} \tag{6}$$

where σ_0 is the lattice friction stress, d the average grain size and k a constant which reflects the ability to transmit slip from one grain to its neighbour. Typical results for strengthening due to grain refinement are shown in Figure 2. In addition to the influence on the initial flow stress refining the grain size can lead to higher initial hardening rates[13][14] due to the creation of plastic gradients in the vicinity of the boundary. With the advent to new methods for processing ultra fine scale microstructures fundamental questions and as to whether as the grain size is refined to the scale of a few nanometers whether plasticity occurs by alternative mechanisms[15].

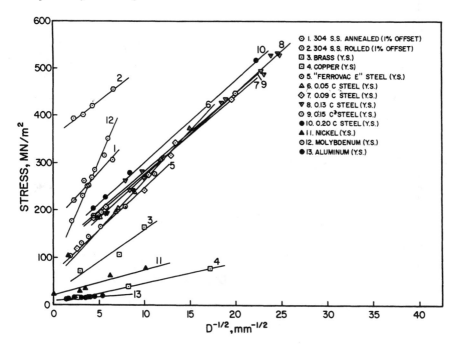

Figure 2   A diagram showing yield strength as a function of grain size plotted according to equation (3) data from ref. 37.

In addition to grain boundaries heterogeneous structures can be produced in the form of a variety of composites and several excellent text books[4,16,17] describe the mechanical behaviour of these materials. In essence when we have a mixture of two

188

materials with markedly different elastic and plastic properties which are bonded by a strong interface we can consider three aspects of the problem.

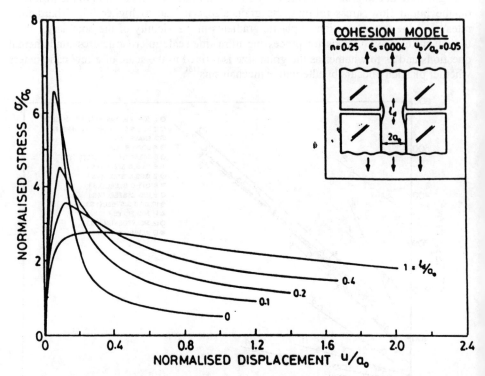

Figure 3   A diagram showing the effect of constraint on the flow of Lead (the degree of constraint is given as Φ/Φ_o).

The first is load transfer to the hard phase and conventional composite strengthening which can be described by an upper and lower bound solution reflecting isostrain or isostress approximations.

Thus the elastic properties have upper and lower bounds given by

$$E_{comp} = E_1 f + E_2 (1 - f) \qquad (7)$$

$$E_{comp} = \left(\frac{f}{E_1} + \frac{1-f}{E_2} \right)^{-1} \tag{8}$$

It should be noted that conventional composite behaviour depends only on the intrinsic properties of the matrix and the embedded phase and their relative volume fractions. No effects of scale enter the description. There is evidence that in many systems particularly those based on eutectic or eutectoid systems some modified rule of mixture which incorporates the scale of the embedded phase is more appropriate[18][19].

The interface may also alter the flow stress in the matrix by virtue of the plastic constraint. This aspect has been considered by a number of authors for model systems and an example from the work of Ashby, Blunt and Bannister[20] is given in Figure 3.

In macroscopic systems the hard phase also serves to alter the pattern of flow in the matrix and this feature and its associated changes in macroscopic texture has been studied by Poole et al[21].

3.   Practical Engineering Materials

The exploitation of grain refinement by controlled hot rolling is the basis of producing the desired mechanical properties in hot rolled plate products as illustrated in Figures 4 and 5. Both the final rolling temperature and the final cooling rate determine the ferrite grain size (Figure 5). As the finishing temperature is reduced to temperatures of order 600°C microstructural features other than the ferrite grain size enter into the overall strength. These effects have not been treated in detail but the work of Branfitt and Marder[22] suggests that if the components are added in a linear manner the results can be expressed in a simplified diagram as shown in Figure 6 in which the influence of substructure and texture can be considered.

The concept of refinement of the microstructural scale can also be applied to cold working operations for both single phase and duplex materials. This is illustrated in the work of Embury and Fisher[23] and others[23] who have examined the evolution of the strength of drawn eutectoid steel (Fe plus Fe_3C) as a function of the initial spacing of the pearlite and the strain level imposed by drawing.

The data can be plotted in a simple scaling law as

$$\sigma = \sigma_0 + k(d_0)^{-1/2} \exp\left(\frac{\varepsilon}{4} \right) \tag{9}$$

where σ_0 is the friction stress, k an unpinning term analogous to the slope in the Hall Petch equation, d_0 is the initial pearlite spacing after transformation and ε is the imposed true strain in the wire drawing process. In Figure 7 data is shown which used this formalism to examine the drawing behaviour of a number of steels. An important feature of this data is the influence of the initial scale of the microstructure on the final properties. This clearly indicates that combinations of methods which produce fine-scale initial structures such as rapid solidification or vapour deposition with deformation

190

processes such as wire drawing or hydrostatic extrusion have great potential in the production of ultra high strength materials in large section sizes.

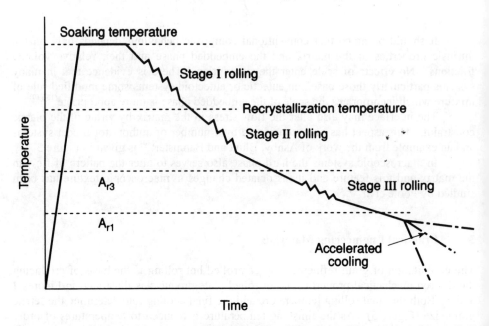

Figure 4 A diagram showing a typical thermal cycle for hot rolling of plate from ref. 3.

It is important to note that drawn pearlitic steel can approach a strength level close to the theoretical strength. Thus the development of the highest strength level requires detailed consideration of the competition between fracture and co-deformation of the hard embedded phase, in this case iron carbide. An excellent example of a process of microstructural refinement and optimization has been given by a group of Japanese researchers[24]. The authors adjusted the composition of the eutectoid steel to promote the refinement of the initial pearlite spacing and suppress the formation of any grain boundary carbides which have a deleterious role. Also the overall section size was reduced to obtain uniform microstructures. After the wire drawing operation it is possible to produce strength levels of close to 6 GPa as shown in Figure 8.

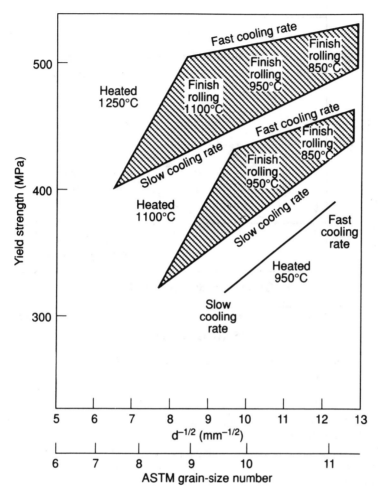

Figure 5   A diagram showing how yield strength and grain size are related to rolling temperature and cooling rate from ref. 3.

As outlined earlier, basic models of precipitation hardening distinguish between particle cutting and the formation of Orowan loops around particles.   The basic assumption is that the microstructural features such as the size and volume fraction of the particles remain constant.

In many practical cases we are concerned with the evolution of some form of precipitation distribution during complex thermal cycles. Thus for microalloyed steels we are concerned with the solubility product as a function of temperature as shown in Figure 9. For microalloyed steels which can be strengthened by the production of carbo-nitrides of Nb and V prediction of the strength level expected based on the Orowan mechanism (and its modified treatments) can be made as a function of particle size and volume fraction as shown in Figure 10.

192

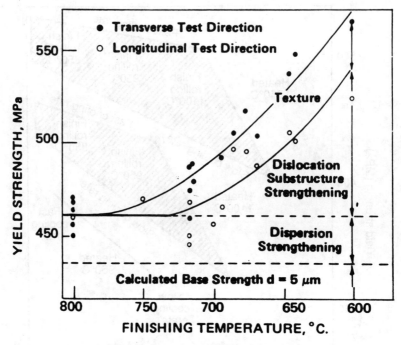

Figure 6   A diagram showing how yield strength and grain size are related to rolling temperature and cooling rate from ref. 22.

In nickel-based alloys a wide range of processes may occur due to the ordered nature of the particles, the ability to modify the particle-matrix mismatch by alloys and the high volume fraction of the precipitated phase. In the regime where particle cutting is dominant the dependence on the size and volume fraction of the particles is given by the expression

$$\Delta\tau = \frac{\gamma}{2b}\left[\frac{4\gamma r_s f}{\pi T} - f\right] \qquad (10)$$

where (is the surface energy based on cutting of ordered particles, f is the volume fraction, r_s the particle radius, and T the line tension. Eq. (10) is in good accord with the experimental data[9].

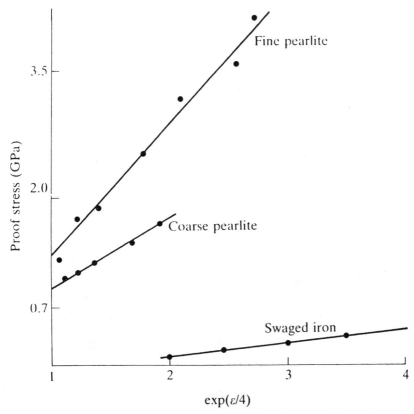

Figure 7   Variation of flow stress as a function of drawing strain for various steels from ref. 23.

In many alloy systems the microstructural development is complex in that it depends on the detailed temperature time history. An example is given in Figure 11 which shows the dependence of strength on aging time at 180°C for an Al-Mg-Si alloy AA 6111 with and without a period of natural aging[25]. The behaviour can be rationalized in terms of the relative development of the two possible metastable precipites β'' and Q'. The number density of these phases can be followed by detailed TEM and the results are shown in Figure 12.

Thus the basic concepts developed to describe the dependence of the yield stress on the character, size and number density of second phase particles can be applied quite successfully to a range of engineering alloys. It provides a rational basis for alloy design and indeed can be extended to describe at least some aspects of the subsequent work hardening of two phase systems, the expected magnitude of path dependent phenomena such as the Bauschinger effects and some aspects of additivity effects[25].

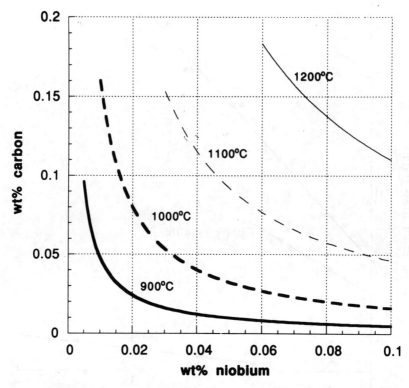

Figure 9   A diagram showing the mutual solubility of carbon and niobium in austenite as a function of temperature from ref. 3.

A number of valuable review articles on topics such as ball milling, equal channel angular extrusion, vapour deposition etc have appeared in the literature and some valuable theoretical models have appeared which deal with the dislocation models appropriate to these structures[28][29].

In general we can consider that ultra fine scale structures can be produced either by mechanical refinement e.g., by wire drawing or extrusion to very large strains and reducing the scale by a similitude principle or by the input of sufficient energy to produce extremely high rates of nucleation coupled with low growth rates.

In processes such as vapour deposition it is possible to define both the scale and detailed architecture of the material so that features such as crystallographic orientation, elastic misfit and scale can be controlled.   The development and analysis of the behaviour of layered structures have received much recent attention. The behaviour of these materials is examined in terms of the Hall-Petch plot relating to the role of grain size.   The results are illustrated in Figure 13.   The change in slope at grain sizes below 50 nm has been interpreted as a change in the mechanism of plasticity to one where the

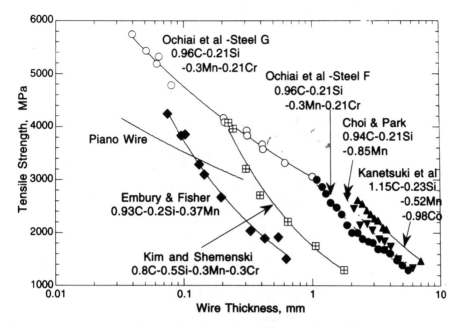

Figure 8   Tensile strength as a function of wire diameter for various steels (Note data from ref. 24).

In general it does not give a comprehensive view of the competition with damage and fracture modes, formability or strain localization but these aspects are being vigorously pursued and no doubt more comprehensive models will soon emerge.

4.   Developments of Ultra High Strength Metallic Materials

In the past three decades a variety of fabrication techniques have emerged for the production of materials in which the controlling microstructural dimensions are of the order of a few nanometers. These include a variety of techniques for the production of nanowires and tubes[26] but also bulk structures with microstructures of nanometer scale[27].

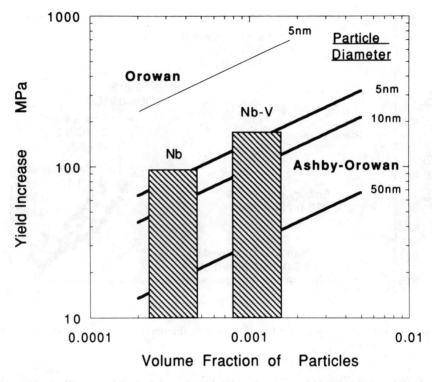

Figure 10 A diagram showing the strengthening due to carbide/nitride precipitation in steel for various volume fractions and particle sizes from ref. 3.

motion of single dislocations between the interfaces becomes the dominant process. Thus the plasticity, both yielding and subsequent dislocation storage, is controlled by the emission of dislocations from an interface and their deposition on an adjacent interface. This has a variety of important consequences, first the scale of events is such that atomistic computational modeling is directly relevant to the scale of the experiments. Thus the detailed processes at the interfaces can be followed and this will yield new descriptions of interface controlled plasticity. Also as dislocation storage occurs only at the interfaces and not within the phases the concept of plastic constraint has to be re-examined at the microscopic level because there is strong evidence that texture development is suppressed in fine scale layered structures resulting in the preservation of the initial Kurdjumov-Sachs relationship between the phases even after large plastic strains. Thus the plastic constraint must be reflected in co-ordination of the various slip systems at the interfaces.

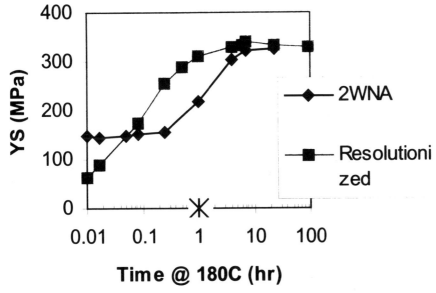

Figure 11   Yield strength as a function of ageing time at 180°C for AA 6111 with and without natural ageing from ref. 25.

Two other aspects of the co-deformation process in layered structures are of considerable interest. The first is the observation that in some cases amorphisation or dissolution processes may occur at the interfaces[30] and the second is that after extensive co-deformation large elastic internal stresses may develop[31]. Clearly, there is a need to document the length scale in which these phenomena occur and their dependence on the physical properties of the constituent phases.

In addition to the work on vapour deposited layer structures much effort has been devoted to the co-deformation of both f.c.c.–f.c.c. and f.c.c.–b.c.c. systems by wire drawing. The interest in these systems is the ability to develop combined structural and functional properties such as strength and electrical conductivity for high field magnet windings[32].

The basic microstructural feature of importance is illustrated in Figure 14 from the work of Raabe et al[19]. In this figure it can be seen that the initial dendritic structure of Nb is deformed and is eventually transformed into a series of long fibres of extremely uniform cross section. This geometric process can only occur if the slip distribution

198

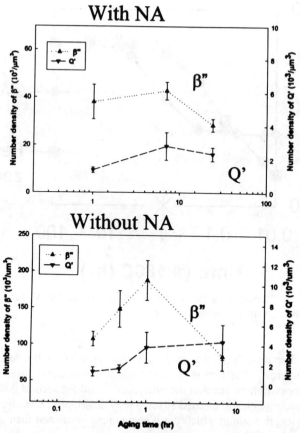

Figure 12 A diagram showing the growth of ∃″ and Q′ phase in AA 6111 on ageing at 180°C with and without natural ageing from ref. 25.

becomes very homogeneous so that the uniform cross-section of the fibres is preserved. This process also results in the production of large internal stresses after deformation as determined by neutron diffraction. This result suggests that the conventional view that phases are either sheared or act as Orowan obstacles needs to be re-examined because shearing can result in the storage of interface dislocations which exert elastic forces on the embedded phase[33]. In essence this means that one aspect of the energy storage process can be via the creation of interface energy[34].

In addition to producing ultra fine scale structure some novel processing methods can produce a variety of non-equilibrium phases. There is much current interest in the production and crystallization of bulk amorphous metals. Recent work has shown that aluminum based metallic glasses can be heat treated to produce materials of exceptionally high strength. These materials may in addition have useful functional properties related to their magnetic behaviour.

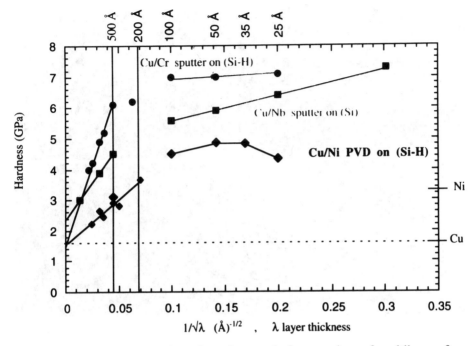

Figure 13 The effect of length scale and strength for a variety of multilayers from ref. 15.

There is much to be gained from considering very rapid heat treatment such as those occurring in femto second laser pulses to produce new structures[35]. This has been explored to a limited extent in semi-conductors but has great potential for metallic systems in terms of modification of surface structures.

Finally a topic which deserves further experimental and theoretical treatment is the deformation of non equilibrium structures such as highly supersaturated solid solutions. The work of Deschamps et al[35] shown in Figure 15 indicates that very high strength levels can be achieved by this process due to the production of fine scale dynamic precipitation.

In conclusion it can be seen that there are a variety of fabrication techniques which enable us to produce materials at strength levels close to the theoretical strength. These will undoubtedly yield rich areas for experimental and the theoretical study and for the direct linkage to atomistic modeling processes.

200

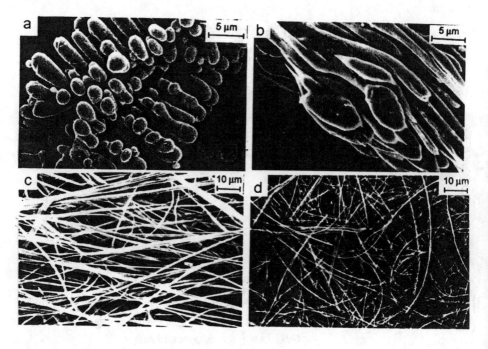

Figure 14 Illustrating the transition from dendrites to fine uniform fibres during wire drawing from ref. 3.

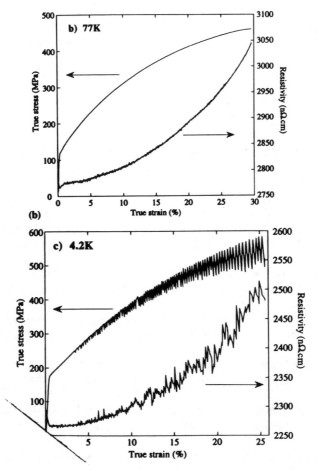

Figure 15 A diagram showing the work hardening and resistances as a function of strain in supersaturated Al-Zn-Mg from ref. 36.

References

1) Kelly A. and Nicholson R.B. 1963 Prog. Mat. Sci., <u>10</u>.
2) Kelly A. and Nicholson R.B. "Strengthening Mechanisms in Crystals", 1971 published by Halsted Press Division of J. Wiley.

202

3) Gladman, T. "The Physical Metallurgy of Microalloyed Steels", 1997 Institute of Metals.

4) Kelly A and MacMillan, N.H. Strong Solids. 1986 3rd edition, Oxford Science Publications.

5) Suresh, S. "Fatigue of Metals". 1982 2nd edition, C.U.P.

6) Ashby, M.F. "Materials Selection and Mechanical Design", 1992 Pergamon Press.

7) Foreman, A.J.E. and Makin, M.J. 1996 Phil. Mag., 14, 911.

8) Kocks, U.F. 1967 Can. J. Phys. 45. 737.

9) Brown, L.M. and Ham, R.K. in "Strengthening Mechanisms in Crystals", edited by Kelly and Nicholson. 1971 published by J. Wiley, p. 22.

10) Nie, J.F., Muddle, B.C. and Polmear, I.J. 1996 Mat. Sci. Forum, 217, 1257.

11) Koppenaal, T.J. and Kuhlmann-Wilsdorf, D. 1964 Appl. Phys. Letters, 4, 59.

12) Kocks, U.F., 1979 Proceedings of I.C.S.M.A. Conference 5, p. 1661.

13) Niewczas. M. and Embury, J.D. "The Integration of Material Process and Product Design". 1999 edited by Zabaras, published by Balkema, The Netherlands, p. 71.

14) Narutani, T. and Takamura, J. 1991 Acta Met. Mat., 59, 2037.

15) Misra, A., Verdier, M., Lu, Y.C., Kung, H., Mitchell, T., Nastasi, M. and Embury, J.D. 1998 Scripta Met., 39, 555.

16) Chawla, K.K. "Composite Materials". 1987 published by Springer-Verlag.

17) Piggott, M.R. Load Bearing Fibre Composites. 1980 published by Pergamon Press.

18) Bevk, J., Harbison, P. and Bell, J.L. 1978 J. Appl. Phys., 49. 6031.

19) Raabe, D., Herringhaus, F., Haugen, U. and Gottstein, G. 1995 Z. Metallkd, 86, 405.

20) Ashby, M.F., Blunt, F.J. and Bannister, M. 1989 Acta Met., 37, 1847.

21) Poole, W.J., Embury, J.D., MacEwan, S. and Kocks, U.F. 1994 Phil. Mag., 69, 667.

22) Bramfitt, R.L. and Marder, A.R. Processing and Properties of Low Carbon Steels. 1967 published by A.I.M.E.

23) Embury, J.D. and Fisher, R.M. 1966 Acta Met., 14, 147.

24) Ochiai, I., Nishida, S., Ohba, H., Serikawa, O. and Takahashi, T. 1994 Bull. Japan Inst. of Metals, 33, 444.

25) Cheng, L.M., Poole, W.J., Wang, X., Embury, J.D. and Lloyd, D.J. 2002 Mat. Science and Engineering, in press.

26) Terrones, M., Grobert, N. and Hsu, W.K. 1999 M.R.S. Bulletin, 24, No. 8, 43.

27) Koch, C.C. 1993 Nanostructured Mat., 2, 109.

28) Rao, S.I. and Hazzledine P.M. 1999 Scripta Met., 41, 1085

29) Nix.W.D. 1997 Mat Sci and Eng., A 234, 37.

30) Languillaume, J., Kapelsk, G. and Baudelet, B. 1997 Acta Met., 46, 1201.

31) Embury, J.D. and Sinclair, C.W. 2001 Mat. Sci. and Eng., A319, 37.

32) Wood, J.T., Griffin, A.J., Embury, J.D., Zhou, R., Nastasi, M. and Veron, M. 1995 J. Mech. Phys. Solids, 44, 737.

33) Sinclair, C.W. 2001 Ph.D. Thesis. McMaster University.

34) Embury, J.D. 1992 Scripta Met., 27, 981.

35)   Cavalleri, A., Toth, C., Siders, C.W., Squier, J.A., Raksi, F., Forge, P. and Kiffer, J.C. 2001 Phys. Rev. Letters, $\underline{87}$, 237401.

36)   Deschamps, A., LeSinq, L., Brechet, Y., Embury, J.D. and Niewczas, M. 1997 Mat. Sci. and Eng., $\underline{A234}$, 477.

MICROSTRUCTURAL ASPECTS OF CYCLIC DEFORMATION AND FATIGUE OF METALS

HAEL MUGHRABI
Universität Erlangen-Nürnberg
Institut für Werkstoffwissenschaften
Martensstr. 5
D-91058 Erlangen, Germany

Abstract

The importance of microstructural effects on the cyclic deformation and fatigue behaviour of metallic materials is discussed and illustrated with reference to selected examples, mainly from the work of the author's research group. The examples concern different single-phase and multi-phase f.c.c. materials, ferritic carbon steels and metastable austenitic stainless steels. Attention is focused in particular on the stability of the microstructures under different cyclic loading conditions, on the fatigue-induced changes of the microstructure and on the effects on fatigue life.

1. Introduction

One of the most important strength properties of metallic engineering materials is the resistance against cyclic loading leading to fatigue failure. The fatigue strength of a particular material depends, like all other strength properties, crucially on the microstructure of the material, both in its initial state and, in particular, on its stability during cyclic loading, compare for example, the reviews by Suresh [1], Klesnil and Lukáš [2], Laird [3], Christ [4] and Mughrabi [5,6].

In this article, some typical examples of fatigue-induced microstructural changes and their effects on the evolution of fatigue damage and on fatigue life will be presented. The examples selected are mainly from work performed in the author's research group under different conditions of cyclic loading on a number of different metals and alloys. In all cases, the cyclic deformation tests were performed in (dry) air in symmetric strain- or stress-controlled tension-compression.

The presentation will be brief, since all examples have been published in more detail earlier, so that reference can be made to the original publications. The materials considered include pure single-phase metals and solid-solution and precipitation-hardened alloys.

A. Finel et al. (eds.), Thermodynamics, Microstructures and Plasticity, 205–214.
© 2003 *Kluwer Academic Publishers. Printed in the Netherlands.*

2. Dislocation pattern evolution in fatigued single-phase materials, cyclic hardening and saturation

Cyclic deformation of annealed single-phase materials leads in general to cyclic hardening, followed by steady-state cyclic saturation and, ultimately, fatigue failure, as indicated schematically in Figure 1 [7]. Cyclic hardening is caused by an increase of the dislocation density. The arrangement of the dislocations is usually more or less heterogeneous and depends on the ease of cross slip, i.e. on whether the slip mode is wavy (easy cross slip) or planar (difficult cross slip), compare Feltner and Laird [8]. In face-centred cubic (f.c.c.) wavy slip materials, such as copper or nickel, dipolar edge dislocation bundles/veins are typical at low amplitudes and dislocation cell structures at higher amplitudes [2-4,7,9]. On the other hand, in planar slip materials, planar arrangements of dislocations on one or, at higher amplitudes, on more than one slip system are observed.

In cyclic saturation, a quasi-steady state situation exists, with small quasi-reversible to-and-fro displacements of dislocations at low amplitudes and a dynamic equilibrium between dislocation multiplication and annihilation at higher amplitudes [10]. Pre-deformed materials undergo cyclic softening, when deformed cyclically subsequently at an amplitude at which the saturation stress of an annealed specimen would be lower than the stress of the pre-deformation [8]. In wavy slip materials, the saturation stress can thus be almost history-independent (unique cyclic stress-strain behaviour). On the other hand, in planar slip materials, the saturation stress is more or less strongly history-dependent, compare Feltner and Laird [8] and Klesnil and Lukáš [2].

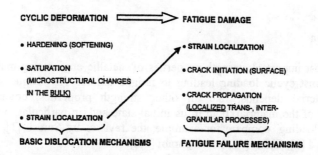

THE FATIGUE PROCESS

Figure 1. Schematic illustration of sequence of events during cyclic deformation and evolution of fatigue damage. After [7].

3. Cyclic strain localization and fatigue failure in single- and multi-phase materials

A common feature of fatigued (f.c.c.) metals and alloys is the development of cyclic strain localization in early saturation in so-called persistent slip bands (PSBs). The latter are thin glide lamellae parallel to the primary glide plane with a thickness of about 1 µm and the so-called "ladder" dislocation wall structure [1-4,7,10] in single-phase material and with much smaller thicknesses (fraction of a micron) in precipitation-hardened alloys with shearable precipitates [3,7,11]. Typically, the local shear strain amplitude in PSBs is about a hundred times larger than in the surrounding "matrix" in single-phase materials and almost an order of magnitude larger in precipitation-hardened alloys [3,7,10,12,13]. In the latter case, the precipitates in the PSBs can be completely destroyed by the extensive localized cyclic slip [7]. In both cases, the fact that the PSBs are softer than the matrix is usually reflected in the observation of cyclic softening as the PSBs develop. In precipitation-hardened alloys, this softening can persist till failure [3,11], whereas, in single-phase materials, cyclic softening usually merges into cyclic saturation. With increasing amplitude, the volume fraction of the PSBs increases. In the case of single-phase materials, the increase is linear and follows Winter's rule [12].

Cyclic strain localization in PSBs leads to fatigue crack initiation at the sites of emerging PSBs, where a rough notch-peak geometry in the form of mainly extrusions and deepenings (intrusions) develops. Earlier studies of the surface topography at the sites of emerging PSBs were made mainly by scanning electron microscopy (SEM) [13, 14]. More recently, finer details have been studied impressively by atomic force microscopy (AFM) [15, 16].

It has been shown that there exist threshold amplitudes of stress and strain below which PSBs do not develop. These thresholds which are closely related to a fatigue limit have been determined with high precision for several mono-crystalline materials [13] and also for some polycrystalline materials. The thresholds for mono- and polycrystalline material can be mutually transformed via an appropriate orientation factor [7,13].

4. Fatigue behaviour and fatigue lives of ultrafine-grained copper produced by equal channel angular pressing (ECAP)

Recently, there has been an increasing interest in the development and characterization of ultrafine-grained (UFG) metals produced by severe plastic deformation techniques such as ECAP [17]. In particular, attention was focused on the extraordinary strength of such UFG materials, also with respect to the potential of developing materials of superior fatigue resistance. While it has been shown, in particular for UFG copper at room temperature [18,19], that the fatigue resistance can be enhanced, certain limitations have been recognized. In a total strain fatigue life diagram in which the total strain amplitude $\Delta\varepsilon_t/2$ is plotted double-logarithmically against the fatigue life (number of cycles to failure, N_f), compare Figure 2, it is apparent that, for long fatigue lives in the low-cycle fatigue (LCF) regime, a large fatigue ductility coefficient ε_f' is required, whereas for long fatigue lives in the high-cycle fatigue (HCF) regime, a high fatigue strength coefficient σ_f' is desirable. Referring to UFG copper produced by ECAP, it is noted that the severe plastic deformation of ECAP enhances σ_f' but reduces the ductility and hence ε_f'. Thus, it is not too surprising that enhanced fatigue lives, compared to lives of conventional grain size material, were found for UFG copper at room temperature in the HCF regime, whereas the fatigue lives in the LCF regime were reduced [18,19]. In the latter case, strong cyclic softening paired with dynamic grain coarsening (Figure 3a) and cyclic strain localization in macroscopic shear bands could be identified as the

208

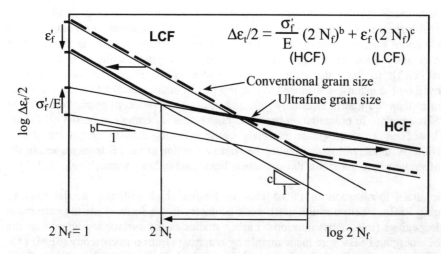

$$\Delta\varepsilon_t/2 = \frac{\sigma_f'}{E}\,(2\,N_f)^b + \varepsilon_f'\,(2\,N_f)^c$$

Figure 2. Schematic total strain fatigue life diagrams for materials of conventional and ultrafine grain size: enhancement and reduction of fatigue life of UFG material in the HCF- and in the LCF-regimes, respectively. E: Young's modulus, N_t: transition fatigue life. The slopes in the limits of LCF and HCF correspond to the fatigue ductility and strength exponents c and b, respectively. After [19].

origin of fatigue damage. The occurrence of strong dynamic grain coarsening at a relatively low homologous temperature (ca. 0.2) reflects the fact that the UFG microstructure in the as-ECAP-processed state is rather unstable. On the other hand, at –50°C the original UFG microstructure was retained, Figure 3b [20]. Höppel et al. [20] have been able to show that the dynamic coarsening is in fact a thermally activated process of dynamic recrystallization. As shown in the author's research group [21], it is possible to render the UFG microstructure more stable by a suitable annealing treatment and thus to improve the initially disappointing LCF behaviour considerably with an enhancement of the fatigue lives by a factor of seven.

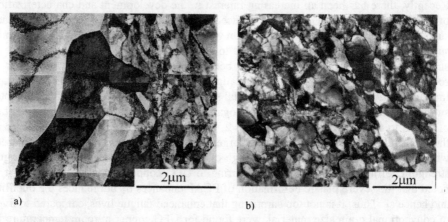

a)   b)

Figure 3. Transmission electron micrographs (TEM) of UFG copper specimens fatigued to failure at a) room temperature and b) –50°C, showing, in the former case, a strongly coarsened microstructure next to the original UFG microstructure, whereas, in the latter case, the UFG microstructure is retained. After [20].

5. Dynamic strain ageing effects in fatigued steels

At higher temperatures, the interaction between glide dislocations and solute atoms can lead to the phenomenon of dynamic strain ageing, accompanied by extra hardening, in a specific range of temperature and strain rate. In this range, the diffusing solute species which are "trapped" in the strain field of the dislocations can follow the gliding dislocations and retard their motion. A lower and higher temperatures, the dislocations are not hindered by the solute atoms, since, at low temperatures, the dislocations are torn loose from their solute atmospheres (upper yield point), whereas, at higher temperatures, the solute atoms diffuse too rapidly to be trapped at the dislocations.

Dynamic strain ageing frequently occurs in steels. In the regime of dynamic strain ageing, a denser (edge) dislocation distribution is observed in the case of ferritic steels. The origin of the higher dislocation density in body-centred cubic (b.c.c.) materials probably lies in the fact that the more mobile edge dislocation segments are hindered more strongly by the solute atoms than the less mobile screw dislocation segments. Then, the dislocation distribution acquires a higher density of dislocation-rich edge dislocation bundles and becomes more similar to that of f.c.c. metals.

Wilson and his co-workers, who studied systematically the effects of dynamic strain ageing on the fatigue behaviour of low carbon steels [22,23], showed that, under conditions of dynamic strain ageing, PSB-like cyclic strain localization occurs which is not usually observed in such clear fashion in (pure) b.c.c. metals [10,24]. The effect of dynamic strain ageing on the fatigue life of carbon steels subjected to cyclic deformation at constant stress amplitude was first studied systematically by the group of Macherauch [25,26]. In particular, these researchers showed convincingly that, in the range of dynamic strain ageing, the fatigue life is enhanced considerably. The reason lies in the extra cyclic hardening which, in a stress-controlled test, leads to a reduction of the plastic strain amplitude which, in turn, then leads to a longer fatigue life, as expected on the basis of the Coffin-Manson law. Figures 4 and 5 show the results of a related study by Weisse et al. [27], compare also [28], on the carbon steel SAE 1045. In addition to the tests at constant axial stress amplitude $\Delta\sigma/2$, also tests at constant plastic strain amplitude $\Delta\varepsilon_{pl}/2$ were carried out. While, at constant stress amplitude, the plastic strain amplitude goes through a minimum in the regime of dynamic strain ageing, the stress amplitude goes through a maximum at constant plastic strain amplitude (Figure 4). In the latter case, the extra cyclic hardening in the regime of dynamic strain ageing (around 300°C) leads to a shorter fatigue life, as shown in Figure 5.

Dynamic strain ageing can occur also in austenitic stainless steels of the types AISI 304 [29] and AISI 316 [30], when fatigued at temperatures above 300°C. In both cases, an interesting change of the cyclic slip mode with increasing temperatures from wavy to planar slip in the regime of dynamic strain ageing and back to wavy slip at higher temperatures is observed. In the case of AISI 304 L, it was shown that the planar slip originates from the formation of small coherent carbides which are cut by the dislocations in planar slip bands [29]. In the case of AISI 316, an interesting observation was that the fatigue lives (in vacuum) were enhanced in strain-controlled tests in the

210

regime of dynamic strain ageing [30], indicating that the beneficial effect of the cyclic slip planarity outweighed the negative effect of extra hardening, compare also [28].

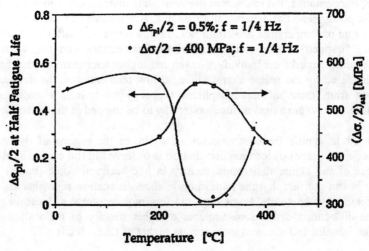

Figure 4. Temperature dependence of stress and plastic strain amplitudes in fatigued carbon steel SAE 1045 for plastic-strain- and stress-controlled tests, respectively. f: frequency of test. After [27].

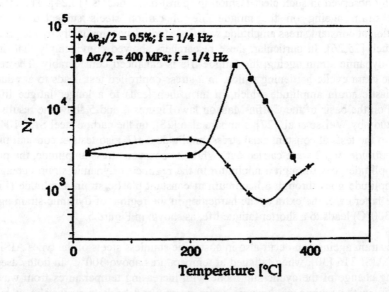

Figure 5. Temperature dependence of fatigue lives, N_f, of carbon steel SAE 1045, at constant stress and constant plastic strain amplitudes, respectively. After [27].

6. Fatigue-induced martensitic transformation in metastable austenitic stainless steels.

In a metastable austenitic stainless steel such as AISI 304 L, plastic deformation can lead to the formation of α'-martensite, larger amounts being formed at lower temperatures, because of the enhancement of the driving force with decreasing temperatures. Under conditions of strain-controlled cyclic deformation, the amount of α'-martensite formed can vary between a few percent and more than 80%, depending on the plastic strain amplitude, the numbers of cycles and, in particular, the temperature. Since the martensite content increases with cumulative plastic strain, the martensite contents achieved in stress-controlled tests are usually somewhat lower. As an example, Figure 6 shows TEM micrographs of fatigue-induced α'-martensite in the steel AISI 304 L [31]. This effect can be exploited technologically, with the aim to enhance the strength of metastable austenitic steels considerably [32,33].

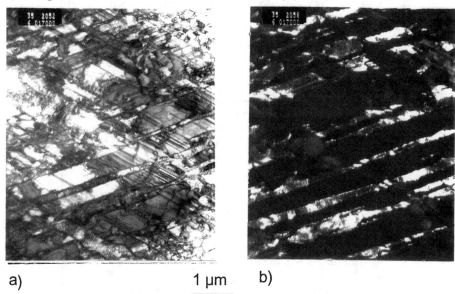

a)   1 µm   b)

Figure 6. Transmission electron micrographs of austenitic stainless steel specimen ($ZA < \overline{2}\overline{1}0 >$) after fatigue at $\Delta\sigma = 500$ MPa, mean stress $\sigma_m = 250$ MPa, 203 K, showing a) bright field image, with α'-martensite and austenite, and b) dark field image of fatigue-induced α'-martensite (volume content: ca. 25%), $1\overline{2}1$ α'-martensite reflection. After [31].

Figures 7 and 8 show two important results from the work of Maier et al. [32,33]. In Figure 7, the dependence of the ultimate tensile strengths measured on specimens of the metastable austenitic stainless steel AISI 304 L that had been pre-fatigued into saturation at a plastic strain amplitude $\Delta\varepsilon_{pl}/2 = 1.26 \cdot 10^{-2}$ at different temperatures is shown as a function of the temperature of pre-fatigue. The result indicates that the introduction of a larger volume fraction (ca. 85%) of α'-martensite by pre-fatigue at a low temperature provides a means to enhance the monotonic strength appreciably.

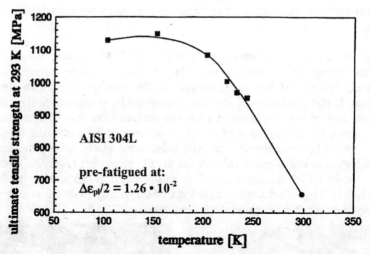

Figure 7. Ultimate tensile strength of the austenitic stainless steel AISI 304 L, measured at room temperature, after introduction of maximum contents of α'-martensite by pre-fatigue with $\Delta\varepsilon_{pl}/2 = 1.26 \cdot 10^{-2}$ at temperatures between 103 K and 293 K. After [32,33].

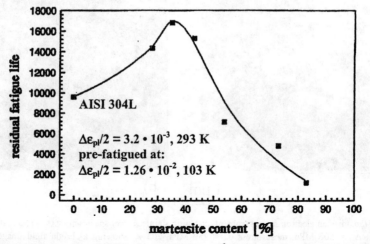

Figure 8. Residual fatigue life after fatigue at $\Delta\varepsilon_{pl}/2 = 3.2 \cdot 10^{-3}$, 293 K, as a function of α'-martensite content introduced by pre-fatigue ($\Delta\varepsilon_{pl}/2 = 1.26 \cdot 10^{-2}$, 103 K) to different numbers of cycles. After [32,33].

In a similar fashion, the martensitic transformation can be exploited in order to enhance the fatigue strength. Figure 8 shows the results of a test series in which different amounts of α'-martensite were introduced by pre-fatigue at $\Delta\varepsilon_{pl}/2 = 1.26 \cdot 10^{-2}$ at a low temperature (103 K) for different numbers of cycles. Subsequently, the residual fatigue lives of the specimens were determined at room temperature for a plastic strain amplitude $\Delta\varepsilon_{pl}/2 = 3.2 \cdot 10^{-3}$. The result makes it evident that an α'-martensite content of ca. 35% is optimal and leads to a maximum residual fatigue life. Obviously, this α'-

martensite content is the best compromise between enhancement of strength and loss of ductility.

7. Concluding remarks

In the preceding sections, a number of examples were given to demonstrate the influence of microstructure on the cyclic deformation and the fatigue resistance of different materials. In addition, the examples underline that the conditions of cyclic loading (stress or strain control, amplitude, temperature) also affect the fatigue behaviours strongly. While the results presented can serve as guidelines for the design of fatigue-resistant materials, it should also be kept in mind that the fatigue life of a component in service can also be governed by a number of other factors that were not discussed here, such as "defects" in the microstructure (pores, inclusions, precipitate-free-zones at the grain boundaries), mean stresses, residual stresses, the state of the surface, environmental effects and, last but not least, the design of the component.

Acknowledgements

The author acknowledges gratefully the support of this research by Deutsche Forschungsgemeinschaft and the competent help of Mrs. Renate Graham and Dr. Heinz Werner Höppel in the preparation of this manuscript.

References

1. Suresh, S. (1998) *Fatigue of Materials*, Cambridge University Press.
2. Klesnil, M. and Lukáš, P. (1992) *Fatigue of Metallic Materials*, Materials Science Monographs, C Laird (ed.), Vol. 71, Elsevier Science Publisher, Amsterdam.
3. Laird, C. (1983) in F.R.N. Nabarro (ed.), *Dislocations in Solids*, Vol. 6, North-Holland Publishing Company, p.1.
4. Christ, H.J. (1996) in *Fatigue and Fracture*, ASM Handbook, Vol. 19, ASM International, Materials Park, Ohio, USA, p.73.
5. Mughrabi, H. (1993) *Journal de Physique IV*, Colloque C7, supplément au Journal de Physique III, **3**, 659.
6. Mughrabi, H. (2002) in A.F. Blom (ed.), *FATIGUE 2002*, Proceedings of the Eighth International Fatigue Congress, Vol. 2, EMAS Ltd., West Midlands, U.K., p. 971.
7. Mughrabi H. (1985) in *Dislocations and Properties of Real Materials*, Book No. 323, The Institute of Metals, London, p. 244.
8. Feltner, C.E. and Laird (C.), (1967) *Acta metall.* **15**, 1621 and 1633.
9. Lukáš, P. and Klesnil, M. (1973) *Mater. Sci.* Eng. **11**, 345.
10. Mughrabi, H., Ackermann, F. and Herz, K. (1979) in J.T. Fong (ed.), *Fatigue Mechanisms*, ASTM STP 675, p. 69.
11. Wilhelm, M. (1981) *Mater. Sci. Eng.* **48**, 91.
12. Winter, A.T. (1974) *Phil Mag.* **30**, 719.
13. Mughrabi, H., Wang, R., Differt, K. and Essmann, U. (1983) in J. Lankford et al. (eds.), *Fatigue Mechanisms: Advances in Quantitative Measurement of Physical Damage*, ASTM STP 811, p. 5.
14. Basinski, Z.S. and Basinski, S.J. (1992) *Progress in Materials Science* **36**, 89.
15. Hollmann, M., Bretschneider, J. and Holste, C. (2000) *Cryst. Res. Technol.* **35**, 479.
16. Villechaise, P., Sabatier, L. and Girard, J.C. (2002) *Mater. Sci. Eng.* A **323**, 377.
17. Valiev, R.Z., Islamgaliev, R.K. and Alexandrov, I.V. (2000) *Progr. in Mater. Sci.* **45**, 103.

214

18. Agnew, S.R., Vinogradov, A-Yu., Hashimoto, S. and Weertman, J. (1999) *J. Electron. Mater.* **28**, 1038.
19. Mughrabi, H. and Höppel, H.W. (2001) *Mat. Res. Soc. Symp. Proc.* **634**, B2.1.1.
20. Höppel, H.W., Zhou, Z.M., Mughrabi, H. and Valiev, R.Z. (2002) *Phil. Mag. A* **82**, 1781.
21. Höppel, H.W., Brunnbauer, M., Mughrabi, H., Valiev, R.Z. and Zhilyaev, A.P., *Proceedings of Materials Week 2000*, http://www.materialsweek.org/proceedings (2001).
22. Wilson, D.V. and Tromans, J.K. (1970) *Acta metall.* **18**, 1197.
23. Wilson, D.V. (1973) *Acta metall.* **29**, 673.
24. Mughrabi, H., Herz, K. and Stark, X. (1981) *Internat. J. Fract.* **17**, 193.
25. Pohl, K., Mayr, P. and Macherauch, E. (1981), *Internat. J. Fract.* **17**, 221.
26. Eifler, D. and Macherauch, E. (1988) in P. Lukáš and J. Polák (eds.), *Mechanisms in Fatigue of Metals*, Academia, Prague, p. 215.
27. Weisse, M., Wamukwamba, C.K., Christ, H.J. and Mughrabi, H. (1993) *Acta metall. mater.* **41**, 2227.
28. Mughrabi, H. and Christ, H.J. (1997) *ISIJ Internat.* **37**, 1154.
29. Petry, F., Christ, H.J. and Mughrabi, H. (1991) in *Microstructure and Mechanical Properties of Materials*, DGM Informationsgesellschaft, Oberursel, p. 79, and F. Petry (1989) *Doctorate Thesis*, Universität Erlangen-Nürnberg.
30. Gerland, M., Mendez, J., Lépinoux, J. and Violan, P. (1993) *Mater. Sci. Eng. A* **164**, 226.
31. Wunschik, W. (1994) Diplomarbeit, Universität Erlangen-Nürnberg.
32. Maier, H.J., Donth, B., Bayerlein, M. and Mughrabi, H. (1993) in J.-P. Bailon and J.I. Dickson (eds.) in *FATIGUE '93*, Proc. of 5th Int. Conf. on Fatigue and Fatigue Thresholds, EMAS Ltd., U.K., p. 85.
33. Maier, H.J., Donth, B., Bayerlein, M., Mughrabi, H., Meier, B. and Kesten, M. (1993) *Z. Metallkde* **84**, 12.

HIGH TEMPERATURE PLASTICITY OF METALLIC MATERIALS

Y. ESTRIN
IWW, Clausthal University of Technology
Agricolastr. 6, D-38678 Clausthal-Zellerfeld, Germany

1. Introduction

The subject of this talk is high temperature plasticity, so we should start by defining what is meant by *high temperature*. Commonly, temperatures above half the melting temperature T_m are regarded as *high*. However, such a definition, though quite practicable, lacks a physical underpinning. Frost and Ashby [1] suggest that the high temperature range – with regard to plasticity – can be associated with temperatures at which the strain rate sensitivity of the flow stress becomes prominent. This is typically the case above $0.3T_m$ for pure metals and above $0.4T_m$ for alloys. We shall adopt this – admittedly somewhat imprecise – definition. Useful guidance can be gained from the Ashby maps [1] that outline the various regions in the normalised stress vs. temperature diagrams where different deformation mechanisms prevail, cf. Fig. 1. The temperature range where specifically high-temperature mechanisms, particularly diffusion controlled ones, produce a noticeable plastic strain rate can be identified on this basis.

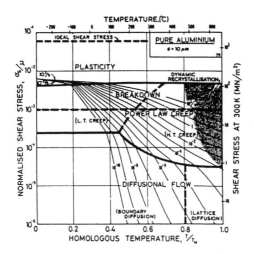

Figure 1. Ashy map for pure Al.

In what follows we shall consider dislocation controlled and diffusion controlled mechanisms of plastic flow with the aim of establishing constitutive equations that describe the mechanical behaviour of single phase and particle reinforced materials. With regard to high temperature deformation, creep behaviour will, of course, be the main focus of this expose. In the high-temperature range, primary creep is of minor

A. Finel et al. (eds.), Thermodynamics, Microstructures and Plasticity, 217–238.

significance, so it is natural to concentrate more on steady state creep. Here, interesting 'master curves' are found which make it possible to describe the creep behaviour of a certain material in a unique way and which apply, it seems, across the borders on the Ashby map. These empirical master curves will be discussed in Section 2. Then, we shall be looking at specific mechanisms of high temperature plasticity, starting with dislocation glide (that corresponds to a horizontal strip of the Ashby map denoted 'BREAKDOWN'). What is meant is that a power law describing the relation between plastic strain rate and stress at lower stresses does not hold within this strip. For this region, we formulate a constitutive model that describes strain hardening and creep in terms of the dislocation density evolution (Section 3). This model will serve as a basis for a constitutive description of primary and secondary creep in particle reinforced alloys: first for a time-independent particle dispersion and then for the case when the particle population evolves during creep. Finally, the effect of particle arrangement on the creep properties will be considered. Climb controlled creep will be considered in Section 4. In Section 5, we then move on the map 'southwards' and investigate how diffusion creep comes in and interacts with dislocation controlled plastic flow. This consideration refers to the region on the Ashby map denoted 'DIFFUSIONAL FLOW'. The focus will be on fine and ultra fine grained materials where diffusional flow is the predominant plasticity mechanism. The shaded region of the Ashby map where dynamic recrystallisation occurs will not be considered.

2. Empirical Correlations

In metals and alloys, the dependence of steady state creep rate, $\dot{\varepsilon}_s$, that is established when strain hardening processes are balanced by dynamic recovery, on stress σ is often represented by a power law [2]

$$\dot{\varepsilon}_s = A \exp\left(-\frac{Q_{CREEP}}{k_B T}\right)\left(\frac{\sigma}{\mu}\right)^n, \tag{1}$$

where the exponent n is a constant, μ is the shear modulus, T the absolute temperature, k_B the Boltzmann constant and A a constant material parameter. Over a wide range of temperatures above $0.6T_m$, say, the exponent n in the power law is about five, and one speaks of 'five-power-law' creep [2]. It is interesting to note that for Al, this exponent is smaller than 5, somewhere between 4 and 4.5. The activation energy for creep, Q_{CREEP}, was consistently found to be close to the activation energy for lattice self-diffusion, Q_{SD}, cf. Fig. 2.

At lower temperatures, in the range below about 0.5-$0.6T_m$, the relation expressed by Eq. (1) breaks down in that n is no longer constant, but rather increases, while the activation energy for creep drops. In other words, the separation of the temperature and stress dependence of $\dot{\varepsilon}_s$ in two unrelated factors does not hold any longer. Another relation (due to Garofalo), designed to embrace both the power-law and the power-law breakdown regimes, employs the hyperbolic sine function,

$$\dot{\varepsilon}_s = B \exp\left(-\frac{Q_{CREEP}}{k_B T}\right)\left[\sinh\left(\psi \frac{\sigma}{\mu}\right)\right]^n \tag{2}$$

where B and ψ are constants. Equation (2) recovers the power law of Eq. (1) in the limit of small stress, while yielding exponential stress dependence for large stresses. These forms of the steady state creep law suggest that by plotting $\dot{\varepsilon}_s$ normalised by the self-diffusion coefficient vs. the modulus-compensated stress σ / μ a degree of universality should be achieved. This is also equivalent to plotting the Zener-Hollomon parameter, $\dot{\varepsilon}_s \exp(Q_{CREEP} / k_B T)$, against σ / μ or σ / E, where E is Young's modulus. It should be noted, however, that in the power-law breakdown range, the activation energy for creep goes down to about $0.6 Q_{SD}$, assuming a level of the activation energy for dislocation pipe-diffusion [2]. Figure 3 does not take this effect into account. It is interesting to note that the curve for high-purity Al exhibits a five-power-law behaviour, while a family of the curves for Al based alloys are consistent with three-power-law - before power-law breakdown sets in.

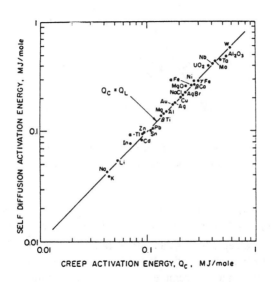

Figure 2. Correlation between Q_{CREEP} and Q_{SD} for a number of metals.

A somewhat different type of 'master curve' was suggested in Ref. 4. It operates with normalised quantities σ / μ, V/b^3 and $\Delta G_o / \mu b^3$, where b is the magnitude of the dislocation Burgers vector, μ is the shear modulus and ΔG_o and V are defined in terms of linearisation of the Gibbs free energy of activation for steady state plastic flow: $\Delta G = \Delta G_o - V\sigma$. It was shown that steady state deformation data for pure Al satisfy the equation

$$\dot{\varepsilon}^* = \dot{\varepsilon}_o \left[\sinh\left(\frac{V}{b^3}\right) \sigma^* \right]^n \qquad (3)$$

where $\dot{\varepsilon}*$, a quantity akin to the Zener-Hollomon parameter, and the normalised stress $\sigma*$ are defined as follows:

$$\dot{\varepsilon}* = \dot{\varepsilon} \cdot \exp\left[\left(\frac{\Delta G_o}{\mu b^3}\right)\left(\frac{\mu b^3}{kT}\right)\right], \quad \sigma* = \frac{\sigma}{\mu} \cdot \frac{\mu b^3}{kT}. \tag{4}$$

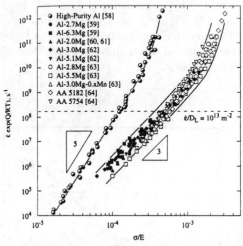

Figure 3. Zener-Hollomon parameter for pure Al and Al-Mg alloys plotted against the modulus-compensated stress (after Ref. 3.)

The best fit of Eq. (3) to experimental data shown in Fig. 4 was obtained with the parameter values of $n = 4$, $V/b^3 = 20$, $\Delta G_o/\mu b^3 = 0.32$, and $\ln\dot{\varepsilon}_o = 14.25$ ($\dot{\varepsilon}_o$ in s^{-1}.) The data stem from both creep and constant strain rate tests conducted by several authors. Thus, in addition to the three- and five-power-law, a *four-power-law* transpires from Eqs. (3) and (4) in the small $\sigma*$ regime. It is worth mentioning that the above value of $\Delta G_o/\mu b^3$ corresponds to ΔG_o equal to about $0.8 Q_{sd}$ and does not appear to be related to the activation energy for self-diffusion, be it bulk or pipe.

In Ref. 4, the effect of solutes, specifically Mg, on the steady state behaviour was investigated as well. The alloying effect on the stress dependence of the strain rate in steady state for a broad range of concentrations c (0.03 – 5.5 wt.% Mg) was shown to be well represented by an additive term in the following equation:

$$\sigma*_{Alloy} = \sigma*_{Al} + a \cdot \frac{c^p}{1 + \left(\dot{\varepsilon}_o' / \dot{\varepsilon}*\right)^q}. \tag{5}$$

The first term on the right-hand side of Eq. (5) is given by Eq. (3); the values of new parameters introduced were obtained from the fit to the experimental curves shown in Fig. 5 ($a = 4$, $p = 0.57$, $\ln(\dot{\varepsilon}_o'/s^{-1}) = 16.22$, and $q = 0.32$). With $p = 0.57$, the concentration dependence of the solid solution hardening term is not far from the common parabolic law. While this term is strain rate independent for sufficiently large strain rates ($\dot{\varepsilon}_o' \gg \dot{\varepsilon}*$), in the opposite limit it assumes the form $ac^p\left(\dot{\varepsilon}*/\dot{\varepsilon}_o'\right)^q$. In the low strain rate limit, when $\dot{\varepsilon}*$ is much smaller than $\dot{\varepsilon}_o$, the smaller of the two reference strain rates, Eq. (5) reduces to the sum of two power-law terms:

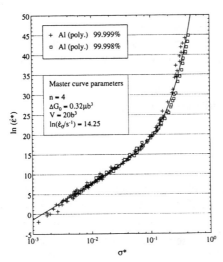

Figure 4. Master curve for high purity Al fitted to experimental data (symbols). (After Ref. 4.)

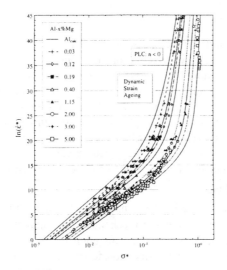

Figure 5. Effect of alloying on steady state plastic flow. Symbols correspond to experimental data, curves to calculations based on Eqs. (3) and (5). (After Ref. 4.)

$$\sigma^*_{Alloy} = \frac{b^3}{V} \left(\frac{\dot{\varepsilon}^*}{\dot{\varepsilon}_o} \right)^{1/n} + ac^p \left(\frac{\dot{\varepsilon}^*}{\dot{\varepsilon}_o'} \right)^q , \qquad (6)$$

with powers of 1/4 and about 1/3, respectively. Obviously, the first term is predominant, and so is the four-power-law, but with increasing $\dot{\varepsilon}^*$ the role of the solid solution contribution gets larger, leading to a departure from the four- towards the three-power-law.

The phenomenological equations presented in this section, though purely empirical, provide useful correlations and possess substantial predictive capability. In what follows, we shall attempt to establish a more physically motivated modelling frame, however.

3. Dislocation Glide Controlled Plasticity

3.1. STRUCTURE OF THE MODEL

For plastic deformation by dislocation glide, a relatively simple constitutive modelling approach that goes back to Kocks and Mecking [5] works remarkably well. The stress required for dislocation glide (DG) at a given plastic strain rate $\dot{\varepsilon}_{DG}$ and absolute temperature T is expressed in terms of the *mechanical threshold stress* $\hat{\sigma}$ in the form of a power law:

$$\sigma = \hat{\sigma} \left(\frac{\dot{\varepsilon}_{DG}}{\dot{\varepsilon}_{REF}} \right)^{1/m} \qquad (7)$$

The mechanical threshold $\hat{\sigma}$ represents the stress required for a gliding dislocation to overcome localised obstacles in its glide plane at absolute zero temperature, i.e. in the absence of thermal activation. Hence, this quantity depends on the obstacle structure and varies as the latter evolves in the course of straining. This evolution has to be included in a constitutive description. The exponent $1/m$ determines the strain rate sensitivity of the flow stress. As the power law, Eq. (7), actually originates from an Arrhenius type relation for thermally activated dislocation glide, $1/m$ can be shown to be linear in T, while for most situations of interest here it can be considered to be stress independent. Finally, $\dot{\varepsilon}_{REF}$ is a reference quantity. For σ to decrease with temperature, $\dot{\varepsilon}_{DG}$ must obviously be smaller than $\dot{\varepsilon}_{REF}$.

While the exponent $1/m$ reflects the nature of the thermally activated process that controls the dislocation glide rate, the mechanical threshold stress $\hat{\sigma}$ is governed by the strength and the concentration of the rate controlling localised obstacles. If these obstacles are associated with the dislocation forest, the density of forest junctions and thus $\hat{\sigma}$ itself being proportional to the inverse of the dislocation spacing, the following expression for the mechanical threshold stress holds:

$$\hat{\sigma} = M\alpha\mu b\sqrt{\rho} \tag{8}$$

Here ρ is the dislocation density, M is the Taylor factor that accounts for polycrystallinity of the material and varies as texture evolves during the deformation process and α a numerical constant.

Now an evolution equation for the dislocation density ρ, the only microstructural variable of the model, needs to be established. The rate of variation of ρ with strain is a result of dislocation storage and concurrent recovery processes. The storage rate is related to the dislocation mean free path L – a distance an average dislocation travels in its glide plane until it gets immobilised at some impenetrable obstacle. The storage is counteracted by dynamic recovery by movement of dislocations out of their glide planes and annihilation with a suitable partner on a different glide plane. The rate of recovery is controlled by cross-slip of screw or – for sufficiently high temperatures - by climb of edge dislocations. The combined effect of dislocation storage and dynamic recovery is summarised in the evolution equation for ρ:

$$d\rho / d\varepsilon_{DG} = M\left(1/(bL) - k_2\rho\right) \tag{9}$$

The coefficient k_2 in the dynamic recovery rate term is associated with the dislocation cross-slip or climb rate and as such is temperature and strain rate dependent. This dependence can be represented as

$$k_2 = k_{20}\left(\dot{\varepsilon}_{DG} / \dot{\varepsilon}_{REF}^*\right)^{-1/n}, \tag{10}$$

where k_{20} is a constant. This form covers both the high and the low temperatures ranges. For the high-temperature, climb controlled regime, the exponent $1/n$ can be taken to be constant. The temperature dependence then resides in the reference quantity $\dot{\varepsilon}_{REF}$ that can be represented by an Arrhenius equation

$$\dot{\varepsilon}_{REF}^* = \dot{\varepsilon}_{REF}^{**} \exp\left(-\frac{Q_{CLIMB}}{k_B T}\right) \tag{11}$$

The activation energy Q_{CLIMB} in this expression is that for the dislocation climb, i.e. the activation energy for self-diffusion. The pre-exponential factor $\dot{\varepsilon}_{REF}^{**}$ can be considered constant.

With the set of Eqs. (7)-(11), the general architecture of the model has been established. Its tenet is that one can separate the dislocation glide processes responsible for plastic strain from the dislocation climb processes. The latter are considered to merely provide a 'background' obstacle structure to gliding dislocations that gradually evolves in the course of plastic deformation. No direct contribution of climb to plastic strain was included so far, but this can, of course, be done.

The recovery processes considered are of dynamic nature, i.e. they are associated with the occurrence of plastic strain. *Time dependent* effects, i.e. static recovery, for which a decrement in the dislocation density is proportional to the corresponding time increment, can also be included [6].

3.2. DEFORMATION OF SINGLE PHASE COARSE-GRAINED MATERIALS

In a coarse-grained single-phase material, the mean free path L of dislocations will be determined by the average dislocation spacing $1/\sqrt{\rho}$ only. Should dislocations be arranged in a cell or subgrain structure, which is usually the case even after a moderate strain, it will be the cell/subgrain size that determines L. However, as the subgrain or dislocation cell size is known to scale with $1/\sqrt{\rho}$, L will be proportional to $\sqrt{\rho}$ in either case. Equation (9) then assumes the form

$$d\rho / d\varepsilon_{GB} = M\left(k_1\sqrt{\rho} - k_2\rho\right) \tag{12}$$

For this case the constitutive equations are so simple that they can be integrated analytically. For the case of constant (plastic) strain rate, $\dot{\varepsilon} = \dot{\varepsilon}_{DG} = const$, the familiar Voce equation is obtained:

$$\frac{\sigma - \sigma_s}{\sigma_i - \sigma_s} = \exp\left(-\frac{\varepsilon - \varepsilon_i}{\varepsilon_{tr}}\right) \tag{13}$$

Here the subscript i refers to some initial point on the stress-strain curve, e.g. the yield point (in which case ε_i is to be set to zero). The quantity

$$\varepsilon_{tr} = \sigma_s / \Theta_{II} \tag{14}$$

determines the rate with which a steady state value of stress,

$$\sigma_s = M\alpha\mu b\left(\frac{k_1}{k_{20}}\right)\left(\frac{\dot{\varepsilon}}{\dot{\varepsilon}_{REF}}\right)^{1/m}\left(\frac{\dot{\varepsilon}}{\dot{\varepsilon}_{REF}^*}\right)^{1/n} \tag{15}$$

is approached. Finally, the quantity

$$\Theta_{II} = \frac{1}{2}M^2\alpha\mu bk_1\left(\frac{\dot{\varepsilon}}{\dot{\varepsilon}_o}\right)^{1/m} \tag{16}$$

is the strain hardening rate corresponding to Stage II hardening [5,6]. In differentiated form, Eq. (13) can be rewritten as

$$\Theta = \Theta_{II}\left(1 - \sigma / \sigma_s\right), \tag{17}$$

where $\Theta = (\partial\sigma / \partial\varepsilon_{\dot\varepsilon})$ is the strain hardening coefficient.

The Voce equation commonly holds for single-phase coarse-grained materials, particularly fcc ones. An example of pure Cu, recently discussed in Ref. 7, is shown in Fig. 6.

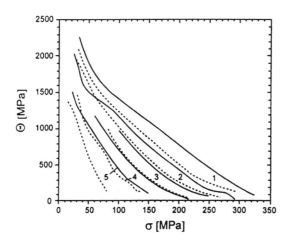

Figure 6. $\Theta - \sigma$ curves for Cu for 5 temperatures (*1:* room temperature; *2:* 100° C; *3:* 200° C; *4:* 300° C; *5:* 400° C) and two strain rates: 1 s^{-1} (solid lines) and 10^{-4} s^{-1} (dashed lines).

The $\Theta - \sigma$ graphs, though not linear in the entire range of data, do show behaviour approximately describable by the above equations over large portions of the curves. This becomes particularly apparent when the $\Theta - \sigma$ diagrams are re-plotted using the normalised quantities Θ / μ and σ / σ_s, cf. Fig. 7. (Note that in the nomenclature of Ref. 7 the quantities σ_v and Θ_o are synonymous with σ_s and Θ_{II} introduced above.) Over a broad range of σ / σ_s or Θ / Θ_{II} all diagrams fall on a common *master curve*: a straight line characterised by two parameters: σ_s and Θ_{II}.

The parameter Θ_{II} is approximately constant - apart from temperature dependence - through the shear modulus μ and a (weak) temperature and strain rate dependence through the factor $(\dot\varepsilon / \dot\varepsilon_o)^{1/m}$. In assessing these dependences it should be noted that the exponent m is large compared to unity, so that for most practical purposes Θ_{II} / μ can be regarded as a constant. By contrast, the steady state stress σ_s given by Eq. (15) is strain rate and temperature dependent. As typically m >> n > 1 holds [6], this dependence resides chiefly in the last factor in Eq. (15). In the low temperature range, when $\dot\varepsilon_{REF}^*$ can be considered as a constant, it is the exponent n that bears a temperature dependence. Kocks and Mecking [7] suggested a compact formula describing this dependence. Their considerations are based on the experimental observation that for a given material (as exemplified by Cu) the steady state stress depends of the quantity $g = (kT/\mu b^3)\ln(10^7 s^{-1}/\dot{\varepsilon})$ in a unique way, as represented by the double-logarithmic plot on Fig. 8. Disregarding a slight 'hump' on the curve (which they associate with a possible solute effect), the authors [7] approximate it by a straight line, obviously corresponding to an Arrhenius type dependence of the strain rate on stress. It can be conveniently represented by a power-law

$$\frac{\sigma_s}{\mu} = \beta\left(\frac{\dot{\varepsilon}}{\dot{\varepsilon}_1}\right)^{k_B T/U} \tag{18}$$

where $\beta, \dot{\varepsilon}_1$ and U are a set of parameters. Obviously, the exponent $U/k_B T$ is synonymous with n in Eq. (15), thus yielding a temperature-dependent exponent in the power-law. The parameter U was shown [7] to depend on the stacking-fault energy. It is believed [5-7], that this behaviour is associated with dynamic recovery controlled by cross-slip. It appears to be relevant up to fairly high homologous temperatures, up to about 2/3 [7].

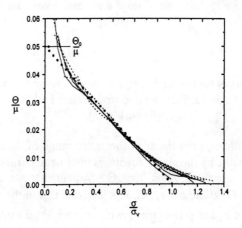

Figure 7. The $\Theta - \sigma$ curves of Fig. 6 re-plotted in terms of the normalised quantities.

In the high-temperature regime proper, when dislocation climb can be expected to control the recovery rate, one can expect $\dot{\varepsilon}^*_{REF}$ to be given by an Arrhenius equation (with the activation energy equal to that for self-diffusion) and exponent n to be constant. It can be set to fit the three-, four-, or five-power-law. Thus, the above constitutive model recovers, in a phenomenological way, both the power-law and the power-law breakdown regimes.

In the above considerations, no distinction was made between steady state flow corresponding to conventional constant strain rate (CSR) testing or constant stress creep. This is an admissible approach as long as static, i.e. time dependent processes do not interfere with the evolution of the microstructure responsible for plastic deformation. Figure 9 shows that the dependence of strain rate on stress for Al obtained under steady state conditions in CSR and creep tests [8] is, indeed, unique: dislocations do not distinguish between creep and CSR tests.

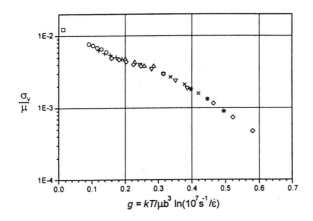

Figure 8. Strain rate and temperature dependence of the steady state stress for Cu. (After Ref. 7)

In materials prone to static recovery, ageing, grain or particle coarsening, *in situ* precipitation, etc. such time effects may be of minor significance under fast strain rates in CSR tests, while under creep conditions, with long times involved, they may become crucial. A unified description is then only possible if time effects are accounted for in an adequate way. Examples of materials exhibiting time dependent structure evolution will be shown below.

An equivalent of the Voce equation for constant stress creep conditions reads [6]

$$\frac{\dot{\varepsilon}^{1/N} - \dot{\varepsilon}_s^{1/N}}{\dot{\varepsilon}_i^{1/N} - \dot{\varepsilon}_s^{1/N}} = \exp\left(-\frac{\varepsilon - \varepsilon_i}{\varepsilon_{tr}^{CREEP}}\right) \tag{19}$$

where $1/N = 1/n + 1/m \approx 1/n$, the steady state creep rate $\dot{\varepsilon}_s$ is given by

$$\dot{\varepsilon}_s = \dot{\varepsilon}_{REF}^* (\sigma/\sigma^*)^n, \quad \sigma^* = M\alpha\mu b(k_1/k_{20}), \tag{20}$$

and

$$\varepsilon_{tr}^{CREEP} = \frac{n}{m}\frac{\sigma}{\Theta_{II}} \tag{21}$$

Equation (19) describes primary (transient) creep, while Eq. (20) gives the steady state creep rate. Again, it recovers the power-law creep or power-law breakdown, depending on the assumptions made with respect to n and $\dot{\varepsilon}^*_{REF}$. The rate of approach of steady state is controlled by the characteristic transient strain ε_{tr}^{CREEP}.

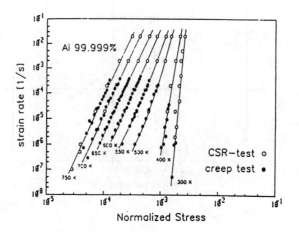

Figure 9. Steady state data for high purity Al [8] for different temperatures. Data from CSR and creep tests fall on a common curve for each test temperature.

3.3. SOLUTE EFFECTS

In discussing solute effects on steady state deformation, we first look at *stationary solutes*. Their role is two-fold. First, they increase the overall flow resistance by an additive term σ_{sol} proportional to some power (typically ½) of the solute concentration c. Second, they may influence the stacking fault energy thus affecting the rate of recovery. As shown in Ref. 6, this effect is reflected in a change of the exponent n that can be expressed as

$$n_{alloy} = n(1 + \eta c) \qquad (22)$$

where η is a (positive) numerical parameter characterising the solute effect on the stacking fault energy. This suggests that the $\lg \dot{\varepsilon}$ *vs.* $\lg \sigma$ diagram for a solid solution can be obtained from the corresponding straight line that represents the steady state behaviour of the solute-free material in two steps. First, this straight line is to be tilted to increase its slope by $(1 + \eta c)$. Second, each point of the resulting graph is to be shifted to the right by an amount σ_{sol} / σ. A curvature is thus introduced which *mimics a decrease of the exponent in the power-law with growing stress*.

For *mobile solutes* capable of diffusing towards dislocations temporarily arrested at localised obstacles and additionally pin them, a negative strain rate sensitivity of the flow stress would result in a certain range of strain rates and temperatures. In addition, dynamic recovery is inhibited. These effects disappear for sufficiently high strain rates, while in the low strain rate limit two effects enhance the steady state stress: (i) the dislocation related contribution is increased due to inhibited recovery and (ii) interaction of dislocations with solute atmospheres segregated on them adds to the overall stress. A detailed description can be found in Ref. 6. As neither of the two effects considered reproduce the empirically found relation given by Eq. (5), the question of the underlying mechanisms is still open.

3.4. EFFECT OF SECOND-PHASE PARTICLES

From among various types of second phases we only discuss non-shearable particles, such as e.g. oxide dispersions, carbides, and the like. These are considered to act as impenetrable obstacles to dislocations and as such they limit the dislocation mean free path L. In the extreme case of very high particle concentrations, L can be replaced with a quantity Λ controlled by and proportional to the particle spacing in the glide plane. Integration of the constitutive equations yields an equation for the evolution of stress with strain similar in structure to the Voce equation, but with σ^2 in place of σ [9]. The strain hardening behaviour is characterised by a linear dependence of the product $\Theta\sigma$ on σ^2. This dependence, shown in Fig. 10, was, indeed, found for Alloy 617 – a Ni based alloy with Mo and Cr carbides as second phase particles [10]. Also shown are creep curves (solid lines) calculated using the parameter values obtained from CSR and strain rate jump tests. The predicted curves are in reasonably good agreement with experimental creep curves.

A next step is to combine dislocation structure related obstacles with particles. This is done by using a superposition rule for inverse mean free paths for the two types of obstacles ('hybrid' model [11]), which leads to the following evolution equation for the dislocation density:

$$d\rho / d\varepsilon_{GB} = M\left(k + k_1\sqrt{\rho} - k_2\rho\right) \tag{23}$$

with $k = 1/(\Lambda b)$. This changes the strain hardening and creep behaviour quite significantly. Without going into detail (which can be found in Ref. 6), we just note that the particle effect reduces the primary creep stage as well as the level of steady state creep rate (Fig. 11, left diagram). The straight line in the $\lg \dot{\varepsilon}$ vs. $\lg \sigma$ diagram turns into a curve whose slope is double that of the matrix diagram in the low strain rate limit, cf. Fig. 11 (right diagram).

Figure 12 that shows the effect of Nb carbides on steady state creep in a Fe-20% Cr alloy illustrates this behaviour. Curve 2 (solid line) was calculated using the constitutive model based on Eq. 23.

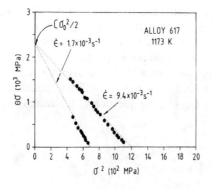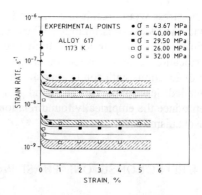

Figure 10. Strain hardening (left) and creep (right) behaviour of Alloy 617.

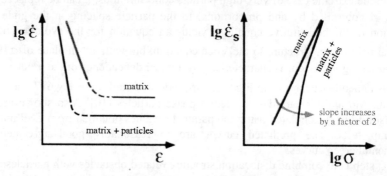

Figure 11. The effect of particles on transient (left) and steady state (right) creep.

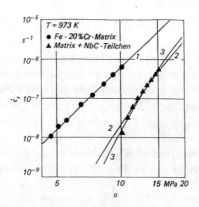

Figure 12. The effect of NbC particles on steady state creep of Fe-20% Cr.
1 – particle-free matrix, *2*- calculation based on Eq. (23), *3* – calculation accounting for attractive dislocation-particle interaction. (Data: courtesy G. Sauthoff, MPIE Düsseldorf).

In the above analysis particles played a 'passive' role, being just immobilisers for gliding dislocations. There are reasons to believe that recovery by cross-slip or glide of immobilised dislocations is inhibited by the attractive interaction between dislocations and particles, cf. [12]. This calls for a modification of the recovery term in Eq. (23). This can be done by multiplying the recovery coefficient k_2 with a factor $\varphi(\sigma, T)$ reflecting the probability of dislocation detachment from a particle. Curve 3 in Fig. 12 represents this modification done in a simplified way. An adequate form of the factor $\varphi(\sigma, T)$ should involve the activation energy for the (thermally activated) detachment process as calculated by Rösler and Arzt [13].

A more complex situation to be modelled is the one when the particle population undergoes *in situ* variations during creep. As an example we consider Alloy 800H undergoing creep deformation at 1173 K when precipitation of Cr-rich and Ti-rich carbides occurs, alongside with Ostwald ripening of carbides [14]. If the specimens are aged prior to creep testing, a normal transient creep is observed, cf. Fig. 13. By contrast, as a result of an interplay between precipitation and ripening, a pronounced minimum on the $\lg \dot{\varepsilon}$ vs. ε curve is found for specimens that did not undergo prior ageing. The steady state creep rate is about the same in both cases. This behaviour is well described by a model [14] that includes the Avrami-type time dependence of the volume fraction of carbides and the Lifshitz-Slyozov-Wagner kinetics of Ostwald ripening.

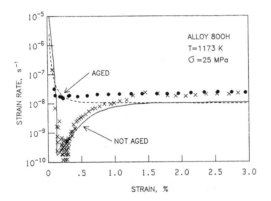

Figure 13. Creep behaviour of Alloy 800H.
The solid and dashed curves represent the model predictions.

A next level of complexity is involved if not only the average particle size and spacing, but also their *arrangement* are to be considered. For that, the Voronoi mesh approach was developed [15]. The effect of clustering calculated in this way is seen in Fig. 14 showing the transient and the steady state behaviour of two similar Ni base ODS alloys, MA 754 and PM 1000, that differ only in the degree of clustering of oxide dispersions. Alloy PM 1000, with clustered particles, exhibits inferior creep properties as compared to its counterpart, PM 1000, that possesses a random particle arrangement. A good accord between the predicted and the experimentally observed curves was reached with this, relatively new, simulation tool.

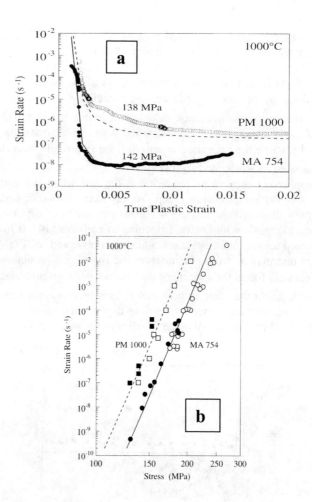

Figure 14. Transient (a) and steady state (b) creep curves
for alloys MA 754 and PM 1000.

3.5. CREEP IN INTERMETALLIC COMPOUNDS

The model description outlined in the previous section can be adopted to describe plastic
flow in various materials whose microstructure provides a geometrical constraint on the
dislocation free path, such as multilayers, Ni base superalloys (where L is determined by
the width of γ 'channels' between γ' particles), pearlitic structures, etc. A recent example
is creep of fully lamellar γ-TiAl [16]. The effect of the average interface spacing on
transient and steady state creep was investigated in the temperature range of 700°C –

800°C. The results were found to be in a very good agreement with the predictions of the hybrid model [11], as illustrated by Fig. 15.

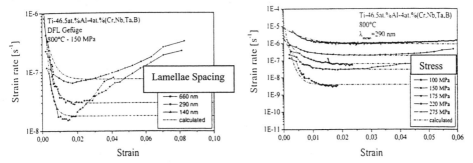

Figure 15. Creep behaviour of fully lamellar γ-TiAl in dependence on the mean interface spacing (left) and stress (right) [16]. The lines correspond to calculations with the hybrid model [11]. (Tertiary creep was not included.)

3.6. TWO INTERNAL VARIABLE MODEL

The above examples show that the model based on just one internal variable (the overall dislocation density ρ) can account for creep behaviour reasonably well, both at low and high temperatures. However, deviations from the model predictions (such as e.g. the occurrence of stage IV hardening precluding the attainment of the steady state stress σ_s) make it necessary to operate with more sophisticated models. A two-internal variable model for dislocation cell-forming materials [17,18] distinguishes between the dislocation densities in the cell walls and cell interiors. The evolution equations for the two dislocation densities, together with respective kinetic equations and a scaling relation between the cell size d_c and the total dislocation density ρ_t,

$$ d_c = K / \sqrt{\rho_t} , \quad \rho_t = f_{wall} \rho_{wall} + (1 - f_{wall}) \rho_{cell} \tag{24} $$

provide a new constitutive frame. It is important that a decrease of the dislocation wall volume fraction f_{wall} is assumed to decrease with (scalar) equivalent resolved shear strain γ^r. The particular form of this decrease is taken to be

$$ f_{wall} = f_\infty + (f_0 - f_\infty) \exp(-\gamma^r / \tilde{\gamma}^r) \tag{25} $$

where f_0, f_∞ and $\tilde{\gamma}^r$ are parameters. Furthermore, crystal plasticity is included linking the scalar equivalent resolved shear strain rate and resolved shear stress to the corresponding quantities for the actual glide systems of a polycrystal. With this model, it was possible to predict the occurrence of Stage IV strain hardening, with all its details, along with the Voce-like behaviour in Stage III preceding the onset of Stage IV, cf. Fig. 16. Texture development in a polycrystal can also be followed using the model, cf. [19].

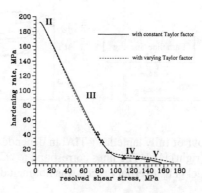

Figure 16. Prediction of all strain hardening stages [17].
The data points refer to pure copper.

4. Dislocation Climb Controlled Plasticity

In the foregoing section, dislocation climb entered only as a possible dynamic recovery mechanism. As such, it controlled steady state plastic flow, but was not considered to provide a direct contribution to plastic strain rate. In Ref. 2, models leading to power law creep, particularly by direct climb mechanism, were reviewed. The latter group of models make use of the relation

$$v_{CLIMB} \propto \frac{D_{SD}}{b} \frac{\Omega \sigma}{k_B T} \tag{26}$$

for climb velocity. Here Ω is the atomic volume. Equation (26) applies for $\Omega \sigma / k_B T \ll 1$. The Orowan equation for plastic strain rate then yields

$$\dot{\varepsilon}_s \propto \rho_m b v_{CLIMB}, \tag{27}$$

where ρ_m is the density of mobile dislocations. Assuming that ρ_m is proportional to the total dislocation density ρ and using a scaling relation between the stress and ρ akin to Eq. (8), one arrives at three-power-law, commonly referred to as the *natural power-law*. However, there are no physical grounds to assume proportionality between the mobile and the total dislocation densities, especially in a situation when a substantial proportion of the dislocation density resides in cell or subgrain walls. Furthermore, the natural power-law is in conflict with numerous observations on pure metals yielding a stress exponent of 4 or 5. Artificial assumptions on the stress dependence of mobile density (like the cubic one suggested by Barrett and Nix) do not appear to resolve the difficulties of the model. The idea of Argon and Moffatt [20] that v_{CLIMB}, in addition to the factors in Eq. (26), should also include the concentration of jogs on edge dislocations appears more plausible. As jogs are formed by forest intersections, their density is linear in the average dislocation spacing $1/\sqrt{\rho}$, which is proportional to stress. Thus, a stress exponent of four in the power-law is obtained. The dependence on the stacking fault energy also comes into play in the right way, at least qualitatively.

5. Diffusion Controlled Plasticity

In a large area in the 'South-East' of Ashby map, plastic flow is controlled by diffusion mass transport, rather than by dislocation movement. We consider a model [21] that makes it possible to follow how the dislocation mechanism of plastic flow yields to diffusion controlled plasticity. The material is partitioned – in much the same way as in Section 2.6 - in two 'phases', but this time in the grain interior (GI) and the grain boundaries (GB). The grain interior may still be sub-divided into cells or subgrains, but these are only considered as agents determining the mean free path of mobile GI dislocations. Plastic flow in the grain interior is taken as a linear superposition of contributions from dislocation glide (as described by the hybrid model, with dislocation mean free path depending on the grain size), diffusion flow through the grain bulk (Nabarro-Herring mechanism) and diffusion flow via grain boundaries (Coble mechanism). The grain boundary phase was assumed to deform by pure diffusion flow. (In this regard the grain boundaries play a dual role, being 'channels' through which their own material flows and, at the same time, the grain interior material is transported by Coble mechanism. The two fluxes thus co-exist in a channel. Stresses in the two phases weighted according to their volume fractions add up to the overall stress following a rule of mixtures. The rate of plastic flow in the grain boundary phase is given by [21]:

$$\dot{\varepsilon}_{GB} = A * \frac{\Omega \sigma_{GB}}{kT} \frac{D_{SD}^{GB}}{d^2}.$$

(28)

This equation offers an interesting blend of a Nabarro-Herring and Coble like features: while the inverse square dependence on the grain size d is akin to that in the Nabarro-Herring formula, the coefficient of self-diffusion, D_{GB}^{SD}, is that for *grain boundary* self-diffusion, as in the Coble formula. $A*$ is a numerical constant. Whereas for sufficiently large average grain size the dislocation glide controlled regime will prevail, for very fine-grained materials (d in the sub micrometer range) the diffusion control terms will be preponderant. The common Hall-Petch relation that gives an inverse dependence of strength on the square root of the grain size will no longer hold in the diffusion controlled case, and softening will be observed. This behaviour for pure copper is seen in Fig. 17.

236

Even for room temperature, the critical grain size d_c below which the Hall-Petch relation does not hold, is in the range of 0.1 μm for the strain rate of 10^{-5} s^{-1}, and it moves to larger values if the strain rate is lowered or temperature is increased. Obviously, for $d < d_c$ viscous flow occurs, characterised by the stress exponent of one.

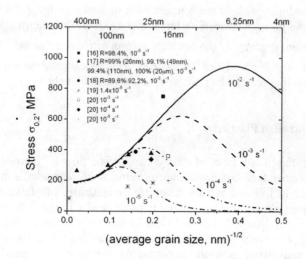

Figure 17. Calculated (lines) and experimental (symbols) yield strength for Cu of different grain size deformed at various strain rates [21].

The model [21] makes it possible to assess the relative roles of the dislocation glide and the diffusion mechanisms of plastic flow and to identify the borderlines between the domains where different mechanisms prevail.

6. Conclusions

High temperature deformation of metallic materials is a vast area, which could not be covered in its entirety in the lecture, of course. These notes rather give a somewhat sketchy, perhaps eclectic, picture that reflects the author's preferences. The largest portion of the paper dealt with dislocation glide controlled plasticity that is believed to be the governing mechanism up to very high homologous temperatures. As steady state deformation gains on significance with increasing temperatures, and as it is largely determined by dynamic recovery by cross slip of screw dislocations and/or climb of edge dislocations, these two processes are particularly important. The growing role of recovery by climb with increasing temperature leads to a change in the character of the stress dependence of steady state creep rate: from an exponential (or hyperbolic sine) to a power-law one, also known as 'Norton' creep. The occurrence of stress exponents higher than 4 still provides difficulties for theoretical models. Possible ways of resolving these difficulties were discussed in recent reviews [3], [23], [24]. Of particular interest is a recent paper [25] in which the significance of dynamic recovery mechanisms is reassessed in the framework of the dislocation glide plasticity model. A crucial role in this approach is ascribed to dislocation dipole collapse by climb, as opposed to that by spontaneous dipole annihilation. The range of applicability of the Kocks-Mecking model is delineated vis-à-vis the Nes-Marthinsen model of climb controlled dipole collapse. However, the position of a 'demarcation line' between the two mechanisms on the modulus-compensated stress vs. normalised temperature ($k_B T / \mu b^3$) diagram hinges on certain assumptions regarding the values of model parameters and cannot be regarded as finally established.

Further down the modulus compensated stress, in the diffusion controlled plastic flow region, some new developments have been taking place lately, particularly with regard to the grain size effects in sub micrometer range. A brief account of these developments was given.

In conclusion, the field of high temperature plasticity, which is of huge significance for applications of structural materials, still provides a number of challenges to researchers, and the coming years will definitely see interesting new results there.

References

1. Frost, H.J., and Ashby, M.F. (1982) *Deformation-Mechanism Maps*, Pergamon Press, Oxford.
2. Kassner, M.E., and Pérez-Prado, M.-T. (2000) Five-power-law creep in single phase metals and alloys, *Progr. Mater. Sci.* **45**, 6-102.
3. Sherby, O.D., and Taleff, E.M. (2002) Influence of grain size, solute atoms and second-phase particles on creep behavior of polycrystalline solids, *Mater. Sci. Eng. A* **322**, 89-99.
4. G. Bermig, A. Bartels, H. Mecking and Y. Estrin (1997) A unique description of steady state deformation of Al and Al-Mg alloys, *Mater. Sci. Eng. A* **234-236**, 904-907.
5. Mecking, H., and Kocks, U.F. (1981) Kinetics of flow and strain hardening, *Acta metall.* **29**, 1865-1877.
6. Estrin, Y. (1996) Dislocation-Density Related Constitutive Modeling, in A.S. Krausz and K. Krausz (eds.), *Unified Constitutive Laws of Plastic Deformation*, Academic Press, San Diego, pp. 69-106.
7. Kocks, U.F., and Mecking, H. (2002) Physics and phenomenology of strain hardening, the fcc case, *Progr. Mater. Sci.* (in press).
8. Mecking, H., Styczynski, A., and Estrin, Y. (1998). Steady state and transient plastic flow of aluminium and aluminium alloys, in P.O. Kettunen, T.K. Lepistö and M.E. Lehtonen (eds.) *Strength of Metals and Alloy (ICSMA 8)*, Pergamon Press, Oxford, pp. 989-994.
9. Estrin, J., (1987), Stoffgesetze der plastischen Verformung und Instabilitäten des plastischen Fließens (Habilitationsschrift) *VDI Forschungsheft* **642**, pp. 1-48.
10. Valsan, M., Ennis, P.J., Estrin, Y., and Schuster, H. (1988), unpublished.
11. Estrin, Y., and Mecking, H. (1984) A unified phenomenological description of work hardening and creep based on one-parameter models, *Acta metall.* **32**, 57-70.
12. Reppich, B. (1993) Particle Strengthening, in R.W. Cahn, P. Haasen, and E.J. Kramer (eds.) *Materials Science and Engineering*, VCH, Weinheim, Vol. 6, pp. 311-357.

238

13. Rösler, J. and Arzt, E. (1990) *Acta metall.* **38**, 671.
14. Reichert, B., Estrin, Y., and Schuster, H. (1998) Implementation of precipitation and ripening of second-phase particles in the constitutive modelling of creep, *Scripta mater.* **38**, 1463-1468.
15. Estrin, Y., Arndt, S., Heilmaier, M., and Brechet, Y. (1999) Deformation behaviour of particle-strengthened alloys: a Voronoi mesh approach, *Acta mater.* **47**, 595-606.
16. Chatterjee, A., Mecking, H., Arzt, E., Clemens H. (2002) Creep behavior of γ-TiAl sheet material with differently spaced fully lamellar microstructures, Mater. Sci. Eng. A **329-331**, 840-846.
17. Estrin, Y., Tóth, L.S., Molinari, A., and Bréchet, Y. (1998) A dislocation based model for all hardening stages at large strain deformation, *Acta mater,* **46**, 5509-5522.
18. Tóth , L.S., Molinari, A., Estrin, Y. (2002) Strain hardening at large strains as predicted by a dislocation based polycrystal model, *J. Eng. Mater. Techn.* **124**, 71-77.
19. Baik, S.C., Estrin, Y., Kim, H.S., Jeong, H.T. , and Hellmig, R.J. (2002) Calculation of deformation behavior and texture evolution during equal channel angular pressing of IF steel using dislocation-based modeling of strain hardening, *Mater. Sci. Forum* **408-412**, 697-702.
20. Argon, A.S., and Moffatt, W.C. (1981) Climb of extended edge dislocations, *Acta mater.* **29** 293-299.
21. Kim, H.S., Estrin, Y., and Bush, M.B. (2000) Plastic deformation behaviour of fine-grained materials, *Acta mater.* **48** 493-504.
22. Yamakov, V., Wolf, D., Phillpot, S.R., Gleiter, H. (2002) Grain boundary diffusion creep in nanocrystalline palladium by molecular-dynamic simulations, *Acta mater.* **50** 61-73.
23. Nes, E., (1998) Modelling of work hardening and stress saturation in fcc metals, Progr. Mater. Sci. **41** 129-193.
24. Blum, W., Eisenlohr, P., and Breutlinger, F. (2002) Understanding creep – a review, *Metall. and Mater. Trans. A* **33** 291-303.
25. Nes, E., Marthinsen, K., and Brechet, Y., (2002) On the mechanisms of dynamic recovery, *Scripta mater.* **47** 607-611.

TRANSFORMATION-INDUCED PLASTICITY IN STEELS

P. J. JACQUES
Université catholique de Louvain, Département des sciences des matériaux et des procédés, PCIM, Place Sainte Barbe 2, B-1348 Louvain-la-Neuve, Belgium

Abstract

The Transformation-induced plasticity (TRIP) phenomenon has been known for quite a long time, especially in steels and iron-based alloys. It was shown to improve in a large way mechanical properties such as strength, ductility and toughness. New applications and a renewed interest for this phenomenon emerged in the last years with the so-called 'TRIP-assisted multiphase steels'. These low-alloy steels consist of metastable fine grained retained austenite dispersed in a ferrite-based microstructure. Thanks to the conjunction of the TRIP effect and a composite strengthening effect, these steels present combinations of strength and ductility never reached before.

The present study focuses on the physics and mechanics of phase transformations occurring during the process and the use of TRIP-assisted multiphase steels. On the one hand, we show how the complex microstructures of these steels are built and optimised during the thermomechanical processing. Thermodynamics and kinetics of phase transformations, especially the bainite transformation were studied. On the other hand, the mechanical properties of the resulting microstructures were scrutinised at the microscopic and macroscopic scales.

1. Introduction

The possibility of formation of martensite during the mechanical loading of austenitic steels has been known for quite a long time. This phenomenon was also shown to bring about large enhancements of the work-hardening rates of these steels and thus to postpone the onset of necking, so delaying the fracture. The acronym 'TRIP' (for TRansformation-Induced Plasticity) was proposed to express the large operative efficiency of the martensitic transformation as a deformation mechanism [1].

A huge literature deals with this topic since the 1950s and very good reviews can be found [2, 3]. The present paper proposes first a short summary of the basic concepts dictating the thermodynamics and mechanics of deformation-activated martensitic transformation. It will then focus more thoroughly on the renewed interest for the TRIP phenomenon in the case of newly developed low-alloy multiphase steels.

A. Finel et al. (eds.), Thermodynamics, Microstructures and Plasticity, 241–250.

2. Mechanical activation of martensitic transformation

In order to proceed, the displacive and diffusionless martensitic transformation needs to be accompanied by a decrease in free energy. As shown by figure 1, the austenite (γ) will not decompose into martensite (α') above the well known Ms temperature even if the γ - α' transformation could proceed for temperatures lower than T_0 (for which the free energy of both phases with identical compositions are equal). A well-defined 'driving force' corresponding to the activation barrier is needed to trigger the transformation. This means that in the absence of any other competing phase transformation, austenite cooled down to temperatures ranging from Ms to T_0 stays in a metastable state. However, it can be easily envisioned that the mechanical state of the surroundings may influence the transformation. A mechanical driving force U, resulting from an externally applied stress may be added to the chemical driving force $\Delta G_{Tl}^{\gamma-\alpha'}$ in order to attain the requested critical driving force $\Delta G_{Ms}^{\gamma-\alpha'}$ at temperature T_l. Consequently, martensitic transformation can be mechanically-induced when a mechanical work supplies the lacking driving force for triggering the transformation [4].

Figure 2 schematically shows the evolution of the critical applied stress for the onset of the martensitic transformation as a function of temperature. Just above Ms, the transformation is said to be *stress-assisted* and the mechanical driving force allows martensitic transformation to occur on the same nucleation sites as during cooling [3]. When temperature is further raised, the critical stress for triggering transformation reaches the austenite elastic limit. Austenite will therefore first be plastically deformed before transformation can occur. This plastic deformation creates potent nucleation sites that activate the martensitic transformation. This regime is called *strain-induced* transformation. The transition temperature Ms^{σ} corresponds to a reversal of the temperature dependence of the macroscopic flow stress [3] and defines the boundary between the temperature regimes where the 2 modes of transformation, i.e. stress-assisted and strain-induced dominate.

2.1. TRANSFORMATION – INDUCED PLASTICITY

Since martensitic transformation can occur during mechanical solicitation, its consequences on the deformation process has been thoroughly studied. It was shown that the aftermath of the mechanically-activated transformation of austenite, such as the dilatational and shear deformations arising from the change of crystal structure, the formation of preferential variants of martensite and the abrupt change of strength [4-7] improve strength and ductility by shaping the work-hardening rate [1, 3, 8]. These effects of Transformation-Induced Plasticity that are associated with the intrinsic characteristics of the martensitic transformation and with the consequences of the transformation on the surrounding microstructure are classically related to two different mechanisms [9, 10]: (i) the stress-assisted nucleation of martensitic variants favourably oriented with respect to the applied stress (Magee mechanism [11]); and (ii) the microscopic plasticity of the phases due to the volume and shape changes associated with the displacive transformation (Greenwood-Johnson effect [12]).

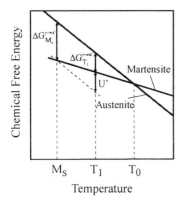

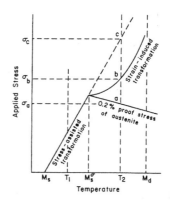

Figure 1: Schematic illustration showing chemical free energies of austenite and martensite as a function of temperature.

Figure 2: Schematic illustration showing the critical stress for martensite formation as a function of temperature.

Figure 3 shows that the combination of mechanical properties such as tensile strength and elongation exhibits an overall optimum when testing is carried out in the temperature range comprised between the Ms temperature of spontaneous martensitic transformation on cooling and the Md temperature of end of mechanically-induced martensitic transformation. Tensile strength increases with decreasing test temperature, and this increase almost corresponds to the increase of the martensite content observed in the specimen after fracture. Total elongation exhibits a maximum at the temperature just above Ms^σ. This enhancement of elongation is attributed mainly to the suppression of necking due to the increase in work-hardening rate induced by the continuous formation of martensite all along plastic straining. As a consequence, this TRIP effect was found to be very beneficial for the deformation properties of a wide variety of fully austenitic Fe-Ni-Cr [13], Fe-Ni [14] or other highly alloyed steels [4].

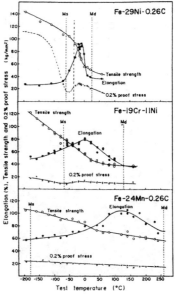

Figure 3: Effect of test temperature on tensile properties of austenitic alloys ([4]).

The TRIP effect has also been found very effective for the improvement of the toughness properties either in fully austenitic steels or in high-strength martensitic and bainitic steels containing dispersed retained austenite [15, 16].

3. TRIP-assisted multiphase steels

The TRIP effect has regained attention recently in the case of low-alloy steels. The studies in this case are motivated by the necessity for the steel industry to develop always better suited high-strength structural steels with low production costs. Beside general strengthening mechanisms such as grain refinement or precipitation strengthening, the efficiency of the martensitic transformation to improve both strength and ductility seems very attractive. The 1990s have thus seen the development and characterisation of new formable high-strength steel grades, the so-called *TRIP-assisted multiphase steels*. As illustrated in figure 4, these steels present complex multiphase microstructures consisting of a ferritic matrix and a dispersion of multiphase grains of bainite, martensite and metastable retained austenite. In the present case, the (meta)stabilisation of austenite at room temperature is ensured by its carbon content. As shown in figure 5, the C enrichment of austenite occurs all along specifically designed thermal or thermomechanical cycles. After cold- or hot-rolling, a first step consisting in an intercritical annealing brings about the general morphology of the microstructure, i.e. the dispersions of ferrite and austenite grains. The first carbon enrichment of the austenite accompanies the nucleation and growth of the phases. Several studies on the formation of the ferrite/austenite mixture during intercritical annealing have been carried out in the 1970s and 1980s [17, 18] in the case of Dual Phase steels. However, the maximum carbon enrichment of the austenite during intercritical annealing is not sufficient for its stabilisation against martensitic transformation at room temperature. A second step combined with particular alloying elements allows further carbon enrichment. Indeed, the carbon redistribution occurring during bainite transformation and the inhibition of carbide precipitation by some alloying elements (Si, Al, ...) allow further C enrichment and the stabilisation of austenite at room temperature. The TRIP-assisted multiphase steels thus classically contain some carbon (from 0.1 to 0.4 wt.%), silicon or aluminium (around 0.5 – 1.5 wt.% to inhibit cementite precipitation) and manganese (around 0.5 to 1.5 wt.%, providing some hardenability).

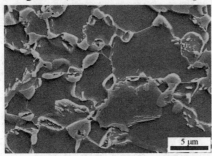

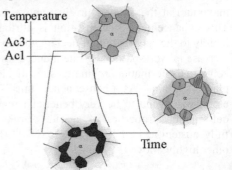

Figure 4: SEM micrograph of the typical microstructure of TRIP - assisted multiphase steels.

Figure 5: Schematic representation of the heat-treatment applied to the TRIP-aided steels.

3.1. AUSTENITE STABILISATION THROUGH BAINITE TRANSFORMATION

Bainite transformation has thus become of primary importance for the optimisation of the mechanical properties of the TRIP-aided steels. Thanks to the redistribution of excess carbon to the austenite occurring after the formation of the bainitic ferrite platelets and to the addition of some alloying element such as silicon or aluminium that inhibits the formation of cementite, residual austenite with C contents as high as 1.2 wt.% can be obtained. However, even if a clear vision of the way by which bainite transformation can be found [19], recent work [20] has shown that the bainite transformation occurring in the case of the TRIP-aided steels presents some peculiarities due to the small grain size of the intercritical austenite present in the TRIP-aided steels [21]. Figures 6 and 7 compare the bainite morphology as a function of the austenite grain size. The classical bainite sheaf structure can been clearly seen in figure 6 which shows that the first bainitic ferrite sub-units nucleate at the austenite grain boundary and grow towards the interior of the austenite grain. New sub-units then nucleate and grow from the tip of the previous ones, bringing about the sheaf structure. This process is valid as long as tip nucleation is possible, i.e. as long as the austenite grains are larger than the platelets length. In contrast, the bainite morphology is completely different when the austenite grain size is only a few µm as in the case of the intercritical austenite of the TRIP-aided steels shown in figure 7. The bainite that forms in very small austenite grains presents adjacent platelets that completely cross the austenite grain. As a consequence, the kinetics of the bainite transformation is strongly modified [22].

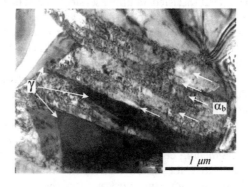

Figure 6: TEM micrograph showing the typical sheaf structure of bainite.

Figure 7: TEM micrograph of the particular morphology of bainite in the case the small austenite grains of the TRIP-aided steels.

Even if the bainite transformation exhibits some peculiarities in the case of the TRIP-aided steels, its main consequence remains the C enrichment of the residual austenite as long as cementite precipitation is inhibited. Figure 8 illustrates the change of the room temperature microstructure of a typical TRIP-aided steel all along the bainitic holding. Such a transformation map clearly shows the progressive change from martensite to austenite accompanying the bainitic holding. As also shown by figure 8, the stabilisation is due to the C enrichment of austenite operating up to a maximum

246

dictated by the T0-curve as shown by figure 9, in agreement with the displacive and diffusionless nature of the bainite transformation [19].

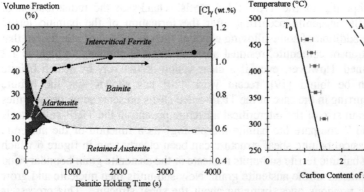

Figure 8: Transformation map and austenite C enrichment during the bainitic holding of a classical TRIP-aided steel [23].

Figure 9: Comparison of the maximum C content of austenite with the T_0 curve of a 1.5Mn-1.5Si TRIP-aided steel.

3.2. MECHANICAL PROPERTIES OF THE TRIP-AIDED STEELS

The TRIP-assisted multiphase steels present the unique feature that the TRIP effect occurs for small austenite grains dispersed in a soft ferritic matrix. The understanding of the hardening properties of these steels thus requires the characterisation not only of the TRIP effect but also of the composite strengthening effect emerging from these complex microstructures.

Figure 10 presents typical true stress – true strain tensile curves of TRIP-aided and Dual Phase steels, as well as the evolution of the retained austenite content with true strain for the specimens TRIP1, TRIP2 and TRIP3. Specimen DP with a microstructure consisting of 25% of martensite dispersed in an intercritical ferrite matrix, exhibits quite a high strength with a reasonable true uniform strain thanks to a composite strengthening effect. However, still better properties are exhibited by specimens TRIP1 and TRIP3 for which martensitic transformation is uniformly distributed all along plastic straining (as illustrated by the evolution of the volume fraction of retained austenite with true strain). Specimen TRIP3 presents an initial microstructure consisting of 8% of retained austenite and around 30% of bainite together with the intercritical ferrite matrix. In the case of specimen TRIP1, the initial amount of retained austenite is about twice the level in specimen TRIP3 while the volume fraction of bainite is 35%. Specimen TRIP2 also contains retained austenite. However, the austenite C content is higher than for specimen TRIP1 so that it does not transform during straining. These results clearly show that the best strength – ductility balance emerges from the continuous strain-induced martensitic transformation and not only from the composite nature of the microstructure.

Figure 11 presents a TEM micrograph of a deformation-induced martensite grain within the ferritic matrix. This figure illustrates that the martensitic transformation in the present multiphase microstructures brings about a large plastic deformation of the surrounding ferrite as shown by the high density of dislocations present at the martensite – ferrite interface [24]. This plastic deformation is furthermore localised at one of the edge of the martensitic variants. The martensitic transformation thus plays the role of an additional source of dislocations, thus leading to an improved work-hardening of the microstructures presenting such thermally- or mechanically-activated martensitic transformation. As a consequence, the increase of the work-hardening capabilities and thus the higher values of strength and uniform strain result not only from the strengthening due to the presence of hard martensite grains within ferrite based microstructures but also from the dislocation strengthening accompanying the progressive martensitic transformation during straining.

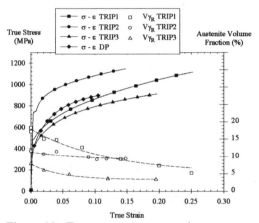

Figure 10: True stress – true strain curves and evolution of V_{γ_r} during plastic straining of the different specimens.

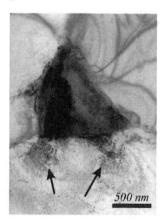

Figure 11: TEM micrograph showing the dislocations generated in the ferrite by the strain-induced marten-sitic transformation .

3.3. MICROMECHANICAL CHARACTERISATION AND MODELLING

Any attempt of developing a physically-based model describing the mechanical properties of the TRIP-aided steels requires the knowledge of the flow properties of each phase of the microstructure. Since it is impossible to obtain the same phases within monophase microstructures (especially retained austenite), their flow properties have to be directly characterised in finely-grained multiphase microstructures. This characterisation was recently started thanks to AFM-nanoindentation [25] and in-situ neutron diffraction.

Thanks to neutron diffraction, the lattice strains of the BCC (ferrite, bainite) and FCC (austenite) phases were measured during uniaxial loading. This lattice strain is calculated by comparing the d-spacing (d) at a given load with the initial d-spacing (d_0) of the unloaded specimen. ($\varepsilon^{hkl} = (d^{hkl} - d_0^{hkl})/d_0^{hkl}$) [25, 26]. As shown by figure 12, these measurements allow to determine the stress partitioning between the phases during loading. Indeed, the elastic lattice strains are converted into stresses thanks to the knowledge of the elastic constants of the diffracting phases. Figure 12 clearly shows

248

that the yield strength and the stress level are higher in the austenite than in the BCC phases (55% ferrite + 30% bainite). This result is quite surprising when compared to fully austenitic steels for which the yield strength of the austenite is quite low [4]. Such measurements have been carried out on different TRIP-aided steels for which the retained austenite presents different C contents. Figure 13 shows that the high C content of austenite together with the small austenite grain size are responsible for its high yield strength.

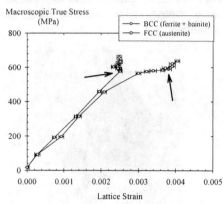
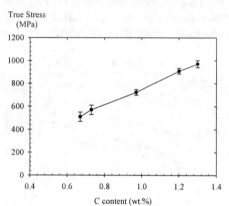

Figure 12: Evolution of the lattice strain of $(211)_\alpha$ and $(311)_\gamma$ during uniaxial tension.

Figure 13: Evolution of the austenite yield strength with its C content in TRIP-aided steels.

Constitutive models for the transformation plasticity in either metastable austenitic steels or multiphase TRIP-assisted steels have been and are still the topic of intensive developments [7, 27, 28]. Micromechanical models may help understanding the physical phenomena occurring during the plastic deformation of metastable multiphase materials. They could also help finding the in-situ mechanical properties of the phases. The current studies on modelling the properties of TRIP-aided steels deal with sophisticated three-phase models accounting for the transformation of austenite to martensite in a ferritic matrix. The selection of the favourably oriented variants of martensite follows the micromechanical criterion originally proposed by Fischer [28].

These models provide a constitutive law for the TRIP-aided steels for given volume fraction, shape, texture of austenite and macroscopic loading conditions. They are for instance based on Finite Element Unit Cell modelling [29], or are developed in the framework of the equivalent inclusion method using the mean field theory for homogenisation [30, 31]. By using these models, the in-situ constitutive law of the phases can be deduced from comparison with experimental values of the stress partitioning between the phases (measured by neutron diffraction). Such models provide satisfactory predictions for the evolution of the austenite volume fraction during loading, showing the strong dependence of the transformation kinetics on the loading conditions [30-32], which was previously experimentally observed.

Implementing the constitutive models into user-defined routines in commercial finite elements codes will help the design of complex structures in which the stress-state usually varies a lot.

4. Conclusion

This paper proposed a short summary of the basic concepts defining the TRIP effect in steels as well as the renewed interest for this phenomenon in the case of the TRIP-assisted multiphase steels. Selected results illustrated how austenite can be retained in low alloy steels thanks to the characteristics of the bainite transformation. The influence of the TRIP effect occurring in a multiphase soft matrix was also exemplified. Finally, the importance of micromechanical characterisation and modelling of these complex steels has been demonstrated.

Acknowledgements

P. J. Jacques acknowledges the Fonds National de la Recherche Scientifique (Belgium). The author is indebted to R&D Cockerill-Sambre (Arcelor group) for continuous support. This work was partly supported by the Belgian State, Prime Minister's Office, Federal Office for Scientific, Technical and Cultural Affairs within the framework of the PAI P5/08 project "From microstructure towards plastic behaviour of single- and multiphase materials".

References

1. Zackay, V.F.; Parker, E.R.; Fahr, D., Busch, R (1967) The enhancement of ductility in high-strength steels, *Trans. Am. Soc. Met.* **60**, 252-260
2. Olson, G.B., Cohen, M. (1986) Martensitic transformation as a deformation process, in S.C. Antolovich, R.O. Ritchie & W.W. Gerberich (eds.), *Mechanical Properties and Phase Transformations in Engineering Materials*, Met. Soc. AIME, New Orleans, 367-390
3. Olson, G.B. (1982) Transformation Plasticity and the stability of plastic flow, in G. Krauss (ed.), *Deformation, Processing, and Structure*, ASM, Metals Park, 391-424
4. Tamura, I., Wayman, C.M., Martensitic transformations and mechanical effects, in G.B. Olson & W.S. Owen (eds.) *Martensite*, ASM, Metals Park, 227-242
5. Patel, J.R., M. Cohen, M. (1953) *Acta Metall.* **1**, 531-538
6. Olson, G.B., Cohen, M. (1982) *Metall. Trans. A* **13A**, 1907-1914
7. Stringfellow, R.G., Parks, D.M., Olson, G.B. (1992) *Acta Metall. Mater.* **40**, 1703-1716
8. Ludwigson, D.C., Berger, J.A. (1969) *J. Iron and Steel Inst.*, 63-69
9. Leblond, J.B., Devaux, J., Devaux, J.C. (1989) *Int. J. Plasticity* **5**, 551-572
10. Marketz, F, Fischer, F.D. (1995) *Metall. Trans. A* **26A**, 267-278
11. Magee, C.L., Paxton H.W. (1968) *Trans. Metall. Soc. AIME* **242**, 1741-1749

250

12. Greenwood, G.W., Johnson, R.H. (1965) *Proc. Roy. Soc. London A* **283**, 403-422
13. Angel, T. (1954) *J. Iron and Steel Inst.*, 165-174
14. Fahr, D. (1971) *Metall. Trans.* **2**, 1883-1892
15. Edmonds, D.V., Cochrane, R.C. (1990) *Metall. Trans. A* **21A**, 1527-1540
16. Haidemenopoulos, G.N., Grujicic, M., Olson, G.B., M. Cohen, M. (1989) *Acta Metall.* **37**, 1677-1682
17. Cai, X.-L., Garatt-Reed, A.J., Owen, W.S. (1985) *Metall. Trans. A* **16A**, 543-551
18. Speich, G.R., Demarest, V.A., Miller, R.L. (1981) *Metall. Trans. A* **12A**, 1419-1428
19. Bhadeshia, H.K.D.H. (2001) *Bainite in Steels*, IOM, London
20. Girault, E., Jacques, P.J., Ratchev, P., Van Humbeeck, J., Verlinden, B., Aernoudt, E. (1999) *Mater. Sc. Eng. A* **273-275**, 471-474
21. Jacques, P.J., Girault, E., Catlin, T., Kop, Th., Geerlofs, N., Van Der Zwaag, S., Delannay, F. (1999) *Mater. Sc. Eng. A* **273-275**, 475-479
22. Jacques, P.J. (2002), Proc. Int. Conf. Martensitic Transformation (ICOMAT02), june 10-13, Helsinki, in press
23. Jacques, P.J., Girault, E., Harlet, Ph., Delannay, F (2001) *ISIJ Int.* **41**, 1061-1067
24. Jacques, P.J., Furnémont, Q., Mertens, A, Delannay, F. (2001) *Phil. Mag. A* **81** 1789-1812
25. Jacques, P.J., Godet, S., Furnémont, Q. (2002), NRC Annual Report, 72
26. Furnémont, Q., Jacques, P.J., Pardoen, T., Lani, F., Godet, S., Harlet, P., Conlon, K.T., Delannay, F. (2002), B.C. De Cooman (ed.), in: Proc. Int. Conf. on TRIP-aided high strength ferrous alloys, *in press*
27. Leblond, J.B., Mottet, G., Devaux, J.C. (1986) *J. Mech. Phys. Solids* **34**, 395-409
28. Fischer, F.D., Reisner, G., Werner, E., Tanaka, K., Cailletaud, G., Antretter, T. (2000) *Int. J. Plasticity* **16**, 723-748
29. Van Rompaey, T., Furnémont, Q., Jacques, P.J., Pardoen, T., Blanpain, B., Wollants, P. (2002) B.C. De Cooman (ed.), in: Proc. Int. Conf. on TRIP-aided high strength ferrous alloys, *in press*
30. Lani, F., Delannay, F., Pardoen, T. (2002), submitted to *Mech. Mater.*
31. Lani, F., Furnémont, Q., Jacques, P.J., Delannay, F., Pardoen, T. (2002) Proc. EMMC6 Conf., Sept. 9-12, Liège, *in press*
32. Reisner, G., Werner, E.A., Fischer, F.D. (1998) *Int. J. Solids Structures* **35**, 2457-2473

IV. Modelling of plasticity and related microstructures

GRAIN SIZE EFFECT OF PLASTICITY MODELLED BY MOLECULAR DYNAMICS

H. VAN SWYGENHOVEN, P.M. DERLET, A. HASNAOUI
Paul Scherrer Institute
CH-5232 Villigen-PSI, Switzerland

1. Introduction

For some polycrystalline metals with grain sizes in the nano regime, experiments have suggested a deviation away from the Hall-Petch relation relating yield stress to average grain size [1]. The debate continues whether or not such deviations are a result of intrinsically different material properties of nanocrystalline (nc) systems, or due simply to inherent difficulties in the preparation of fully dense nc-samples and in their microstructural characterization. Nevertheless, it suggests that the traditional work hardening mechanism of pile-up of dislocations originating from Frank-Read sources may no longer be valid at the nanometer scale. In-situ deformation testing in the transmission electron microscope (TEM), performed on Cu and Ni3Al nc samples, reveals a limited dislocation activity in grains below 50nm [2,3]. However, due to the presence of large internal stresses which make grain boundaries (GB) in TEM images difficult to observe, and also possible artifacts induced by thin-film geometry such as dislocations emitted from the surface [4], in-situ tensile tests did not until now, bring convincing evidence for abundant dislocation activity.

Shear bands have been observed in nc materials after deformation at low and high strain rates [5,6], where in the vicinity of the shear bands, large sized grains, eventually showing texture are present. These are indications that cooperative processes are active in the nc regime, an example of which being the formation of planar interfaces at a scale greater than the grain size, leading to collective sliding. The model of Hahn [7] suggests that grain boundary sliding controls the deformation mechanism, and that, in order to overcome grain boundary obstacles, two or more grain boundaries must cooperate to form a plane interface, which then by further interconnection with other plane interfaces, will lead to long-range sliding.

Mechanical testing also revealed the issue of the "GB state" by means of a property dependence on thermal history and internal strains. It is shown that a substantial strengthening can be obtained by a short heat treatment. The cause of the strengthening is possibly associated with a reduction in internal strains and/or dislocation content produced by the annealing [8]. The effect of strengthening has been measured both on nc materials obtained by grain refinement techniques and those obtained by

A. Finel et al. (eds.), Thermodynamics, Microstructures and Plasticity, 253–263.

consolidation of clusters. These observations strongly suggest that there is still a lack in understanding of the relationship between the structure of GB and triple junctions (TJ) and the overall mechanical properties.

Large-scale molecular dynamics are used successfully to study the structural and mechanical properties of nc metals. Despite the limitations in time and length scales, which impose high strain rates and short deformation times, as is discussed by the author in [9], the MD technique is providing an invaluable detailed picture of the atomic scale processes during plastic deformation. The present paper gives an overview of the key observations on structural and mechanical properties observed in MD simulations that are performed on fully 3D nanocrystalline fcc metals, synthesized using the Voronoi construction [10-13]. These samples are not comparable with the 2D columnar network structures of Yamakov et al [14] since, in fully 3D, the resulting shear across a grain can be accommodated by surrounding grains, whereas this is not possible in a planar network [15].

All deformation simulations are performed at room temperature or, in the special case of collective processes, at 800K, corresponding to 0.45Tm. These temperatures are considerably lower than the one used in [16] where Coble creep was found as the dominating mechanism at T>0.7Tm. Temperature control during all simulations is via a velocity re-scaling every 100fs.

2. Simulation technique

All nc samples are created using the Voronoi construction with random nucleated seeds and random crystallographic orientations: the simulation cell volume was filled with nanograins grown from seeds with random location and crystallographic orientation; the space was filled according to the Voronoi construction. Molecular dynamics at 300K is then performed to anneal the sample for 50ps to allow the system to find a more relaxed configuration. It has been previously shown that the samples resulting from this type of synthesis, have grain boundaries that are fundamentally not different from their coarse grained counterparts [17]. For all types of missorientation, a large degree of structural coherence is observed and misfit accommodation occurs in quiet regular patterns.

For the present work we consider the following samples: (1) a 12nm model Ni sample containing 15 grains. Other samples used for comparison have the same general microstructure, but scaled to three characteristic grain diameters: 5nm, 10nm and 20nm are considered in [10-13,17]. This comparison allowed us to study the same grain boundary for different grain sizes, isolating the effect of grain size on grain boundary structure. (2) A so-called "annealed" sample, made by annealing the 12nm sample at 800K via velocity re-scaling for 100ps in order to relax the GB and TJ regions. Due to computer time limitations, a higher annealing temperature was chosen in the computer compared to the annealing temperatures in experiment, in order to accelerate the relaxation processes. However, due to the shorter annealing times applied in simulations, no grain growth was observed. The sample is then cooled down to 300K prior to deformation. (3) A sample containing 125 grains with a mean grain size of 5nm in order to study collective processes. This sample is deformed at 800K.

All molecular dynamics are performed within the Parrinello-Rahman approach with periodic boundary conditions and fixed orthorhombic angles. Samples are deformed under uniaxial tensile loading. We used the second moment (tight binding) potential of Cleri and Rosato for "model" fcc Ni [18]. Atomic visualization is aided by determining the local crystalline order according to the Honneycutt analysis. For more details about this procedure and the simulations we refer to [9,15,17]. Using this classification scheme, we define four classes of atoms via a color-code system: gray=fcc, red=hcp, green=other twelve, and blue non-twelve coordinated atoms. Fig. 1 shows a view of a nc sample, containing 100 10nm size grains. Using this colour coding scheme, the grain and grain boundary regions can clearly be identified.

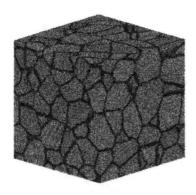

Figure 1. General view of a modelled nanocrystalline sample containing 100 grains with a mean grain diameter of 10nm.

3. Grain boundary sliding

In all samples with mean grain sizes up to 20nm, grain boundary sliding is observed as being the main contribution to the observed plasticity. Careful analysis of the GB structure during sliding under constant tensile load shows that sliding includes a significant amount of discrete atomic activity, either through uncorrelated shuffling of individual atoms or, in some cases, through shuffling involving several atoms acting with a degree of correlation [10]. In all cases, the excess free volume present in the disordered regions plays an important role. In addition to the shuffling, we have observed hopping sequences involving several GB atoms. This type of atomic activity may be regarded as stress assisted free volume migration. Together with the uncorrelated atomic shuffling they constitute the rate controlling process responsible for the GB sliding (GBS).

The atomic shuffling is shown in Fig.2 where a section of the grain boundary between grain number 1 and grain number 14 of the 12 nm sample is shown. The plane of the paper represents a (011) plane for grain 1 and a (-112) plane for grain 14. The atoms are shown at their positions prior to loading and the thickness of the section (perpendicular to the sheet of paper) is 4 atomic layers. In grain 1 the unit cell is highlighted in yellow, in grain 14 the (111) planes are indicated and also the crystallographic directions [110] and [1-11]. To visualize the relative displacement of atoms in the two grains resulting from approximately 2% plastic deformation, the displacement vectors have been calculated with respect to the local center-of-mass coordinate system of the atoms in grain 1. As can be observed GB sliding is approximately parallel to the GB plane and with a significant sliding component in the (111) plane. Within the GB some single events of much larger displacements across the grain boundary are observed, which can be identified as shuffling of atoms across the interface.

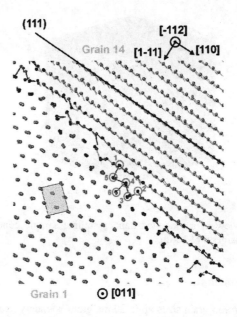

Figure 2. A section of the GB between grains 1 and 14 in the 12nm sample. Displacement vectors are shown indicating the change in position between two levels of strain during plastic deformation. Atomic shuffling between the grains can be observed.

In order to resolve the atomic activity in time, numbers are attributed to those columnar regions (with respect to the viewing direction) taking part in the atomic shuffling. When a load is applied, the atoms occupying regions 1 and 2 (constituting lattice positions in grain 14) slide away increasing the excess free volume at the connecting region between grains 14 and 1. When the atoms in grain 14 have slid by an amount equal to about 50% of the indicated sliding vector, two atoms sitting in region 3 of an (011) plane of grain 1, move in the direction of region 2, transferring free volume into grain 1. The new positions taken by these atoms are again lattice positions of an (011) plane in grain 1. Then a shuffle takes place to refill one of the vacant sites in region 3 by an atom sitting in region 4. Region 4 constitutes lattice positions common to both grain 1 and grain 14. Only a very short time later, a similar process takes place: an atom sitting in region 5 shuffles to region 1. This position is filled up with another atom in region 4, which then in turn is filled up by a shuffle of an atom in region 6. This atom sits closer to the GB plane since the GB plane is slightly inclined relative to the perpendicular of the viewing plane.

Sliding is accompanied by stress build up across neighboring grains. A local stress increase across a grain will lead indirectly to accommodation by GB and triple junction migration, as is shown in other grain boundaries studied in detail in [10].

4. Dislocation activity

At larger grain sizes, dislocation activity is observed. In fully three dimensional GB networks, which have been modelled now up to 20nm grain sizes, only partial dislocations have been observed. Only in quasi 2D-samples with columnar diameters greater than 20 nm full dislocations have been observed. A detailed study of the difference between a 2D columnar and a fully 3D network has been given in [15].
MD simulations have shown that a GB dislocation emits a partial lattice dislocation meanwhile changing the grain boundary structure and its dislocation distribution [12]. This mechanism is the reverse of what is often observed during absorption of a lattice dislocation, where the impinging dislocation is fully or partially absorbed in the GB, creating local changes in the structure and GB dislocation network .

Fig.3 shows the dislocation activity occurring in grain 13 at the interface with grain 12. Fig.3a shows a section of the GB 12-13, including part of the triple junction (TJ) involving grain 1 just at the onset of plastic deformation. Such a configuration will be referred to as the elastically deformed case. The view is along a [1-10] direction of grain 13, where for this grain the unit cell has been highlighted in yellow. The grain boundary plane is close to a (1,-1,13) plane of grain 13 and the tilt angle between the observed (111) planes in grain 13 and 12 is approximately 24° and a twist angle of approximately 18° is found.

The GB structure has to accommodate the above-mentioned misfit through a GB dislocation (GBD) network. Two of these GBDs are indicated in fig.3a by yellow circles. Others are not shown, since they can only be observed in other viewing orientations. For detailed information we refer to [12].

During deformation of this sample the GBD, which is closest to the TJ, dissociates into a Shockley partial, travelling through grain 13. This can be seen in fig.3b, showing the same GB after 2.3% plastic strain, in which this GBD has annihilated. The two (111) planes of red HCP atoms indicate the intrinsic stacking fault left behind when the partial is travelling through grain 13. Since the Burgers vector of the lattice dislocation is not the same as the Burgers vector of the GBD, the emission of the partial occurs along with local changes in GB orientation, GB structure and GBD distribution.

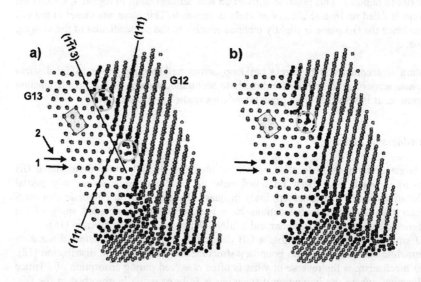

Figure 3. View of a grain boundary in the 12nm sample for a) the configuration at elastic loading and b) the configuration at a plastic strain of 2.3%. Grain boundary dislocations (of type A) that accommodate the misfit between the grains are highlighted by yellow circles.

The atomic mechanism behind the dislocation emission is the following: during deformation sliding is observed and free volume migrates from the nearby triple junction, and diffuses under the applied stress towards the GBD observed in fig.3a, or, equivalently, two atoms from the core region of the GBD migrate to nearby positions within the GB. Local shuffling around the GBD allows the creation of the necessary Burgers vector for the partial dislocation. The nucleation and propagation induces changes in the resultant GBD distribution and additional structural relaxation is observed in the GB and nearby TJ.

It is shown in [12] that for increased grain diameters, an increase in partial dislocation activity is seen, but no full dislocations are observed probably due to subsequent structural relaxation after the emission of the partial. For the model Ni potential used, the room temperature stacking fault energy is 280 mJ/m^2 [13]. Such a value, together with a lack of full dislocation activity, seems to suggest that the stacking fault energy plays only a minor role in the issue concerning whether or not a full dislocation is seen.

At smaller grain sizes, the same GBDs are observed. They undergo significantly more atomic scale activity resulting in climb of the GBD without the generation of partials, suggesting that GBD are more delocalised and therefore motion of GBDs by atomic shuffling is facilitated and association of GBDs into a partial dislocation is hindered.

5. Collective processes

In order to investigate collective processes such as cooperative grain boundary sliding via the formation of shear planes spanning several grains, a sample with 125 grains and a mean grain size of 5nm was deformed. A large number of grains are necessary in order to minimize the effects imposed by the periodicity used to simulate bulk conditions. The small grain size is chosen to reduce the total number of atoms in the sample (to 1.2 million) so that longer deformation times are possible at acceptable strain rates. In order to increase grain boundary activity the deformation was done at 800K.

A typical example of formation of common shear planes during deformation is presented in fig.4. Fig.4a displays the atomic positions prior to deformation and fig.4b shows the atomic positions after 3.6% plastic deformation. Comparison of the two plots demonstrates that after plastic deformation this section of the sample underwent a reorganization of the GB regions (in particular GB(85,108), GB(10,17) and GB(8,117)) resulting in an alignment of multiple GBs to form two common shear planes indicated by the large black arrows in fig.4b. The nature of the atomic activity that facilitates such migration is predominantly of the form of atomic shuffling with some free volume migration resulting in an increase in GB structural order similar to what we have seen in past work [10-12]. The orientation of shear plane 1 is inclined at about 53° to the tensile axis, whereas shear plane 2 is orientated at 46° to the tensile axis, which is close to the maximum resolved applied shear at 45°.

It is also observed that some grains move collectively relatively to some others. Arrows in fig.4b represent the relative sliding of some grains involved in this collective motion. The directions of the three nearly vertical arrows in fig.4b for shear plane 1, are derived by the displacement of atoms in grains 108, 17 and 117 relative to the centre of mass of the respective grains 85, 10 and 8. The length of arrows does not represent the absolute degree of sliding, which for all grains was typically between 1.5 and 2 Å.
The underlying mechanisms that have been observed for the formation of shear planes are (1) Pure GBS induced migration of parallel and perpendicular GBs to form a single shear interface (2) Coalescence of neighboring grains that have low angle GB facilitated

by the propagation of Shockley partials. (3) Continuity of the shear plane by intragranular slip. A detailed description of the processes is given in [13].

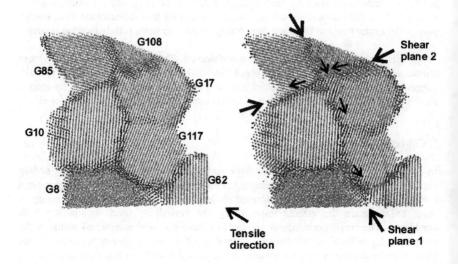

Figure 4. Region of nc-Ni sample (5nm mean grain size) in which two shear planes (1 and 2) have been identified. The small arrows display the direction of the relative sliding between grains. a) is the atomic configuration before loading, b) after 3.6% plastic deformation

6. Role of grain boundary structure.

It has been observed that with increased structural order within the GB region, or equivalently the extent to which the GB is in equilibrium, there is a corresponding decrease in the level of plasticity under uniaxial tensile conditions and thus an effective increase in the strength of the nc material. Fig.5 shows the deformation curve for a 12nm sample prepared according to the Voronoi construction (called as-prepared), and a sample that has been annealed at 800K prior to deformation in order to relax more the GBs and TJ's. Both are deformed at 300K. The annealed sample exhibits significantly less deformation compared to the as-prepared sample. To confirm that the decrease in strain does indeed arise from a reduction in plastic deformation activity at the GB, we unloaded the as-prepared and annealed samples to determine the residual strain. The difference in residual strain is 0.43%, which is only slightly, less than the 0.56% difference in strain under load conditions. Inspection of this figure also reveals that the strain rate for the as-prepared sample is 2.2×10^{7} /s, whereas for the annealed sample it is 9.4×10^{6} /s, which is a reduction of more than half. These strain rates where calculated using the last 30ps of deformation data before unloading and should be regarded as applicable only to the time scale of fig. 4. The deformation simulations were continued up to 0.3ns, by which time the strain rates became closer in value to approximately 3.8×10^6 for both the as-prepared and annealed samples.

When GB structure is more disordered, an increase in plastic strain occurs in the early stages of deformation. This seems however to be a transient regime since after a while, the strain rates begin to converge to the same value for the three sample types considered. In this transient regime, significant GB relaxation contributes to the strain. Alternatively such relaxation pathways can be removed by an annealing treatment prior to deformation. In a more detailed study [11] we also demonstrated that the corresponding structural relaxation processes under applied temperature are functionally equivalent to that under applied stress. The transient structural relaxation processes involve atomic processes the nature of which, does not seem to differ fundamentally from that identified in the steady state regime, apart from the level of activity. This observation hints at the general hypothesis that GB's further from equilibrium exhibit increased atomic shuffling and atomic migration activity, and that upon annealing this can be reduced

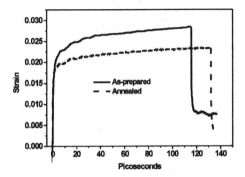

Figure 5. The deformation curves for the as-prepared and the annealed in which, all have been unloaded to determine the residual strain.

7. Conclusion

Uniaxial tensile loading of nc-Ni samples reveals the presence of three deformation mechanisms: (1) GB sliding triggered by atomic shuffling and to some extent stress assisted migration (2) the emission of partial dislocations from GB that contain GBD, resulting in a redistribution of GBD and additional structural relaxation (3) cooperative GBS via the formation of shear planes that extend over a number of grains. It is also observed that increased structural order within the GB region results in a decrease in the level of plasticity under uniaxial tensile conditions and thus an effective increase of

262

strength of the nc material, suggesting a beneficial effect of structural disorder from the perspective of grain boundary sliding.

8. Acknowledgments

Work is supported by the Swiss NSF (2000-056835.99)

9. References

1. Weertman, J.R. (2002) Mechanical behaviour of nanocrystalline metals, *Nanostructured Materials: Processing, Properties, and Potential Applications*, William Andrew Publishing, Norwich.
2. Youngdahl, C.J., Hugo, R.X., Kung, H. and Weertman, J.R. (2002), TEM observation of nanocrystalline copper during deformation, *Structure and Mechanical Properties of Nanophase Materials- Theory and Computer Simulation vs. Experiment*, MRS Symposium Series Vol. **634**, B1.2.
3. McFadden, S.X., Sergueeva, A.V., Kruml, T., Martin, J-L. and Mukherjee, A.K. (2000), Superplasticity in nanocrystalline Ni3Al and Ti Alloys, *Structure and Mechanical Properties of Nanophase Materials- Theory and Computer Simulation vs. Experiment*, MRS Symposium Series Vol. **634** , B1.3.
4. Derlet, P.M. and Van Swygenhoven, H. (2001) The role played by two parallel free surfaces in the deformation mechanism of nanocrystalline metals: a molecular dynamics simulation, *Phil. Mag. A*. **82**, 1-15.
5. Dalla Torre, F., Van Swygenhoven, H., Victoria, M. (2002) Nanocrystalline electrodeposited Ni: microstructure and tensile properties, *Acta Materialia* **50**, 3957-3970.
6. Jia, D., Ramesh, K.T. and Ma, E. (2000) Compressive,Tensile,and Dyamic Behaviour of Nanophase Iron, *Ultrafine Grained Materials* Edited By R. S. Mishara, S. L. Semiatin,C. Suryanarayana,N. N. Thadhani,T. C. Löwe, TMS , 309-318.
7. Hahn, H. and Padmanabhan, K.A. (1997) A model for the deformation of nanocrystalline materials, *Phil. Mag. B* **76**, 559-571.
8. Volpp, T., Goring, E., Kuschke, W.M. and Arzt, E. (1997), Grain size determination and limits to Hall-Petch behaviour in nanocrystalline NiAl powders, *Nanostruct. Mater.* **8**, 855-865.
9. Van Swygenhoven, H. (2002) Polycrystalline materials: grain boundaries and dislocations, *Science* **296**, April 4, 66-67.
10. Van Swygenhoven, H. and Derlet, P.M. (2001) Grain-boundary sliding in nanocrystalline fcc metals, *Phys. Rev. B* **64**, 224105-9.
11. Hasnaoui, A., Van Swygenhoven, H. and Derlet, P.M. (2002) On non-equilibrium grain boundaries and their effect on thermal and mechanical behaviour: a molecular dynamics computer simulation, *Acta Mater.* **50**, 3927-3939.
12. Van Swygenhoven, H,. Derlet, P.M. and Hasnaoui, A. (2002) Atomic mechanism for dislocation emission from nanosized grain boundaries, *Phys. Rev. B* **66**, 024101-8.
13. Hasnaoui, A., Van Swygenhoven, H. and Derlet, P.M. (2002) Cooperative processes during plastic deformation in nanocrystalline fcc metals - a molecular dynamics simulation, *Phys. Rev. B (in press)*.

14. Yamakov. V., Wolf, D., Salazar, M., Phillpot, S.R.and Gleiter, H. (2001) Length-scale effects in the nucleation of extended dislocations in nanocrystalline al by MD simulations, *Acta Mater.* **49**, 2713-2722.

15. Derlet, P.M. and Van Swygenhoven, H. (2002) Length scale effects in the simulation of deformation properties of nanocrystalline metals, *Scripta Mater.* **47**, 719-724.

16. Yamakov, V., Wolf, D., Phillpot, S.R. and Gleiter, H. (2002) Grain-boundary diffusion creep in nanocrystalline palladium by molecular dynamics simulation, *Acta. Mater.* **50**, 61-73.

17. Van Swygenhoven, H., Farkas, D. and Caro, A. (2000) Grain-boundary structures in polycrystalline metals at the nanoscale, *Phys. Rev. B* **62**, 831-838.

18. Cleri F. and Rosato, V. (1993) Tight bindng potentials for transition metals and alloys, *Phys. Rev. B* **48**, 22-33

MIXED ATOMISTIC/CONTINUUM METHODS :
STATIC AND DYNAMIC QUASICONTINUUM METHODS

D. RODNEY
Génie Physique et Mécanique des Matériaux (ESA CNRS 5010)
101, rue de la Physique BP46   38402 Saint Martin d'Hères   FRANCE

Introduction

The motivation for mixed atomistic/continuum methods in material science modeling is similar to the motivation for statistical physics: given the number of atoms in a typical grain ($(5 \ \mu m)^3 \sim 10^{13}$ atoms), it is *impossible* and *useless* to simulate this grain directly from the atomic scale. It is *impossible* because, even with the rapid development of computational technologies, the largest molecular dynamics simulation to date [1] contains *only* 10^9 atoms, which represents 1/10000 of the number of atoms in a real grain. It would also be *useless* because (1) it would be extremely difficult to extract the relevant information from this huge amount of data and (2) atomic resolution is required only in very limited regions of space and most of the crystal can be accurately modeled with lower resolution using continuum techniques, such as the finite element method (FEM).

The objective of the coupling methods is to use a resolution adapted to the region considered, i.e. preserve atomic resolution near defects while using a lower resolution (treat collectively large numbers of atoms) in conjunction with continuum techniques in regions where the deformation field varies slowly compared to the lattice parameter. This allows to reduce drastically the number of degrees of freedom to account for and thus allows to simulate larger cells for equivalent computational overheads.

In the first coupling methods proposed in the literature, the system was decomposed in atomistic/continuum regions, each region was simulated with a classical well-developed method (molecular statics in the atomistic region, Green's function [2,3] or the finite element method in the continuum region (FEAt method [4])) and all the art was in coupling both methods in a *seamless fashion*. The philosophy behind more recent methods is to start with a system entirely modeled at the atomic scale, i.e. described by a potential energy which depends on the position of all the atoms in the system, and to construct an effective potential energy which depends on a reduced set of atomic positions. The system is then often said *coarse grained*. We will describe here one such method, the Quasicontinuum method (QCM).

The QCM has been developed and successfully applied to static T=0K simulations, i.e. the coarse-grained effective energy is minimized in order to obtain equilibrium configurations. On the other hand, the issue of accounting for finite temperature in coarse-grained systems remains unclear. Two types of simulations may be performed: study either systems equilibrated at finite temperature or dynamic systems driven out-of-equilibrium. In the first case, an effective free energy must be constructed and the solutions proposed up to now involve strong approximations. In the second case, spurious wave reflections due to inhomogeneties in the finite element mesh imply that heat cannot flow away from atomistic

A. Finel et al. (eds.), Thermodynamics, Microstructures and Plasticity, 265–274.

266

regions.

In this review article, we first present the QCM in its static T=0K formulation. Since the later has been described over a series of articles [5-8], we present in Section 1 and 2, only its main ingredients and discuss one illustrative application. In Section 3, we insist on the less discussed aspects of finite temperature and dynamic simulations.

1. The quasicontinuum method: static T=0K formulation

The QCM is best described as a *method to approximate an atomic scale potential energy*, with 2 objectives: (1) retain, in a systematic and automatic way, only those degrees of freedom (atomic positions) needed to describe the deformation field of the system and (2) simplify (and accelerate) the calculation of the potential energy.

Here, we restrict ourselves to a monocrystal with energetics described by an additive interatomic potential (f.e. Embedded Atom Method (EAM) potentials). More general presentations of the QCM can be found in Refs. 6-8. The crystal contains N atoms with positions given by their displacements $\{u_i\}_N$ with respect to a reference configuration. The atomic scale potential energy is:

$$E\left(\{u_i\}_N\right) = \sum_{i=1}^{N} E_i\left(\{u_i\}_N\right) - \sum_{i=1}^{N} f_i^{ext} \cdot u_i \qquad (1)$$

The degrees of freedom are the atomic displacements. In a static T=0K formulation, we are interested in equilibrium configurations determined as sets of displacements which minimize the above energy functional.

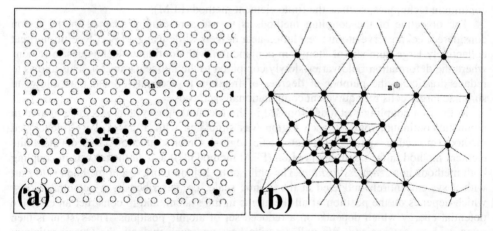

Figure 1: Section perpendicular to a Lomer dislocation in a FCC crystal.
In (a), black atoms are representative atoms, the corresponding finite element mesh in shown in (b)

1.1 REDUCTION IN THE NUMBER OF DEGREES OF FREEDOM

The first step of the QCM, illustrated in figure 1, is to select a reduced set of atoms, called *representative atoms* and construct a finite element mesh based on these atoms. From there on, the position of any atom (f.e. atom B in figure 1(b)) is linearly interpolated from the displacements of the representative atoms, using the shape functions of the finite element mesh. *The only remaining degrees of freedom of the potential energy are thus the displacements of the representative atoms.*

Since the displacements are linearly approximated, the displacement gradient $F=\nabla_x u$ is constant within each element, i.e. *deformations are constant (homogeneous) within the finite elements.* All atoms embedded in a given element see therefore the same environment and have the same energy $E(F_e)$, which depends only on the displacement gradient in the element and is computed very efficiently from the interatomic potential.

1.2 AUTOMATIC ADAPTION

Since deformations are constant inside the finite elements, the density of representative atoms (which controls the size of the finite elements) must depend on the spatial variations of the deformation: in regions where the deformation varies rapidly (f.e. near dislocation cores), atomic resolution must be preserved; inversely, far from lattice defects, the deformation varies slowly and a low density of atoms is sufficient to interpolate the displacements. In the course of a simulation, the configuration may evolve, implying to adapt the number and geometry of the representative atoms. This is achieved by *automatic adaption techniques* typical of the FEM (see Ref. 6 for details).

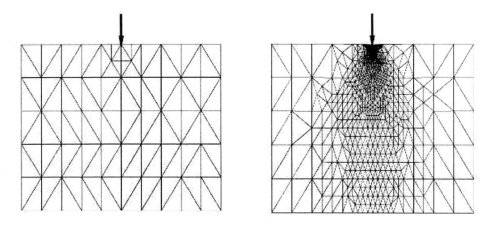

Figure 2: Automatic adaption during a 2D nanoindentation simulation

An example of adaption in the case of nanoindentation is presented in figure 2. Another illustrative example is given in figure 3(a) which shows a typical mesh used in the 2D simulation of a Lomer dislocation in a FCC crystal. The dislocation is introduced by displacing all representative atoms according to the elastic displacement of the dislocation. The mesh has atomic resolution near the dislocation glide plane in order to capture the displacement jump across this plane.

The deformation is very inhomogeneous near the glide plane of the dislocation since elements between the glide plane and the plane just above are sheared by b/d (b the Burgers vector, d the interplanar distance), while the other elements are only weakly deformed. This region is therefore triggered for atomic resolution. However, at the atomic scale, the deformation is everywhere close to homogeneous since the shear b/d brings the crystal in correspondence with itself. In order to coarsen this region, the adaption criterion must be based on the *elastic deformation* and not on the total deformation which includes the inhomogeneous *plastic deformation*. Such a procedure has been developed [9] and results in figure 3(b) which shows that, after adaption, the only atomistically refined regions are close to the dislocation core and to the step left by the dislocation on the left surface.

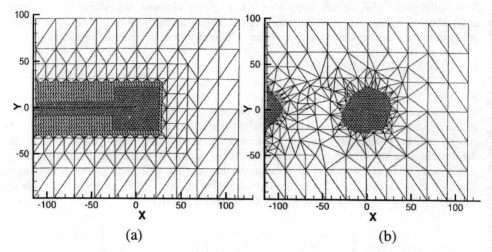

(a)   (b)

Figure 3: Mesh coarsening along the glide plane of a Lomer dislocation in a FCC crystal (see text for details)

1.3 POTENTIAL ENERGY APPROXIMATION

Evaluation of the potential energy (equation 1) is computationally expensive since it requires computing the energy of all the atoms, representative or not. This calculation is simplified using a *quadrature approximation*: only the energy of the representative atoms is computed and an approximated potential energy is constructed as a weighted sum of these energies (R, number of representative atoms):

$$E_{QC}\left(\{\mathbf{u}_i\}_R\right) = \sum_{\alpha=1}^{R} n_\alpha E_\alpha\left(\{\mathbf{u}_i\}_R\right) - \sum_{\alpha=1}^{R} n_\alpha \mathbf{f}_\alpha^{ext} \cdot \mathbf{u}_\alpha \qquad (2)$$

The weights $\{n_\alpha\}_R$ correspond physically to the number of atoms represented by each representative atom. They may be computed in different ways [6,8].

The energy calculation can be further simplified using the fact that all atoms embedded in a given element have a same energy $E(\mathbf{F})$. The energy of a representative atom α may be approximated by the average energy of the atoms in its M surrounding elements (n_α^e, number of atoms represented by atom α in element e):

$$E_\alpha^L = \frac{1}{n_\alpha} \sum_{e=1}^{M} n_\alpha^e E(\mathbf{F}_e) \qquad (3)$$

Atoms whose energy is computed using this approximation are said *local*. The resulting error depends on the homogeneity of the deformation around the representative atom. Local atoms are therefore those for which the maximum difference between the displacement gradients in the surrounding elements is below a given threshold. Local atoms are essential for computation efficiency because evaluation of their energy is fast.

Atoms that do not meet the locality criterion are said *non local*. Their energy is computed by explicitly accounting for their neighbors, as in classical molecular statics. In order to avoid unphysical forces, non local atoms are forced to be in atomistically refined regions, such that $n_\alpha = 1$ for all non local atoms and the QC potential energy is rewritten:

$$E_{QC}\left(\{\mathbf{u}_i\}_R\right) = \sum_{\alpha=1}^{R_{NL}} E_\alpha\left(\{\mathbf{u}_i\}_R\right) + \sum_{\beta=1}^{R_L} \sum_{e=1}^{M_\beta} n_\beta^e E(\mathbf{F}_e) \qquad (4.a)$$

$$E_{QC}\left(\{\mathbf{u}_i\}_R\right) = \sum_{\alpha=1}^{R_{NL}} E_\alpha\left(\{\mathbf{u}_i\}_R\right) + \sum_{e=1}^{M} n^e E(\mathbf{F}_e) \qquad (4.b)$$

where R_{NL} (resp. R_L) is the number of non local (resp. local) atoms, M is the number of finite elements and $n^e = \sum_\beta n_\beta^e$ is the number of atoms in element e.

The final expression of E_{QC} in equation 4.b is obtained by inverting the sum signs in equation 4.a. In this expression we recognize the *usual decomposition between atomistic and continuum regions*: the first sum is the energy of the atomistic non local regions, where the QCM is equivalent to classical molecular statics, while the second sum is the energy of the finite element local regions. In the local regions, the QCM is equivalent to the FEM except that the constitutive law is computed directly from an interatomic potential. This implies that the approximation made in the local regions does not depend on the norm of the deformation (which is therefore not limited to the linear regime) but only on the spatial variations of the deformation compared to the size of the finite elements (since the deformation is assumed constant inside the elements).

The effective potential energy may then be minimized using standard methods such as the conjugate gradient in order to obtain equilibrium configurations. Quasi-static evolutions may also be obtained, for example, by increasing in steps the applied force and minimizing the effective energy between each step.

2. Applications of the static T=0K quasicontinuum method

The QCM was first applied to the static 2D simulation of nanoindentation [5,10], of grain boundaries interacting with cracks and dislocations [11,12], and of crack propagation [13]. The generalization of the QCM to 3D allowed to study dislocation junctions [14] and dislocation/precipitate interactions [15]. A slightly modified version of the QCM was recently proposed and applied to the simulation of nanoindentation in 3D [8]. All previous simulations concern FCC crystal described with semi-empirical EAM potentials. A QCM formulation adapted to complex Bravais lattices and tight-binding functionnals has also been developed [7] and applied to silicon crystals [16]. An extension to finite temperatures has been proposed [17] and will be discussed in Section 3.2.

The aim of this review paper is to insist more on methods than on applications. We will therefore present only one illustrative application: the 3D simulation of dislocation interactions in FCC crystals. We considered pairs of glissile dislocations on intersecting {111} glide planes. Using the QCM, we studied the structure of the different three-

270

dimensional dislocation junctions that may form in a FCC crystal. Our aim was to analyze the role played by non-linear effects in the structure and strength of dislocation junctions.

Equilibrium configurations of these junctions are shown in figure 4. With the use of the QCM, only limited regions near the glide planes are atomistically refined (bands made of 3 {111} planes on both sides of the glide planes) while the rest of the cell is modeled with a lower resolution using the local approximation. This lead to a significant reduction in the number of degrees of freedoms to account for: the number of representative atoms is only of the order of one tenth of the total number of atoms in the system (which is of the order of $1.5\ 10^6$ atoms). Because of the assumed important role of the Lomer-Cottrell junctions in FCC hardening, we studied the strength of these particular junctions [14]: shear stresses were applied by application on the boundary of the simulation cell of increasing shear deformations and lead to the study of the breaking mechanism and the evaluation of the resistance of the junction.

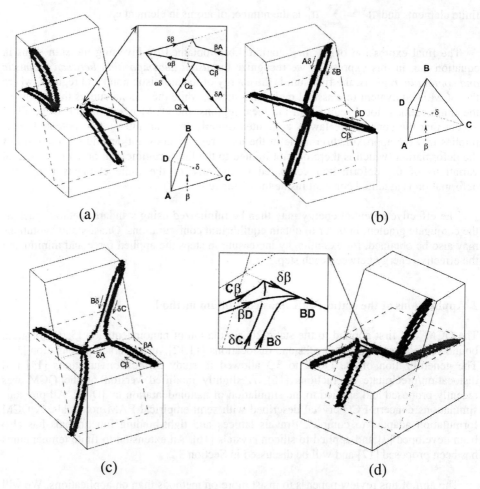

Figure 4: FCC dislocation junctions: (a) Colinear interaction (dislocations with same **AC** Burgers vector), (b) Hirth junction, (c) Glissile junction, (d) Lomer-Cottrell junction.
The Burgers vectors are given with respect to the Thomson tetrahedron shown in (a) and (b).
Dislocation core atoms are shown by representing only the most energetic atoms in the cells.

3. Dynamics and Temperature in coarse-grained systems

Firstly, methods to apply dynamics are discussed i.e. how to study the time evolution of a coarse-grained system as would be done by Molecular Dynamics. Secondly, we discuss equilibrium properties of coarse-grained systems at finite temperature. The first approach is limited to short time scales but can treat non-equilibrium configurations while the second approach deals with long time scales and systems at equilibrium.

3.1 DYNAMIC SIMULATIONS

Dynamics may be applied straightforwardly by adding to the QC potential energy (equation 4.b) the kinetic energy of the atoms and applying the same approximations as presented in previous section: the velocity of any atom is obtained by linear approximation from the velocity of the representative atoms. Defining an effective mass $M_\alpha = n_\alpha m$ and effective momentum $P_\alpha = n_\alpha p_\alpha$ for the representative atoms (n_α: weights of equation 2, m: atomic mass, p_α: atomic momentum), the QC Hamiltonian is:

$$H_{QC}\left(\{u_i\}_R, \{p_i\}_R\right) = E_{QC}\left(\{u_i\}_R\right) + \sum_{\alpha=1}^{R} \frac{P_\alpha^2}{2M_\alpha} \qquad (5)$$

Equations of motion for the representative atoms are derived using Hamilton's equations $\left(\dfrac{\partial P_\alpha}{\partial t} = -\dfrac{\partial E_{QC}}{\partial x_\alpha} \text{ and } \dfrac{\partial x_\alpha}{\partial t} = \dfrac{P_\alpha}{M_\alpha} = \dfrac{p_\alpha}{m}\right)$. These equations may then be integrated to obtain time evolutions using standard finite difference schemes, such as Verlet algorithm.

This method, similar to the *lumped mass approach* of FEM, was implemented by Shenoy [18] and applied to the dynamic simulation of nanoindentation and crack propagation. The main limitation of this method, which is the main problem to solve in order to account properly for the dynamics of coarse-grained systems, is the *unphysical reflection of waves in the case of a non-uniform mesh*. Said in practical words, the finite elements can transmit only waves of wavelength larger than their size while they reflect waves with smaller wavelength. This point was clearly illustrated in Ref. 18 on a non-uniform 1D mesh. The short wavelength waves generated in atomistic regions and representing the thermal motion of the atoms will be reflected at the boundary with the coarser continuum region and will be trapped in the atomistic region. In other words, *heat cannot flow away from atomistic regions*.

Recently, Cai *et al.* [19] proposed a method to minimize boundary reflections in the case of systems composed of an atomistic region (called P) surrounded by a linear region (called Q). As for methods using Green's function [2,3], Q-Q and P-Q interactions are assumed linear. The influence of Q is then be given by a *generalized Langevin equation*:

$$m\ddot{x}_i(t) = -\frac{\partial E}{\partial x_i} + \int_0^t d\tau \sum_{j \in P} \beta_{ij}(\tau)\dot{x}_j(t-\tau) + \sum_{j \in P} \beta_{ij}(t)x_j(0) \qquad (6)$$

where E is the potential energy of the entire system with Q atoms at equilibrium positions.

The response functions $\beta_{ij}(t)$ (i,j: atoms in P) characterize the influence of the atoms in

Q in the form of correlation functions between P atoms. They are at the heart of the method and may be computed from test simulations. The method was on a 1D chain and a 2D square lattice of harmonic oscillators where very low wave reflections were found.

This method is promising and may be generalized to more complex interatomic potentials. However, it suffers from one difficulty: its computational need. The $\beta_{ij}(t)$ functions must be stored for all atomic pairs along the boundary of the simulation cell within a cut-off distance and for all time intervals within a cut-off time interval. In the example discussed in Ref. 19 which is 2D and contains 3600 atoms, the $\beta_{ij}(t)$ functions represent about 10^6 elements. This is an excessively large matrix, which may prohibit large 3D simulations.

3.2 EQUILIBRIUM CONFIGURATIONS AT FINITE TEMPERATURE

The objective is to build an effective energy for a system maintained at fixed temperature which depends only on the position of the representative atoms, the influence of the other atoms (called here *slave atoms*) being averaged out. This is in fact reminiscent of *Renormalization Group (RG)* theories, where effective hamiltonians describing the interactions of a decreasing number of atoms are obtained by iterative partial summations of the partition function of the system.

One possible strategy is, as in RG theories, to define the effective energy E^{eff} by imposing that the *partition function of the system is unchanged when averaging out the slave atoms*:

$$Z = \int d\mathbf{u}^R d\mathbf{u}^S \exp\left(-\beta E\left(\{\mathbf{u}_i\}_R, \{\mathbf{u}_i\}_S\right)\right) = \int d\mathbf{u}^R \exp\left(-\beta E^{\text{eff}}\left(\{\mathbf{u}_i\}_R, T\right)\right) \quad (7)$$

E is the atomic scale energy (equation 1) which depends on both the position of the representative atoms $\{\mathbf{u}_i\}_R$ and of the slave atoms $\{\mathbf{u}_i\}_S$. Equation 7 describes the partial summation of the motion of the slave atoms and serves to define the effective energy:

$$\exp\left(-\beta E^{\text{eff}}\left(\{\mathbf{u}_i\}_R\right), T\right) = \int d\mathbf{u}^S \exp\left(-\beta E\left(\{\mathbf{u}_i\}_R, \{\mathbf{u}_i\}_S\right)\right) \quad (8)$$

Note that the summation of the motion of the slave atoms is performed while keeping the representative atoms fixed. The effective energy thus defined is temperature dependent, as noted explicitly in equation 8. In the case of pair potentials, it leads to temperature-dependent interaction potentials between atoms.

The effective energy cannot be computed exactly, except in special cases such as nearest-neighbor polynomial pair potentials, as recently shown by Curtaloro and Ceder [20].

In order to make the calculation tractable, Shenoy et al. [17] proposed to compute equation 8 approximately using the *Local Harmonic (LH) approximation* [21,22]: slave atoms are assumed to have harmonic displacements around their equilibrium position; as in the static T=0K formulation of the QCM, these equilibrium positions are given by linear interpolation from the equilibrium position of the representative atoms.

It can then be easily shown that the effective energy (which may be noted $E_{QC}(\{\mathbf{u}_i\}_R, T)$) is the difference between the internal energy of the system (equal to the QC potential energy of equation 4.b, noted $E_{QC}(\{\mathbf{u}_i\}_R, 0)$) and the entropy contribution due to averaging the motion of the slave atoms. In the LH approximation, this entropy is the sum of the entropy of each atom, which depends on the determinant of the local dynamical matrix of the atom moving in the potential of the other atoms kept at their equilibrium position. In a given finite element, all atoms see the same environment and have the same dynamical matrix that depends only on the displacement gradient of the element and can be computed very efficiently. The effective energy is thus written:

$$E_{QC}\left(\{u_i\}_R, T\right) = E_{QC}\left(\{u_i\}_R, 0\right) + 3k_B T \sum_{e=1}^{M} (n_e - 1) \ln\left(\frac{\hbar D^{\frac{1}{6}}(F_e)}{k_B T}\right) \quad (9)$$

where $n_e - 1$ is the number of slave atoms in element e and $D(F_e)$ is the determinant of the dynamical matrix of an atom moving in an infinite crystal deformed by F_e.

This effective energy can be used in Monte Carlo simulations to study finite temperature equilibrium properties of deformed solids. A free energy can also be constructed by adding to equation 9 the entropy of the representative atoms, which can also be computed in the Local Harmonic approximation. This *coarse-grained free energy* may then be minimized to determine finite temperature equilibrium configurations. Note that, as in the static T=0K case, this finite temperature QCM is equivalent in atomistic regions to fully atomistic simulations, as performed in Refs. 21 and 22.

In conclusion, this approach have the advantage of being computationally tractable, requiring computational resources comparable to static T=0K simulations. It is also applicable to complex potentials such as EAM potentials as done in Ref 17. On the other hand, it has limitations related to the Local Harmonic approximation (as reviewed in Ref. 22): it is applicable only at low temperatures and it underestimates the temperature dependence of defect free energies.

Finally, it is tempting to apply Hamilton's equations to the effective energy of equation 9 in order to obtain equations of motion for the representative atoms and simulate dynamics at finite temperature. However, this would be incorrect for 3 reasons (these remarks also apply to the effective Hamiltonian discussed in section 4.1, which is simply a particular case of equation 9 with T=0K): (1) equation 9 was obtained by explicitly assuming that the system is in canonical equilibrium, (2) the issue of wave reflections discussed in previous Section is not handled here and (3) the position and momentum of the representative atoms are macroscopic variables (they are averages) and thus Hamilton's equations, which apply to microscopic degrees of freedom, cannot be applied to these variables. Equations of motion for the representative atoms should be adapted to the evolution of macroscopic variables. Up to now, there exists no proper theory to obtain these equations.

Conclusion

The methodology to construct effective potential energies for static T=0K simulations is now well developed and has been applied successfully to various multiscale problems. Comparatively, methodologies to account for finite temperatures and dynamics are far less clear. In the first case, a computationally tractable method has been proposed but is based on the Local Harmonic approximation which implies strong limitations, in particular to low temperatures. In the second case, the method recently proposed by Cai *et al.* is very

274

efficient but computationally expensive. Moreover, up to now, none of these methods has been applied to large-scale simulations. Therefore, most of the work is still to be done in order to account properly for temperature in coarse-grained systems.

References

1. http://www.research.ibm.com/resources/news/20020429_fracture_simulation.shtml.
2. Sinclair, J.E. (1971) Improved atomistic model of a bcc dislocation core, *Journal of Applied Physics* **42**, 5321-5329; Sinclair, J.E., Gehlen, P.C., Hoagland, R.G. and Hirth, J.P. (1978) Flexible boundary conditions and nonlinear geometric effects in atomic dislocation modeling, *Journal of Applied Physics* **49**, 3890-3897.
3. Thomson, R., Zhou, S.J., Carlsson A.E. and Tewary, V.K. (1992) Lattice imperfection studied by use of lattice green's functions, *Physical Review* **B46**, 10613-10622.; Zhou, S.J., Carlsson, A.E. and Thomson, R. (1993) Dislocation nucleation and crack stability: lattice green's functions treatment of cracks in a model hexagonal lattice, *Physical Review* **B47**, 7710-7718.
4. Kohlhoff,S. , Gumbsch, P. and Fischmeister, H.F. (1991) Crack propagation in bcc crystals studied with a combined finite-element and atomistic model, *Philosophical Magazine* **64**, 851-878.
5. Tadmor, E.B., Ortiz, M. and Phillips, R. (1996) Quasicontinuum analysis of defects in solids, *Philosophical Magazine* **73**, 1529-1563.
6. Shenoy, V.B., Miller, R., Tadmor, E.B., Rodney, R., Phillips, R. and Ortiz, M. (1999) An adaptive finite element approach to atomic-scale mechanics, *Journal of the Mechanics and Physics of Solids* **47**, 611-642.
7. Tadmor, E.B., Smith, G.B., Bernstein, N. and Kaxiras, E. (1999) Mixed finite element and atomistic formulation for complex crystals, *Physical Review* **B59**, 235-245.
8. Knap, J. and Ortiz, M. (2001) An analysis of the quasicontinuum method, *Journal of the Mechanics and Physics of Solids* **49**, 1899-1923.
9. Miller, R. and Rodney, D. (1998) *unpublished*.
10. Tadmor, E.B., Miller, R., Phillips, R. and Ortiz, M. (1999) Nanoindentation and incipient plasticity, *Journal of Materials Research* **14**, 2233-2250.
11. Shenoy, V.B., Miller, R., Tadmor, E.B., Phillips, R. and Ortiz, M. (1998) Quasicontinuum models of interfacial structure and deformation, *Physical Review Letters* **80**, 742-745.
12. Miller, R., Ortiz, M., Phillips, R., Shenoy, V. and Tadmor, E.B. (1998) Quasicontinuum models of fracture and plasticity, *Engineering Fracture Mechanics* **61**, 427-444.
13. Miller, R., Tadmor, E.B., Phillips, R., and Ortiz, M. (1998) Quasicontinuum simulations of fracture at the atomic scale, *Modeling and Simulations in Material Science and Engineering* **6**, 607-638.
14. Rodney, D. and Phillips, R. (1999) Structure and strength of dislocation junctions: an atomic level analysis, *Physical Review Letters* **82**, 1704-1707.
15. Hardikar, K., Shenoy, V., Phillips, R. (2001) Reconciliation of atomic level and continuum notions concerning the interaction of dislocations and obstacles, *Journal of the Mechanics and Physics of Solids* **49**, 1951-1967.
16. Smith, G.S., Tadmor, E.B. and Kaxiras, E. (2000) Multiscale simulation of loading and electrical resistance in silicon nanoindentation, *Physical Review Letters* **84**, 1260-1263.
17. Shenoy, V., Shenoy, V. and Phillips, R. (1999) Finite temperature quasicontinuum methods, in V.V. Bulatov, T. Diaz de la Rubia, R. Phillips, E. Kaxiras and N. Ghoniem (eds.), *Multiscale Modeling of Materials*, Mater. Res. Soc. Symp. Proc. 538, Warrendale, PA, pp. 465-471.
18. Shenoy, V.B. (1999) Quasicontinuum models of atomic-scale mechanics, PHD Thesis, Brown University
19. Cai, W., de Koning, M., Bulatov, V.V. and Yip, S. (2000) Minimizing boundary reflections in coupled-domain simulations, *Physical Review Letters* **85**, 3213-3216.
20. Curtaloro, S., Ceder, G. (2002) Dynamics of an inhomogeneously coarse grained multiscale system, *Physical Review Letters* **88**, 255504.
21. LeSar, R., Najafabadi, R., Srolovitz, D.J. (1989) Finite-temperature defect properties from free-energy minimization, *Physical Review Letters* **63**, 624-627.
22. Foiles, S.M. (1994) Evaluation of harmonic methods for calculating the free energy of defects in solids, *Physical Review* **B49**, 14930-14938.

BOUNDARY PROBLEMS IN DD SIMULATIONS

B. DEVINCRE[1], A. ROOS[2] AND S. GROH[1]
[1] *LEM, CNRS-ONERA, 29 av. de la division Leclerc, 92322 Chatillon Cedex, France*
[2] *DMSE/LCME, ONERA, 29 av. de la division Leclerc, 92322 Chatillon Cedex, France*

Abstract. Over the years, different approaches have been developed to calculate the state of mechanical equilibrium in a dislocated finite body. The purpose of this paper is to show the common aspects among the approaches used with Dislocation Dynamics simulations, as well as their distinctitive features. Given the uniqueness of the solution, an attempt is made to explicitly illustrate how it is translated within the different existing frameworks. Two approaches are distinguished according to whether the stress singularity along the dislocations lines is homogenized or not. It is shown that the solution to be preferred depends on the problems at hand.

1. Introduction

Over the years, different methods have been developed to calculate the state of mechanical equilibrium in a dislocated finite body. Recently, such studies were found to be useful for the development of exact 3D simulations of dislocation dynamics (DD). In this paper, a critical review of the existing numerical solutions is carried out. Such solutions are based on the *superposition method* [1, 2, 3, 4] and the *Discrete-Continuous Model* (DCM) [5, 6]. Although the underlying physical problem is the same, and therefore also its mechanical solution, these methods follow different strategies. As a consequence, it is not always clear how and where exactly they differ or are similar. Our purpose is here to establish their common features and their differences in a concise manner, as well as to determine their preferential domain of application. By lack of space, the case of simulations making use of periodic boundary conditions, as well as the recent phase-field approaches used to model dislocation dynamics, are not considered.

275

A. Finel et al. (eds.), Thermodynamics, Microstructures and Plasticity, 275–284.

Throughout this study, use is made of normal type (a) to denote scalar quantities and of bold characters (**a**) for vectors. Second order tensors are underlined ($\underline{a}$), and fourth order tensors are doubly underlined $\underline{\underline{a}}$.

2. Statement of the problem

2.1. DISLOCATIONS IN A CONTINUUM MEDIUM

Following the seminal work of Mura [7], dislocations can be regarded as line defects causing eigenstresses in an elastic medium. For each dislocation, a volume C is defined, which surrounds the line and its slip trace (see figure 1). Within this volume (of penny shape with constant thickness h in the direction **n**, normal to the dislocation slip plane) an eigenstrain field, $\underline{\varepsilon}^p$, exists that leaves into the medium a plastic deformation in the form of a displacement shift (or shear). The direction and the magnitude of this plastic shear are characterized by the Burgers vector, **b**. Outside C and at the boundary ∂C, the eigenstrain $\underline{\varepsilon}^p$ is by definition equal to zero. When a small volume element v included in V is sheared by several dislocation segments i, the associated eigenstrain is denoted by $\underline{\varepsilon}^p_{,v} = \sum_i \underline{\varepsilon}^{p,i}$. One must realize that the dislocation core properties are not accounted for in this framework, but this is not a problem in what follows since the dislocation core structure has a weak influence on the boundary value problem.

In a few ideal cases, such as dislocations in an infinite elastic medium and with a Heaviside step function as eigenstrain solution, the mechanical problem of the equilibrium state of a dislocated body can be solved analytically. This development is reported in many textbooks (see for instance [7]) and is, therefore, not be reproduced here. One must only keep in mind that the solutions for $\underline{\sigma}^\infty$ and $\underline{\varepsilon}^\infty$ are relatively simple and are commonly used to predict the elastic properties of dislocations [8].

2.2. MECHANICAL EQUILIBRIUM IN A FINITE BODY

Van der Giessen and Needleman [1] were the first to emphasize the importance of the boundary value problem in mesoscopic simulations of plastic deformation. Indeed, DD codes, in their standard formulation, use for reason of simplicity and code efficiency analytical forms justified only in an isotropic infinite body. When a finite body is concerned, or when there are internal interfaces, this simple solution is *a priori* no longer valid and the true boundary value problem (BVP) has to be solved.

Consider, as in figure 1, a computational volume V, which may contain elastic inclusions of volume V^* and dislocations (line segments i) in a matrix phase $V^M = V - V^*$. The elastic properties of the matrix are governed by the fourth-order tensor $\underline{\underline{L}}^M$, while the elastic modulus tensor of the inclusion

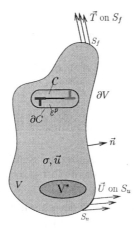

Figure 1. Statement of the mechanical problem. A description of all quantities can be found in the text. In mechanical equilibrium all fields must satisfy equation (1).

is denoted $\underline{\underline{L}}^*$. Tractions $\mathbf{T}$ can be prescribed at the boundary S_f, and displacements $\mathbf{U}$ at the boundary S_u. The latter may also include a plastic displacement $\mathbf{u}^p$ induced by dislocations that moved out of the volume V in an earlier deformation stage. The outward normal to the surface is denoted by $\mathbf{n}$.

In the absence of overall body forces, the equilibrium state of the total volume is described by the stress field $\underline{\sigma}$ and the total strain field $\underline{\varepsilon}$ that must satisfy:

$$\begin{cases} \nabla \cdot \underline{\sigma} &= \mathbf{0} \\ \underline{\sigma} \cdot \mathbf{n} &= \mathbf{T} \text{ at } S_f \\ \mathbf{u} &= \mathbf{U} \text{ at } S_u \\ \nabla \mathbf{u} &= \underline{\varepsilon} \\ \underline{\sigma} &= \underline{\underline{L}}^M : (\underline{\varepsilon} - \underline{\varepsilon}^p) \text{ in } V^M \\ \underline{\sigma} &= \underline{\underline{L}}^* : \underline{\varepsilon} \text{ in } V^* \end{cases} \qquad (1)$$

3. Solution strategies for DD simulations

Depending upon the precision required and on the nature of the problem to be solved, different numerical methods are available for computing field that satisfy equations (1). Two approaches are particularly well adapted to DD simulations and have been the object of several applications. Both of them make use of a coupling with a Finite Element (FE) code, but the resolution of the algebraic equations involved is not attached to any particular method.

For instance, Boundary Elements, Finite Volume or Element Free Galerkin methods could be a good alternative to the classical FE approach.

3.1. THE DISCRETE-CONTINUOUS MODEL

The Discrete-Continuous Model (DCM) was the first solution effectively used in conjunction with 3D simulations of dislocation dynamics [5, 6]. In this method, a FE code computes directly the displacement field satisfying equations (1), making use of the plastic strain, $\underline{\varepsilon}^p$, yielded by the DD simulation. Thus the DD code serves as a substitute for the constitutive form used in standard FE frameworks. The most difficult part of the coupling consists in setting conditions that leave the possibility for the FE mesh to capture the complexity of the elastic fields involved during plastic deformation. Two important steps of the computational method must be distinguished: a homogenization procedure for the calculation of $\underline{\varepsilon}^p$ and an interpolation procedure for the calculation of $\underline{\sigma}$.

The homogenization procedure is certainly the most critical part of the method, and a publication has been dedicated to it[9]. In consistency with the framework recalled in § 2.1, the plastic shear associated with the motion of dislocations is extended over a slab of thickness h. This is formally equivalent to replacing one dislocation by a continuum distribution of parallel infinitesimal dislocations, or to distributing the eigenstrains $\underline{\varepsilon}^{p,i}$ of each dislocation i over a large slab, rather than on an infinitely thin plate. Within the slab of typical thickness $h = 3\,v_G^{1/3}$ (v_G is the elementary volume associated to the Gauss points in the FE mesh), the Burgers vector can be split linearly or with the help of a shape function that helps localizing the dislocation slip plane in the FE mesh. In what follows, only the linear solution will be considered for the sake of simplicity, although a polynomial solution is preferred in the practice.

When a dislocation segment i belonging to the slip system of normal $\mathbf{n}$ and Burgers vector $\mathbf{b}^i$ moves, it generates increments of resolved plastic shear. These increments, $\Delta\gamma_{n,e}^i$, are homogenized within the elementary volumes $v_{G,e}$ of the Gauss points e according to:

$$\Delta\gamma_{n,e}^i = \frac{(b^i/h)v_{int,e}^i}{v_{G,e}} \tag{2}$$

where $v_{int,e}^i$ is the volume of intersection between the sheared slab of segment i and a volume $v_{G,e}$. Notice that C^i and $v_{int,e}^i$ are related by:

$$C^i = \sum_e v_{int,e}^i \tag{3}$$

Then, the plastic strain increment at each Gauss point of the FE mesh is the sum of the contributions of all the segments i of unit shear direction $\mathbf{l}^i = \mathbf{b}^i/b^i$ and moving in $v_{G,e}$:

$$\Delta\underline{\varepsilon}^p_{,e} = \sum_i \Delta\gamma^i_{n,e}(\mathbf{l}^i \otimes \mathbf{n}^i)^{sym} \tag{4}$$

Finally, to obtain the total plastic strain at step k, the sum of all increments from step 0 to step k is needed:

$$\underline{\varepsilon}^p_{,e} = \sum_0^k \Delta\underline{\varepsilon}^p_{,e} \tag{5}$$

Once $\underline{\varepsilon}^p$ is defined at each Gauss point of the FE mesh, the problem of the equilibrium state (Eq. 1) can be solved in a conventional manner. The only modification made with respect to an ordinary FE explicit scheme is that the increments of total deformation $\Delta\underline{\varepsilon}_{,e}$ need to be computed simultaneously everywhere in the FE mesh. This modification is necessary for the DD simulation.

The second critical procedure of the DCM is the interpolation step. This operation is not specific to the DCM and exists in the other approach discussed below. Nevertheless, this procedure is more critical in the DCM case where the FE mesh is strongly deformed. To compute the Peach-Koehler force on each dislocation, the stresses calculated at the Gauss points, $\underline{\sigma}_{,e}$, must be interpolated at reference points along the dislocation line. The quality of this interpolation strongly influences the dynamics of the dislocations. It is directly related to the flexibility offered by the polynomial shape functions associated with the FE mesh. For this reason, mesh elements containing a large number of nodes and Gauss points are required. Typically, the DCM calculations make use of elements consisting of 20 nodes - 27 Gauss points.

3.2. THE SUPERPOSITION METHOD

This approach was first used by Van der Giessen and Needleman [1] for 2-D simulations, but is now applied to 3-D simulations [2, 3, 4]. The basic objective of this method is to enable an accurate description of the dislocation-dislocation interactions at short distances, while, at the same time, simplifying as much as possible the computations delivered to the FE code. In rough terms, the idea is to eliminate from the FE mesh the elastic singularity associated to the dislocation fields. This can be done by extracting the singular solutions for dislocations in an infinite body, $\underline{\sigma}^\infty$ and $\mathbf{u}^\infty$ (see § 2.1) from the whole mechanical problem.

More precisely, the simulated problem (cf. fig. 1) is decomposed into two sub-problems. The first sub-problem is that of interacting dislocations in a homogeneous, isotropic, infinite solid, the ($\sim$) fields, and the second one is the complementary problem of accounting for the initial non-homogeneous body, but without dislocations and with modified boundary conditions, the ($\wedge$) fields. The state of the simulated body is then re-written as the superposition of two fields:

$$
\begin{aligned}
\mathbf{u} &= \hat{\mathbf{u}} + \tilde{\mathbf{u}} \\
\underline{\varepsilon} &= \underline{\hat{\varepsilon}} + \underline{\tilde{\varepsilon}} \\
\underline{\sigma} &= \underline{\hat{\sigma}} + \underline{\tilde{\sigma}}
\end{aligned}
\tag{6}
$$

Note that the fields of the ($\sim$) sub-problem may be decomposed again into two contributions:

$$
\begin{aligned}
\tilde{\mathbf{u}} &= \mathbf{u}^\infty + \mathbf{u}^b \\
\underline{\tilde{\varepsilon}} &= \underline{\varepsilon}^\infty + \underline{\varepsilon}^b \\
\underline{\tilde{\sigma}} &= \underline{\sigma}^\infty
\end{aligned}
\tag{7}
$$

where the additional fields $\mathbf{u}^b$ and $\underline{\varepsilon}^b$ exist only at the boundary and account for the body shape transformation. Indeed, each time a dislocation moves out of the finite body, it vanishes at the surface leaving a step of magnitude b, and the shape of the finite body is changed. Accounting for this dislocation-boundary interaction is one of the most difficult parts of the superposition method. Among possible solutions, the concept of virtual dislocations developed by Weygand *et al.* [10] is extremely useful. To avoid the cumputation of $\mathbf{u}^b$ and $\underline{\varepsilon}^b$, it is suggested that the outgoing dislocations should never be eliminated, but accumulated outside the simulated body, at virtual coordinates consistent with the dislocation slip planes. The benefit of that procedure is that displacements at the boundary are computed in the same manner as in the volume.

Then, the governing equations for the ($\sim$) fields are, by construction, free of boundary conditions and can be summarized as follows:

$$
\left\{
\begin{aligned}
\nabla \cdot \underline{\tilde{\sigma}} &= \mathbf{0} \\
\underline{\tilde{\sigma}} &= \underline{L} : \underline{\tilde{\varepsilon}} \\
\underline{\tilde{\varepsilon}} &= \nabla \tilde{\mathbf{u}}
\end{aligned}
\right.
\quad \text{in } V = V^M \cup V^*
\tag{8}
$$

With the help of equations (1), (6) and (8), we can now define the equations governing the sub-problem for the ($\wedge$) fields:

i) in V

$$\nabla \cdot \underline{\sigma} = \mathbf{0} \quad <=> \quad \nabla \cdot (\underline{\widehat{\sigma}} + \underline{\widetilde{\sigma}}) = \nabla \cdot \underline{\widehat{\sigma}} + \nabla \cdot \underline{\widetilde{\sigma}} = \nabla \cdot \underline{\widehat{\sigma}}$$
$$=> \quad \nabla \cdot \underline{\widehat{\sigma}} = \mathbf{0} \tag{9}$$

and

$$\varepsilon = \widehat{\varepsilon} + \widetilde{\varepsilon} = \nabla \mathbf{u} \quad <=> \quad \varepsilon - \widetilde{\varepsilon} = \nabla \mathbf{u} - \nabla \widetilde{\mathbf{u}}$$
$$=> \quad \widehat{\varepsilon} = \nabla \widehat{\mathbf{u}} \tag{10}$$

ii) at the boundary we have on S_f:

$$\underline{\sigma} \cdot \mathbf{n} = \mathbf{T} \quad <=> \quad \underline{\widehat{\sigma}} \cdot \mathbf{n} = \mathbf{T} - \underline{\widetilde{\sigma}} \cdot \mathbf{n}$$
$$=> \quad \underline{\widehat{\sigma}} \cdot \mathbf{n} = \mathbf{T} - \widetilde{\mathbf{T}} \tag{11}$$

and on S_u:

$$\mathbf{u} \quad = \quad \widehat{\mathbf{u}} + \widetilde{\mathbf{u}} = \mathbf{u}^{app}$$
$$=> \quad \widehat{\mathbf{u}} = \mathbf{u}^{app} - \widetilde{\mathbf{u}} \tag{12}$$

iii) finally, keeping in mind that $\underline{\varepsilon}^p$ is non-zero only inside C and that the thickness of C tends to zero when the analytical solutions hold, we have in V^M:

$$\underline{\sigma} = \underline{\underline{L}} : (\varepsilon - \underline{\varepsilon}^p) \quad = \quad \underline{\underline{L}} : (\widehat{\varepsilon} + \widetilde{\varepsilon} - \underline{\varepsilon}^p) = \underline{\underline{L}} : (\widehat{\varepsilon} + \widetilde{\varepsilon}) = \underline{\underline{L}} : \widehat{\varepsilon} + \widetilde{\underline{\sigma}}$$
$$=> \quad \underline{\widehat{\sigma}} = \underline{\underline{L}} : \widehat{\varepsilon} \tag{13}$$

and in V^*:

$$\underline{\sigma} = \underline{\widehat{\sigma}} + \underline{\widetilde{\sigma}} \quad = \quad \underline{\underline{L}}^* : \varepsilon = \underline{\underline{L}}^* : (\widehat{\varepsilon} + \widetilde{\varepsilon})$$
$$=> \quad \underline{\widehat{\sigma}} = \underline{\underline{L}}^* : \widehat{\varepsilon} + (\underline{\underline{L}}^* - \underline{\underline{L}}) : \widetilde{\varepsilon} \tag{14}$$

In summary, the equilibrium state equations for the ($\widehat{}$) sub-problem are:

$$\begin{cases} \nabla \cdot \underline{\widehat{\sigma}} &= \mathbf{0} \\ \underline{\widehat{\sigma}} \cdot \mathbf{n} &= \mathbf{T}^{app} - \underline{\sigma}^\infty \cdot \mathbf{n} \quad \text{at } S_f \\ \widehat{\mathbf{u}} &= \mathbf{U}^{app} - \widetilde{\mathbf{u}} \quad \text{at } S_u \\ \nabla \widehat{\mathbf{u}} &= \widehat{\varepsilon} \\ \underline{\widehat{\sigma}} &= \underline{\underline{L}} : \widehat{\varepsilon} \quad \text{in } V^M \\ \underline{\widehat{\sigma}} &= \underline{\underline{L}}^* : \widehat{\varepsilon} + (\underline{\underline{L}}^* - \underline{\underline{L}}) : \widetilde{\varepsilon} \quad \text{in } V^* \end{cases} \tag{15}$$

By resolving this elastic problem with a FE meshing, and adding the analytical solutions of the ($\widetilde{}$) fields problem to that solution, i.e. by adding σ^∞, $\varepsilon^\infty[+\varepsilon^b]$ and $\mathbf{u}^\infty[+\mathbf{u}^b]$, one can determine solutions for the total problem as defined in equations (1).

282

4. Discussion

In this section the strengths and weaknesses of the BVP solutions presented in §3 are compared. The discussion is based on the authors' recent evaluations and validation tests. We tried, as much as possible, to restrict our comments to general features independent of the coding details of the simulations.

In terms of CPU, the comparison of the DCM and the superposition method is usually in favor of the last one. Indeed, as mentioned in §3.1, the DCM requires a more detailed meshing (i.e. more elements with many nodes and Gauss points) and, therefore, larger computations. Also, the superposition method requires only data transfer between the DD and FE codes at the boundary elements of the mesh, whereas, the DCM imposes data transfer everywhere. From a practical viewpoint, the implementation of DD-FE coupling based on the superposition approach is easier to realize, but this last point may depend on the FE code that is used. From a theoretical point of view, the two approaches are perfectly equivalent with one exception; calculations in an elastically anisotropic medium can be realized much more easily using the DCM approach. Indeed, for the superposition solution to be efficient, analytical forms are required for the displacement field of dislocation segments and for anisotropic media a general solution does not exist. In contrast, elastic anisotropy is easily taken into account with the DCM by only changing one input of the DCM calculation: the tensor of elastic moduli of the considered material. As a result, computing times are virtually the same for isotropic and anisotropic simulations.

To obtain the stress at an arbitrary position in the simulated body, FEs are using shape functions which interpolate the stresses calculated at Gauss points. A basic hypothesis of the classical FE method imposes that these shape functions are continuous within the elements. Since the dislocation fields vary as the inverse of the distance to the line, the resulting singularities at the dislocation line cannot be exactly accounted for. For instance, our tests show that quadratic elements (i.e. with a polynomial shape function of order two) become inaccurate at distances from a dislocation line of the order of $v_G^{1/3}$. This is why, without correction, the DCM cannot precisely account for dislocation-dislocation contact and short distance interactions. An immediate solution to this problem consists in refining the mesh and, consequently, decreasing the homogenization thickness h around the dislocation lines. The latter is directly linked to the size of finite element mesh and to v_G. However, this leads rapidly to useless simulations in term of computation times, especially in 3D. Alternatively, we can use in the DCM analytical forms for segment-segment elastic interactions at short distances (i.e. when two segments are at a distance smaller than $1.7\,v_G^{1/3}$).

This additional force contribution to the dislocation dynamics simulation can be regarded as a constitutive rule caring of the lack of accuracy of the FE mesh at short distance.

It is interesting to note that the above problem of the DCM approach turns into an advantage when dislocation segments are close to boundaries. Indeed, in the superposition approach the boundary conditions on S_f and/or S_u contain an image contribution proportional to minus σ^∞ and $\mathbf{u}^\infty$, respectively. When a dislocation approaches the boundary, these contributions diverge, and in fact become infinite when dislocations touch a Gauss point at the boundary. To precisely account for this surface effect requires an accurate subtraction of the large ($\tilde{}$) and ($\hat{}$) fields, hence a remeshing of the boundaries close to the points where segments are emerging [11]. Without such a procedure, important errors arise on the amplitude of the image force (the nodes at the surface elements cannot accommodate the important elastic displacement imposed by the emerging dislocations). To the best knowledge of the authors, no satisfactory solution to this problem has been published yet. On the other hand, as a result of the homogenization procedure, such difficulty is not met within the DCM approach. The only limitation is that the amplitude of the image force close to the surface may be smaller than the elastic prediction because of the smoothing of the shape functions.

Lastly, as emphasized by equation (14), the existence of elastic inclusions in the simulated body impose additional computations to the superposition approach and makes it less attractive by comparison to the DCM scheme, since the latter is transparent to this problem. As discussed in reference [1] and [12], in the case of the superposition approach, two-phase problems impose the calculation of the so-called polarization stresses in the inclusions, $\hat{\underline{p}} = (\underline{L}^* - \underline{L}) : \tilde{\underline{\varepsilon}} = (\underline{L}^* : \underline{L}^{-1} - \underline{I}) : \tilde{\underline{\sigma}}$. Hence, the number of analytical calculations of the $\tilde{\underline{\sigma}}$ solutions at each step of the simulation is significantly increased (in addition to the boundary calculation, the stress field of each segments must be calculated at each Gauss points in V^*).

5. Conclusion

In this paper, we provide a non-exhaustive list of advantages and disadvantages of the two numerical approaches used to solve boundary value problems in DD simulations, respectively the DCM and the superposition method. During these early stages of development of these simulation techniques, our aim was to draw the contour of their domains of excellence. Based on the comments made in §4, our recommendation to those starting in this field could be to try first the superposition method. This approach is easy to implement, delivers precise results and is efficient in terms of

CPU for simple problems. Then, for those interested in problems with a large density of dislocations and complex materials (e.g. multiphase materials, materials with strong elastic anisotropy materials, materials exhibiting plastic strain localization, ...) our preference would go to the second approach. Indeed, the DCM is numerically competitive (even faster) with large numbers of segments. Furthermore, with its concept of plastic strain homogenization,it provides an interesting self-consistent description of plasticity compatible with mesoscopic (discrete) and macroscopic(continuous) modeling of solid mechanics.

6. Acknowledgments

The authors wish to thank J-L. Chaboche, F. Feyel and L.P. Kubin for helpful discussions.

References

1. E. Van der Giessen and A. Needleman, Model. Simul. Mater. Sci. Eng. **3**, 689 (1995).
2. D. Weygand, E. Van der Giessen and A. Needleman, Mat. Sci. Eng. **A309-310**, 420 (2001).
3. C.S. Shin, M.C. Fivel and K.H. Oh, J. Phys. IV France **11**, Pr5-27 (2001).
4. H. Yasin, H.M. Zbib and M.A. Khaleel, Mat. Sci. Eng. **A309-310**, 294 (2001).
5. C. Lemarchand, B. Devincre, L.P. Kubin and J.L. Chaboche, MRS Symposium Proceedings **538**, 63 (1999).
6. C. Lemarchand, Ph.D. Thesis, University of Paris VI (1999).
7. T. Mura,"Micromechanics of defects in solids", Second edition, Kluwer Academic Publishers (1993).
8. J.P. Hirth and J. Lothe,"Theory of dislocations", Second edition, John Wiley and Sons (1982).
9. C. Lemarchand, B. Devincre and L.P. Kubin, Journal of the Mechanics and Physics of Solids **49**, 1969 (2001).
10. D. Weygand, L.H. Friedman, E. Van der Giessen, A. Needleman, Model. and Simu. in Mat. Sci. and Eng. **10**, 437 (2002).
11. M.C. Fivel, T.J. Gosling and G.R. Canova Model. Simul. Mater. Sci. Eng. **4**, 581 (1996).
12. C.S. Shin, M.C. Fivel and K.H. Oh J. Phys. IV France **11**, 27 (2001).

DISCRETE DISLOCATION PLASTICITY

Numerical simulation of dislocation dynamics

E. VAN DER GIESSEN
Netherlands Institute for Metals Research/
Dept. of Applied Physics, University of Groningen
Nyenborgh 4, 9747 AG Groningen

Abstract. This article provides a brief overview of the developments in discrete dislocation plasticity, with emphasis on solving boundary-value problems. It reviews the superposition technique that was proposed in 1995 and presents a number of applications of two-dimensional simulations that give new insights. One class of applications relates to fracture problems, and highlight the dual role that dislocations can play in the fracture of materials with limited plasticity. The second group of simulations is used to help the development of nonlocal continuum crystal plasticity theories.

1. Introduction

With reference to the schematic of pertinent length scales in plasticity in Fig. 1, I will start by pointing out that there is a seemingly growing class of plasticity problems where the relevant length scale (determined either by geometry or by the wavelength of the deformation pattern) is such that there are too few dislocations in order that a continuum plasticity description is meaningful, but at the same time there are too many dislocations for atomistic modeling to be practical. This class of problems is in the realm of discrete dislocation plasticity (DDP), in which dislocations are treated as line singularities in an elastic continuum. The solution of discrete dislocation plasticity problems involves two essential ingredients: (i) the determination of the fields inside a dislocated body; (ii) the evolution of the dislocation structure on the basis of the current fields. The basic idea of the approach goes back a long time, and seminal work on the first aspect can be found in the famous textbook by Hirth and Lothe [1]. The drawback in this classical view is that for each boundary value problem the solution of the elasticity problem with a number of dislocations in the studied geometry subject to its boundary conditions needs to be found. Because of the singularity along the dislocation line

A. Finel et al. (eds.), Thermodynamics, Microstructures and Plasticity, 285–298.

286

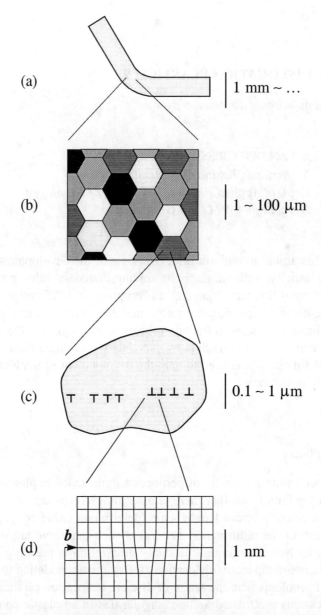

Figure 1. Schematic of the various pertinent length scales in between a single dislocation and a plastically deforming polycrystalline metal at the macro-scale. Each length scale requires its own type of model: (a) macroscopic phenomenological; (b) crystal plasticity; (c) discrete dislocation plasticity; (d) atomistics.

this is a horrifying task, and in fact only a finite number of relatively simple cases have been solved. In 1995, Alan Needleman and I [2], devised a more versatile approach in which we use the singular part of a known solution (e.g. in infinite space) and employ superposition to augment it with a finite element solution to comply with the actual boundary conditions. Since then a few other approaches have been proposed in the literature, where, after some sort of coarse-graining of the dislocations, the elasticity problem is solved directly using finite elements (e.g. [3]) or phase-field methods (e.g. [4]).

In this article, the essentials of the superposition approach will be summarized, including the additional ingredients, so-called constitutive rules, needed to analyze dislocation dynamics (limiting attention to two-dimensional situations). The subsequent section will discuss what we have learned from discrete dislocation plasticity in relation to fracture problems under monotonic or fatigue loading. A very important characteristic of discrete dislocation plasticity is that it is inherently size dependent, contrary to classical continuum models. The subsequent section will therefore discuss a number of model problems that we have considered in helping the development of nonlocal continuum plasticity theories by comparing the discrete dislocation results with the solutions of a few recent nonlocal theories.

1.1. GENERAL APPROACH

In discrete dislocation plasticity, a dislocation is treated as a line singularity in a linear elastic continuum, whose motion produces what we observe as permanent, plastic strain. Such a description obviously cannot capture the core structure of a dislocation, but it does capture the fields further away than five to ten times the atomic spacing. Within the linear elastic approximation, the fields around a dislocation have the typical structure that (i) the displacement component parallel to the slip plane on which it lives is discontinuous across the slip plane and that (ii) the stress and strain fields decay as $1/r$ away from the dislocation. Because of this $1/r$ decay, dislocations have long-range effects and interactions with other dislocations. In addition, these interactions depend in a rather complex manner on the orientation relative to the other dislocations. Owing to these characteristics, dislocations can organize in dislocation structures, such as walls and cells.

For the determination of the fields in an arbitrary dislocated body subject to boundary conditions, the basic idea put forward in [2] was to exploit the known singular solutions, notably in infinite space, and to use superposition to correct for the proper boundary conditions, as illustrated in Fig. 2. The displacement, strain and stress fields are decomposed as

$$u_i = \tilde{u}_i + \hat{u}_i, \quad \varepsilon_{ij} = \tilde{\varepsilon}_{ij} + \hat{\varepsilon}_{ij}, \quad \sigma_{ij} = \tilde{\sigma}_{ij} + \hat{\sigma}_{ij}. \quad (1)$$

The ($\tilde{}$) fields are the superposition of the singular fields of the individual dislocations in their current configuration, but in infinite space. Identifying the fields for

288

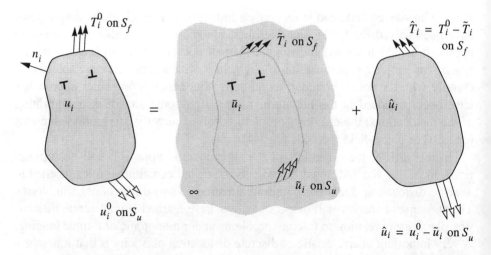

Figure 2. Decomposition into the problem of interacting dislocations in the infinite solid (˜ fields) and the complementary problem for the finite body without dislocations (ˆ or image fields).

dislocation k by a superscript (k), the (˜) stress field, for example, is obtained as

$$\tilde{\sigma}_{ij} = \sum_k \sigma_{ij}^{(k)}.$$

The actual boundary conditions, in terms of prescribed displacements u_i^0 or tractions $T_i^0 = \sigma_{ij} n_j$, are imposed through the (ˆ) fields, in such a way that the sum of the (˜) and the (ˆ) fields in (1) gives the solution that satisfies all boundary conditions. It is important to note that the solution of the (ˆ) problem does not involve any dislocations. Therefore, the (ˆ) fields (often called 'image' fields) are smooth and the boundary value problem for them can conveniently be solved using a finite element method.

2. Constitutive rules

Once the fields in the dislocated solid are known, the second ingredient is to determine the instantaneous change of the dislocation structure. Materials scientists have discovered a variety of different ways in which this may happen, such as (i) (predominantly) dislocation glide; (ii) climb; (iii) cross slip; (iv) annihilation; (v) junction formation with other dislocations and (v) pinning at obstacles. Each of these mechanisms is controlled by atomistic processes, which, by definition, are not resolved in discrete dislocation plasticity. These have to be incorporated by a set of constitutive equations or rules, just like in plasticity theories at higher size scales (Fig. 1a, b). These constitutive rules have to be inferred from experiments or from atomistic simulations.

The important point to note first is that the key quantity involved in constitutive rules for dislocation evolution is the so-called Peach-Koehler force. It is a configurational force acting on the dislocation (per unit length) that is work-conjugate to motions of this dislocation that leave the total length of the dislocation unchanged. It can be shown [2] that in the approach outlined above, the component of the Peach-Koehler force in the slip plane can be expressed as

$$f^{(k)} = n_i^{(k)} \left(\hat{\sigma}_{ij} + \sum_{l \neq k} \sigma_{ij}^{(l)} \right) b_j^{(k)}. \tag{2}$$

This expression highlights the long-range contribution of all other dislocations, through the second term in parentheses, as well as the image stresses.

It would take too far to discuss the constitutive rules in general (a now classical account can be found in [5]); we will confine ourselves here to those used in the two-dimensional simulations to be presented later. All these problems involve only edge dislocations, for which the glide component of the Peach-Koehler force reduces to $f^{(k)} = \tau^{(k)} b^{(k)}$ where $\tau^{(k)}$ is the resolved shear stress on the plane. The following ingredients to the evolution of the dislocation structure are incorporated: the motion of dislocations along their slip plane, pinning of dislocations at obstacles, annihilation of opposite dislocations, and generation of new dislocation pairs from discrete sources.

Glide of a dislocation is accompanied by drag forces due to interactions with electrons and phonons. The simplest models of this lead to drag forces that can be expressed as $Bv^{(k)}$ where B is the drag coefficient. A value of $B = 10^{-4} \, \text{Pa s}$ is representative for aluminum [5]. When inertia effects of dislocations are ignored, the magnitude of the glide velocity $v^{(k)}$ of dislocation k becomes linearly related to the Peach-Koehler force through $f^{(k)} = Bv^{(k)}$.

New dislocation pairs are generated by simulating Frank-Read sources. The initial dislocation segment of a Frank-Read source bows out until it produces a new dislocation loop and a replica of itself. The Frank-Read source is characterized by a critical value of the Peach-Koehler force, the time it takes to generate a loop and the size of the generated loop. In two dimensions, this is simulated by point sources which generate a dislocation dipole when the magnitude of the Peach-Koehler force at the source exceeds a critical value $\tau_{\text{nuc}} b$ during a period of time t_{nuc}. The distance L_{nuc} between the dislocations is specified so that the dipole does not collapse onto itself under an applied force of $\tau_{\text{nuc}} b$. In the examples shown later, the strength of the dislocation sources is chosen at random from a Gaussian distribution with mean strength $\bar{\tau}_{\text{nuc}}$ and standard deviation of $0.2\bar{\tau}_{\text{nuc}}$. The nucleation time for all sources is typically taken as $t_{\text{nuc}} = 0.01 \, \mu s$.

Annihilation of two dislocations with opposite Burgers vector occurs when they are sufficiently close together. This is modeled by eliminating two dislocations when they are within a material-dependent, critical annihilation distance L_e, which is taken as $L_e = 6b$ [5] in the examples later.

In some calculations, obstacles to dislocation motion are included that are modeled as fixed points on a slip plane. Such obstacles can represent either small precipitates or forest dislocations. Pinned dislocations can only pass the obstacles when their Peach-Koehler force exceeds an obstacle dependent value $\tau_{obs}b$.

3. Applications to crack problems

Plastic deformation near crack tips can be considered at different length scales. Under the assumption of small-scale yielding, continuum plasticity representations of near-tip fields were established on the basis of isotropic models halfway the last century, and more recently for anisotropic crystal plasticity by Rice [6]. When strain hardening is neglected, the latter fields are predicted to have a remarkable geometry, with the stress state being uniform in distinct sectors around the tip, and with either slip bands or kink bands in between the sectors. Evidently, this analysis ignores the discreteness of dislocations inside the plastic zone.

Cleveringa et al. [7] carried out a discrete dislocation analysis of a mode I crack. Crack propagation was modeled by a cohesive surface with fracture properties approaching atomic (de)bonding. Depending on the densities and strengths of the sources and obstacles that were randomly introduced around the crack tip, these calculations showed that (i) crack growth was brittle if insufficient dislocation activity was triggered; (ii) crack blunting without fracture occurred when dislocations could move sufficiently far away from the top to effectively shield the tip; and (iii) in intermediate situations, the crack could grow in spurts due to the fact that dislocation structures formed near the tip that raised the local opening stress state to the cohesive strength over a sufficiently large region. This is demonstrated for a particular case with two slip systems at $\pm 60°$ from the crack plane in Fig. 3. The salient feature of the findings in [7] is that dislocations can play a dual role: On the one hand they are the carriers of plastic relaxation, while at the same time they move with them a singularity.

The other observation made in [7] was that beyond some distance ($\approx 0.5\mu m$) away from the tip, the stresses appear to be uniform on average in three sectors around the tip, consistent with Rice's [6] continuum crystal plasticity analysis for this orientation of two slip systems. This connection was studied in more detail in [8] for crystals with three slip systems. One of the orientations studied was so that in addition to the two $\pm 60°$ slip systems considered in Fig. 3 there was a slip system with its slip planes parallel to the crack plane. In this case, Rice's analysis predicts four fully plastic, uniform stress sectors in the absence of hardening, with kink bands rather than slip bands in between two of these. The average stresses from the discrete dislocation analysis were found to agree fairly well with the continuum predictions, but no sign of kink bands was found. However, the non-hardening continuum slip solutions are not unique and Drugan [9] has constructed alternative elastic-ideally plastic solutions in which kink bands do not form. In

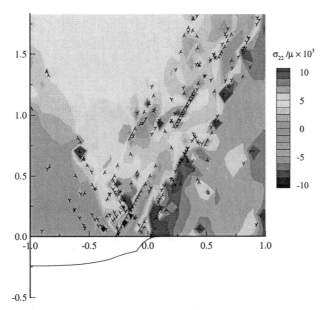

Figure 3. Distribution of dislocations and the opening stress σ_{22} in the immediate neighborhood $(2\mu\mathrm{m} \times 2\mu\mathrm{m})$ of the crack tip for the case with $\rho_{\mathrm{nuc}} = 49/\mu\mathrm{m}^2$ ($\bar\tau_{\mathrm{nuc}} = 50$ MPa) and $\rho_{\mathrm{obs}} = 98/\mu\mathrm{m}^2$ (τ_{obs}) at the onset of crack growth. The corresponding crack opening profiles (displacements magnified by a factor of 10) are plotted below the x_1-axis. From [7].

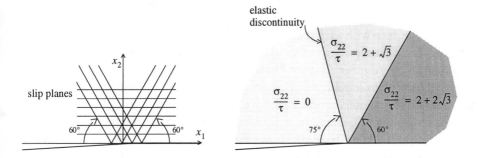

Figure 4. Opening stress states (σ_{22} normalized by the critical resolved shear stress τ) in the three sectors of Drugan's [9] solution.

fact, there are several solution families without kink bands. For the same configuration analyzed by discrete dislocation plasticity, one of his solution families has a slip band at $\theta = 60°$, consistent with our discrete dislocation simulations. Within this family, the solution that seemed to agree best with the discrete dislocation results, see Fig. 4, has one of the sectors being elastic (instead of fully plastic) and collapsing to a line discontinuity.

The duality in the role of near-tip dislocations mentioned above is not re-

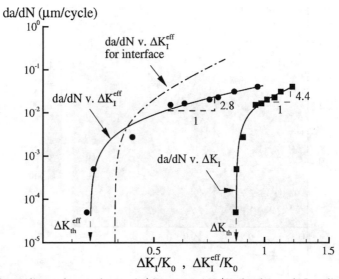

Figure 5. The cyclic crack growth rate da/dN versus $\Delta K_I/K_0$ for the mode I cyclic loading of a single crystal and for an interface crack. The same data is also plotted versus the effective loading amplitude ΔK_I^{eff} which accounts for crack closure in the assumed irreversible cohesive surface. The slopes of the curves marked correspond to the Paris law exponents for the curves fitted through the numerical results. From [10].

stricted to the mode I crack problem, but is probably generic for many fracture issues. For example, Deshpande et al. [10] have considered the growth of a crack along the interface with a rigid material under cyclic remote mode I loading, where the near-tip enhanced stress state was again found to determine when the crack would advance. Another discrete dislocation effect that is critical in fatigue is the irreversibility of dislocation motion. A summary of some salient results in [10] shown in Fig. 5, demonstrates that the cyclic crack growth rate $\log(da/dN)$ versus applied stress intensity factor range $\log(\Delta K_I)$ curve that emerges naturally from the discrete dislocation solution shows distinct threshold and Paris law regimes. It is emphasized that no special features were introduced in the discrete dislocation plasticity analysis for fatigue loading. However, the cohesive law used in this calculation was modified to be irreversible, in a simple attempt to account for an oxidizing environment.

4. Bridging scales to strain gradient continuum theories

With reference to the scale transitions illustrated in Fig. 1, a discrete dislocation description should be able to provide a true foundation for crystal plasticity. The continuum description of plastic deformation in the latter implies an averaging of the behavior of a sufficiently large ensemble of dislocations. Statistical approaches are now starting to be developed, but the link between the two descriptions of plasticity is currently done indirectly through constitutive rules. One of the most important constitutive laws in a crystal plasticity theory is that for hardening of slip systems. From the point of view of discrete dislocations, hardening is largely due to the interactions between the dislocations on the slip system under consideration with those on intersecting slip systems, i.e. the so-called forest dislocations. Three-dimensional discrete dislocation models are capable of simulating this forest hardening mechanism and thereby to provide input to the hardening laws in crystal plasticity models (see, e.g., [11]).

Such simulations, however, deliberately ignore other interactions and are therefore relevant for the behavior in the interior of a grain. The interaction with boundaries, such as interfaces with second-phase particles and grain boundaries gives rise to additional effects. Cleveringa et al. [12], for example, performed a discrete dislocation analysis of plastic flow in a model composite material containing hard elastic particles, under single slip conditions. They demonstrated that, depending on the particle shape and size, the material may develop geometrically necessary dislocations (see e.g. [16]). In such cases, the overall response was also found to be dependent on the size of the reinforcements (at the same volume fraction). The corresponding deformation field, shown in Fig. 6a, reveals rotation of the central particle as well as localized shearing near the ends of the particles.

Since the predicted dislocation densities, even for overall strains as small as 1%, reach values on the order of $10^{14} m^{-2}$, one may ask the question if the problem could have been analyzed by continuum plasticity. Although standard continuum plasticity theories do predict additional hardening due to the particles [12], they cannot predict size-dependent behavior. For this, one needs nonlocal or strain gradient theories.

Aifantis [13] was the first to suggest a plasticity model that not only depends on plastic strain but also on plastic strain gradients. Just from dimensional considerations this introduces a length scale in the model, which will give rise to size effects. At the early stages of strain gradient plasticity, there was some debate in the literature about which gradient to use, but the paper by Fleck et al. [14] on size effects in thin wires under torsion reminded the community of the concept of geometrically necessary dislocations. Nye [15] in the early 50's showed that the essential element is the existence of a net Burgers vector averaged over a small region when there is a strain gradient. In a continuum setting of plasticity, he further showed that the net Burgers vector B_i can be described by a second-order tensor α_{ij}, which can be directly related to the gradient of the plastic strain ε_{ij}^{p} through

$$\alpha_{ij} = e_{jlk}\varepsilon_{ik,l}^{p},$$

where e_{jlk} is the alternating tensor and $(),j$ denotes differentiation with respect to coordinate x_j. When this is used in the framework of continuum crystal plasticity, where the plastic strain is given by

$$\varepsilon_{ij}^{p} = \sum_{\alpha}\gamma^{(\alpha)}\frac{1}{2}\left[s_i^{(\alpha)}m_j^{(\alpha)} + s_j^{(\alpha)}m_i^{(\alpha)}\right], \tag{3}$$

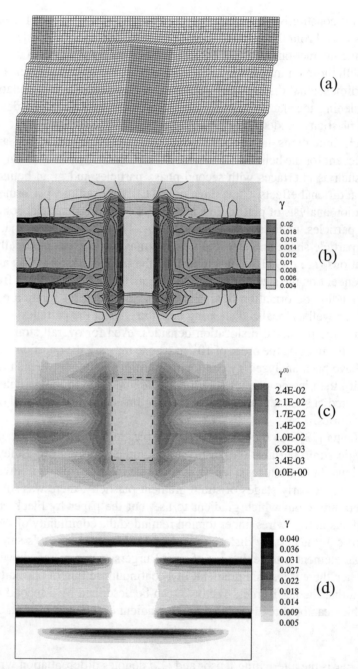

Figure 6. Deformation fields for the composite under simple shear (see inset) according to various theories: (a) discrete dislocation plasticity (from [12]); (b) Acharya-Bassani theory (from [21]); (c) Gurtin's nonlocal theory [22] and (d) Groma's theory [20].

in terms of the slip $\gamma^{(\alpha)}$ on slip system α (with unit normal $m_i^{(\alpha)}$ and slip direction $s_i^{(\alpha)}$, it can be shown that α_{ij} depends on $\gamma_{,k}^{(\beta)}$ at most through derivatives of $\gamma^{(\beta)}$ in the β-th slip-plane [14]. There are a number of strain gradient crystal plasticity models available in the literature at the moment which are based on the notion that the presence of geometrically necessary dislocations, measured through the dislocation density tensor, give rise to additional hardening.

We have re-analyzed the composite problem of [12] with three of these: (i) by Acharya and Bassani [21]; (ii) by Gurtin [18]; (iii) by Yefimov et al. [19, 20]. These theories differ in several ways. The Acharya–Bassani theory is the simplest one with the strain gradient only entering as an additional contribution to the hardening of a slip system. The Gurtin theory incorporates this effect too, but in addition includes kinematic hardening; the latter involves an internal variable that is governed by a separate differential equation with associated boundary conditions. While these mentioned theories are entirely phenomenological, the third theory considered is based on a statistical description of an ensemble of interacting edge dislocations on parallel glide planes by Groma [19]. This leads to two dislocation density fields: one being the total dislocation density, the other being the dislocation-difference density (directly related to the density of geometrically necessary dislocations). These two fields are governed by two coupled transport equations with associated boundary conditions. Making use of the Orowan relationship, Yefimov et al. [20] coupled this to a continuum slip theory based on the plastic strain definition (3).

First it should be pointed out that all three nonlocal theories are capable of picking up the size dependence found in the discrete dislocation simulations, by appropriate selection of the material parameters (notably, the length scale parameters entering in the Acharya–Bassani [21] and Gurtin theories [22]). The aspect of the overall response that the Acharya–Bassani theory did not capture was the strong Bauschinger effect seen in [23], because this theory only involves slip system hardening and no kinematic hardening contribution as in the Gurtin theory. The current Groma theory only has kinematic hardening. The deformation fields predicted by these nonlocal theories are also shown in Fig. 6. The Acharya–Bassani prediction, in Fig. 6b, still resembles the local solution [12, 21] but it somewhat smoother and has increased slip above and below the central particle. The slip distribution according to Gurtin's theory, in Fig. 6c, differs from it by the slip on the left and right hand sides of the particle rapidly dropping to zero at the interface, because of the boundary condition imposed their. It seems that Groma's theory predicts a deformation field that is closest to the discrete dislocation result, Fig. 6d vs Fig. 6a, with deformation concentrating into slip bands near the top and bottom of all particles and no slip on the particle sides.

The second example to be briefly discussed here is that of the simple shear of single crystal between rigid walls. Assuming the crystal to have two slip systems with slip planes at $\pm 60°$ from the shearing direction, discrete dislocation simula-

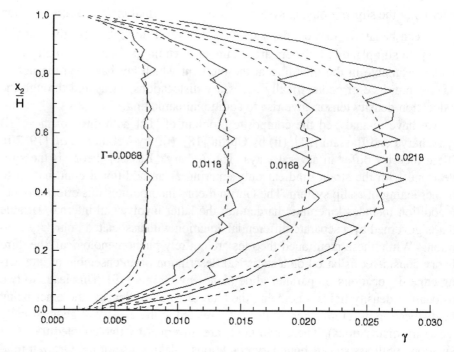

Figure 7. Shear strain profiles at various values of the applied shear Γ for a single-crystalline strip of height $H = 1\mu$m with two symmetric slip systems. The dashed lines are fitted exponential strain profiles. From [25].

tions predict the development of boundary layers because of the constraint on slip imposed by the impenetrable boundaries. This is illustrated in Fig. 7 showing the shear γ, averaged over $-\infty < x_1 < \infty$, across the thickness for four values of the overall shear Γ. The development of these boundary layers is accompanied by the development of geometrically necessary dislocations, which again induces a size effect [24] Standard continuum theories not only fail to pick up the boundary layers, they also do not capture the size effects. In fact, nonlocal theories that do not have additional (higher-order) boundary conditions are also not able to pick up these effects, because deformation will remain uniform when starting from a homogeneous state, just like in a local theory. The Shu-Fleck [25] and the Gurtin theory [22] both can capture the boundary layer formation and the size effect.

5. Concluding remarks

Discrete dislocation plasticity (DDP) applies to problems that are neither amenable to atomistics nor to continuum theories of plasticity. From this position, it holds promises in two directions.

One is the vertical direction in the length scale picture in Fig. 1: DDP can help to bridge the gap between atomistic descriptions of dislocations and continuum descriptions of crystal plasticity. An obvious route is to fine-tune DDP models on the basis of atomistic studies and to use DDP simulations to provide quantitative input for phenomenological constitutive rules in crystal plasticity. This assumes the existence of a theory. However, the form of crystal plasticity theories that account for size effects is not known; several attempts are being made, but this subject leaves challenges for the future.

The second direction in which DDP is expected to become a major player is the horizontal direction in Fig. 1, i.e. as a tool to analyze plasticity problems at the micron scale. With the continued miniaturization of components that is expected in this new century, this may become a major application area for DDP. Quantitative predictions, evidently, require a three-dimensional implementation and this is well underway now (e.g. [5, 26], but there are a number of technical difficulties to be solved to guarantee accuracy of the solutions [27, 28].

References

1. Hirth, J.P. and Lothe, J. (1968) *Theory of Dislocations*, McGraw-Hill, New York.
2. Van der Giessen, E. and Needleman, A. (1995) Discrete Dislocation Plasticity: A Simple Planar Model, *Model. Simul. Mater. Sci. Eng.* 3, 689–735.
3. Lemarchand, C., Devincre, B. and Kubin, L.P. (2001) Homogenization method for a discrete-continuum simulation of dislocation dynamics, *J. Mech. Phys. Solids.* 49, 1969–1982.
4. Wang, Y.U., Jin, Y.M., Cuitiño, A.M. and Khachaturyan, A.G. (2001) Nanoscale phase field microelasticity theory of dislocations: model and 3D simulations, *Acta Mat.* 49, 1847–1857.
5. Kubin, L.P., Canova, G., Condat, M., Devincre, B., Pontikis, V. and Brechet, Y. (1992) Dislocation Microstructures and Plastic Flow: A 3D Simulation, in *Nonlinear Phenomena in Materials Science* II, eds. G. Martin and L.P. Kubin, Sci-Tech, Vaduz, p. 455.
6. Rice, J.R. (1987) Tensile crack tip fields in elastic-ideally plastic crystals, *Mech. Mater.* 6, 317–335.
7. Cleveringa, H.H.M., Van der Giessen, E. and Needleman, A. (2000) A discrete dislocation analysis of mode i crack growth, *J. Mech. Phys. Solids* 48, 1133–1157.
8. Van der Giessen, E., Deshpande, V.S., Cleveringa, H.H.M. and Needleman, A. (2001) Discrete dislocation plasticity and crack tip fields in single crystals, *J. Mech. Phys. Solids* 49, 2133–2153.
9. Drugan, W.J. (2001) Asymptotic solutions for tensile crack tip fields without kink-type shear bands in elastic-ideally plastic single crystals, *J. Mech. Phys. Solids* 49, 2155–2176.
10. Deshpande, V.S., Needleman, A. and Van der Giessen, E. (2002) Discrete dislocation modeling of fatigue crack propagation, *Acta Mat.* 50, 831–846.
11. Fivel, M., Tabourot, L., Rauch, E. and Canova, G. (1998) Identification through mesoscopic simulations of macroscopic parameters of physically based constitutive equations for the plastic behaviour of fcc single crystals, *J. Phys. IV France* 8, 151–158.
12. Cleveringa, H.H.M., Van der Giessen, E. and Needleman, A. (1997) Comparison of discrete dislocation plasticity and continuum plasticity predictions, *Acta Mater.* 45, 3163–3179.
13. Aifantis, E.C. (1984) On the microstructural origin of certain inelastic models, *J. Eng. Mater. Tech.* 106, 326-334.
14. Fleck, N.A., Muller, G.M., Ashby, F. and Hutchinson, J.W. (1994) Strain gradient plasticity: theory and experiment, *Acta Metall. Mater.* 42, 475–487.
15. Nye, J.F. (1953) Some geometrical relations in dislocated solids, *Acta Metall.* 1, 153–162.
16. Ashby, M.F. (1970) The deformation of plastically non-homogeneous materials, *Phil. Mag.* 21, 399–424.

298

17. Acharya, A. and Bassani, J.L. (2000) Incompatibility and crystal plasticity, *J. Mech. Phys. Solids* **48**, 1565–1595.
18. Gurtin, M.E. (2002) A gradient theory of single-crystal viscoplasticity that accounts for geometrically necessary dislocations, *J. Mech. Phys. Solids* **50**, 5–32.
19. Groma, I. (1997) Link between the microscopic and mesoscopic lenght-scale description of the collective behaviour of dislocations, *Phys. Rev. B* **56**, 5807–5813.
20. Yefimov, S., Groma, I. and Van der Giessen, E. (2001) Comparison of a statistical-mechanics based plasticity model with discrete dislocation plasticity calculations, *J. Phys. IV* **11**, Pr5/103–110.
21. Bassani, J.L., Needleman, A. and Van der Giessen, E. (2001) Plastic flow in a composite: a comparison of nonlocal continuum and discrete dislocation predictions, *Int. J. Solids Struct.* **38**, 833–853.
22. Bittencourt, E., Needleman, A., Gurtin, M.E., Van der Giessen E. (2002) A comparison of nonlocal continuum and discrete dislocation plasticity predictions, *J. Mech. Phys. Solids* (in print).
23. Cleveringa, H.H.M., Van der Giessen, E. and Needleman, A. (1999) A discrete dislocation analysis of residual stresses in a composite material, *Phil. Mag.* **A79**, 893–920.
24. Shu J.Y., Fleck N.A. (1999) Strain gradient crystal plasticity: size-dependent deformation of bicrystals, *J. Mech. Phys. Solids* **47**, 297–324.
25. Shu J.Y., Fleck N.A., Van der Giessen E. and Needleman, A. (2001) Boundary layers in constrained plastic flow: comparison of nonlocal and discrete dislocation plasticity, *J. Mech. Phys. Solids* **49**, 1361-1395.
26. Weygand, D.M., Friedman, L.H. Van der Giessen, E. and Needleman, A. (2001) Discrete dislocation modeling in three-dimensional confined volumes, *Mat. Sci. Engrg.* **A309-310**, 420–424.
27. Weygand, D.M., Friedman, L.H., Van der Giessen, E. and Needleman, A. (2002) Aspects of boundary-value problem solutions with three-dimensional dislocation dynamics, *Model. Simul. Mater. Sci. Eng.* **10**, 437–468.
28. Bulatov, V.V. (2002) Crystal Plasticity from Defect Dynamics, in *Microstructures 2002*, ed. A. Finel (this volume).

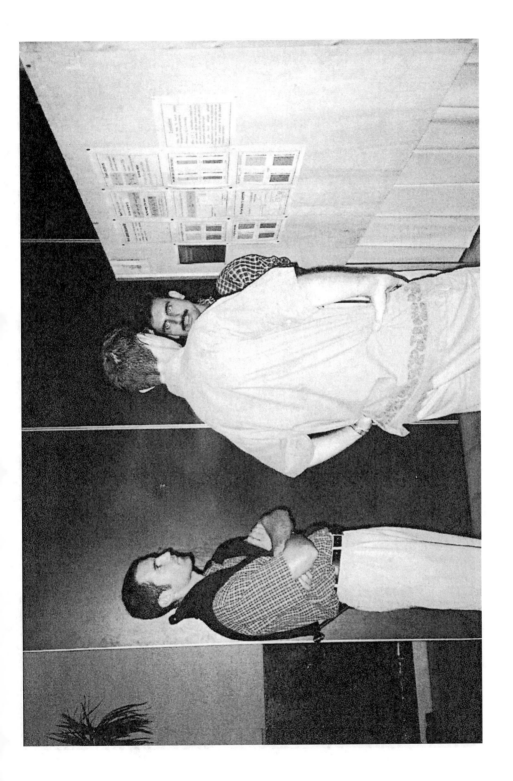

SURFACE INSTABILITIES AND MISFIT DISLOCATIONS IN ANNEALED HETEROEPITAXIAL FILMS

MIKKO HAATAJA (mhaataja@princeton.edu)
Princeton Materials Institute and Department of Mechanical and Aerospace Engineering, Princeton University, Princeton NJ 08544, USA

MARTIN GRANT (grant@physics.mcgill.ca)
Centre for the Physics of Materials, Department of Physics, McGill University, Rutherford Building, 3600 rue University, Montréal, Québec H3A 2T8 Canada

Abstract.
    Heteroepitaxial films are typically under large strains due to a mismatch in the bulk lattice constants between the film and the substrate. Theoretically, strain relaxation in these systems is attributed either to a morphological instability at the surface leading to a grooved profile, or to the nucleation of misfit dislocations which partially destroy the coherency of the film-substrate interface. However, it has been experimentally demonstrated that both of these mechanisms operate and interact simultaneously. By employing a recently introduced continuum model, we show that dislocations compete with the stress-induced instability of the film-vapor interface as a strain relief mechanism. We show that mean-field theory provides an accurate description at early times, where the dislocations simply suppress the initial instability by relaxing the misfit strain partially. The late-time morphology, however, is dictated by the strong interaction between the dislocations and the localized stress build-ups at the bottom of the grooves. We demonstrate that decreasing the dislocation density or mobility leads to enhanced island formation.

Key words: heteroepitaxy, misfit dislocations, strain relaxation

1. Introduction

The precise control of the morphology of thin films under strain is an important ingredient in the fabrication of microelectronic devices. For example, various semiconductor alloys are commonly used to engineer devices with particular values of the band gap. Due to the misfit in the constituent's lattice constants and silicon, the most commonly used substrate, the film is typically under a compressive misfit stress. It is now understood that

A. Finel et al. (eds.), Thermodynamics, Microstructures and Plasticity, 301–310.

such strained films may undergo a morphological instability owing to the stress in the film. This instability, known as the Asaro-Tiller-Grinfeld [1–3] (ATG) instability, allows the film to partially relax its elastic energy by becoming corrugated ("buckling"), thereby making the growth of planar films difficult to achieve. An alternative strain relief mode can be important for a sufficiently thick ("supercritical") film [4, 5]: misfit dislocations can nucleate and climb to the film-substrate interface, thereby partially relaxing strain. This latter mechanism leads typically to a large number of threading dislocations which can deteriorate the electrical properties of the film.

Experiments suggest [6–8] that both of these mechanisms — the buckling of the film and the nucleation of dislocations — take place in supercritical films. For example, it has been demonstrated recently that groove alignment in $Si_{1-x}Ge_x$ films grown on a Si surface depends crucially upon whether misfit dislocations are present or not [7]. Hence, even a qualitative understanding of the morphology requires an approach that explicitly treats both mechanisms at the same time. In this paper, we examine the morphology evolution of misfitting films by employing a continuum model introduced recently [9, 10] that treats both strain relief mechanisms on an equal footing.

The interaction between nonuniform stresses and plasticity is theoretically challenging to address, since the former constitutes a free-boundary problem, while the latter involves singular contributions to the strain. Dong et al. [11] carried out Molecular Dynamics (MD) simulations and studied the growth and relaxation of two-dimensional misfitting films. They demonstrated that the surface morphology plays an important role in the nucleation of dislocations, in which the formation of deep valley structures acted as preferential nucleation sites for dislocations. However, due to small (microscopic) length and time scales amenable to MD simulations, they did not address the physics in the nonlinear regime where both coarsening and misfit dislocations and their interactions contribute to the morphology of the film.

A different type of approach to studying dislocation dynamics in three dimensions under external stresses was undertaken by Schwarz [12]. In his approach, the motion of dislocation lines is driven by the net local stress through the Peach-Koehler force. While treating the interactions and topology of the dislocation network physically, the motion of the dislocations was not coupled to the evolution of a free surface, and therefore the response of the free surface to both non-singular and singular stresses could not be addressed.

Recently, an interesting approach to studying heteroepitaxial growth has been undertaken by Elder et al. [13]. In this approach, one simulates the atoms themselves using a phase-field model, thereby having access to diffusive time and atomic length scales.

2. Model equations

Our approach [9, 10] is based on the phase-field method, which is particularly well-suited for studying free-boundary problems. This method was previously employed by Müller and Grant [14] and Kassner *et al.* [15, 16] to model the ATG instability, in the absence of dislocations. A scalar vapor-film-substrate order parameter, which describes the morphology of the film-substrate system, is coupled to vectorial displacement $\vec{u}(\vec{r})$ and dislocation density $\vec{b}(\vec{r})$ fields, where the latter gives the local Burger's vector density. Dislocations give rise to and interact through singular stresses, and they are coupled to the external stresses via standard elastic interaction; this provides a thermodynamic driving force for the nucleation of dislocations.

The free-energy functional is constructed such that it describes a strained solid-vapor system with coexisting phases, and dislocations interacting with each other and with the nonsingular strains in a finite film. We write $\mathcal{F}$ as the sum of six terms [10, 17, 18]:

$$\mathcal{F} = \mathcal{F}_\phi + \mathcal{F}_{\text{ext}} + \mathcal{F}_{\text{el}} + \mathcal{F}_{\text{int}} + \mathcal{F}_{\text{loc}} + \mathcal{F}_{\text{coup}}. \tag{1}$$

Phase coexistence of solid, film, and vapor is described by the term involving the order parameter ϕ:

$$\mathcal{F}_\phi = \int d\vec{r} \left[\frac{\epsilon^2}{2} |\nabla \phi|^2 + f(\phi) + \frac{\eta_0^2}{2\kappa} \Phi_{\text{mis}}^2 \right], \tag{2}$$

where $f(\phi) = \frac{1}{a}\phi^4(\phi-1)^2(\phi-2)^2$, a is a constant, κ is the compressibility modulus, and η_0 is proportional to the externally applied stress. Hence the homogeneous equilibrium phases of the vapor, film, and solid substrate correspond to $\phi_{\text{eq}} = 0$, 1, or 2, respectively. For an inhomogeneous system, these coexisting phases will be separated by diffuse interfaces of thickness $w \sim \epsilon\sqrt{a}$. The quantity Φ_{mis} ensures that the misfit stress only occurs within the film; the functional form employed here can be found in [10]. The external (misfit) stress enters through

$$\mathcal{F}_{\text{ext}} = \int d\vec{r} \left[\eta_0 \Phi_{\text{mis}} \nabla \cdot \vec{u}^{ns} \right]. \tag{3}$$

This "pre-stress" leads to non-zero stresses in the epilayer, the strength of which is determined by the coupling constant η_0. Smooth nonsingular elastic strains u_{ij}^{ns} due to the misfit are described by

$$\mathcal{F}_{\text{el}} = \int d\vec{r} \left[\frac{1}{2}\kappa(\nabla \cdot \vec{u}^{ns})^2 + \mu\Phi_{\text{sol}} \left(u_{ij}^{ns} - \frac{\delta_{ij}}{2}\nabla \cdot \vec{u}^{ns} \right)^2 \right], \tag{4}$$

where μ is the shear modulus. Φ_{sol} ensures that only the two solid phases can support shear, and its functional form can be found in [10].

The singular part of the elastic energy involves several terms. Non-local interactions between dislocations are conveniently mediated by introducing the complex Airy stress function ξ through [10, 19]

$$\mathcal{F}_{\text{int}} = \int d\vec{r} \left[\frac{1}{2\mathcal{Y}} \xi \nabla^4 \xi + i\xi\eta + \frac{1}{2\mathcal{Y}\ell^4}(1 - \Phi_{\text{sol}})\xi^2 \right], \tag{5}$$

where $\mathcal{Y}$ is the Young's modulus, ℓ is a microscopic length entering into the boundary conditions for ξ at the free surfaces, and

$$\eta(\vec{r}) = \Phi_{\text{sol}} \left(\nabla_x b_y - \nabla_y b_x \right). \tag{6}$$

In particular, it can be shown that if ξ equilibrates instantaneously, it gives rise to the correct non-local interaction between the dislocations [10, 19].

The local energy of a dislocation enters through

$$\mathcal{F}_{\text{loc}} = \int d\vec{r} \left[c\vec{b}^2 (\vec{b}^2 - b_0^2)^2 + E_c \vec{b}^2 \right], \tag{7}$$

where b_0 is the size of the dislocation, which is set by the anharmonic term, E_c is the core energy, and c is a constant which determines the barrier for the nucleation of defects. Tuning the core energy E_c provides a convenient way of controlling the equilibrium dislocation density in the solid phases. In particular, increasing E_c leads to smaller dislocation densities, and letting $E_c \to \infty$ forces the dislocations to disappear altogether. Finally, the crucial coupling between singular strain u_{ij}^s and smooth stress σ_{ij}^{ns} is accomplished by

$$\mathcal{F}_{\text{coup}} = \int d\vec{r} \, \sigma_{ij}^{ns} u_{ij}^s. \tag{8}$$

The origin of the elastic coupling terms is straightforward. The total strain u_{ij}^{tot} of the system is the sum of smooth (non-singular) and singular strains, i.e., $u_{ij}^{\text{tot}} = u_{ij}^{ns} + u_{ij}^s$. Also, the total stress can be decomposed to the sum of smooth and singular stresses: $\sigma_{ij}^{\text{tot}} = \sigma_{ij}^{ns} + \sigma_{ij}^s$. The elastic free energy can be written as $\frac{1}{2} \int d\vec{r} \, \sigma_{ij}^{ns} u_{ij}^{ns} + \frac{1}{2} \int d\vec{r} \, \sigma_{ij}^s u_{ij}^s + \int d\vec{r} \, \sigma_{ij}^s u_{ij}^{ns}$. Therefore the energy breaks up naturally into three contributions: smooth strain energy ($\mathcal{F}_{\text{el}}$), singular strain energy ($\mathcal{F}_{\text{int}}$), and their coupling ($\mathcal{F}_{\text{coup}}$).

An additional term can be included in the free energy functional of Eq. (1) to describe external flux, permitting the study of films with deposition [10].

With regard to the dynamics, we assume that they are driven by the minimization of the free energy of Eq. (1). As our interest lies in the slow collective dynamics giving rise to the film morphology, we invoke mechanical equilibrium conditions and solve for the smooth and singular elastic fields in terms of ϕ and $\vec{b}$ [10, 14, 20, 21]. Substituting the elastic fields back into Eq. (1) yields an effective free-energy $\mathcal{F}_{eff}$ which implicitly contains the (non-singular and singular) elastic interactions. The time dependence of the order parameter is then assumed to obey

$$\frac{\partial \phi}{\partial t} = -\Gamma \frac{\delta \mathcal{F}_{eff}}{\delta \phi}, \tag{9}$$

where the constant mobility Γ is an inverse time scale related to the attachment and detachment kinetics of atoms at the interface. Note that this implies that the main matter transport mechanism is via evaporation-condensation. For convenience we will set this mobility to unity hereafter (hence time is in units of Γ^{-1}). Dislocations are conserved due to their topological nature; they can only appear and disappear in pairs and hence the total Burger's vector is conserved. We thus write the equation of motion for b_x as

$$\frac{\partial b_x}{\partial t} = \left(m_g \frac{\partial^2}{\partial x^2} + m_c \frac{\partial^2}{\partial y^2} \right) \frac{\delta \mathcal{F}_{eff}}{\delta b_x}, \tag{10}$$

where m_g and m_c denote the glide and climb mobilities. The equation for b_y is of the same form with, however, $m_g \partial^2/\partial x^2 + m_c \partial^2/\partial y^2 \rightarrow m_c \partial^2/\partial x^2 + m_g \partial^2/\partial y^2$, in the equation above. Typical parameters employed in the simulations were $N_x = N_y = 256$, $a = 1.0$, $\mu = 0.25$, $\kappa = \eta_0 = 1.0$, $b_0 = 0.025$, $\Delta x = \epsilon = 1.0$, and $\Delta t = 0.01$.

3. Dynamics of annealed heteroepitaxial thin films

We are now ready to explore the effect of dislocations on the morphology evolution by solving the coupled dynamics for the order parameter ϕ and the local dislocation density $\vec{b}$. We first report on the growth of sinusoidal perturbations of the film-vapor interface at early times, which we have studied numerically. In the absence of dislocations, perturbation analysis predicts [2, 3] that a perturbation $\delta \xi \exp(iqx + \omega(q)t)$ with wave-number q grows or shrinks according to

$$\frac{d\delta \xi \exp(iqx + \omega(q)t)}{dt} = \omega(q)\delta \xi \exp(iqx + \omega(q)t), \tag{11}$$

where the growth rate $\omega(q)$ satisfies

$$\omega(q) = Aq - Bq^2, \tag{12}$$

appropriate for evaporation-condensation mass transport. In the above, $A \propto \sigma_{ext}^2/\mathcal{Y}$ and $B \propto \gamma$, where γ denotes the surface tension. In particular, $\omega(q) > 0$ for small q (or large length scales), and $\omega(q) < 0$ for

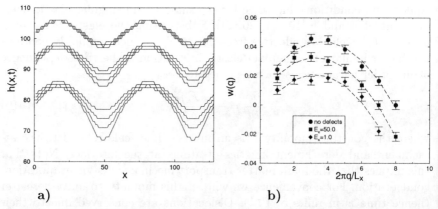

Figure 1. (a) Growth of a sinusoidal perturbation of wavelength $\lambda = L_x/2$ for a film devoid of dislocations (bottom), dislocations with $E_c = 50.0$ (middle), and $E_c = 1.0$ (top). Profiles (shifted for clarity) were obtained at times $t = 10.0, 15.0, 20.0, 25.0$. (b) Numerically evaluated perturbation growth rates $\omega(q)$ vs. wave-number q with and without dislocations. The dashed lines are second degree polynomial fits through the data and serve as guides to the eye. Notice how the data with dislocations is shifted downwards, consistent with an effectively smaller external misfit stress.

large q, implying the unstable growth of perturbations with a sufficiently long wave-length $2\pi/q^*$ such that $\omega(q^*) > 0$. This behavior can be understood physically as follows. A perturbation of the planar films leads to non-uniformities in the stress state of the film [2, 3]; in particular, the atoms at protrusions become less strained, while those at the grooves experience an enhanced local stress. This leads to preferential evaporation from the grooves, providing a positive feedback mechanism leading to unstable growth. Surface tension, however, resists the growth of perturbations, and completely suppresses the instability on small scales (large q with $\omega(q) < 0$), as described by Eq. (12). This interplay between the de-stabilizing misfit strain and the stabilizing surface tension gives rise to a length scale $\lambda \propto (\mathcal{Y}\gamma)/\sigma_{ext}^2$ which roughly determines the typical lateral scale of the resulting morphology (decreasing σ_{ext} leads to larger λ).

Physically, misfit dislocations appear in order to screen (renormalize) the misfit stress in the film [5]. Thus, one expects that dislocations which equilibrate quickly in the external stress field, lead to an effectively smaller stress at early times. Indeed, our numerical results support this mean-field picture. In Figure 1 (a) we show typical interface profiles, starting from a sinusoidal perturbation, for $E_c = \infty$ (i.e., no dislocations), $E_c = 1.0$, and $E_c = 50.0$. Dislocation mobility was set to $m_c = m_g = 0.01$. It can be seen that the presence of dislocations leads to a slower growth of the perturbation, the more so for $E_c = 1.0$ which corresponds to a larger dislocation

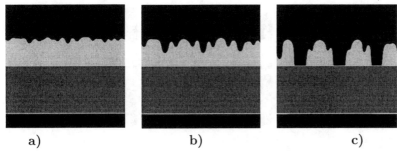

a)   b)   c)

Figure 2.   Typical configurations for the evolving film in the absence of dislocations from a random initial condition. The gray scale is such that dark gray corresponds to vapor, light gray to film, and intermediate gray to substrate, respectively. (a) $t = 20.0$, (b) $t = 40.0$, (c) $t = 80.0$. Notice how the initially flat surface with small fluctuations develops into a grooved profile with a well-defined lateral scale.

density in the film. The growth rates $\omega(q)$ can be obtained by evolving the system according to Eqs. (9) and (10) and numerically evaluating the amplitude of the surface perturbation corresponding to the initial perturbation with wave-numbers q. In Figure 1 (b) we plot the numerically evaluated growth rates $\omega(k)$ for the aforementioned three cases. It is noteworthy that increasing the dislocation density leads to slower growth of perturbations and stabilization of the surface on longer scales. This is due to the fact that misfit dislocations partially screen the stress in the film, thereby reducing the driving force for the instability.

Similar considerations apply for annealed films where the initial surface corrugation corresponds to uniformly random fluctuations with a small amplitude. In Figure 2 (a) we show an early-time configuration of a film devoid of dislocations, and the subsequent morphology evolution is depicted in Figure 2 (b) and (c). It is noteworthy that the small-amplitude initial fluctuations are amplified during strain relaxation, and lead to a film which has broken up into islands with a characteristic lateral scale λ, in qualitative agreement with experimental observations [7]. Upon allowing dislocations to appear, the morphology evolution can change dramatically, depending on the dislocation core energy and mobility. In Figures 3 (a)-(c) we show the configurations corresponding to Figure 2 (c), starting from the same initial configuration as in Figure 2. In (a) $E_c = 1.0$ and $m_c = 0.1$, (b) $E_c = 1.0$ and $m_c = 0.025$, and (c) $E_c = 10.0$ and $m_c = 0.01$. It is noteworthy that either decreasing the dislocation density in the film by increasing E_c or decreasing the dislocation mobility m_c tends to make the surface more rough via island formation. Physically, it is easy to see why the dislocations decouple from the system, as either $E_c \to \infty$ or $m_c \to 0$: in the first case, dislocations are energetically unfavorable in the the solid, whereas in the second case, the

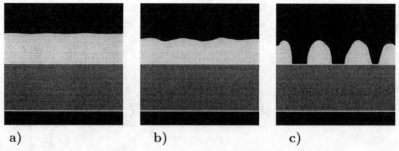

a)   b)   c)

Figure 3. Late-time configurations ($t = 80.0$) for the evolving film in the presence of dislocations. (a) $E_c = 1.0$ and $m_c = 0.1$, (b) $E_c = 1.0$ and $m_c = 0.025$ (c) $E_c = 10.0$ and $m_c = 0.01$. Notice how the initial perturbation decays in the presence of a sufficiently large density of mobile dislocations (a), leading to a flat surface. Furthermore, decreasing either the dislocation density or mobility enhances island formation, as shown in (b) and (c).

time for defect distribution build-up is too long, on the time scale of the buckling, to be effective in relaxing the strain.

In Figures 4 (a) and (b) we show the dislocation distributions corresponding to the film configuration shown in Figure 3 (d). In particular, in (a) we show b_x while in (b), b_y is shown. It is noteworthy, that there is a localized concentration of dislocations with $b_x > 0$ at the film-substrate interface. It is precisely these dislocations which partially screen the external stress within the film, in agreement with the Matthews-Blaskeslee theory and experimental observations [7]. We would like to emphasize that the initial dislocation distribution was uniformly random, and that our choice of free-energy and dynamics naturally leads to the localized dislocation concentration at the film-substrate interface. Interestingly, there is also a build-up of dislocations near the film-vapor interface around depressions. This is due to the fact that large stresses are localized in regions around depressions, and the dislocations attempt to screen these stresses. The interplay between the mobility of these dislocations and the dynamics of the groove gives rise to a non-equilibrium brittle-to-ductile transition [9]: low-mobility dislocations are outrun by the grooves, while high-mobility dislocations rapidly screen the large stresses at the bottom of the groove and inhibit the propagation of the grooves. Finally, as shown in Figure 4 (b), the y-component of the Burger's vector b_y is non-zero everywhere except at the film-vapor interface. This is due to the fact that these dislocations do not relax the external stress within the film, and therefore the driving force for nucleating such dislocations in the bulk is small.

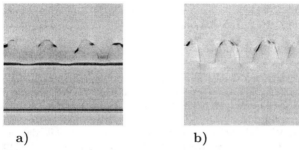

a) **b)**

Figure 4. Dislocation density corresponding to the film morphology shown Figure 3 (c). (a) x-component of the Burger's vector b_x. It is noteworthy, that there is a localized concentration of dislocations with $b_x > 0$ at the film-substrate interface. (b) y-component of the Burger's vector b_y. We note that b_y is non-zero everywhere except at the film-vapor interface.

4. Conclusions

In this paper we have examined the coupled dynamics of strained heteroepitaxial films and dislocations. Our results show that dislocations renormalize the misfit stress in the film in agreement with the (mean-field) Matthews-Blakeskee theory. Moreover, we have shown that the film undergoes a "buckling" instability, driven by the residual stress in the film. We have quantified this by numerically evaluating the perturbation growth rates $\omega(q)$. We find that increasing core energy and/or decreasing mobility lead to less effective screening of the external stress. In agreement with our previous results [9], this implies that the film will undergo a ductile-to-brittle transition for sufficiently low dislocation mobility, where the dislocations are unable to keep up with the evolving film and thus become ineffective in screening the misfit strain.

Acknowledgements

This work was supported by the Natural Sciences and Engineering Research Council of Canada, *le Fonds pour la Formation de Chercheurs et l'Aide à la Recherche du Québec*, and, in part, by the Academy of Finland (MH). We thank Judith Müller, Andrew Rutenberg, Ken Elder, and Dong-Hee Yeon for useful discussions.

References

1. Asaro R. J. and Tiller W. A. (1972) Stress corrosion cracking: Part I. Via surface diffusion, *Metall. Trans.* **3**, 1789-1796.

310

2. Grinfeld M. (1986) Instability of the separation boundary between a non-hydrostatically stressed elastic body and a melt, *Dokl. Akad. Nauk. SSSR* **265**, 831-834.

3. Srolovitz D. J. (1989) On the stability of surfaces of stressed solids, *Acta Metall.* **37**, 621-625.

4. Frank F. C. and van der Merwe J. H. (1949) One-dimensional dislocations. II. Misfitting monolayers and oriented overgrowth, *Proc. R. Soc. London, Ser. A* **198**, 216-225; van der Merwe J. H. (1962) Crystal Interfaces. Part II. Finite Overgrowths, *J. Appl. Phys.* **34**, 123-128.

5. Matthews J. W. and Blakeslee A. E. (1974) Defects in epitaxial multilayers. I. Misfit dislocations, *J. Cryst. Growth* **27**, 118-125; Matthews J. W. (1975) Defects associated with the accommodation of misfit between crystals, *J. Vac. Sci. Technol.* **12**, 126-133.

6. Gao H. and Nix W. D. (1999) Surface roughening of heteroepitaxial thin films, *Annu. Rev. Mat. Sci.* **29**, 173-209.

7. Ozkan C. S., Nix W. D., and Gao H. (1997) Strain relaxation and defect formation in heteroepitaxial $Si_{1-x}Ge_x$ films via surface roughening induced by controlled annealing experiments, *Appl. Phys. Lett.* **70**, 2247-2249.

8. Gray J. L., Hull R., and Floro J. A. (2002) SiGe Epilayer Stress Relaxation: Quantitative Relationships Between Evolution of Surface Morphology and Misfit Dislocation Arrays, *Mat. Res. Soc. Symp. Proc.* **696**, N8.3.1.

9. Haataja M., Müller J., Rutenberg A. D., and Grant M. (2002) Dynamics of dislocations and surface instabilities in misfitting heteroepitaxial films, *Phys. Rev. B* **65**, 035401.

10. Haataja M., Müller J., Rutenberg A. D., and Grant M. (2002) Dislocations and morphological instabilities: Continuum modelling of misfitting films, *Phys. Rev. B* **65**, 165414.

11. Dong L., Schnitker J., Smith R. W., and Srolovitz D. J. (1998) Stress relaxation and misfit dislocation nucleation in the growth of misfitting films: A molecular dynamics simulation study, *J. Appl. Phys.* **83**, 217-227.

12. Schwarz K. W. (1999) Simulation of dislocations on the mesoscopic scale. I. Methods and examples, *J. Appl. Phys.* **85**, 108-119; Simulation of dislocations on the mesoscopic scale. II. Application to strained-layer relaxation, *J. Appl. Phys.* **85**, 120-129.

13. Elder K. R., Katakowski M., Haataja M., and Grant M. (2002) Modeling elasticity in crystal growth, *Phys. Rev. Lett.* **88**, 245701.

14. Müller J. and Grant M (1999) Model of Surface Instabilities Induced by Stress, *Phys. Rev. Lett.* **82**, 1736-1739.

15. Kassner K. and Misbah C. (1999) A phase-field approach for stress-induced instabilities, *Europhys. Lett.* **46**, 217-223.

16. Kassner K., Misbah C., Müller J., Kappey J., and Kohlert P. (2001) Phase-field modeling of stress-induced instabilities, *Phys. Rev. E* **63**, 036117.

17. Landau L. and Lifshitz E. M. (1986), *Theory of Elasticity*, Pergamon Press, Oxford.

18. Nabarro F. R. N. (1967) *Theory of crystal dislocations*, Dover Publications, New York.

19. Aguenaou K. (1997) Modeling solidification, PhD thesis, McGill University.

20. Onuki A. and Nishimori H. (1991) Anomalously slow domain growth due to a modulus inhomogeneity in phase-separating alloys, *Phys. Rev. B* **43**, 13649-13652.

21. Sagui C., Somoza A. M., and Desai R. C. (1994) Spinodal decomposition in an order-disorder phase transition with elastic fields, *Phys. Rev. E* **50**, 4865-4879.

ON THE ROLE OF THE STRAIN-RATE SENSITIVITY IN COLLECTIVE DISLOCATION EFFECTS

S. BROSS[1], P. HÄHNER[2,3]

[1] *Institut für Allgemeine Mechanik, TU Braunschweig, D-38106 Braunschweig*
[2] *Institut für Metallphysik und Nukleare Festkörperphysik, TU Braunschweig, D-38106 Braunschweig*
[3] *Institute for Energy, Joint Research Centre of the European Commission, PO Box 2, NL-1755 ZG Petten*

Abstract. The plastic instability due to strain-rate softening, which is associated with the Portevin–Le Chatelier effect, results from an interplay between dynamic strain ageing by solute diffusion, on the one hand, and long-range dislocation interactions, on the other hand, while neither of the effects alone induces instability. This assertion, which has been suggested by analytical modelling results, is confirmed by numerical simulation of the dislocation dynamics in 2D at constant stress rate.

1. Introduction

Collective dislocation effects underlying, for instance, the spontaneous emergence of mesoscopic dislocation patterns, or the macroscopic plastic instabilities associated with the Portevin–Le Chatelier (PLC) effect [1] have attracted much interest, both from the experimental and the theoretical point of view. The theoretical description of those collective effects, however, is complicated by the fact that dislocations are subject to long-range interactions, which is why averageing (coarse graining) procedures in going from the microscopic dynamics of individual dislocations to the meso- or macroscopic evolution of dislocation densities are anything but straightforward.

313

A. Finel et al. (eds.), Thermodynamics, Microstructures and Plasticity, 313–330.

In principle, the dynamics of a dislocation ensemble can be described by a hierarchical set of kinetic equations with dislocation interactions to be traced back to higher-order correlation functions. The formulation of such a correlation hierarchy and its self-consistent truncation, however, is a tremendous task, which seems out of reach at the present time. A much easier, semi-phenomenological alternative consists in formulating the dislocation dynamics in terms of a stochastic process. In this case, dislocation correlations are taken into account by random fluctuations acting on the evolution of the dislocation densities, and the self-consistent formulation of the fluctuation intensities becomes the important issue.

In recent years, such a stochastic dislocation dynamics (SDD) approach has been proposed and applied to various dislocation patterning phenomena [2-4]. The pivot of the SDD consists in fluctuation–dissipation relationships between the plastic shear strain rate fluctuations and the mechanical work dissipated during plastic flow. Let $\langle \dot{\gamma} \rangle$ denote the ensemble average of the plastic shear strain rate that is observed macroscopically. Then variations in $\dot{\gamma}$ can be estimated if one notes that the work done by the internal shear stresses τ_{int} is completely dissipative to a good approximation: $\langle \tau_{int} \dot{\gamma} \rangle = 0$. Adopting the self-consistency requirement, according to which the macroscopic dynamic response function also governs the fluctuations at the mesoscopic scale, we introduce the strain-rate sensitivity of the flow stress τ_{ext}, $S = \partial \tau_{ext} / \partial \ln \langle \dot{\gamma} \rangle$ to translate mesoscopic fluctuations of the effective driving stress into strain-rate fluctuations $\delta \dot{\gamma}$. This leads to the following fluctuation intensity of the plastic strain rate [2,4]:

$$\frac{\langle \delta \dot{\gamma}^2 \rangle}{\langle \dot{\gamma} \rangle^2} = \frac{\langle \tau_{int} \rangle}{S} \tag{1}$$

which illustrates the crucial role of the dynamic response function S in controlling the degree of intermittency of mesoscopic slip. As to the origin of strain-rate fluctuations, we note that, owing to the discrete nature and the long-range interactions of dislocations, *noise* is inherent to the plastic deformation by dislocation glide. Hence, it escapes external control.

As to their relative size (1), one concludes that the strain-rate fluctuations are appreciable, in particular, in low rate-sensitivity f.c.c. metals, where $\langle \tau_{int} \rangle / S$ is of the order of 10^2. It is important to stress that the spatial extent of those fluctuations is limited by a correlation length, which behaves as $\xi \sim S^{-1/2}$, see [3,4,6] for more details. Consequently, fluctuations average out on the macroscope scale, as long as S remains finite. However, we expect critical fluctuations to appear on all scales (up to the macroscopic scale which is limited by the specimen size), if the strain-rate sensitivity S tends to zero.

In fact, this is what is observed with the macroscopic plastic instabilities at the onset of the PLC effect. This instability is due to dynamic strain ageing (DSA) which causes S to become a function of temperature T and strain rate $\dot{\gamma}$ in a way that $S < 0$ (strain-rate softening) may occur within some intermediate range of $\dot{\gamma}$ and T. Consequently, DSA alloys constitute interesting model systems in that S can be tuned over a wide range by the choice of appropriate deformation conditions. While this has been well established, there is still lack of understanding how DSA may induce strain-rate softening. This is the subject of the present paper.

Let us proceed on the well accepted idea that dislocation motion is a discontinuous process, as glide dislocations become temporarily arrested at obstacles, notably forest dislocations, which are then overcome by the aid of thermal fluctuations. In DSA alloys containing mobile solute atoms, a glide dislocation segment may be subjected to an additional pinning due to solute atoms migrating towards the dislocation core during the waiting time at an obstacle. This solute accumulation in the glide dislocation core leads to an increasing activation energy for the unpinning of the dislocation segment. This is expressed through a Gibbs' free activation enthalpy which depends explicitly on the waiting time t_{w}

$$G(t_{\mathrm{w}}) = G_0 + \Delta G_{\max} \cdot \left(1 - \exp\{-(\eta \, t_{\mathrm{w}})^n\}\right), \tag{2}$$

where G_0 is the basic activation enthalpy in the absence of DSA (characteristic of the pinning strength of the forest dislocation), $\Delta G_{\max}$ is the additional activation enthalpy contribution of a completely aged dislocation (saturated solute cloud), and η is the ageing rate. The characteristic exponent n, which governs the initial ageing behaviour at small waiting times t_{w}, is usually observed to be $n \approx 1/3$, [5]. Using equation (2), the evolution of the plastic shear strain rate with time t can then be written as a generalized Arrhenius law [6][1]

$$\dot{\gamma}(t) = \nu_0 \Omega \int_0^\infty f(t_{\mathrm{w}}, t) \, \exp\left\{-\frac{G(t_{\mathrm{w}}) - V\tau_{\mathrm{eff}}(t)}{kT}\right\} \mathrm{d}t_{\mathrm{w}} \tag{3}$$

Here ν_0 denotes the attempt frequency, Ω is the elementary strain associated with an activation event. Thermal activation is assisted by the effective shear stresses τ_{eff} scaled by the activation volume V. The Boltzmann constant and absolute temperature are denoted by k and T, respectively.

An exact analytical expression for the distribution function of the waiting times, $f(t_{\mathrm{w}}, t)$ can be obtained for the case of a 'non-interacting dislocation

---

[1]For simplicity, only one glide system is considered.

ensemble' [6]. This artificial case, however, is interesting in itself, since it has been shown that it does not give rise to plastic instability, contrary to what was commonly believed. While DSA is confirmed to lower the strain-rate sensitivity S, it is not that efficient as to make S negative and, hence, give rise to strain-rate softening instability. It has therefore been concluded that long-range dislocation interactions are another necessary ingredient to conceive the plastic instability associated with the PLC effect [6]. In the present work, we are going to address this issue from a numerical modelling point of view.

2. Two-dimensional cellular automaton model

The numerical model simulates the dynamics of an ensemble of $N_\perp$ discrete straight parallel edge dislocations on a single slip system during tensile deformation. Figure 1 gives a schematic of the simulation area. The cellular automaton contains $N_X \cdot N_Y$ cells of size $(cb)^2$, where b is the length of the Burgers vector, and c a scaling factor. Hence, the automaton simulates an area $A_{sim} = N_X \cdot N_Y \cdot (cb)^2$.

The dashed lines in figure 1 indicate the cells neighbouring a given cell (highlighted in grey) along the glide plane. The arrow on top illustrates periodic boundary conditions. So, any dislocation leaving the simulation area is re-inserted at the opposite side, ensuring a constant number of dislocations and, hence, statistically steady-state conditions. The vertical segments represent forest dislocations obstructing free dislocation glide. The mean distance between these segments gives the mean free glide path of the glide dislocations, L (cf. figure 1), which translates into the elementary strain increment

$$\Omega = \frac{N_\perp}{A_{sim}} L c b^2 = \frac{L N_\perp}{c \cdot N_X \cdot N_Y}. \tag{4}$$

Since a cell cannot contain more than one mobile dislocation at a time, six different cell states result, which are displayed in figure 2. The share of the occupied cells in the model amounts to less than 10^{-5}. Thus, for the sake of efficiency the code concentrates on the occupied cells in the field instead of keeping a huge number of empty cells in memory. In addition, each mobile dislocation is characterized by a set of individual variables as specified in table 1. The automaton is initialized by distributing the $N_\perp$ positive and negative mobile dislocations evenly on $N_\perp$ equally spaced parallel glide planes, while forest dislocation segments are distributed randomly on each glide plane in a way that the same mean free glide path L results in each glide plane.

There are two types of short range interactions implemented in the model. The first one is associated with the release of a dislocation by cutting the

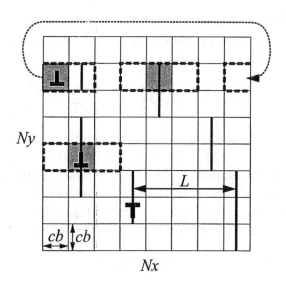

Figure 1: Schematic of the simulation field.

Figure 2: Illustration of the six possible cell states of the automaton: empty, forest dislocation, free dislocation of either sign, arrested dislocation of either sign.

Table 1: Individual variables of mobile dislocations.

(X_i, Y_i)	Current cell position
$\tau_{\text{int},i}$	Interaction shear stress
$\tau_{\text{eff},i}$	Effective shear stress
$t_{\text{w},i}$	Waiting time at obstacle
$\Delta t_{\text{att},i}$	Time elapsed since last attempt
$n_{x,i}$	Number of jumps in x-direction

forest dislocation it had been arrested at (states $5 \to 3$ or $6 \to 4$). Assuming a Boltzmann distribution of energy fluctuations the probability of thermal activation is given by

$$P = \nu_0 \Delta t_{\text{att}} \; \exp\left\{ -\frac{G(t_{\text{w}}) - V\tau_{\text{eff}}}{kT} \right\} \qquad (5)$$

with the time Δt_{att} elapsed since the last attempt, and the other quantities as introduced in equations (2,3). Thermal activation is then simulated by random number generation in combination with the activation probability (5).

The second type of short range interaction refers to the opposite case that a free dislocation runs into a forest dislocation segment resulting in cell states 5 or 6, cf. figure 2. The glide condition for a free dislocation is given by

$$\tau_{\text{eff},i} > \tau_{\text{P}} \qquad (6)$$

with the individual effective shear stress of dislocation i and the lattice friction stress τ_{P}. By choosing a sufficiently high basic activation enthalpy G_0 in equation (2) that glide condition is fulfilled for virtually all dislocations.

The effective shear stress which governs both free glide and the release of pinned dislocations by thermal activation, is obtained by summing three different stress contributions, namely external, hardening and interaction stresses:

$$\tau_{\text{eff},i} = \tau_{\text{ext}} - \tau_{\text{hard},i} + \tau_{\text{int},i}. \qquad (7)$$

The first contribution stems from the external tensile stress σ_{ext}, according to $\tau_{\text{ext}} = \sigma_{\text{ext}}/M$, with the Taylor factor $M = 2$ in the present case. In order to avoid that the system gets locked in a periodic solution, we introduce a strain hardening component $\tau_{\text{hard},i}$ which reduces the effective shear stress depending on the plastic strain. For the sake of simplicity a straightforward ansatz is adopted from stochastic material models [7]:

$$\tau_{\text{hard},i} = \frac{h}{M} N_\perp \varepsilon_{\text{pl},i}, \qquad (8)$$

according to which the back stress due to hardening is expressed by the individual plastic strain accommodated by the dislocation, and the tensile strain hardening coefficient h. The total number of dislocations $N_\perp$ relates the individual strain contribution $\varepsilon_{\text{pl},i}$ to the total tensile plastic strain accommodated by the dislocation ensemble.

Finally, the interaction stress $\tau_{\text{int},i}$ encompasses the long-range dislocation interactions. In Cartesian coodinates the shear stress around an edge dislocation segment [8] is given by

$$\hat{\tau}_{xy} = \frac{\mu b}{2\pi(1-\nu)} \frac{x(x^2-y^2)}{(x^2+y^2)^2}, \tag{9}$$

where μ and ν denote the shear modulus and the Poisson ratio, respectively. The Peach–Koehler equation [8] then gives the interaction force $\underline{F}_{ij}$ per unit length on a dislocation segment i with line vector $\underline{s}$ and Burgers vector vector $\underline{b}$

$$\frac{1}{l}\,\underline{F}_{ij} = \left(\underline{\underline{\sigma}}\,\underline{b}\right) \times \underline{s}. \tag{10}$$

with the stress tensor $\underline{\underline{\sigma}}$ containing the shear stress contribution of dislocation j according to equation (9). Summation of all interaction forces $\underline{F}_{ij}$ between dislocation i and dislocations $j \neq i$, gives the resulting interaction force

$$\underline{F}_i = \sum_{j=1}^n \underline{F}_{ij} \;\; \text{for} \left\{ \begin{array}{l} |\Delta x_{ij}| \leq R_{\text{int}} \\ |\Delta y_{ij}| \leq R_{\text{int}}. \end{array} \right. \tag{11}$$

It should be noted that a finite interaction radius $R_{\text{int}} = N_X/2$ has been introduced. To avoid artifacts from a discontinuous cut-off, eigenstresses are smoothly trimmed to zero using a cosine function.

During a simulation run, the system time is stopped at the beginning of each time step and all cell states are frozen. The individual effective stresses are calculated from the interaction forces for all members of the dislocation ensemble. A loop, which addresses all members, decides on moving or arresting a dislocation depending on the friction or the activation probability criterion applied to free or arrested dislocations, respectively. Subsequently, all strains and stresses are updated. Then the simulation time is increased by a certain time increment Δt_{g}, which is defined by the average glide time of a free dislocation between two obstacles divided by the average number of cells separating the obstacles. Finally, the code updates all cells, releases the automaton and starts the next cycle.

3. Results

The following numerical results illustrate the isolated and combined effects of DSA and dislocation interactions on the plastic deformation behaviour. The model parameters used for all simulations are listed in table 2. While a quantitative parameter fit has not been attempted here, the parameters adopted are suggested by experimental findings, see [9–11].

Table 2: Parameter values used for the simulations.

Shear modulus	μ	$4.37 \cdot 10^4$ MPa
Burgers vector	b	$2.56 \cdot 10^{-8}$ cm
Poisson ratio	ν	0.340
Diffusion exponent	n	0.33
Ageing rate	η	$0.20\,\mathrm{s}^{-1}$
Basic activation enthalpy	G_0	$108\,kT$
Saturated DSA-related enthalpy	$\Delta G_{\max}$	$10\,kT$
Attempt frequency	ν_0	$10^5\,\mathrm{s}^{-1}$
Temperature	T	400 K
Friction stress	τ_P	1.0 MPa
Instantaneous macroscopic SRS	$S_0 = kT/V$	1.38 MPa
Taylor factor	M	2
Strain hardening coefficient	h	10^3 MPa
Number of cells in x- and y-direction	N_X, N_Y	3400
Area of simulation	A_sim	$19\,\mu\mathrm{m}^2$
Number of mobile dislocations	$N_\perp$	100
Mobile dislocation density	ρ_m	$5.3 \cdot 10^8\,\mathrm{cm}^{-2}$
Average cell spacing between obstacles	L	425
Elementary strain increment	Ω	$3.68 \cdot 10^{-4}$
Average glide time between obstacles	t_g	$10^{-4}\,\mathrm{s}$

3.1 SYSTEM OF INDIVIDUAL DISLOCATIONS

Before we investigate the combined influence of DSA and elastic interactions on the plastic behaviour of the discrete dislocations ensemble, let us first consider an artificial case where both interactions and solute diffusion are switched off. This is achieved by setting

$$\Delta G_{\max}/kT = 0, \quad R_{\mathrm{int}} = 0. \tag{12}$$

Besides providing a reference to the results to be presented in the sections below, this section introduces the methods used for analysing the simulation results.

Figures 3 and 4 show numerical results from a stress-controlled tensile test at a constant stress rate $\dot{\sigma} = 9,26 \cdot 10^{-2}$ MPa/s under the conditions given in equation (12). With the hardening coefficient h the nominal plastic strain rate is

$$\dot{\varepsilon}_{\mathrm{pl,nom}} = \frac{\dot{\sigma}}{h} = 9,26 \cdot 10^{-5}\,\mathrm{s}^{-1}, \tag{13}$$

while the nominal plastic strain $\varepsilon_{\text{pl,nom}}$ is proportional to the time t. As a measure quantifying the stability of plastic flow figure 3 shows the ratio $Q_{\varepsilon_{\text{pl}}}$ of actually accommodated and nominally expected plastic strain increments

$$Q_{\varepsilon_{\text{pl}}}(\varepsilon_{\text{pl,nom}}) = \frac{\Delta\varepsilon_{\text{pl}}(\varepsilon_{\text{pl,nom}})}{\Delta\varepsilon_{\text{pl,nom}}}. \tag{14}$$

In the limiting case of an infinitely large system containing an infinite number of dislocations, $Q_{\varepsilon_{\text{pl}}}$ tends to unity as actual and nominal strain are equal during stable plastic flow, while the fluctuations revealed by figure 3 constitute the finite size effect (100 dislocations).

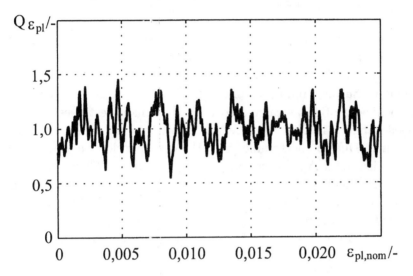

Figure 3: Ratio of actual and nominal strain increment; $\Delta G_{\text{max}}/kT = 0$, $R_{\text{int}} = 0$, $\dot\sigma = 9.26 \cdot 10^{-2}\,\text{MPa/s}$.

As an advantage of the present numerical approach, we have access to a record of all thermal activation events with corresponding time and space coordinates $(t_{\text{act},i}, x_{\text{act},i}, y_{\text{act},i})$. We may thus consider the following dichotomous random variable E

$$E(t, x, y) = \begin{cases} 1 & \text{for } (t, x, y) = (t_{\text{act},i}, x_{\text{act},i}, y_{\text{act},i}) \\ 0 & \text{for } (t, x, y) \neq (t_{\text{act},i}, x_{\text{act},i}, y_{\text{act},i}) \end{cases}, \tag{15}$$

which quantifies the local plastic activity. Correlation analysis of E then provides a quantitative measure of the degree of collectivity in the dislocation dynamics. The auto-correlation function describes the spatio-temporal decay of dislocation correlations as a function of space shift Δy and time delay Δt, respectively:

$$R_{EE}(\Delta t, \Delta y) = \frac{1}{(T_0 - \Delta t)(Y - \Delta y)} \int_0^{T_0 - \Delta t} \int_0^{Y - \Delta y} E(t, y) E(t - \Delta t, y - \Delta y) \, dy \, dt$$

(16)

where T_0 is the simulation time and Y the height of the simulation field. Normalization of equation (16) gives the auto-correlation coefficient function

$$\rho_{EE}(\Delta t, \Delta y) = \frac{R_{EE}(\Delta t, \Delta y)}{R_{EE}(0, 0)}$$

(17)

which is in the range $0 \leq \rho_{EE} \leq 1$. In the present context, ρ_{EE} denotes the conditional probability of thermal activation events at a distance $(\Delta t, \Delta y)$. The numerical realization of the auto-correlation coefficient function requires a binning of the time and space intervals, $\Delta \Delta t$ and $\Delta \Delta y$. The discrete correlation coefficient of E, as calculated from the tensile test presented above, is displayed in figure 4. The time interval of $0.01\,s$ is subdivided into 100 equally spaced bins. Due to the symmetry in the Δy-direction only absolute values of spatial-distances Δy are considered. The limit of $2.2\,\mu m$ of the Δy-axis corresponds to half the system height devided into 50 equally spaced bins. As expected for the present non-interacting dislocation ensemble without ageing, thermal activation is completely uncorrelated resulting in $\rho_{EE}(\Delta t, \Delta y)$ values which do not exceed $1.3 \cdot 10^{-4}$. The correlation coefficient does not reveal any spatio-temporal structure at all.

3.2 DSA OF NON-INTERACTING DISLOCATIONS

This section summarizes the sole influence of dynamic strain ageing on the deformation behaviour, while elastic interactions are still being suppressed. The corresponding parameters are

$$\Delta G_{\max}/kT = 10, \quad R_{\text{int}} = 0.$$

(18)

It should be noted that the deformation conditions and material parameters (table 2) were chosen such that the system is in the centre of the PLC regime, as predicted by simple linear stability analysis. Nevertheless, deformation is still stable under those conditions, although the fluctuations are slightly bigger as compared to the previous case. Figure 5 shows the ratio of actual and nominal plastic strain increments. This observation already supports the

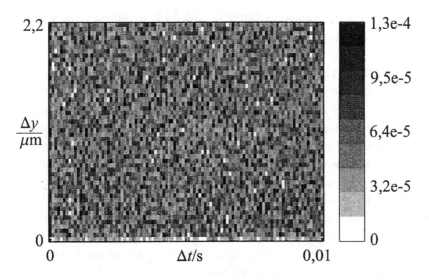

Figure 4: Auto-correlation function $\rho_{EE}(\Delta t, \Delta y)$ of thermal activations; $\Delta G_{\max}/kT = 0$, $R_{\text{int}} = 0$, $\dot{\sigma} = 9.26 \cdot 10^{-2}$ MPa/s.

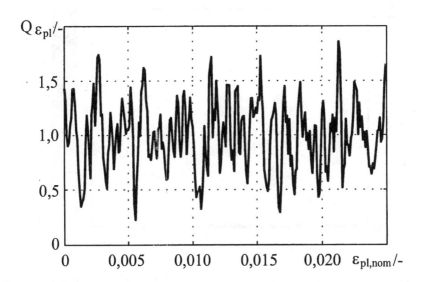

Figure 5: Ratio of actual and nominal strain increment; $\Delta G_{\max}/kT = 10$, $R_{\text{int}} = 0$, $\dot{\sigma} = 9.26 \cdot 10^{-2}$ MPa/s.

previous theoretical assertion [6] that dislocation interactions should play a fundamental role in destabilizing plastic flow of DSA alloys.

The correlation analysis of the activation events in figure 6 substantiates that impression. While the maximum value of $\rho_{EE}(\Delta t, \Delta y)$ exceeds the values reported in the previous section by about two magnitudes, correlations are restricted to $\Delta y = 0$ and decay very slowly in time Δt. The strong temporal correlations are caused by the fact, that once a dislocation has been activated it will be prone for repeated activation, since it posseses a lower activation enthalpy than those dislocations that have been arrested and aged for a long time. So, the probability of beeing activated again remains close to unity, until the individual effective shear stress is reduced significantly by strain hardening. Then, the waiting time starts to increase and so does the DSA-related additional activation enthalpy. Now, this dislocation will be excluded from plastic deformation process until the global stress level compensates the ageing effects as well as the back stresses due to strain hardening. These repeated activations are responsible for the slightly stronger fluctuations of the ratio Q_{pl} as seen in figure 5, without creating spatial correlation of plastic activity. On a macroscopic scale, deformation will be perfectly smooth again.

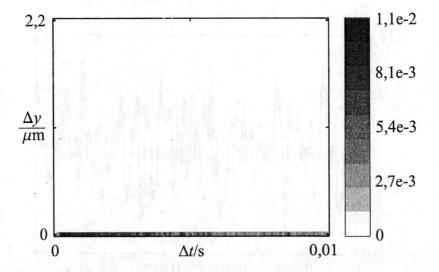

Figure 6: Auto-correlation function $\rho_{EE}(\Delta t, \Delta y)$ of thermal activations; $\Delta G_{max}/kT = 10$, $R_{int} = 0$, $\dot{\sigma} = 9.26 \cdot 10^{-2}$ MPa/s.

3.3 THE INFLUENCE OF LONG-RANGE INTERACTIONS

The sole effect of long-range interactions, in the absence of DSA, is simulated using

$$\Delta G_{max}/kT = 0, \quad R_{int} = N_X/2. \tag{19}$$

The interaction range R_{int} covers all dislocations in the field, whereas ageing is suppressed by setting the additional activation enthalpy equal to zero. Therefore, the activation probability is not affected by the waiting time. On the other hand, the effective shear stresses experienced by individual dislocations differ due to the internal stresses and the strain hardening, cf. equation (7).

As it is apparent from figure 7, long-range interactions affect the mesoscopic uniformity of plastic deformation to a much larger extent than DSA does. The plot of the ratio of actual and nominal plastic strain increments emphasizes this in terms of a pronounced intermittency with periods when plastic activity is virtually absent. The serrated curve denotes deformation behaviour close to instability. The strain rate fluctuations amount to several times the nominal strain rate.

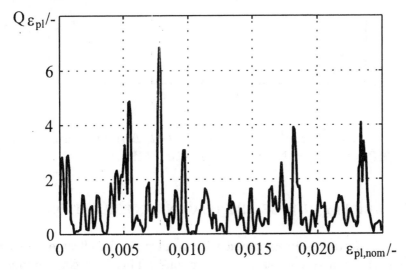

Figure 7: Ratio of actual and nominal strain increment; $\Delta G_{max}/kT = 0$, $R_{int} = N_X/2$, $\dot{\sigma} = 9.26 \cdot 10^{-2}$ MPa/s.

To elucidate this behaviour, the correlation analysis is again helpful. The plot of $\rho_{EE}(\Delta t, \Delta y)$ in figure 8 now reveals appreciable correlations with respect to Δy, while temporal correlations are less pronounced than in the previous case. In particular, this is true for in the class $\Delta y = 0$, meaning that

we are no more talking about repeated activation of the very same dislocation, but correlated activation on adjacent glide planes. The maximum value
exceeds the correlation coefficient of the previous case (cf. figure 6) by about
one magnitude which explains the serrations mentioned above. The collectivity is caused mainly by interaction-induced thermal activation which will be
discussed in more detail at the end of the next section.

Anyway, it should be stressed that the plastic deformation behaviour under the conditions described in equation (19) will be macroscopically stable,
as in the previously discussed cases.

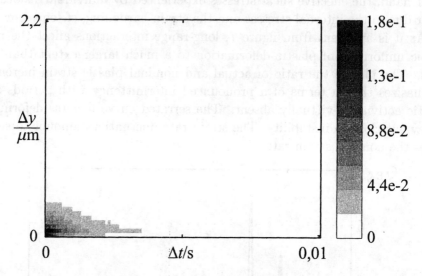

Figure 8: Auto-correlation function $\rho_{EE}(\Delta t, \Delta y)$ of thermal activations;
$\Delta G_{\max}/kT = 0$, $R_{\mathrm{int}} = N_X/2$, $\dot{\sigma} = 9.26 \cdot 10^{-2}\,\mathrm{MPa/s}$.

3.4 INSTABILITY INDUCED BY DSA AND INTERACTIONS

Finally, the system is investigated with respect to the effect of DSA in conjunction with long-range elastic interactions. This case corresponds to the
following parameters:

$$\Delta G_{\max}/kT = 10, \quad R_{\mathrm{int}} = N_X/2 \tag{20}$$

Thus, the members of the dislocation ensemble do not only differ in their
individual effective shear stresses $\tau_{\mathrm{eff},i}$ but also in their waiting time dependent
activation enthalpy $G(t_{w,i})$. As expected the deformation behaviour is now
accompanied by developed instability causing plastic bursts followed by long

quiescent periods without any plastic activity (zero strain rate). The plot of the ratio $Q_{\dot{\varepsilon}_{pl}}$ in figure 9 gives a good impression of the strongly intermittent plastic deformation. Each peak of the strain rate gives rise to a significant step in the strain vs. time curve (strain burst). During a peak the actual strain rate exceeds the nominal rate by a factor of up to 20 which emphasizes the serrated character of plastic flow, in good qualitative agreement with experimental findings.

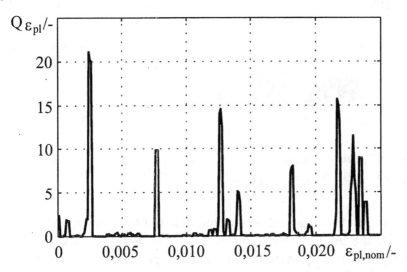

Figure 9: Ratio of actual and nominal strain increment; $\Delta G_{max}/kT = 10$, $R_{int} = N_X/2$, $\dot{\sigma} = 9.26 \cdot 10^{-2}\,\mathrm{MPa/s}$.

The correlation analysis (figure 10) shows that plastic activity is extremely correlated in space *and* time. The maximum value of 0.6 exceeds the values of the previously discussed types of model systems by a factor of three. In other words for any given activation event at a certain time and position there is an excessive probability of an impending activation event in the vincinity of the previous event. So, we conclude that only the combination of DSA and long-range interactions gives rise to a synchronized interaction-induced thermal activation, a collective effect which manifests itself on the macroscopic scale in terms of system-spanning plastic strain bursts. To illustrate this, we have analysed the strain bursts more closely. Figure 11 shows all thermal activation events in the initial phase of a strain burst at a high resolution in time. As one can see, there is a clear structure in the sequence of activation events. The plastic deformation starts with the activation of the dislocation ①. After several repetitive activation events the neighbouring dislocation ② starts

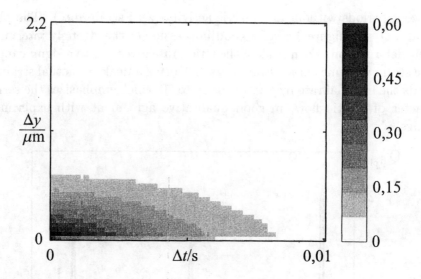

Figure 10: Auto-correlation function $\rho_{EE}(\Delta t, \Delta y)$ of thermal activations; $\Delta G_{\max}/kT = 10$, $R_{\mathrm{int}} = N_X/2$, $\dot{\sigma} = 9.26 \cdot 10^{-2}\,\mathrm{MPa/s}$.

contributing to the plastic deformation, followed by dislocation ③ and so on. Eventually the plastic activity propagates in an avalanche-like way along the positive y direction (perpendicular to the glide plane) at some finite speed. Similar avalanches are found to underly the other strain burst, confirming the idea that the PLC effect may go along with plastic waves propagating macroscopically all along the tensile axis.

4. Conclusions

A mesoscopic 2D computer model has been presented for the simulation of the stability properties of plastic flow during tensile deformation, in terms of an ensemble of discrete edge dislocations subjected to dynamic strain ageing (DSA) and long-range elastic interactions. The separate numerical investigation of dislocation interaction effects, on the one hand, and of the influence of DSA, on the other hand, confirms a previous assertion made on analytical grounds [6], that plastic instability of the strain-rate softening type arises only by the combined effect of DSA and interactions, while either effect taken alone does not manifest itself on the macroscopic scale.

That combined effect can be envisaged in terms of a positive feedback mechanism: DSA tends to lower the strain-rate sensitivity S of the flow stress. As a result, the strain-rate fluctuations increase, as described by equation (1).

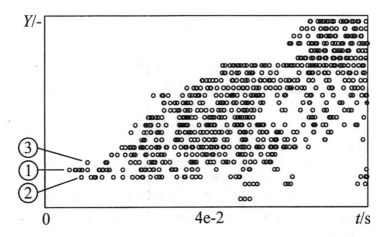

Figure 11: Activation events $E(t, y)$ at the beginning of a plastic strain burst displayed at a high temporal resolution; $\Delta G_{max}/kT = 10$, $R_{int} = N_X/2$, $\dot{\sigma} = 9.26 \cdot 10^{-2}\,\mathrm{MPa/s}$.

This implies that the dislocation motion becomes more correlated owing to more efficient interactions which, in turn, make DSA more efficiently act on synchronized dislocations. Consequently, S is further reduced until a critical point is reached where strain-rate softening becomes appreciable on the macroscopic scale.

330

5. References

1. Zaiser, M. and Hähner, P. (1997) Oscillatory modes of plastic deformation *phys. stat. sol. B* **199**, 267–330.
2. Hähner, P. (1996) A theory of dislocation cell formation, *Acta Mater.* **44**, 2345–2352.
3. Hähner, P., Bay, K. and Zaiser, M. (1998) Fractal dislocation patterning during plastic flow, *Phys. Rev. Lett.* **81**, 2470–2473.
4. Hähner, P. (1996) Stochastic dislocation patterning during cyclic plastic deformation, *Appl. Phys. A* **63**, 45–55.
5. Nortmann, A. and Schwink, Ch. (1997) Characteristics of dynamics strain ageing in binary f.c.c. copper alloys, I and II, *Acta Mater.* **45**, 2043 and 2051.
6. Hähner, P. (1996) On the physics of the Portevin–Le Chatelier effect, Part 1 and Part 2, *Mater. Sci. Eng. A* **207**, 208 and 216.
7. Steck, E. (1985) A stochastic model for the high-temperature plasticity of metals, *Int. J. Plasticity* **1**, 243–258.
8. Hirth, J.P. and Lothe, J. (1982) *Theory of Dislocations*, McGraw-Hill, New York.
9. Schwink, Ch. and Nortmann, A. (1997) The present experimental knowledge of dynamic strain ageing in binary f.c.c. solid solutions, *Mater. Sci. Eng. A* **234–236**, 1–7.
10. Neuhäuser, H. (1990) Plastic instabilities and the deformation of metals, in D. Walgraef and N.M. Ghoniem (eds.), *Patterns, Defects and Material Instabilities*, Kluwer, Dordrecht, pp. 241–276.
11. Hähner, P., Ziegenbein, A., Rizzi, E. and Neuhäuser, H. (2002) Spatio-temporal analysis of Portevin–Le Chatelier deformation bands: Theory, simulation, and experiment, *Phys. Rev. B* **65**, 134109–134128.

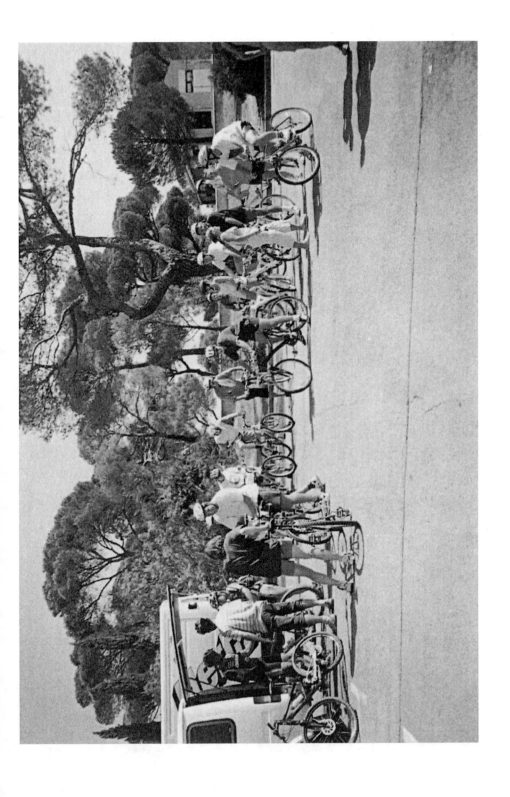

Statistical properties of dislocation ensembles

I. Groma and F.F. Csikor
Department of General Physics, Eötvös University, POB 32, 1518 Budapest, Hungary
(groma@metal.elte.hu)

Abstract. The problem of the collective behavior of straight parallel dislocations is investigated. A continuum description is derived from the equation of motion of individual dislocations. It is shown that the influence of the short range dislocation-dislocation interactions on the dislocation dynamics can be described either by a diffusion-like gradient or a stochastic term. The properties of the stochastic stress component are discussed in details.

Keywords: dislocations, theory and modeling of defects, mechanical properties

1. Introduction

In order to study the plastic properties of macroscopic (larger than $1\mu m$) systems a continuum theory of crystal plasticity needs to be developed. The traditional crystal plasticity models operate with phenomenological constitutive relations between the stress, the strain and the strain rate. Recently several models have been published in which nonlocal terms (in most cases gradient terms) are also introduced (Fleck et al., 1994; Zaiser and Aifantis, 2002). It is essential, however, that the underlying dynamics of discrete dislocations is taken into account.

During the past 30 years several continuum dislocation dynamics description have been elaborated. In the models of Holt (1970) and Ricman & Viñals (1997) an irreversible thermodynamics analogy was used, Walgraef and Aifantis (1985) proposed an approach where the mobile and immobile dislocations evolve according to reaction-diffusion equations, Kratochvil et. al. (1993) developed the idea of non-local hardening based on particular mechanisms such as dislocation sweeping, and Hähner and Zaiser (1996; 1998) proposed a stochastic description of dislocation dynamics in terms of nonlinear stochastic processes. Most of these models are based on analogies with other physical problems like spinodal decomposition, oscillating chemical reactions and chemical patterning, etc. As a consequence of this, the properties of individual dislocations are taken into account only in a very indirect way. The goal of the investigations presented in this paper is to derive a continuum description of dislocation dynamics from the dynamics of discrete dislocations by a rigorous procedure. Due to the large complexity of the discrete dislocation systems the problem cannot be tackled in general. We consider only a system of straight parallel dislocations. This is certainly an extremely

A. Finel et al. (eds.), Thermodynamics, Microstructures and Plasticity, 333–342.

simplified dislocation configuration but a useful model system to determine the structure of the evolution equations of the dislocation system.

The general structure of a continuum theory of dislocation evolution were investigated in detail in (1997; 1999; 2001; 2002). Recently, the approach has been generalized for curved dislocations by El-Azab (2000). The goal of the present paper is to study the influence of short range dislocation-dislocation correlations in details. It is shown that as a first order approximation short range correlation effects can be taken into account by a gradient term. However, for strongly inhomogeneous systems this is obviously not satisfactory. In order to step beyond the limitations related to the gradient approximation the concept of stochastic dislocation dynamics is outlined.

2. Linking discrete and continuum dislocation dynamics descriptions

Let us consider N straight parallel edge dislocations with positions $\{\vec{r}_i, i = 1..N\}$ in the (x, y) plane perpendicular to the line vector $\vec{l}$. For the sake of simplicity we assume that the Burgers vector of the ith dislocation can have only the values $\vec{b}_i = \pm \vec{b}$ (i.e. single glide configuration is considered). With the commonly applied over-damped dislocation dynamics approximation, the velocity of the ith dislocation is given by

$$\vec{v}_i = B\vec{b} \left(\sum_{j \neq i}^{N} s_i s_j \tau_{\text{ind}}(\vec{r}_i - \vec{r}_j) + s_i \tau_{\text{ext}} \right), \tag{1}$$

where $s_i = \vec{b}_i \vec{b}/(\vec{b}^2)$ is the sign of the dislocation, $\tau_{\text{ind}}(\vec{r}_i - \vec{r}_j)$ is the shear stress created at $\vec{r}_i$ by a positive dislocation located at $\vec{r}_j$, τ_{ext} is the external stress, and B is the dislocation mobility. Since B can always be absorbed into the time variable it is dropped in the following.

After carrying out an averaging procedure on Eq. (1) (for details see (Groma, 1997; Groma and Balogh, 1999)) we obtain the following balance equations for the time evolution of the averaged positive and negative dislocation densities denoted by $\rho_+(\vec{r})$ and $\rho_-(\vec{r})$ respectively:

$$\frac{\partial \rho_+(\vec{r}_1, t)}{\partial t} = \tag{2}$$
$$-\vec{b}\frac{\partial}{\partial \vec{r}_1} \left[\rho_+(\vec{r}_1, t)\tau_{\text{ext}} + \int \{\rho_{++}(\vec{r}_1, \vec{r}_2, t) - \rho_{+-}(\vec{r}_1, \vec{r}_2, t)\} \tau_{\text{ind}}(\vec{r}_1 - \vec{r}_2)d\vec{r}_2 \right]$$

$$\frac{\partial \rho_-(\vec{r}_1, t)}{\partial t} = \tag{3}$$
$$+\vec{b}\frac{\partial}{\partial \vec{r}_1} \left[\rho_-(\vec{r}_1, t)\tau_{\text{ext}} - \int \{\rho_{--}(\vec{r}_1, \vec{r}_2, t) - \rho_{-+}(\vec{r}_1, \vec{r}_2, t)\} \tau_{\text{ind}}(\vec{r}_1 - \vec{r}_2)d\vec{r}_2 \right]$$

where ρ_{++}, ρ_{--}, ρ_{+-} and ρ_{-+} are two-particle density functions. They may be interpreted as follows: $\rho_{++}(\vec{r}_1,\vec{r}_2,t)dV_1 dV_2$ is the joint probability to find at time t a positive dislocation in a volume element dV_1 at $\vec{r}_1$ and another positive dislocation in a volume element dV_2 at $\vec{r}_2$. The two-particle density functions ρ_{+-}, ρ_{-+} and ρ_{--} are interpreted accordingly.

By adding and subtracting the above two equations, the following evolution equations can be obtained for the total dislocation density $\rho(\vec{r},t) = \rho_+(\vec{r},t) + \rho_-(\vec{r},t)$ and the sign dislocation density $\kappa(\vec{r},t) = \rho_+(\vec{r},t) - \rho_-(\vec{r},t)$:

$$\frac{\partial \rho(\vec{r}_1,t)}{\partial t} + \vec{b}\frac{\partial}{\partial \vec{r}_1}[\kappa(\vec{r}_1,t)\tau_{ext} + \int \{\rho_{++}(\vec{r}_1,\vec{r}_2,t) + \rho_{--}(\vec{r}_1,\vec{r}_2,t) \tag{4}$$
$$-\rho_{+-}(\vec{r}_1,\vec{r}_2,t) - \rho_{-+}(\vec{r}_1,\vec{r}_2,t)\}\tau_{ind}(\vec{r}_1 - \vec{r}_2)d\vec{r}_2] = 0,$$

$$\frac{\partial \kappa(\vec{r}_1,t)}{\partial t} + \vec{b}\frac{\partial}{\partial \vec{r}_1}[\rho(\vec{r}_1,t)\tau_{ext} + \int \{\rho_{++}(\vec{r}_1,\vec{r}_2,t) - \rho_{--}(\vec{r}_1,\vec{r}_2,t) \tag{5}$$
$$-\rho_{+-}(\vec{r}_1,\vec{r}_2,t) + \rho_{-+}(\vec{r}_1,\vec{r}_2,t)\}\tau_{ind}(\vec{r}_1 - \vec{r}_2)d\vec{r}_2] = 0.$$

We have to mention at this point that the connection with plasticity is established by taking into account that the signed dislocation density $\kappa(\vec{r},t)$ is related to the plastic shear strain $\gamma(\vec{r},t)$ as $\kappa = \vec{b}(\partial \gamma/\partial \vec{r})/(\vec{b}^2)$ (cf. e.g. (Landau and Lifshitz, 86)). By comparing the above expression with Eq. (5) one can find that

$$\dot{\gamma}(\vec{r}_1) = b^2\rho(\vec{r}_1)\tau_{ext} + b^2 \int \{\rho_{-+}(\vec{r}_1,\vec{r}_2) - \rho_{+-}(\vec{r}_1,\vec{r}_2) \tag{6}$$
$$+\rho_{++}(\vec{r}_1,\vec{r}_2) - \rho_{--}(\vec{r}_1,\vec{r}_2)\}\tau_{ind}(\vec{r}_1 - \vec{r}_2)d\vec{r}_2$$

It is important to note that Eqs. (4) and (5) are exact, i.e. no assumption is required to derive them from Eq. (1). Since, however, they depend on the two-particle density functions they do not form a closed set of equations. Although equations can be derived for the two-particle densities, these depend on the three-particle densities and so on, resulting an infinite hierarchy of equations (Groma, 1997; Zaiser et al., 2001). So, in order to get usable equations the ρ and κ dependence of the two particle density functions need to be determined.

3. Separation of short and long range effects

The two particle density function $\rho_{ss'}(\vec{r}_1,\vec{r}_2,t)$ can always be given in the form

$$\rho_{ss'}(\vec{r}_1,\vec{r}_2,t) = \rho_s(\vec{r}_1)\rho_{s'}(\vec{r}_2)(1 + d_{ss'}(\vec{r}_1 - \vec{r}_2)) \qquad s,s' \in \{+,-\} \tag{7}$$

where $d_{ss'}$ is the dislocation-dislocation correlation function. ($d_{ss'}$ might depend on $(\vec{r}_1 + \vec{r}_2)/2$ too, but this is not indicated for simplicity.) The actual form of $d_{ss'}$ can be determined by discrete dislocation dynamics simulations.

336

Fig. 1 shows the variation of the function $d(\vec{r}) = 1/4[d_{++}(\vec{r}) + d_{--}(\vec{r}) + d_{+-}(\vec{r}) + d_{-+}(\vec{r})]$ obtained on a dislocation configuration which was relaxed from an initially random configuration. Its most important feature revealed is that it decays to zero within a few dislocation distances. This opens the possibility to separate the short and long range type terms in the evolution equations. Namely, the terms depending on the correlation functions are determined only by the short range properties. With expression (7), Eqs. (4,5)

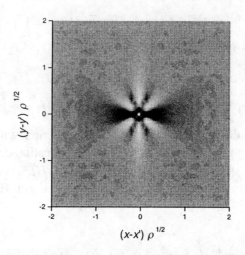

Figure 1. Pair correlation function $d(\vec{r})$ of a dislocation system relaxed at zero external stress. The x and y coordinates have been scaled by the mean dislocation spacing $1/\sqrt{\rho}$. Light areas correspond to positive pair correlations, dark areas to negative correlations.

can be written as

$$\frac{\partial \rho(\vec{r},t)}{\partial t} + \vec{b}\frac{\partial}{\partial \vec{r}}[\kappa(\vec{r},t)\{\tau_{sc}(\vec{r}) + \tau_{ext} - \tau_f(\vec{r}) + \tau_b(\vec{r})\}] = 0, \qquad (8)$$

$$\frac{\partial \kappa(\vec{r},t)}{\partial t} + \vec{b}\frac{\partial}{\partial \vec{r}}[\rho(\vec{r},t)\{\tau_{sc}(\vec{r}) + \tau_{ext} - \tau_f(\vec{r}) + \tau_b(\vec{r})\}] = 0, \qquad (9)$$

where

$$\tau_{sc}(\vec{r}) = \int \kappa(\vec{r}_1,t)\tau_{ind}(\vec{r}-\vec{r}_1)d\vec{r}_1 \qquad (10)$$

is the long range (self-consistent) internal-stress field created by the dislocations,

$$\tau_f(\vec{r}) = \frac{1}{4}\int \rho(\vec{r}_1)[d_{+-}(\vec{r}-\vec{r}_1) - d_{-+}(\vec{r}-\vec{r}_1)]\tau_{ind}(\vec{r}-\vec{r}_1)d\vec{r}_1, \qquad (11)$$

and

$$\tau_b(\vec{r}) = \int \kappa(\vec{r}_1) d(\vec{r} - \vec{r}_1) \tau_{\text{ind}}(\vec{r} - \vec{r}_1) d\vec{r}_1. \tag{12}$$

From the scaling properties of the correlation function one can find that τ_f is proportional to $\sqrt{\rho}$, so it can be associated with the flow stress.

Because of short range nature of d in the integral in Eq. (12) the function $\kappa(\vec{r}_1)$ can be approximated by their Taylor expansion around the point $\vec{r}$. Keeping only the first non-vanishing term we arrive at

$$\tau_b(\vec{r}) = -\frac{D}{\rho(\vec{r})} \frac{\partial \kappa(\vec{r})}{\partial \vec{r}} \tag{13}$$

where D is a constant. For the actual value of D the reader is referred to (Groma et al., 2002).

The balance equations (8,9) together with expressions (10,11,12) form a closed set of dislocation evolution equations. It is important to stress, however, that the scheme outlined above is applicable only if the gradient of the dislocation densities are small. If it is not the case one cannot stop the Taylor expansion of $\kappa(\vec{r}_1)$ at the first nontrivial term. It needs to be mentioned that dislocation annihilation and/or multiplication can be incorporated into this framework by adding a source term on the right-hand side of Eq. (8). However, Eq. (9) has to remain unchanged, reflecting the fact that the net Burgers vector of the system is conserved.

4. Statistical properties of the internal stress

As it is mentioned above if the dislocation system is strongly inhomogeneous τ_b cannot be approximated by a gradient term. However, discrete dislocation dynamics simulations indicate that the time variation of the internal stress has a slowly varying and a strongly irregular component (see Fig. 2). This

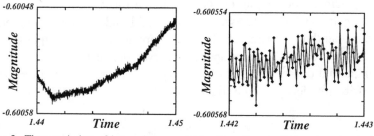

Figure 2. Time variation of internal stress obtained by discrete dislocation simulation. The right stress-time curve is a magnified part of the left one.

numerical finding supports the idea that the time variation of τ_b can be well described by a stochastic process. (The self-consistent field τ_{int} accounts for the slowly varying component.)

In order to obtain the stochastic properties of τ_b first we have to determine the steady state probability distribution of the internal stress. In precise mathematical terms this means that knowing the probability of a dislocation configuration we ask the probability of finding the dislocation created stress τ in the interval $[\tau_0 - \delta\tau/2, \tau_0 + \delta\tau/2]$ at a given point $\vec{r}$. According to Groma and Bakó (1998) for a system of straight parallel dislocations with uniform Burgers vectors the Fourier transform of the stress probability density can be given as follows:

$$P_s(q) = \int w_N(\vec{r}_1, \vec{r}_2, \ldots, \vec{r}_N) \prod_{j=1}^{N} \exp\left\{ iq\tau_{ind}(\vec{r} - \vec{r}_j) \right\} d\vec{r}_1 d\vec{r}_2 \ldots d\vec{r}_N$$

$$= \exp[Q(\vec{r}, q)], \tag{14}$$

where $w_N(\vec{r}_1, \vec{r}_2, \ldots, \vec{r}_N)$ is the probability density of the dislocation configuration. By introducing the notation

$$B(\vec{r}, q) = 1 - \exp\{i\tau_{ind}(\vec{r})q\} \tag{15}$$

the quantity $Q(\vec{r}, q)$ introduced above can be given as a polynomial of $B(\vec{r}, q)$:

$$Q(\vec{r}, q) = - \int \rho(\vec{r}_1) B(\vec{r} - \vec{r}_1, q) d\vec{r}_1 \tag{16}$$

$$+ \frac{1}{2} \int \rho(\vec{r}_1) \rho(\vec{r}_2) d(\vec{r}_1, \vec{r}_2) B(\vec{r} - \vec{r}_1, q) B(\vec{r} - \vec{r}_2, q) d\vec{r}_1 d\vec{r}_2 + \ldots$$

As it can be seen the first term is determined by the dislocation density, while the second one depends on the dislocation-dislocation correlation function $d(\vec{r}_1, \vec{r}_2)$. As it was found by Groma and Bakó (1998) for small enough q values $P_s(\vec{r}, q)$ has the form:

$$P_s(q, \vec{r}) \approx \exp\left[C\rho(\vec{r})q^2 \ln(q/q_{eff}) + \ldots \right] \tag{17}$$

where C is a constant determined by the type of dislocations considered and q_{eff} is a parameter depending on the correlation function $d(\vec{r}_1, \vec{r}_2)$. It is important to note here that expression (17) is valid under much more general conditions than those considered above. It is applicable for multiple slip too, as well as under certain circumstances for curved dislocations.

An important consequence of expression (17) is that the tail of the probability distribution decays as (Groma and Bakó, 1998)

$$P_s(\tau_b)|_{\tau_b \to \infty} = C\rho(\vec{r}) \frac{1}{|\tau_b|^3} \tag{18}$$

A remarkable feature of this tail behavior is that it is independent from the actual form of the dislocation distribution, it is determined only by the dislocation density. This is clearly observable on the stress probability distributions obtained numerically on a random and a relaxed dislocation configuration. The two probability distributions are plotted in Fig.3. As it can be seen the

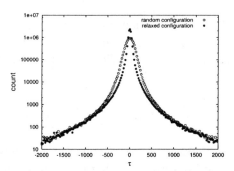

Figure 3. Internal stress distribution obtained on a random and a relaxed dislocation configuration.

tails of the two curves are the same while the center part of the distribution is much narrower for the relaxed configuration. According to Fig.4 the inverse cubic decay of the tail is clearly fulfilled.

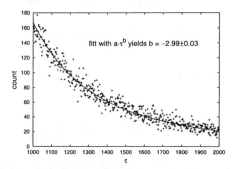

Figure 4. Tail of the internal stress distribution obtained numerically on a relaxed dislocation configuration, together with the fitted power function.

After carrying out a similar procedure for dislocation dipoles one can find that for large q values

$$P_s(q) = \exp(-G|q|) \tag{19}$$

As a consequence of this for a dipole system the central part of the probability distribution is Lorentzian. However, we need to emphasize that the tail always decays as $1/\tau_b^3$. This also means that in contrast with many other physical systems for dislocation ensembles the steady state probability distribution of the relevant stochastic variable is never Gaussian.

5. Time evolution of stress probability distribution

In order to obtain the governing equation of the time evolution of the stochastic component of the internal stress τ_b we have to determine the evolution equation of $P(\tau,t)$. This cannot be done as rigorously as the steady state distribution function was obtained, but as a simplest possible nontrivial possibility we can assume that the evolution of $P(\tau,t)$ is described by the following Fokker-Planck equation:

$$\frac{\partial}{\partial t}P(\tau_b,t) = \frac{\partial}{\partial \tau_b}\left\{-b(\tau_b)P(\tau_b,t) + \frac{1}{2}\frac{\partial}{\partial \tau_b}\sigma^2(\tau_b)P(\tau,t)\right\} \qquad (20)$$

To indicate the type of approximation involved in the above assumption we need to emphasize that Eq. (20) is nothing more but assuming that the time derivative of $P(\tau_b,t)$ is a linear functional of $P(\tau_b,t)$ (which is certainly valid if we are not far from equilibrium) and we stop at the quadratic term in the series expansion of the linear functional.

In the general form of the Fokker-Planck equation the so called "dissipation" function $b(\tau_b)$ and the "diffusion" coefficient $\sigma^2(\tau_b)$ are arbitrary functions. However, we require that $P_s(\tau_b)$ determined above is the steady state solution of Eq.(20), i.e. $P_s(\tau_b)$ has to fulfill the equation

$$b(\tau_b)P_s(\tau_b) = \frac{1}{2}\frac{\partial}{\partial \tau_b}\sigma^2(\tau_b)P_s(\tau_b) \qquad (21)$$

This means that $b(\tau_b)$ and $\sigma^2(\tau_b)$ are not independent from each other. From Eq. (21) we get that

$$\sigma^2(\tau_b) = \left(2\int_0^{\tau_b} b(\tau_b)P_s(\tau_b)d\tau\right)\frac{1}{P_s(\tau_b)} \qquad (22)$$

which is the fluctuation-dissipation relation of the problem considered.

6. Stochastic dislocation dynamics

As it is well known the Fokker-Planck equation (20) is equivalent with the stochastic differential equation

$$\frac{d\tau_b(\vec{r})}{dt} = -b(\tau_b)(\vec{r}) + \sigma(\tau_b(\vec{r}))\eta \qquad (23)$$

where η is white noise. According to Eq. (22) $b(\tau_b)$ and $\sigma(\tau_b)$ are related to each other, but to obtain a closed set of equations $b(\tau_b)$ should be determined. This requires further investigations, but since dissipation cannot occur in the absence of plastic deformation $b(\tau_b)$ needs to be proportional to the average shear rate. As a simplest possible approximation we can assume that $b(\tau_b)$ is proportional to τ_b, i.e.

$$b(\tau_b) = \dot{\gamma} M \tau_b \tag{24}$$

where M is a constant.

After these, the framework of stochastic dislocation dynamics can be summarized as follows:

- The evolution of the total and sign dislocation densities are determined by Eqs. (8) and (9) respectively.

- The long range stress field is given by Eq. (10).

- The flow stress can be approximated by the following form:

$$\tau_f = \begin{cases} \tau_{ext} + \tau_{sc} & \text{if } \tau_{ext} + \tau_{sc} < \beta\sqrt{\rho} \\ \beta\sqrt{\rho} & \text{otherwise} \end{cases} \tag{25}$$

   where β is a constant.

- The evolution of the back stress is determined by the stochastic differential equation (23).

It has to be mentioned that the generalization of the model to multiple slip in 2D is straightforward. Even more, it seems to be feasible to extend the model to 3D.

Dislocation distribution obtained by the method in 2D with two slip systems having equal Schmidt factors can be seen in Fig.5. The detailed analysis of the dislocation distribution revealed that there is no characteristic cell size, the pattern obtained is a fractal.

7. Conclusions

We have derived a continuum model of dislocation dynamics by statistically averaging the equations of motion of discrete dislocations. In this model the dislocation fluxes are governed by different stress contributions, viz., the external stress, a long-range stress created by 'geometrically necessary' excess dislocations related to strain gradients, and a stress contribution which characterizes short-range interactions between individual dislocations. This stress contribution is related to the presence of non-vanishing short-range correlations in the dislocation arrangement. It can be approximated as a sum of two contributions, namely a local flow stress which scales as the square root of the dislocation density, and a term which can be called back stress. It is shown that as a first order approximation the back stress can be approximated

342

Figure 5. Fractal type dislocation cell structure obtained by stochastic dislocation dynamics.

with a gradient term. For strongly inhomogeneous systems, however, it is not satisfactory. In this case the time variation of the back stress can be described by a stochastic process. The stochastic differential equation governing the evolution of the back stress is determined. Numerical investigations reveal that the stochastic dislocation dynamics is able to reproduce the formation of fractal type dislocation cell structure.

Acknowledgements

The authors are grateful to Professors M. Zaiser, E. Von der Giessen and E. Aifantis for valuable discussions. The financial support of the Hungarian Scientific Research Fund (OTKA) under contract number T030791 is gratefully acknowledged.

References

El-Azab, A.: 2000. *Phys. Rev. B* **61**, 11956.
Fleck, N., G. Muller, M. Ashby, and J. Hutchinson: 1994. *Acta Metall. Mater.* **42**, 475.
Groma, I.: 1997. *Phys. Rev. B* **56**, 5807.
Groma, I. and B. Bakó: 1998. *Phys. Rev. B* **58**, 2964.
Groma, I. and P. Balogh: 1999. *Acta Mater.*
Groma, I., F. Csikor, and M. Zaiser: 2002. *Acta Materialia* **submitted**.
Hähner, P.: 1996. *Acta mater.* **44**, 2345.
Hähner, P., K. Bay, and M. Zaiser: 1998. *Phys. Rev. Lett.* **81**, 2470.
Holt, D.: 1970. *Journal Appl. Phys.* **41**, 3179.
Kratochvil, J. and M. Saxlova: 1993. *Scripta Met. Mater.* **26**, 113.
Landau, L. and E. Lifshitz: 86. In: *Theory of Elasticity, Course in Theoretical Physics Vol. 7. 3rd ed.* Oxford: Pergamon.
Rickman, J. and J. Viñals: 1997. *Phil. Mag. A* **75**, 1251.
Walgraef, D. and E. Aifantis: 1985. *J. Appl. Phys.* **15**, 688.
Zaiser, M. and E. Aifantis: 2002. *Scripta mater.* **in press**.
Zaiser, M., M. Miguel, and I. Groma: 2001. *Phys. Rev.* **64**, 224102.

COLLECTIVE BEHAVIOR OF DEFECT ENSEMBLES AND SOME NONLINEAR ASPECTS OF FAILURE

O.B. NAIMARK

Institute of Continuous Media Mechanics, Russian Academy of Sciences

1, Academician Korolev Str., 614013 Perm, Russia

Abstract.

The multifield statistical approach for the description of collective properties of mesodefect ensembles is developed that allowed the determination of specific nonlinear form of the evolution equation for the macroscopic tensor parameter of the defect density. Characteristic self-similar solutions of these equations were found that responsible for the transition from damage to damage localization stage, change of symmetry properties of system due to the generation of collective modes of mesodefects. This approach was applied for experimental and theoretical study of stochastic crack dynamics and the resonance excitation of failure (failure waves).

1. Introduction

During the last few decades the interrelation between structure and properties of solids has been the key problem in physics and mechanics. An extensive study of transformation of structure stimulates bridging the gap between general approaches of physics and mechanics of solids. This tendency provides the deeper insight into general laws of plasticity and failure, which can be also treated as structural transitions induced by defects. Real solids are complex in structure defined as a hierarchy of different scale levels. Solids under loading demonstrate changes on all structural levels. On analyzing, these changes have been qualified as plastic deformation and damage processes, realized by nucleation, evolution and interaction of defects on appropriate structural levels as well as by the interaction of defects between levels. Until recently, no unified multifield theory of solids has been developed to describe a variety, complexity and interaction of processes

A. Finel et al. (eds.), Thermodynamics, Microstructures and Plasticity, 345–361.

commonly observed on all levels of structure. Thus, to construct the model allowing for the material structure and its variation, the prime attention should be given to the choice of physical level of microstructure and, consequently to the type of defects. Furthermore, each structural level involves the results of processes occured on the smaller scale levels. Experimental study of material responses in a large range of loading rate reveals some unresolved puzzles in failure and plasticity, and shows the linkage of solid behavior with evolution of the ensemble of typical mesoscopic defects (dislocation substructures, microcracks, microshears). This situation has pronounced features for dynamic and shock wave loading, when the internal times of the ensemble evolution for different structural levels are approaching to characteristic loading times. As the consequence, the widely used assumption in phenomenology of plasticity and failure concerning the subjective role of structural variables to stress-strain variables (adiabatic limit) can not be generally applied. However, in this case another fundamental problem arises concerning description of the defect ensemble behavior taking in view the multifield nature of the defect interaction and specific role of defects. These problems are related to the role of the defect ensemble that could be considered in the local as the localized change of the symmetry of the displacement field, in the global, as the change of the system symmetry under the generation of new collective modes. These modes could subject the system behavior and determine the relaxation ability (plasticity) and failure.

2. Mesodefect Properties

2.1. DISLOCATION SUBSTRUCTURES

It is well-known, that the dislocation density increases due to the plastic deformation and the consequent changes of dislocation substructures are observed. These phenomena occur under the active loading, fatigue, creep, dynamic and shock wave loading. Despite the variety of the deformed materials the limited types of the dislocation substructures are observed. The reason of such universality is related to the inherent property of dislocation ensembles as the essentially non-equilibrium system which reveals self-similarity features in the sense of characteristic nonlinear responses. This universality in the behavior of dislocation systems appears in the experiments as the low sensitivity of the evolution of dislocation structures to the external stress, but the high sensitivity to structural stresses induced by dislocation interaction. The increase of the dislocation density is accompanied by the decrease of the distance between the dislocations and the stresses of the dislocation interaction generated on the corresponding dislocation substructures. The collective properties in dislocation ensembles begin to play the leading role in these transitions and the substructure formation. The driving force of the reconstruction of dislocation ensembles is the tendency to reach the relative minimum of total energy due to the creation of dislocation substructures. The energy of the dislocation substructure includes two parts: the own dislocation energy and the energy of the dislocation interaction. The reconstruction of the dislocation substructure leads to the change of both parts. As the consequence, the energy of new formed dislocation substructure is less than the energy of the preceding substructure. The main part in the energy of the dislocation substructure belongs to the own dislocation energy [1]

$$\Delta U = \frac{\rho G b^2}{2\pi} \ln \left(\frac{L}{r_0} \right) \tag{1}$$

where ρ is the dislocation density in the dislocation substructure, **b** is the Burgers vector, G is the shear elastic modulus, r_0 is the radius of the dislocation nucleus, L is the screening radius of the elastic field created by the dislocations. The last scale plays the important role in the evolution of dislocation substructures: the increase of the dislocation density leads to the decrease of the -scale in the order of the substructure succession. The typical dislocation substructures that are experimentally observed are chaotic, tangle, walls of the cells, sub-boundaries of strip substructures. The value of r_0 is close to the grain size of chaotic substructures and has the order of the width of cell walls or high density dislocation area in strip substructures.

2.2. MICROCRACK (MICROSHEAR) ENSEMBLE

Of all structural levels of sub-dislocation defects, the level of microcracks and microshear may be considered as the representative for developed stage of plastic deformation and failure. The rest of defects (point defects, dislocations, dislocation pile-ups) have the smaller values of intrinsic elastic fields and energies in the comparison with microcracks and microshears. Moreover, the nucleation and growth of these defects (that are closest to the macroscopic level) are some final acts of the previous rearrangement of the dislocation substructures, when all the defects take part in the present local volume of the material. The density of these defects reaches $10^{11}\ cm^3$, but each mesoscopic defect consists of a dislocation ensemble and exhibits the properties of this ensemble. Scenarios of the evolution of ensembles of these mesoscopic defects show features of non-equilibrium kinetic transitions, and experimental data obtained in a wide range stress intensities and rates of strain confirm the universality of structural evolution and its effect on relaxation properties and failure. The important features of the quasi-brittle fracture were established for the understanding of various stages of failure: damage, damage localization, crack nucleation and propagation. It was shown that microcracks have the dislocation nature and represent the core nuclei of the dislocation pile-ups. The model representation of microcrack as dislocation pile-up allowed the estimation of the own microcrack energy [2]

$$E \simeq \left[\frac{G}{V_0} \ln \left(\frac{R}{r_0} \right) \right] s^2 \tag{2}$$

where $\mathbf{B} = n\mathbf{b}$ is the total Burgers vector; $s = S_D B$ is the penny-shape microcrack volume; S_D is the microcrack base; $V_0 = \frac{4}{3} r_0^3$ is the volume of the defect nuclei, r_0 is the characteristic size of the dislocation core (defect nuclei); R is the characteristic scale of the elastic field produced by microcrack. The estimation given in [3] showed that the power of the dislocation pile-up is close to $n \approx 20$. Two reasons are important for the dislocation representation of microcracks (microshears). The first one is the determination of the microcrack energy as the energy of the dislocation pile-up. The second reason is the determination of microscopic parameters for the microcracks as the consequence of the symmetry change of displacement field due to the microcrack nucleation and growth. Study of the microcrack (microshear) size distribution for the different spatial scales revealed the self-similarity of the mesodefect pattern [4]. The statistical self-similarity reflects the invariant form of the distribution function for the mesodefects of different structural levels. This fact has important consequence for the development of the statistical multifield theory of the evolution of the defect ensemble.

3. Order Parameters of Continuum with Defects

3.1. MICROSCOPIC AND MACROSCOPIC VARIABLES FOR MICROCRACK (MICROSHEAR) ENSEMBLE

Structural parameters associated with microcracks and microshears were introduced [5] as the derivative of the dislocation density tensor. These defects are described by symmetric tensors of the form

$$s_{ik} = s \nu_i \nu_k \tag{3}$$

in the case of microcracks and

$$s_{ik} = \frac{1}{2} \left(\nu_i l_k + l_i \nu_k \right) \tag{4}$$

for microshears. Here $\vec{\nu}$ is unit vector normal to the base of a microcrack or slip plane of a microscopic shear; $\vec{l}$ is a unit vector in the direction of shear; s is the volume of a microcrack or the shear intensity for a microscopic shear.

The change of the diffeomorphic structure of the displacement field due to these defects has also important consequences from point of view of the symmetry change of the system "solid with defects". This symmetry aspect can be used to model arbitrary defects both in crystalline and amorphous materials without the assumption concerning the dislocation nature of the defects that originally is the property of crystalline materials. The average of the "microscopic" tensor gives the macroscopic tensor of the microcrack or microshear density

$$p_{ik} = n \langle s_{ik} \rangle \tag{5}$$

that coincides with the deformation caused by the defects, n is the defect concentration.

4. Statistical Model of Continuum with Defects

4.1. EFFECTIVE FIELD METHOD

The effective field method is frequently used to refer to any auxiliary field (real or virtual) introduced into a theoretical model in order to construct a simplified way of taking into account the effect of complicated factors like interparticle interactions, which are either too difficult to evaluate rigiriously or are even not yet clear in detail. In our consderation the reference to the effective field method means that we use the concept of an auxiliary external multicomponent field, constructed in such a way that in addition of the corresponding term to the Hamiltonian of the system under consideration makes it state an equilibrium one at any given instant in time. This simple idea has proved itself as a useful approximation or the treatment of a number of problems. The effective field method was reintroduced into statistical physics a fruitful physical idea put forward in thermodynamics by Leontovich [6]. According to Leontovich for an arbitrary nonequilibrium state of any thermally uniform system, that is characterized by definite values of internal parameters, the transition into the equilibrium state with the same values of those internal parameters may be performed by introducing an additional force field. By definition, the entropy of this nonequilibrium state is equal to the entropy of the equilibrium (being that due the presence of the additional force field) state characterized by the same values of the considered material parameters. The microscopic kinetics for the parameter s_{ik} is determined by the Langevin equation

$$\dot{s}_{ik} = K_{ik}(s_{lm}) - F_{ik} \tag{6}$$

where $K_{ik} = \partial E / \partial s_{ik}$, E is the energy of the defect and F_{ik} is a random part of the force field and satisfys the relations $\langle F_{ik}(t) \rangle = 0$ and $\langle F_{ik}(t') F_{ik}(t) \rangle = Q\delta(t-t')$. The parameter Q characterizes the mean value of the energy relief of the initial material structure (the energy of defect nuclei). Statistical model of the defect ensemble was developed in the terms of the solution of the Fokker-Plank equation in [7]

$$\frac{\partial W}{\partial t} = -\frac{\partial}{\partial s_{ik}} K_{ik} W + \frac{1}{2} Q \frac{\partial^2}{\partial s_{ik} \partial s_{ik}} W \qquad (7)$$

According to the statistical self-similarity hypothesis the distribution function of defects can be represented in the form $W = Z^{-1} \exp(-E/Q)$ where Z is the normalization constant. As it follows from (7) the statistical properties of the defect ensemble can be described after the determination of the defect energy E and the dispersion properties of the system given by the value of Q. In the term of the microscopic and macroscopic variables and according to the presentation of these mesodefects as the dislocation substructure the energy of these defects (the Lagrangian) can be written in the form

$$E = E_0 - H_{ik} s_{ik} + \alpha s_{ik}^2 \qquad (8)$$

where the quadratic term represents the own energy of defects (2) and the term $H_{ik} s_{ik}$ describes the interaction of the defects with the external stress σ_{ik} and with the ensemble of the defects in the effective field approximation:

$$H_{ik} = \sigma_{ik} + \lambda p_{ik} = \sigma_{ik} + \lambda n \langle s_{ik} \rangle \qquad (9)$$

where α, λ are the material constants. The average procedure gives the self-consistency equation for the determination of the defect density tensor

$$p_{ik} = n \int s_{ik} W(s, \vec{\nu}, \vec{l}) ds_{ik} \qquad (10)$$

For the dimensionless variables

$$\hat{p}_{ik} = \frac{1}{n} \sqrt{\frac{\alpha}{Q}} p_{ik}, \quad \hat{s}_{ik} = \sqrt{\frac{\alpha}{Q}} s_{ik}, \quad \hat{\sigma}_{ik} = \frac{\sigma_{ik}}{\sqrt{Q\alpha}} \qquad (11)$$

self-consistency equation has the form

$$\hat{p}_{ik} = \int \hat{s}_{ik} Z^{-1} \exp\left((\hat{\sigma}_{ik} + \frac{1}{\delta} \hat{p}_{ik}) \hat{s}_{ik} - \hat{s}_{ik}^2 ds_{ik} \right) d\hat{s}_{ik} \qquad (12)$$

that includes the single dimensionless material parameter $\delta = \alpha/\lambda n$. The dimension analysis allowed us to estimate that $\alpha \sim G/V_0$, $\lambda \sim G$, $n \sim R^{-3}$. Here G is the elastic modulus, $V_0 \approx r_0^3$ is the mean volume of the defect nuclei, R is the distance between defects. Finally we obtain for δ the value $\delta \approx (R/r_0)^3$ that is in the correspondence with the hypothesis concerning the statistical self-similarity of the defect distribution on the

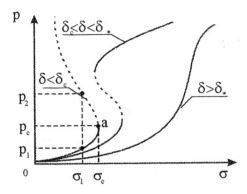

Figure 1. Characteristic responses of matrials to defect growth

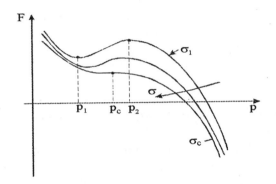

Figure 2. Free energy dependences

different structural level. The solution of the self-consistency equation (11) was found for the case of the uni-axial tension and simple shear [8], (Fig. 1,2).

The existence of characteristic nonlinear behavior of the defect ensemble in the corresponding ranges of δ ($\delta > \delta_* \approx 1.3$, $\delta_c < \delta < \delta_*$, $\delta < \delta_c \approx 1$) was established, where δ_c and δ_* are the bifurcation points. It was shown [8] that the above ranges of δ are characteristic for the quasi-brittle ($\delta < \delta_c \approx$ 1.3), ductile ($\delta_c < \delta < \delta_*$) and nanocrystalline ($\delta > \delta_* \approx 1.3$) responses of materials. It is evidence from this solution that the behavior of the defect ensemble in the different ranges of δ is qualitative different. The replace of the stable material response for the fine grain materials to the metastable one for the ductile materials with the intermediate grain size occurs for the value of $\delta = \delta_* \approx 1.3$, when the interaction between the orientation modes of the defects has more pronounced character. It means that the metastability has the nature of the orientation ordering in the defect ensemble.

5. Collective Properties of Ensembles of Defects

5.1. PHENOMENOLOGY OF SOLIDS WITH DEFECTS. FREE ENERGY

The statistical description allowed us to propose the phenomenology of solid with defects based on the appropriate presentation of the free energy form F. Taking in view that Eqn.10 corresponds to the equation $\partial F / \partial p = 0$, the simple phenomenological form of the part of the free energy caused by defects (for the uni-axial case $p = p_{zz}$, $\sigma = \sigma_{zz}, \epsilon = \epsilon_{zz}$) is given by the six order expansion, which is similar to the well-known Ginzburg-Landau expansion [8].

$$F = \frac{1}{2} A \left(1 - \frac{\delta}{\delta_*}\right) p^2 - \frac{1}{4} B p^4 + \frac{1}{6} C \left(1 - \frac{\delta}{\delta_c}\right) p^6 - D\sigma p + \chi \left(\nabla_l p\right)^2 \quad (13)$$

The bifurcation points δ_*, δ_c play the role that is similar to the characteristic temperatures in the Ginzburg-Landau expansion in the phase transition theory. The gradient term in (12) describes the non-local interaction in the defect ensemble in the so-called long wave approximation; A, B, C, D and χ are the phenomenological parameters. The defect kinetics is determined by the evolution inequality

$$\frac{\delta F}{\delta t} = \frac{\delta F}{\delta p} \frac{dp}{dt} \leq 0 \qquad (14)$$

that leads to the kinetic equation for the defect density tensor

$$\frac{dp}{dt} = -\Gamma \left[A \left(1 - \frac{\delta}{\delta_*}\right) p - Bp^3 + C \left(1 - \frac{\delta}{\delta_c}\right) p^5 - D\sigma - \frac{\partial}{\partial x_l} \left(\chi \frac{\partial p}{\partial x_l}\right) \right], \qquad (15)$$

where Γ is the kinetic coefficient. Kinetic equation (14) and the equation for the total deformation

$$\epsilon = \hat{C}\sigma + p \qquad (16)$$

($\hat{C}$ is the component of the elastic compliance tensor) represent the system of the constitutive equations of solid with considered types of the defects.

5.2. COLLECTIVE PROPERTIES OF DEFECT ENSEMBLES

As it follows from the solution of (10), presented in Fig.1, transitions through the bifurcation points δ_c and δ_* lead to a sharp change in the symmetry of the distribution function as a result of the appearance of some orientationally pronounced macroscopic modes of the tensor p_{ik}. The effect of transitions on the evolution of the defect ensemble is determined by the type of bifurcation — the group properties of the kinetic equation for the tensor p_{ik} for different domains of δ: $\delta > \delta_*$, $\delta_c < \delta < \delta_*$, $\delta < \delta_c$. The qualitative relationships governing the changes in the behavior of the system are reflected in Fig.2 in the form of families of heteroclines, which are the solutions of equation

$$A\left(1 - \frac{\delta}{\delta_*}\right)p - Bp^3 + C\left(1 - \frac{\delta}{\delta_c}\right)p^5 - D\sigma - \frac{\partial}{\partial x_l}\left(\chi\frac{\partial p}{\partial x_l}\right) = 0. \quad (17)$$

In the region $\delta > \delta_*$ this equation is of the elliptic type with periodic solutions with spatial scale Λ and possesses p anisotropy determined mainly by the applied stress. This distribution of p gives rise to weak pulsations of the strain field. As $\delta \to \delta_*$ the solution of Eqn.13 passes the separatrix S_2, and the periodic solution transforms into a solitary-wave solution. This transition is accompanied by divergence of the inner scale Λ: $\Lambda \approx -\ln(\delta - \delta_*)$. In this case the solution has the form $p(\zeta) = p(x - Vt)$. The wave amplitude, velocity and the width of the wave front are determined by the parameters of non-equilibrium (orientation) transition:

$$p = \frac{1}{2}p_a\left[1 - \tanh\left(\zeta l^{-1}\right)\right], \quad l = \frac{4}{p_a}\left(2\frac{\chi}{A}\right)^{\frac{1}{2}} \quad (18)$$

The velocity of solitary wave is $V = \chi A(p_a - p_m)/2\zeta^2$, where $p_a - p_m$ is the jump in p in the course of an orientational transition. A transition through the bifurcation point δ_c (separatrix S_3) is accompanied by the appearance of spatio-temporal structures of a qualitatively new type characterized by explosive accumulation of defects as $t \to t_c$ in the spectrum of spatial scales (blow-up regime) [9]. In this case the kinetics of p is determined by the difference of the power of the terms in the expansion (12).

It is shown in [10] that for this type of the equations the developed stage of kinetics of p in the limit $t \to t_c$ can be described by a self-similar solution

$$p(x,t) = \psi(t)f(\zeta), \quad \zeta = \frac{x}{L_c}, \quad \psi(t) = \Psi_0\left(1 - \frac{t}{t_c}\right)^{-m} \quad (19)$$

354

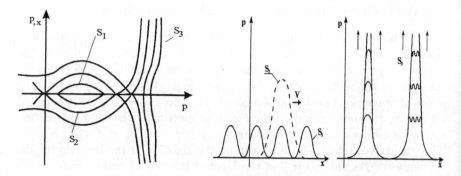

Figure 3. Types of heteroclines and the corresponding characteristic forms

where $m > 0$, $\Psi_0 > 0$ are the parameters related to the nonlinear term of Eqn.14; L_c and t_c are the scaling parameters. The function $f(\zeta)$ is determined by solving the corresponding eigenvalue problem. The self-similar solution (18) describes the blow-up damage kinetics for $t \to t_c$ on the set of spatial scales $L_H = kL_c$, $k = 1, 2, \ldots K$. The blow-up kinetics of damage localization allowed us to link the hotspots of failure with the above mentioned self-similar structures.

6. Collective Behavior of Cracks and Defects

6.1. INTRODUCTORY REMARKS

The interaction of the main crack with the ensemble of the defects is the subject of intensive experimental and theoretical studies that revealed some unresolved puzzles in the quasi-brittle failure. The rebirth of interest in the issue of dynamic fracture is observed during last decade due to the variety of new experimental results which are not explainable within the prediction of classical fracture mechanics, where it was shown that the crack in infinite plane specimen has two steady-state velocities: zero and the Rayleigh speed [11]. The recent experimental study revealed the limiting steady state crack velocity, a dynamical instability to micro-branching [12], the formation of non-smooth fracture surface [13], and the sudden variation of fracture energy (dissipative losses) with a crack velocity [14]. This renewed interest was the motivation to study the interaction of defects at the crack tip area (process zone) with a moving crack.

6.2. ORIGIN OF CRACK INSTABILITY

The classic theory of fracture treats a cracks as mathematical branch cuts which begin to move when an infinitesimal extension of the crack releases more energy then it is needed to create fracture surface. This idea is successful in some cases in practice but conceptually incomplete. The experiments fail to confirm this idealized picture. The surface created by the crack is not necessarily smooth and flat. In a series of experiments on the brittle fracture the simultaneous propagation of an ensemble of microcracks, instead of a single propagating cracks, was observed [15]. The fracture process was viewed as a coalescence of defects situated in the crack path; the mean acceleration drops, the crack velocity develops oscillations and a structure is formed on the fracture surface [11-15]. As the branches grow in size, they evolve into macroscopic large scale crack branches. A theoretical explanation of the limited steady-state crack velocity and the transition to branching regime was proposed in [16-18] due to the study of collective behavior of the microcrack ensemble in the process zone. It was shown by the solution of evolution equation for the defect density tensor that the kinetics of microcracks accumulation at the final damage stage includes the generation of spatial-temporal structures (dissipative structures with blow-up damage kinetics) that is the precursor of the nucleation of the "daughter" cracks. The kinetics of the daughter crack generation is determined by two parameters, which are given by the self-similar solution (18). The parameters L_c and t_c are the spatial scales of the blow-up damage localization and the so-called "peak time", which is the time of damage localization in the self-similar blow-up regime. The velocity limit V_c of the transition from the steady-state to the irregular crack propagation is given by the ratio: $V_c \approx L_c/t_c$. The set of spatial scales $L_H = L_c k$ (daughter crack sizes) represents new set of independent coordinates (collective modes of the defect ensemble) of the nonlinear system for $p > p_c$ ($\sigma > \sigma_c$). These coordinates characterize the property of a second attractor that could subject the behavior of the nonlinear system. The first attractor corresponds to the well-known self-similar solution for the stress distribution at the crack tip that is the background for the stress intensity factor conception. This solution is available in the presence of the metastability (local minimum) for p in the range $\sigma < \sigma_c$. The steady-state crack propagation is realized in the case when the stress rise in the process zone provides the failure time $t_f > t_c = L_c/V_c$ for the creation of the daughter crack only in the straight crack path. The failure time t_f follows from the kinetic equation (14) and represents the sum of the induction time t_i (the time of the approaching of the defect distribution to the self-similar profile on the $L_H = kL_c$ scales) and the peak time t_c: $t_f = t_i + t_c$. For the velocity $V < V_c$ the induction time $t_i >> t_c$ and

356

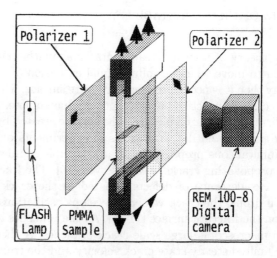

Figure 4. Scheme of experiment

the daughter crack appears only along the initial main crack orientation. For the crack velocity $V \approx V_c$ there is a transient regime $(t_i \approx t_c)$ of the creation of number of the localization scales (daughter cracks) in the main crack path. The crack velocity growth in the area $V > V_c$ leads to the sharp decrease of the induction time $t_i \to 0$, $t_f \to t_c$ that is accompanied by the extension of the process zone in both (tangent and longitudinal) directions where the multiple blow-up structures (daughter cracks) and, as the consequence, the main crack branching appears.

6.3. EXPERIMENTAL STUDY OF NONLINEAR CRACK DYNAMICS

6.3.1. *Experimental setup*

Direct experimental study of crack dynamics in the preloaded PMMA plane specimen was carried out with the usage of a high speed digital camera *Remix REM 100-8* (time lag between pictures $10\,\mu s$) coupled with photo-elasticity method, Fig. 4 [17,19].

The pictures of stress distribution at the crack tip is shown in Fig.5 for slow $(V < V_c)$ and fast $(V > V_c)$ cracks. The experiment revealed that the path of the critical velocity V_c is accompanied by the appearance of a stress wave pattern produced by the daughter crack growth in the process zone. Independent estimation of critical velocity from the direct measurement of crack tip coordinates and from pronounced stress wave Doppler pattern gives a correspondence with the Fineberg data $(V_c \approx 0.4V_R)$ [12].

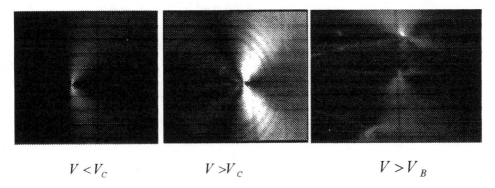

$V < V_C$ $V > V_C$ $V > V_B$

Figure 5. Different regimes of crack dynamics

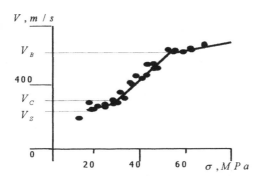

Figure 6. Crack velocity versus stress

6.3.2. *Characteristic crack velocity*

The dependence of crack velocity on the initial stress is represented in Fig.6.

Three portions with different slopes can be shown. The existence of these portions determines three characteristic velocities: the velocity of the transition from the steady-state to the non-monotonic straight regime $V_s \approx 220\,m/s$, the transient velocity to the branching regime $V_c \approx 330\,m/s$ and

358

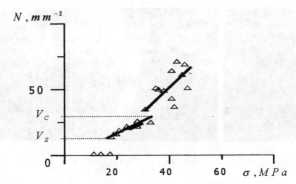

Figure 7. Mirror zone concentration versus stress

the velocity $V_B \approx 600\,m/s$ when the branches behave autonomous. The characteristic velocity $V_c \approx 330\,m/s$ allowed us to estimate the peak time t_c to measure the size of the mirror zone $L_c \approx 0.3\,mm$: $t_c = L_c/V_c \approx 10^{-6}\,s$. This result allowed also the explanation of the linear dependence of the branch length on the crack velocity [15]. Actually, since the failure time for $V > V_c$ is approximately constant $(t_f \approx t_c \approx 1\,\mu s)$, there is a unique way to increase the crack velocity to extend the size of the process zone. The crack velocity V is linked with the size of the process zone L_{pz} by the ratio $V = L_{pz}/t_c$. In our experiments the dependence of the density of the localized damage zone on the stress was observed (Fig.7). Since the branch length is limited by the size of the process zone, we obtain the linear dependence of branch length on the crack velocity. This fact explains the sharp dependence (quadratic law) of the energy dissipation on the crack velocity [15].

7. Resonance Excitation of Failure

7.1. DELAYED FAILURE PHENOMENON. FAILURE WAVES

Rasorenov et al [20] were the first to observe the phenomenon of delayed failure behind the elastic wave in glass. The existence of failure wave was established by considering a small recompression signal in the VISAR record of the free surface velocity of K19 glass (similar to soda-lime). This recompression signal resulted from a release returning after reflection of the shock at the glass rear surface reflecting again in compression at the lower

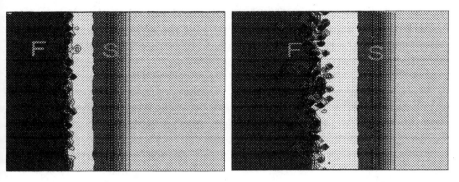

Figure 8. Propagation of stress (S) and failure (F) waves

impedance failure front. Recent studies have suggested that a wave of failure propagates behind the elastic wave in glass with a velocity in the range 1.5-2.5 km/s. Failure wave appeared in shocked brittle materials (glasses, ceramics) as a particular failure mode in which they lose strength behind a propagating front [21-23]. The description of failure wave phenomenon as the consequence of the generation of collective burst modes of mesodefects was proposed in [24,25] in the course of study of non-equilibrium transition in defect ensemble. It was shown the existence of the self-similar solution for the microshear density tensor, which describes the qualitative changes in the microshear density kinetics in the course of the non-equilibrium transition. The failure waves represent the specific dissipative structures (the "blow-up" dissipative structures [26,27]) in the microshear ensemble that could be excited due to the pass of the elastic wave.

7.2. SIMULATION OF FAILURE WAVES

The study of failure wave initiation and propagation was carried out on the basis of constitutive equations (14), (15) coupled with momentum transfer equation [28]. The defect density tensor in the compression stress wave represents the microshear density.

The system of equations was solved using original finite element code. The initial flaws in materials were introduced as the random field of the kinetic coefficient Γ in (14). The simulation confirmed the delayed propagation of the failure front behind the stress wave and the finger-like structure of the failure wave front, Fig.8. We observed also that the shear stress vanishes in the failure wave. The failure wave propagation leads to qualitative change in the transverse stress when this stress is approaching to the longitudinal stress level [28].

8. Acknowledgement

This work was supported in part by the Russian Foundation of Basic Research (grant 02-01-00736).

References

1. Hansen, N. and Kuhlmann-Wilsdorf, D., Materials Science and Engineering, 81, 141 (1986).
2. Betechtin, V.I. and Vladimirov, V.I., In: Problems of Strength and Plasticity of Solids, Zhurkov, S.N. (Ed.). Nauka, Leningrad, 142 (1979).
3. Betechtin, V.I, Naimark, O.B. and Silbershmidt, V.V., In: Proceedings of Int. Conf. of Fracture (ICF 7), 6, 38 (1989).
4. Barenblatt, G.I. and Botvina, L.R., Izv. AN. SSSR, Mech. Tv. Tela, 4, 161 (1983), (in Russian).
5. Naimark, O.B. and Silbershmidt, V.V., Eur.J.Mech., A/Solids, 10, 607 (1991).
6. Leontovich, M.A., Introduction to Thermodynamics, Statistical Physics, Chapter 3, Nauka, Moscow (1983).
7. Naimark, O.B., In:Proceedings of the IUTAM Symposium on Nonlinear Analysis of Fracture, J.R.Willis (Ed). Kluver Academic Publishers, Dordrecht, 285-298 (1997).
8. Naimark, O.B., JETP Letters, 67, 9, 751 (1998).
9. Kurdjumov, S.P., In: Dissipative Structures and Chaos in Non-Linear Space, Utopia, Singapure, 1, 431 (1988).
10. Naimark, O.B., Davydova, M.M. and Plekhov, O.A., In: Proceedings of NATO Workshop Probamat 21 Century, G.Frantziskonis (Ed.). Kluwer, 127 (1998).
11. Freund, L.B., Dynamic Fracture Mechanics, Cambridge University Press, Cambridge, England (1990).
12. Fineberg, J., Gross, S.P. and Sharon, E., In: Proceedings of the IUTAM Symposium on Nonlinear Analysis of Fracture, J.R.Willis (Ed). Kluver Academic Publishers, Dordrecht, 177 (1997).
13. Sharon, E., Gross, S.P. and Fineberg, J., Phys.Rev.Lett., 74, 5096 (1995).
14. Boudet, J.F., Ciliberto, S. and Steinberg, V., J. de Physique, 6, 1493 (1993).
15. Sharon, E., Gross, S.P., Fineberg, F., Phys.Rev.Lett , 76, 2117 (1996) .
16. Naimark, O.B., Davydova, M.M. and Plekhov, O.A., In: Proceedings of NATO Workshop Probamat 21 Century, G.Frantziskonis (Ed.). Kluwer, 127 (1998).
17. Naimark, O.B., Davydova, M.M., Plekhov O.A. and Uvarov, S.V., Physical Mesomechanics, 2, 3, 47 (1999).
18. Naimark, O.B., Davydova, M.M. and Plekhov, O.A., Computers and Structures, 76, 67 (2000).
19. Naimark, O.B., In: Proceedings of EUROMAT 2000. Advances in Mechanical Behavior. Plasticity and Damage (plenary lecture), D Miannay, P.Costa, D. Francois, A.Pineau (Eds.), Elsevier, 1, pp.15-28 (2000).
20. Rasorenov, S.V., Kanel, G.J., Fortov V.E. and Abasenov, M.M., High Press. Res., 6, 225 (1991).
21. Brar, N.K. and Bless, S.J., High Press.Res., 10, 773 (1992).
22. Bourne, N., Rosenberg, Z., Field, J.E., J. Appl. Phys. 78, 3736 (1995).
23. Bourne N.K., Rosenberg, Z., Field, J.E. and Crouch, I.G., J. Physique IV, Colloq.C

8, 635(1994).

24. Naimark, O.B., In: Proceedings of IX Int. Conference of Fracture, Sydney (key-note lecture), B. Karihaloo (Ed.), 6, 2795 (1997).

25. Naimark, O.B., Collombet, F. and Lataillade, J.-L., J. Physique IV Colloq. C. 7, 773 (1998).

26. Beljaev, V.V.and Naimark O.B., Sov. Phys. Doklady., 312, 2, 289 (1990).

27. Bellendir, E., Beljaev, V.V. and Naimark, O.B., Sov. Tech. Phys. Lett., 15, 3, 90 (1989).

28. Plekhov, O.A., Eremeev, D.N. and Naimark, O.B., J. Physique IV Colloq C. 10, 811 (2000).

FIELD THEORY OF CRYSTAL DEFECT STRUCTURE

A.I. OLEMSKOI AND A.V. KHOMENKO

Sumy State University
2, Rimskii-Korsakov St., 40007 Sumy, Ukraine

Abstract. A description of single defects is carried out within the gauge theory where the compensating field represents the elastic shear and the material field, being a complex order parameter, describes long–range correlations in variation of potential relief of structural units (atoms). Dislocation as a localized material field of the shear and disclination as a localized rotation field are represented on the basis of simple variant of U(1) theory. A description of a point defect leads to non Abelian SU(2) theory with components of gauge field being relevant to different polarizations of elastic wave.

1. Introduction

At present time the geometrical models of crystal defects based on obvious or topological considerations are worked out and generally accepted [1, 2]. They allow the continual description of elastic fields created by these defects [2]. Nevertheless, these models do not describe the real structure of defects kernels. For example, there is no explanation for anomalous high values of a diffusion coefficient along dislocation tubes.

Thus, the problem appears, within the framework of which the material field accounting the atomic structure of defects should be described and a force field created by these defects has to be reproduced. It turns out that the basis for representation of the former is the conception of rearrangeable potential relief [3]. It allows self–consistent picture representing defect structure within field approach.

A. Finel et al. (eds.), Thermodynamics, Microstructures and Plasticity, 363–373.

2. Field Theory of Crystal Defects

Let us consider the spatial distribution of a field that characterizes the structure level of defects at steady state. For this purpose, it is necessary to introduce a field $U(\mathbf{x})$ conjugated to a material field $\rho(\mathbf{x})$ that characterizes distribution of structural units on the initial (atom) level. The later field realizes a basis of irreducible representations of initial group of the system symmetry. The distribution of the material field specifies thermodynamic potential $\Phi\{\rho(\mathbf{x})\}$ whose variation gives conjugate field [3]

$$U(\mathbf{x}) = \frac{\delta\Phi(\mathbf{x})}{\delta\rho(\mathbf{x})} \tag{1}$$

being determined analogously to energy of elementary excitations of many–particle system [4]. Physically, this field represents the potential relief of structural units on the initial level (for example, the field $U(\mathbf{x})$ gives the Paierls relief for dislocation distribution [5]).

A system deviation from equilibrium results in excitation of ensemble of structural units, which is displayed in a smearing of the potential relief. Related probability of steady state distribution $U(\mathbf{x})$ is determined by Boltzmann–like form

$$P\{U(\mathbf{x})\} \propto \exp\left(-\frac{V\{U(\mathbf{x})\}}{\Theta}\right) \tag{2}$$

which is given by the corresponding functional of the synergetic potential $V\{U(\mathbf{x})\}$ and excitation intensity Θ [3]. By definition, the functional order parameter that characterizes new defect level is given by long-range correlations of variation $\delta U(\mathbf{x}') = U(\mathbf{x}') - \langle U(\mathbf{x}')\rangle$ with respect to a deviation $\delta U(\mathbf{x}) = U(\mathbf{x}) - \langle U(\mathbf{x})\rangle$ fixed at $\mathbf{x} = \mathbf{const}$:

$$\lim_{|\mathbf{x}'-\mathbf{x}|\to\infty} \frac{\langle \delta U(\mathbf{x})\, \delta U(\mathbf{x}')\rangle}{\langle|\delta U(\mathbf{x})|^2\rangle} \equiv \varepsilon^2(\mathbf{x}) \tag{3}$$

where the averaging is performed over distribution (2). Here, the change in characteristic length on new level is displayed as follows: if the potential relief $U(\mathbf{x})$ oscillates over small distances $x \sim a$ being microscopic, then macroscopic order parameter $\varepsilon(\mathbf{x})$ varies at much more distances $x \geq \xi$ which scale $\xi \gg a$ corresponds to new level. To take into account such scaling, it is convenient to introduce complex field of order parameter

$$\epsilon(\mathbf{x}) = \varepsilon(\mathbf{x})e^{i\varphi(\mathbf{x})} \tag{4}$$

where variation of the phase $\varphi(\mathbf{x})$ is observed at microscopic distances $x \sim a$. Then, transition from initial scale a to new one ξ implies spontaneous breaking conformal invariance being realized below within the framework of a gauge field scheme.

The simplest case of a scalar parameter (4) is characterized by the symmetry group $U(1)$ that makes *external* four–dimensional coordinate x_μ, $\mu = 0, 1, 2, 3$ ambiguous for new level (here $x_0 = ct$, c is velocity of transversal sound) [3]. This ambiguity reflects the appearance of defects and can be compensated by prolongation of the derivative:

$$\partial^\mu \equiv \frac{\partial}{\partial x_\mu} \Rightarrow \nabla^\mu \equiv \partial^\mu + \Gamma^\mu, \quad \mu = 0, 1, 2, 3 \tag{5}$$

where Γ^μ is the 4–potential of the corresponding gauge field. From geometrical point of view Γ^μ is the connectivity in the stratification, which is a nontrivial generalization of the direct product of a manifold of initial structural units and gauge group $U(1)$ [6]. To study new structure level, it is convenient to use associated stratification, considering at each point x^μ not the transformations of the gauge group, but the corresponding potential Γ^μ whose value is associated with 4–vector of displacements $A^\mu \equiv (\varphi, \mathbf{u})$, $\mu = 0, 1, 2, 3$ by the equality $\Gamma^\mu = -ig A^\mu$ where g is elastic charge [3].

Within four–dimensional representation, we consider the simplest Ginzburg–Landau scheme whose material Lagrangian

$$\mathcal{L}_m \equiv L_m - L_c \tag{6}$$

is determined by usual term

$$L_m = \frac{\beta}{2} |\nabla^\mu \epsilon|^2 - V(\epsilon) \tag{7}$$

with gradient constant $\beta > 0$ and synergetic potential $V(\epsilon)$. The diminution L_c takes into account a specific integral condition for field distribution being relevant to a defect:

$$\oint \Gamma_\mu dx^\mu = w \oint d\varphi = 2\pi w n, \quad n = 0, \pm 1, \ldots \tag{8}$$

According to this condition, defects play the role of elementary carriers of gauge fields to make a manifold of defect crystal multiply connected. It is easily to see that the constraint (8) is a result of spontaneous breaking of conformal invariance related to the Lagrangian

$$L_c = \nu_\mu \left(\nabla^\mu \varepsilon - w \varepsilon \partial^\mu \varphi \right) \tag{9}$$

with respect to variation of Lagrange multiplier ν_μ. Actually, relevant differential constraint reads:

$$(\partial^\mu + \Gamma^\mu)\varepsilon = w \varepsilon \partial^\mu \varphi. \tag{10}$$

Since the external coordinate x^μ corresponds to a change in the phase $\varphi(x^\mu)$ at distances $x^\mu \sim a$ while the internal coordinate should describe a variation of the amplitude $\epsilon(x^\mu)$ at $x^\mu \geq \xi$, one obtains the condition $\xi = a/w$ which fixes the scale ξ by the assignment of the parameter $w \ll 1$ (a is given *a priori*). On the other hand, the gauge choice

$$\Gamma^\mu = \partial^\mu(-\ln\varepsilon + w\varphi), \tag{11}$$

reducing Eq. (10) to identity, gets many–valued field Γ^μ obeying to the constraint (8) needed.

The equation for material field following from Eq. (6) takes the form

$$\partial^\mu\partial_\mu\epsilon = \Gamma^\mu\Gamma^*_\mu\epsilon + \beta^{-1}\frac{\partial V}{\partial\epsilon^*} \tag{12}$$

where the gauge condition $\partial^\mu\Gamma_\mu = 0$ has been taken into account. Here, the order parameter ϵ varies at correlation length ξ, whereas the 4–displacement $A^\mu \equiv (\varphi, \mathbf{u})$ related to the gauge field $\Gamma^\mu = -igA^\mu$ does at microscopic distance a. Then, elastic charge g and correlation length ξ are determined by equalities

$$g = \frac{1}{\xi a}; \qquad \xi^2 = \frac{\beta}{|A|}, \quad A \equiv \left.\frac{\partial^2 V}{\partial\varepsilon^2}\right|_{\varepsilon=0} \tag{13}$$

and the scale ratio $w \equiv a/\xi \ll 1$ takes the magnitude

$$w = ga^2. \tag{14}$$

3. Description of Disclination and Dislocation

Within the framework of the field approach developed, let linear defects be represented as autolocalized regions possessing rearranged potential relief. A defect kernel is described by distribution $\epsilon(\mathbf{r})$ of a relief rearrangement parameter over spatial components of radius–vector $\mathbf{r}$. A conjugated field is specified by an elastic component of a 4-vector of static displacement $A^\mu_e = (\varphi_e, \mathbf{u}_e)$ reduced to 4-potential of a gauge field. Corresponding strengths

$$\chi_e = -\frac{\partial\mathbf{u}_e}{c\partial t} - \mathrm{grad}\varphi_e, \quad \omega_e = \mathrm{curl}\,\mathbf{u}_e \tag{15}$$

represent elastic components of shift and rotation vectors. The material component $A^\mu_m = (\varphi_m, \mathbf{u}_m)$ is related to coherent displacement of minima and smearing of a potential relief. The gauge symmetry is connected with invariance of medium characteristics relative to translation and rotation of a specimen as a whole. A charge (13) is determined by inherent atomic distance a and correlation length ξ to be characteristics of the material field. Moreover, there is the third scale

$$\lambda = \frac{\nu}{c}; \quad \nu \equiv \frac{\eta}{\rho}, \quad c^2 \equiv \frac{\mu}{\rho} \tag{16}$$

to determine a length of elastic field smearing (here ρ, η and μ are density, shear viscosity and related elastic modulus of medium, respectively). The system behavior is determined by value of Landau–Ginzburg parameter

$$\kappa = \frac{\lambda}{\xi} \equiv \frac{\eta}{\xi\sqrt{\rho\mu}}. \tag{17}$$

In weakly excited state the crystal possesses so large shear viscosity η that one has the values $\kappa \gg 1$. Under such a condition distributions of material and elastic fields have soliton–like form type of Gross–Pitayevskii soliton [3]. Here, rotation field $\omega(\mathbf{r})$ corresponds to disclination whereas shear field $\chi(\mathbf{r})$ does to dislocation. Let study conditions for realization of the first type solutions.

In real crystal there are stress concentrators always. Let one of them creates a homogeneous rotation field ω_{ext} over distances $x \leq \lambda$. If, in addition, the local rearrangement of atomic system takes place over distances $x \leq \xi$, at critical value $\omega_{\text{ext}} = \omega_c$ the situation is realized when variation of thermodynamic potential

$$\Phi = \mathcal{L}_m + L_f \tag{18}$$

becomes decreasing. Here, material component $\mathcal{L}_m$ is given by Eqs. (6), (7), (9) with synergetic potential

$$V = \frac{A}{2}|\epsilon|^2 + \frac{B}{4}|\epsilon|^4 \tag{19}$$

where A and B are material constants. The field addition is specified as

$$L_f = \frac{1}{4}\left(\partial^\mu A^\nu - \partial^\nu A^\mu\right)^2 - \frac{1}{c}A^\mu j_\mu \tag{20}$$

where $A^\mu(\mathbf{r}, t)$ is a 4-potential of the gauge field, c is sound velocity, $j^\mu = (\rho c, \mathbf{j})$ is a 4–current.

Using expressions (6), (7), (9), (18) — (20), it is easy to show that a condition $\Delta\Phi = 0$ for variation of thermodynamic potential is realized if external field is reduced to critical one ω_c defined by relation

$$\omega_c^2 = \frac{1}{2}|a|\varepsilon_0^2 \tag{21}$$

where ε_0 is the order parameter at steady state. The variation of thermodynamic potential connected with creation of new phase

$$\Delta\Phi \equiv \int [\Phi(\epsilon) - \Phi(\epsilon = 0)]\, d\mathbf{r} \qquad (22)$$

takes the explicit form [4]

$$\Delta\Phi = \rho c^2 \omega_c^2 \int \left[-\left(1 - \mathbf{u}_e^2\right)\varepsilon^2 + \frac{1}{2}\varepsilon^4 + \kappa^{-2}\left(\frac{d\varepsilon}{d\mathbf{r}}\right)^2 + \left(\frac{1}{\sqrt{2}} - \operatorname{curl}\mathbf{u}_e\right)^2 \right] d\mathbf{r}. \qquad (23)$$

Here $\mathbf{u}_e$ is elastic displacement, the coordinate $\mathbf{r}$ is measured in units of the length λ. Then, static Euler equations

$$\kappa^{-2}\nabla^2\varepsilon = -(1 - \mathbf{u}_e^2)\varepsilon + |\epsilon|^2\varepsilon, \qquad (24)$$

$$-\operatorname{curl}\operatorname{curl}\mathbf{u}_e = |\epsilon|^2\mathbf{u}_e \qquad (25)$$

and variation of thermodynamic potential (23) are equivalent to corresponding equations of Ginzburg – Landau – Abrikosov theory for a vortical (mixed) state of a superconductor [4] where an elastic component $\mathbf{u}_e$ of displacement vector is meant as vector potential and a rotation vector $\omega_e = \operatorname{curl}\mathbf{u}_e$ is understood as magnetic field. Having in mind this analogy, let the main results of theory [4] apply to the case under consideration.

The mixed state is realized provided $\kappa > 1/\sqrt{2}$ within the interval of rotation field $\omega_{c1} < \omega_{ext} < \omega_{c2}$ where

$$\omega_{c1} = \frac{\ln\kappa}{\sqrt{2}}\omega_c, \quad \omega_{c2} = \sqrt{2}\kappa\omega_c. \qquad (26)$$

Near ω_{c2} the closed packed lattice of vortical threads is formed: at $\omega_{ext} = \omega_{c2}$ threads density N per unit of area is maximum and amounts to value $N_{max} = 1/\pi\xi^2$; at ω_{ext} decreasing in a region $\omega_{c2} - \omega_{ext} \ll \omega_{c2}$ it varies according to the equality

$$\frac{N}{N_{max}} = \frac{\omega_{ext}}{\kappa} - \frac{\overline{\varepsilon^2}}{2\kappa^2} \qquad (27)$$

where average over volume $\overline{\varepsilon^2}$ is connected with a rotation vector magnitude ω_{ext} by equality

$$\overline{\varepsilon^2} = \frac{2\kappa}{\beta(2\kappa^2 - 1)}(\kappa - \omega_{ext}), \quad \beta \equiv \overline{\varepsilon^4}/(\overline{\varepsilon^2})^2 = 1.1596. \qquad (28)$$

The average value

$$\overline{\omega_e} = \omega_{ext} - \overline{\varepsilon^2}/2\kappa = \omega_{ext} - (\kappa - \omega_{ext})/\beta(2\kappa^2 - 1) \qquad (29)$$

is smaller than applied field in a quantity being equal to the average of a medium polarization

$$\overline{\omega_m} = -\overline{\varepsilon^2}/2\kappa = -(\kappa - \omega_{ext})/\beta(2\kappa^2 - 1). \qquad (30)$$

The maximum value of a rotation vector field is reached in threads cores, and the minimum one $\omega_{min} = \omega_{ext} - \sqrt{2}(\kappa - \omega_{ext})/(2\kappa^2 - 1)$ — in the centers of triangles formed by threads. The average variation of thermodynamic potential (18) caused by medium rotation

$$\overline{\Delta\Phi} = \rho c^2 \omega_c^2 \left(\frac{1}{2} + \overline{\omega_e^2} - \frac{\overline{\varepsilon^4}}{2} \right) = \rho c^2 \omega_c^2 \left[\frac{1}{2} + \overline{\omega_e^2} - \frac{(\kappa - \overline{\omega_e})^2}{1 + \beta(2\kappa^2 - 1)} \right] \qquad (31)$$

is the function of the average turning $\overline{\omega_e}$ differentiation with respect to which results in Eq. (29).

Near the lower critical value ω_{c1} threads density $N = (\kappa/2\pi)\overline{\omega_e}$ is not large and they can be treated independently. Taking into account that $u_e(r)$ varies at distances $r \sim 1$ and $\varepsilon(r)$ does at $r \sim \kappa^{-1}$, the displacements are determined by Eq. (25) with $|\epsilon|^2 \approx 1$ and $\kappa \gg 1$:

$$u_e = -\kappa^{-1} K_1(r) \qquad (32)$$

where $K_1(\mathbf{r})$ is the Hankel function of imaginary argument. Respectively, the order parameter is determined by Eq. (24) with $u_e = 1/\kappa r$:

$$\begin{aligned} \varepsilon &\simeq c_1 r \quad \text{at} \quad r \ll \kappa^{-1}, \\ \varepsilon^2 &\simeq 1 - (\kappa r)^{-2} \quad \text{at} \quad r \gg \kappa^{-1} \end{aligned} \qquad (33)$$

where c_1 is positive constant. According to Eq. (32) one has $u_e \approx -1/\kappa r$ at $r \ll 1$ and $u_e \approx -\sqrt{\pi/2\kappa^2}\, r^{-1/2} e^{-r}$ at $r \gg 1$. The dependence $\overline{\omega_e}(\omega_{ext})$ is of steadily increasing nature: at $\omega_{ext} = \omega_{c1}$ it has the vertical tangent and with ω_{ext} growth it asymptotically approaches to the straight line $\overline{\omega_e} = \omega_{ext}$. Thermodynamic potential per thread length unit is $(2\pi/\kappa^2) \ln \kappa$. The ω_e value in a thread center is twice as large as ω_{c1}.

The described system of vortical threads related to rotation field ω_{ext} corresponds to periodical distribution of rectilinear disclinations. In the same way, it can be shown that the mixed state formed by threads in a shear field χ_{ext} is possible as well to be corresponded to a system of rectilinear dislocations. To prove the above, it is necessary to determine the law of strength field decrease near the threads. Within the framework of the field scheme the function of this field components is performed by material components of strength vectors: ω_m — for disclination and χ_m — for dislocation. Thus, it is necessary to reestablish dependencies $\omega_m(r)$,

$\chi_m(r)$ by the obtained dependence $\mathbf{u}_e(r)$. In order to do this, let the elastic part of current be written down:

$$\mathbf{j}_e = -\beta g^2 c |\epsilon|^2 \mathbf{u}_e. \tag{34}$$

As it is evident from the motion equation written in the form

$$\triangle \mathbf{u} = \text{curl } \omega_m + \frac{\partial \chi_m}{c \partial t} + \frac{\mathbf{j}}{c}, \tag{35}$$

the material fields ω_m, χ_m relevant to defects causes the current

$$\mathbf{j}_m = c \text{ curl } \omega_m + \frac{\partial \chi_m}{\partial t}. \tag{36}$$

At steady state one has $\mathbf{j}_e + \mathbf{j}_m = 0$ and for the case of simple rotation ($\chi_m = 0$) within actual region $\kappa^{-1} < r < 1$ where $u_e \propto r^{-1}$ it follows from Eqs. (34), (36) that creation of a localized rotation thread results in variation of the external field ω_{ext} by value $\omega_m \propto \ln r$. Respectively, in a shear field one has $\chi_m \propto r^{-1}$. Since it is in this way that the stress of the elastic field decreases near disclinations and dislocations [1], it can be concluded that the described mixed state represents an ensemble of crystal structure linear defects.

In real case the appearance of defects, being localized carriers of plastic deformation, is caused by stress concentrators providing the system transition into the region $\omega_{c1} < \omega_{\text{ext}} < \omega_{c2}$ (or $\chi_{c1} < \chi_{\text{ext}} < \chi_{c2}$) of mixed state even with lack of external load. The main condition of realization of this inequalities is utmost large values of the parameter κ ($1 \ll \ln \kappa \ll \kappa$) giving a very small lower critical field and the large upper one in accordance with Eqs. (26). Therefore, in a real crystal there is a stable system of lattice defects (with the exception of thread type crystals where in view of potential relief stiffness resulting in large values of a gradient parameter β the quantity κ is small).

The consideration carried out above shows that a linear defect represents an autolocalized formation corresponding to a small domain of rearranged potential relief $U(\mathbf{r})$ and to a considerable region of the elastic field distribution. The coordinate dependence of the corresponding fields is shown in Fig. 1a. Since relief $U(\mathbf{r})$ rearrangement in the region of a defect kernel is reduced to smearing into ensemble $\{U(\mathbf{r})\}$ and a field switching results in contribution of energy $-\chi_m(\mathbf{r})$ to the total potential, the potential relief picture near the defect has the form shown in Fig. 1b. It is smeared within a kernel region that causes increase of a tube diffusion coefficient due to considerably decrease of effective height of the barrier. The elastic field is expressed in variation of a reference level and quick oscillations of potential $U(\mathbf{r})$ at interatomic distances.

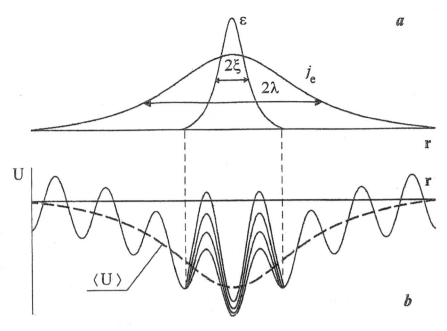

Figure 1. a: Coordinate dependence of material deformation field ε and elastic stress field j_e near a defect. b: Corresponding form of the potential relief

4. Non Abelian Field Theory of Point Defect

It seems at the first glance that the point defect, being the simplest one in geometrical form, have to be described by a simple variant of the field theory. Such insight is based on naive representation that the point defect is a result of shrinking three orthogonal dislocation loops [2]. But it needs to keep in mind that this model implies junction of dislocation kernels which arrives at a new type of the field singularity being fundamentally non-linear in nature. It turned out that, besides above solutions that belong to Abelian gauge groups, the SU(2)–symmetry leads to new solution related to topology change in distribution of plastic deformation being relevant to the point defect [3]. Let us consider main statements of corresponding field scheme.

In this case, the order parameter ϵ_a has three components numbered by polarizations $a = 1, 2, 3$ of related displacement waves. Moreover, two-component material field $\mathbf{\Psi}_a$ describes a structural unit distribution for each polarization a to realize a basis of the gauge group (the components correspond to excited and unexcited states — in accordance with dimensionality of generators $\hat{\tau}^\mu$ of the group SU(2) which are the Pauli matrices [6]). Material component of generic Lagrangian has the form [7]

$$L_m = i\beta^{1/2}\overline{\mathbf{\Psi}}_a\hat{\tau}^\mu\nabla_\mu^{ab}\mathbf{\Psi}_b + \frac{\beta}{2}\left|\nabla_\mu^{ab}\epsilon_b\right|^2 - \omega j^a\epsilon_a - \frac{B}{4}|\epsilon_a|^4 \qquad (37)$$

where β, ω and B are constants, $\nabla_\mu^{ab} = \partial_\mu\delta_{ab}+\varepsilon_{abc}\Gamma_\mu^c$, $j^a = i\varepsilon_{abc}\overline{\Psi}^b\Psi^c$ is the current of structural units, ε_{abc} is structural constant which is reduced to the antisymmetric tensor. Similar to all non Abelian models, the Lagrangian (37) results in both asymptotic freedom and confinement. These facts reflect the long–range order in distribution of structural units over excited and unexcited states.

Variation of the action corresponding to Eq. (37) with respect to the fields $\mathbf{\Psi}_a(x^\mu)$ results in equation of Weyl–type

$$(\hat{\tau}^\mu\partial_\mu\delta_{ab} - i\varepsilon_{abc}\hat{\tau}^\mu\Gamma_\mu^c + \omega\beta^{-1/2}\varepsilon_{abc}\hat{I}\epsilon^c)\mathbf{\Psi}^b = 0 \qquad (38)$$

where $\hat{I}$ is 2×2 unit matrix. It is typical that the term that contains the order parameter (the plastic deformation field $\epsilon^c(x^\mu)$) performs the function of the mass of bare "fermions" that are reduced to the structural units distributed over excited and unexcited states. Using Eq. (38) allows to eliminate the field $\mathbf{\Psi}_a(x^\mu)$ in the Lagrangian (37) to reduce the latter to the form (6) where potential $V(\epsilon^a)$ is given by Landau–type expansion with a minimum at the point $|\epsilon_0^a| = i\omega|\mathbf{\Psi}^a|$. This means that dispersion law (the mass) ω of bare "fermion" has to be of imaginary nature. As a result of exchange with Higgs bosons, corresponding to the field of plastic deformation, between the fermions, which are the structural units, the gauge symmetry is spontaneously broken. Longitudinal component ϵ^1 of the plastic deformation takes on a fixed value ϵ_0^1, and the two transversal components ϵ^2, ϵ^3 transform into Goldstone bosons of restoration of SU(2)–symmetry, i.e., they become the elastic components e^2 and e^3 of the strain field. Conversely, for the corresponding components Γ_μ^2 and Γ_μ^3 of the potential of the stress $\hat{\sigma}$, the dispersion law acquires a plastic character, while for the longitudinal component Γ_μ^1 it remains elastic.

References

1. Katsnelson, A.A. and Olemskoi, A.I. (1990) *Microscopic theory of inhomogeneous structures*, MIR Publishers, Moscow.

2. de Witt, R. (1977) *Continual Theory of Dislocations*, MIR Publishers, Moscow (Russian translation).
3. Olemskoi, A.I. (1999) *Theory of Structure Transformations in Non-equilibrium Condensed Matter*, NOVA Science, N.-Y.
4. Lifshitz, E.M. and Pitaevskii, L.P. (1978) *Statistical Physics*, Part 2, Nauka, Moscow (in Russian).
5. Hirth, J.P. and Lothe, J. (1968) *Theory of Dislocations*, McGraw–Hill, N.Y.
6. Dubrovin, B.A., Novikov, S.P., and Fomenko, A.T. (1979) *Modern Geometry*, Nauka, Moscow (in Russian).
7. Olemskoi, A.I. and Sklyar, I.A. (1992) *Sov.Phys.Uspekhi* **35**, 455-480.

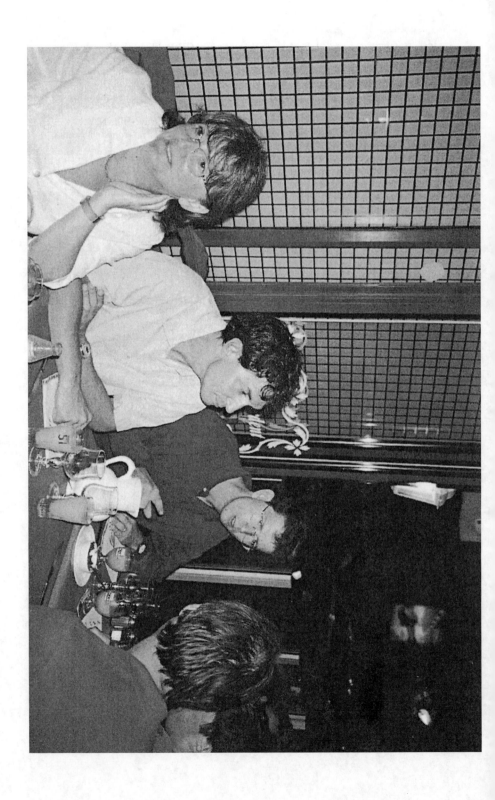

V. Kinetics and phase transformations at the atomic scale

MONTE-CARLO METHODS

V. PONTIKIS
Centre d'Etudes de Chimie Métallurgique, CNRS-UPR2801
15, rue Georges Urbain, F- 94407 Vitry-sur Seine Cedex, FRANCE

ABSTRACT. Numerical techniques for computing integrals by using pseudo-random numbers are generically referred to as Monte Carlo methods. After an elementary introduction of the basic concepts, explanations are given on how these techniques apply to the computation of statistical mechanical integrals, that is to say of thermodynamic quantities in various statistical mechanical ensembles. The implementation of atomistic Monte Carlo is considered with emphasis on pseudo-random number generators. Finally, we present the principles of few extended Monte Carlo techniques together with few examples taken from the literature.

1. Introduction

The complexity of real phenomena is partly due to the large number of degrees of freedom of the studied systems, the non-linear superimposition of the underlying elementary causes and/or the non-trivial initial and boundary conditions. These are among reasons that make the analytic formulation of complex phenomena difficult, if not impossible, and hinder understanding and prediction. Facing the complexity of the real world, numerical methods are often more successful than is analytic theory in de-convoluting experimental observations, making predictions and improving understanding. When the phenomenon under study is by essence probabilistic e.g. traffic jams, particle beam/target interactions, radioactive decay, its modeling is straightforward by using series of random numbers with probability density distributions adapted to each of the elementary constitutive causes. The same approach is also applicable for the study of deterministic phenomena, provided a stochastic process exists featuring the physical property of interest. «Experimental» mathematical approaches of this kind, used in modeling probabilistic or deterministic complex phenomena, are what we customarily call, since the early 1940's, the Monte Carlo methods [1].

Over half a century of use of Monte Carlo in different areas of science, has led to a large number of published articles and, periodically, to valuable reviews of methods, applications and results. A biased selection among these provides a possible starting point for the reader who is interested in historical aspects [1,2], in a tutorial introduction in the matter [1, 3] or in applications [4, 5, 6]. In the following, because we are principally interested in atomistic scale modeling of condensed matter systems, the Monte Carlo method is introduced via the example of the numerical computation of simple integrals by using pseudo-random numbers. Such numbers are generally produced by deterministic algorithms and often do not meet all the qualities of

A. Finel et al. (eds.), Thermodynamics, Microstructures and Plasticity, 377–392.

randomness. The consequences of such a drawback on the results of the Monte Carlo method are illustrated, in the case of correlated series of such numbers, with the objective of fixing the readers attention on this crucial ingredient of Monte Carlo methods. Computing thermodynamical properties of condensed matter is a difficult and complicated problem that perfectly highlights the benefits of using the numerical approach. This is why a presentation is further given of the principles of atomistic Monte Carlo with the Metropolis selection rule of configurations. Finally, a brief presentation is made of extensions of the Monte Carlo method to other thermodynamical ensembles and to transformations aiming to improve the computational performance. For illustration purposes, some examples from the literature are reproduced and commented on.

2. Monte Carlo methods

2.1 COMPUTING SIMPLE INTEGRALS

Through an appropriate scale transformation, any given definite integral can be put in the following simple form:

$$A = \int_0^1 f(x)dx \tag{1}$$

with, $0 \le f(x) \le 1$. The value of this integral i.e. the area delimited by the abscissa axis and the graph of f(x) (Fig. 1), can be computed by means of random numbers as follows: we first define three independent random variables, X, Y and Z, with the two first uniformly distributed over the interval, [0,1] and Z, given by:

$$Z = \begin{cases} 1 & if \quad Y \le f(X) \\ 0 & if \quad Y > f(X) \end{cases} \tag{2}$$

Z, is a Bernoulli random variable with values relating with the position of the random point (X,Y), below or above the graph of f(x). The probability, $P(Z=1)$, the average value of Z and its standard deviation, σ, are given by [7]:

$$\begin{cases} P(Z = 1) = P(Y \le f(X)) \\ \langle Z \rangle = A \\ \sigma^2 = A(1 - A) \end{cases} \tag{3}$$

Considering now N values Z_k, generated by N random couples (X_k, Y_k) and Eqs. (2), the following relation holds because of the law of large numbers (Khinchine's theorem) [7]:

$$\langle Z \rangle = \lim_{k \to \infty} \left\{ \frac{1}{k} \sum_{i=1}^k Z_i \right\} = A \tag{4}$$

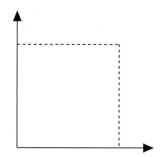

Figure 1. The hashed area represents $I = \int\limits_0^1 f(x)dx$.

Proceeding as indicated above, the problem of computing the integral of Eq. 1, is transformed to the problem of computing the average value of a random variable characteristic of a stochastic process.

The error committed in estimating the average value of Z, and thus of the integral, A, reflects not only rounding, because of the discrete representation of numbers in the computer, but also finite statistics. Consider a finite size estimate of <Z>, Z(N), made of N terms contributing in Eq. 4. The variance of Z(N), Σ^2, is given by [6]:

$$\Sigma^2 = \frac{A(1-A)}{N} \tag{5}$$

so that the dispersion of Z(N) with respect to the average value, <Z>=A, decreases as $1/\sqrt{N}$. In addition, Z(N), is a random variable with a normal distribution [6]. The probability that the also normally distributed reduced variable, R_N:

$$R_N = \sqrt{N} \frac{[Z(N)-A)]}{\sqrt{A(1-A)}} \tag{6}$$

be bounded in absolute value by λ, $|R_N| \le \lambda$, is given by [6]:

$$P[|R_N| \le \lambda] = \frac{2}{\sqrt{\pi}} \int\limits_0^\lambda \exp(-x^2)dx \tag{7}$$

With λ=1.821, this probability amounts P 99% for the integral being estimated with an error, ε:

$$\varepsilon = |Z(N)-A| \le \frac{2\sqrt{A(1-A)}}{\sqrt{N}} \tag{8}$$

Equation (8) highlights one major difficulty in estimating integrals by Monte Carlo, namely that the calculation converges very slowly with N, the number of evaluations of the integrand. However, it is worth noting that whatever is the complexity of this last, the expressions established above still remain very simple.

2.2 COMPUTING MULTIPLE INTEGRALS

Computing d-dimensional integrals goes along the procedure above sketched: N randomly distributed points are produced in the d-dimensional space by using the uniformly distributed random variables, $X_1, ..., X_n$. The integral is then estimated by:

$$Z(N) = \frac{1}{N} \sum_{k=1}^{N} f_k(X_1, ..., X_d) \tag{9}$$

while

$$\begin{cases} I = <Z> = \lim_{k \to \infty} \left(\frac{1}{k} \sum_{i=1}^{k} f_i(X_1^i, ..., X_d^i) \right) \\ \sigma_I^2 = \sigma_Z^2 = \frac{1}{N} \sigma^2 \left(f_i(X_1^i, ..., X_d^i) \right) = \frac{1}{N} \int ... \int (f - I)^2 \, dx_1 ... dx_d \end{cases} \tag{10}$$

with I and σ_I^2 the values of the integral and the error committed evaluating it, $<Z>$ and σ_Z^2, the average and the variance of Z and, N, the number of evaluations of the integrand. It is worth noting that the error does not depend on the dimension, d, of the integral. This is indeed the main interest in using Monte Carlo for the computation of multi-dimensional integrals, given that for a given error tolerance, the computational effort with traditional integration schemes increases extremely fast with d.

2.3 BIAS MONTE CARLO

The computational effort increases with increasing the size of the statistical sample, a compromise should be found between this last and the related to it error affecting the estimation of the integral. The difficulty in reaching this objective is often considerable, especially when the integrand takes significant values only within a reduced portion of its domain of definition. This is schematically illustrated in figure 2.

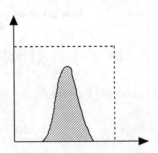

Figure 2. Bias Monte-Carlo decreases the variance.

This difficulty can be circumvented if, instead of the *uniform sampling* of the domain of definition of the integrand, *importance sampling* is performed such as to evaluate more frequently values which significantly contribute to the integral. To this end, a non-uniform probability distribution function should first be chosen, p(x), such as to mimic f(x) as close as possible. The integral is then transformed as follows:

$$I = \int_0^1 f(x)dx \equiv \int_0^1 \left[\frac{f(x)}{p(x)} \right] p(x)dx \tag{11}$$

where:

$$\int_0^1 p(x)dx = 1 \tag{12}$$

Sampling f/p from the distribution, p, leads to the desired estimate of I:

$$E(I) = E\left(\frac{f}{p}\right) = \frac{1}{N}\sum_{i=1}^{N} \frac{f(X_i)}{p(X_i)} \tag{13}$$

Importance sampling improves dramatically the efficiency of the Monte Carlo estimation even for the simplest one-dimensional integral. It is always required in order to get meaningful results in Monte Carlo computations of statistical mechanical integrals. Reducing variance in Monte Carlo calculations is a topic that has attracted much attention. A review of classical variance-reducing techniques can be found in [2].

Comparison of Monte Carlo integration with other available methods helps deciding which one should be adopted. The simple trapezoidal integration rule converges faster for one-dimensional integrals ($1/n^2$) than Monte Carlo ($1/\sqrt{n}$) [2]. However, with increasing dimensionality the efficiency of Monte Carlo does not change while the computational effort associated with classical quadrature methods increases as a power law of the dimensionality. For statistical mechanics integrals in 3N dimensions, where N is the number of particles, m^{3N} evaluations of the integrand should be made if m regularly spaced values of each coordinate had to be considered. This number diverges rapidly with N so that numerical quadrature becomes unfeasible well before the difference in uncertainty with Monte Carlo favors the use of this last [2].

3. Monte Carlo in statistical mechanics

3.1 THE CANONICAL ENSEMBLE

The canonical average of a configurational property, $X(r^N)$, of a system made of N point particles is given by:

$$\langle X \rangle = \int_\Omega X(r^N)p(r^N)dr^N = \int_\Omega X(r^N)\frac{\exp[-\beta U(r^N)]}{Z_N}dr^N \tag{14}$$

where, Ω, is the volume of the phase space, $U(\mathbf{r}^N)$, is the configuration energy, $\beta=1/k_b T$, with, k_b, the Boltzmann constant and Z, is the partition function:

$$Z_N = \int_\Omega \exp\left[-\beta U(r^N)\right] dr^N \tag{15}$$

Estimating $<X>$ by importance sampling could be made possible provided a probability distribution function, $Q(\mathbf{r}^N)$, can be found that closely resembles to the integrand. Eq. (14) is then transformed into:

$$\langle X \rangle = \frac{\sum_{i=1}^{n} X(r_i^N) \exp(-\beta U(r_i^N))/Q(r_i^N)}{\sum_{i=1}^{n} \exp(-\beta U(r_i^N))/Q(r_i^N)} \tag{16}$$

and, by choosing for $Q(\mathbf{r}^N)$ the normal distribution of states itself, the simple expression below is obtained:

$$\langle X \rangle = \frac{1}{n}\sum_{i=1}^{n} X(r_i^N) \tag{17}$$

However, only the shape, not the value of the equilibrium distribution function is known, since the partition function, Z_N, is unknown and therefore the direct application of the methods exposed in section 2 is not possible here.

This difficulty can be bypassed if the condition of random sampling of the integrand is relaxed and successive configurations from a *Markov chain* of states are used instead. Such a chain is made of states each of which is obtained from the immediate precedent state though independent of all others. It can be described by a one-step transition probability matrix, p_{ij}, such element of which represents the probability that the step next to i will be j. This is a stochastic matrix with $p_{ij}>0$ and $\sum_j p_{ij} =1$ which is said be *irreducible*, if any state can be reached from any other. Provided that a set of positive numbers, π_i, exists such as:

$$\sum_j \pi_j = 1 \quad (18a) \quad \text{and} \quad \sum_i \pi_i p_{ij} = \pi_j \tag{18b}$$

the frequency of occurrence of a state j in a *irreducible* Markov chain becomes proportional to π_j and the limiting distribution of the chain is $(\pi_1, \pi_2, \dots)$.

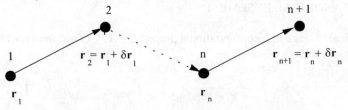

Figure 3. Markov chain of states of a n-particle system.

It is important to notice that Eq. (18b) is homogeneous in π_i, so that only ratios π_i/π_j, are needed in practice. Therefore ignoring the value of the partition function, Z_N, is not an obstacle anymore for sampling the canonical distribution and computing thermodynamical averages by Monte Carlo. More generally, any sampling distribution can be chosen provided Eq. (18b) is satisfied. A particular realization of the above is the Metropolis Monte Carlo method [8, 9].

Metropolis et al. [8], have made the choice, $\pi(\mathbf{r}^N) \propto Q(\mathbf{r}^N)$, where, Q, is the normal, equilibrium distribution of states and obtained the fulfillment of Eq. (18b), by imposing the micro-reversibility condition which postulates that every pair of states, i and j:

$$\pi_i p_{ij} = \pi_j p_{ji} \tag{19}$$

indeed:

$$\sum_i \pi_i p_{ij} = \sum_i \pi_j p_{ji} = \pi_j \sum_i p_{ji} = \pi_j \tag{20}$$

Moreover, starting from an initial configuration of the system of interest, trial transitions are produced by means of a stochastic transition matrix, t_{ij}, arbitrarily chosen to be symmetric. The trial move is accepted with a probability a_{ij} and otherwise it is rejected, thus leading to:

$$p_{ij} = t_{ij} a_{ij} \qquad p_{ii} = 1 - \sum_{j \neq i} p_{ij} \tag{21}$$

and since matrix T is symmetric:

$$\frac{p_{ij}}{p_{ji}} = \frac{a_{ij}}{a_{ji}} \tag{22}$$

The ratio of transition probabilities above must be equal to the ratio of probabilities of states j and i, π_j/π_i, condition easily realized by taking:

$$a_{ij} = \min\left(1, \frac{\pi_j}{\pi_i}\right) \qquad \forall (i, j) \tag{23}$$

Given that $\pi_j/\pi_i = \exp[-\beta\Delta U]$, ΔU, the energy difference between states j and i, is the only quantity needed in order to calculate the acceptance probability.

3.2 IMPLEMENTATION FOR SYSTEMS OF PARTICLES

The initial state is often made of interacting particles contained in a computational box with positions, $\mathbf{r}_i$, compatible with the appropriate crystalline lattice. Usually, the potential energy, $U(\mathbf{r}_i)$ is computed with periodic boundary conditions acting in the three directions of space, excepted if free surfaces are desired or the calculations focus on bulk defects with long range deformation and stress fields, such as dislocations. The common trial consists in choosing at random a particle and moving it, within a small volume centered on its original position, along a random direction. The resulting change of the total potential energy of the system, ΔU, defines the transition probability a_{ij}.

384

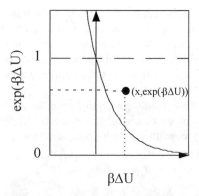

Figure 4. Acceptation rule of trial configurations in Metropolis Monte Carlo by comparison between exp(-βΔU) and a pseudo-random number in the interval [0,1].

The decision of keeping or rejecting the new state is taken as follows: (i) $\Delta U < 0$, the new state is accepted because $a_{ij}=1$ (ii) $\Delta U \; 0$, a random number, x, uniformly distributed over the interval [0,1] is chosen and is compared to $a_{ij}=\pi_j/\pi_i$. If $a_{ij} \; x$, the new state is accepted, otherwise it is rejected. An empirical rule with little justification is that convergence is best when $<a_{ij}> \; 0.5$. When possible, the acceptance should be fixed at a value such as to minimize the variance. In calculations of thermodynamical properties, this obtained by empirically choosing the maximum amplitude allowed for atomic moves. Trial moves of particles can be either, individual or collective [10]. However, the variance is strongly influenced by this choice in a way specific of the system and/or the studied property. On the other hand, if particles are individually moved, these *must* be chosen at random. Sequential moves could violate the micro-reversibility condition thus yielding incorrect average values.

4. Pseudo-random numbers

In Monte Carlo calculations, one would like to rely on truly random numbers such as those produced by taking advantage of random physical processes such as natural radioactivity or the thermal white noise produced by electronic devices (see Ref. [2] and references therein). However, using such numbers requires the existence in the computer of large databases, not easy to store and handle. Thus the so-called, pseudo-random number generators became very popular and are used nowadays quasi-systematically. These are algorithms generating sequences of numbers that resemble those drawn by repeated independent trials from a probability distribution uniform on a given interval. For the sake of clarity we present here the classical multiplicative congruential method attributed to Lehmer. This generator produces a sequence of I_k integers uniformly distributed over the interval $[0, 2^m-1]$, with m, the number of bits used in the computer for integer representation. To this end, a seed, I_0, (I_0: an odd integer e.g. $I_0=103$) is arbitrarily chosen and the relation below is used to generate pseudo-random numbers:

$$I_{k+1} = SI_k \quad \mathrm{mod} \quad \left(M = 2^m - 1\right) \tag{24}$$

S=65539, is usually the value assigned to the multiplicative constant. The sequence of random numbers obtained by Eq. 24 is periodic, with period long enough (2^{m-1}-1) to suggest that not major disturbance could exist for most of the practical uses of this generator. Random numbers in the interval [0,1] are easy to obtain, dividing the pseudo-random integers by the period of the generator. For several computers, m=32, since integer arithmetic is 32-bit based. However, exhausting the period is likely in large calculations affecting the quality of the results. Therefore, in such cases, generators with longer period must be employed [2, 11].

Besides the errors related to the exhaustion of the period of the random number generator, other effects exist that can dramatically affect the quality of Monte Carlo calculations. By testing the generator above, very popular because it was part of the standard software implementation in IBM 360 series computers and also simple to implement, Coldwell [12] has remarked that correlations existed between successive pseudo-random numbers thereby produced. This can be revealed by computing the two-particle correlation function, $g_2(r)$, for the randomly distributed particles of an ideal gas, contained in a cubic box with unit linear dimensions. Figure 5, displays the pair correlation function for such a system, set up by assigning to N=4000 particles as coordinates, x_i, y_i and z_i, sequential pseudo-random numbers in the interval [0,1] obtained from Eq. 24. Periodic boundary conditions along the three directions in space have been used in this calculation. It can be seen in this figure, that $g_2(r)$ values depart strongly at short distances from those trivially expected for an ideal gas, $g_2(r)$=1, where full lines represent the confidence limit expected for the average number of particles per neighbor shell in a truly random distribution.

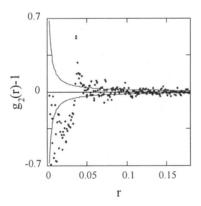

Figure 5. Pair correlation function of N=4000 particles contained in a unit cubic box, with coordinates obtained as triplets of consecutive pseudo-random numbers in the interval [0,1] generated by using Lehmer's method.

It should be remarked that at short distances, several points lie outside the confidence intervals for the average number of particles per shell, corresponding to the area delimited by the full lines drawn in Fig. 5. This is because consecutive triplets of pseudo-random numbers are correlated and is not due to the variance implied by the finite statistics in atomic shells located close to the origin. The method above, illustrating the unsatisfactory behavior of Lehmers's random number generator, is identical to that used by Coldwell [12], but for the number of particles (N=32) in the system he studied. This is the difference explaining why correlations in Fig. 5 appear somehow different from those he has published.

Subtle deviations from randomness of pseudo-random numbers and their influence on the results of Monte Carlo calculations are difficult to identify. The safest method to succeed in such an attempt is that of a systematic comparison between Monte Carlo estimations of physical properties with the exact theoretical values in models that can be successfully solved analytically. In doing so, the random number generator is tested together with the specific algorithm used in the Monte Carlo calculations. This is the conclusion drawn from the results by Ferrenberg and Landau [13] reporting that the best cluster-flipping Monte Carlo algorithm for circumventing *critical slowing down*, used in calculations on Ising square lattices with periodic boundary conditions, led to systematic errors. The last have been identified by comparing, exact analytical results available for this system with these obtained in the Monte Carlo calculations. Adopting this method, Vattulainen, Ala-Nissila and Kankaala [14] have tested several random number generators by examining self-correlations in the energy of a 2D Ising model and the deviations of mean square displacements from the value expected for a random walk process. Their results show that many of these generators display short-range correlations and that the errors reported by Ferrenberg et al. [13] are due to local correlations introducing strong bias in the cluster formation process.

5.   Extended Monte Carlo methods

There exists an abundant literature devoted, on one hand, to novel, new, fast or faster, Monte Carlo methods all aiming to improve computational performance and/or decrease the variance and, on the other hand, to extensions of the standard Monte Carlo to thermodynamical ensembles other than the canonical ensemble. The former are often specific to the systems studied and their presentation is beyond the purposes of this overview. Among the latter, the following examples are representative of the guidelines that led to the development of existing methods and could serve as a basis for the development of new, forthcoming implementations of Monte Carlo.

5.1 SIMULATED ANNEALING

Determining the N-dimensional vector, x, which corresponds to a minimum of a real function, $G(x)$, is a difficult task when N is large because the configuration space is correspondingly extended. Also, the localization of the absolute minimum of such a function is still today an unsolved problem. Searching for minima of $G(x)$, has led to the development of adapted heuristic algorithms which allow for localizing minima, generally with the help of a computer. Among these methods, Simulated Annealing [15] has proven to be quite successful in atomistic simulations, when searching for total energy minima in the studied system. Mention of this method here is justified because, to a large extent, it is analogous to the Monte Carlo method.

In simulated annealing, the function under study is viewed as the energy functional of a fictitious system with configuration phase space defined by the set of coordinates $x_1, x_2, ..., x_N$ of vector $\mathbf{x}$. In this system, the probability of an equilibrium state is given by the canonical distribution at the temperature, T. If T is lowered according to a given quenching schedule, $T_1 > T_2 > ... > T_i > 0$, the probability of high energy states decreases progressively and the system, passing through barriers in the configuration space, evolves toward a state, x_0 of minimum energy. Successive configurations at each stage of simulated annealing at $T=T_i$, are produced via the standard Metropolis Monte Carlo algorithm. By fixing the 'cooling' rate at low values, the system can escape from possible traps in the configuration space, which correspond to local minima or to flat potential energy regions. It is this feature that constitutes the major advantage of simulated annealing over traditional minimization schemes. Several variants of simulated annealing exist for some of which an informative presentation has been given by Salazar and Toral [16].

5.2 FORCE BIAS AND SMART MONTE CARLO

The convergence rate is a determining factor in Monte Carlo, which can be improved by introducing slight modifications in the usual Metropolis procedure. In atomistic Monte Carlo, two such modifications are based on particle moves guided by the forces acting on the particles and are customarily referred to as force bias [17] and smart [18] Monte Carlo methods respectively. In the former, the transition probability a_{ij}, from state i to j, is given by:

$$a_{ij} = \frac{1}{C} \exp(\lambda \beta \mathbf{F}_i \circ \delta \mathbf{r}_{ij}) \qquad (25)$$

where, C, is a normalizing factor, λ, a constant, $\mathbf{F}_i$ the force acting in state i on the particle being moved and, $\delta \mathbf{r}_{ij}$, the corresponding random displacement vector in the trial move. In smart Monte Carlo, a_{ij} is given by:

$$a_{ij} = \frac{1}{C} \exp\left(-\frac{(\delta \mathbf{r}_{ij} - \beta A \mathbf{F}_i)^2}{4A} \right) \qquad (25a) \quad \text{with} \quad \delta \mathbf{r}_{ij} = \beta A \mathbf{F}_i + \delta \mathbf{r}_G \quad (25b)$$

where, A, is an adjustable parameter, $\delta \mathbf{r}_G$, a random displacement chosen from a Gaussian distribution with zero mean and variance, 2A. Both methods provide an appreciable improvement over the standard Monte Carlo method, though the need of computing atomic forces makes them as complicated and computationally heavy as is Molecular Dynamics. A tutorial presentation of these methods, well adapted for the needs of the reader interested in their practical implementation can be found in [19].

5.3 GRAND CANONICAL MONTE CARLO

In addition to the pressure, temperature and density, equilibrium properties of multi-component systems are realistically determined by mastering the chemical potential of the species in presence. However, the number of particles in the standard Monte Carlo method is fixed and this constraint would lead to unrealistic equilibrium configurations when studying structurally heterogeneous systems. Indeed, segregation at extended defects such as, dislocations, grain boundaries or surfaces will result in a position-dependent composition strongly size-dependent in systems like these studied in the computer. Similarly, in structurally homogeneous systems, phase separation renders the system heterogeneous in the later stages when phase interfaces appear with local composition possibly differing from that of bulk phases. Amongst others, these are important motivations that justify the development of the grand-canonical Monte Carlo method [3,19].

The starting point for the implementation of a grand canonical Monte Carlo simulation is the probability distribution of states in this ensemble:

$$P_{\mu VT}(N, \mathbf{r}^N) \propto \frac{\exp(\beta \mu N) V^N}{\lambda^{3N} N!} \exp\left(-\beta U(\mathbf{r}^N)\right) \qquad (26)$$

where, N, is the number of particles, μ, the chemical potential, V, the volume, $\beta = 1/k_B T$ and λ, the thermal de Broglie wavelength. Similar to the standard procedure employed in the canonical Monte Carlo, this distribution is sampled by submitting to the Metropolis test trial configurations obtained by moving, inserting or removing particles. The acceptance probability for particle moves is given by Eq. (23) above, while insertions or removals are accepted with probabilities given below by Eqs. 27a and 27b respectively:

$$a_{ij} = \min\left(1, \frac{V}{\lambda^3 (N+1)} \exp\left(\beta[\mu + U(N-1) - U(N)]\right)\right) \qquad (27a)$$

$$a_{ij} = \min\left(1, \frac{\lambda^3 N}{V} \exp\left(-\beta[\mu + U(N-1) - U(N)]\right)\right) \qquad (27b)$$

In principle, this algorithm slightly modified, allows for calculations of phase diagrams and studies of phase transitions in multi-component systems to be made. However, in dense condensed systems, such as crystalline or amorphous solids, acceptance of trial insertion or removal of particles is very low, this fact limiting the applicability range of the algorithm. Though, in multi-component systems the equilibrium chemical composition can be calculated by using the pseudo-grand canonical Monte Carlo variant. This consists in trial moves of particles, as above, and of attempts to transform the chemical nature of particles, picked at random in the studied system. For a binary mixture, AB, the acceptance probability is given by:

$$a_{ij} = \min\left(1, \exp[-\beta(\delta U - (\mu_B - \mu_A)\delta N)]\right) \qquad (28)$$

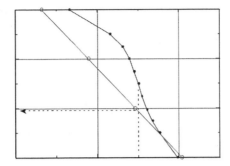

Figure 6. Chemical potential difference calculated as a function of the composition in a Cu-Au crystalline system with the L1$_2$ structure: Monte Carlo calculations with (full dots) and without (open circles) particle moves. Dashed horizontal and vertical lines mark the value, $\delta\mu_{AuCu}$ for the compound Cu$_3$Au (after Ref. [20]).

where, dN=dN$_A$=-dN$_B$,is the number of particles candidates to chemical identity changes, usually chosen equal to unity, whereas, μ_A, μ_B are the chemical potentials of species A and B. In general, the values of the last are unknown, so that the variation of the composition as a function of the chemical potential difference should be empirically established for any particular cohesion model being used. Figure 6, shows that, in a binary Cu-Au alloy with the L1$_2$ structure, two significantly different values are obtained for the difference in chemical potential between the two species at the nominal composition of the compound Cu$_3$Au, depending on whether or not atomic moves were included in the Markov chain [20].

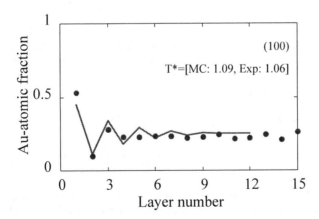

Figure 7. Gold content of (100) atomic planes in Cu$_3$Au in the disordered state ($T_c^* > 1$) given as a function of the distance from the free surface: pseudo-canonical Monte Carlo with atomic moves (full circles, T*=1.09)[20] and experiment (full line, T*=1.06) [21].

There is strong indication that adding atomic vibrations can significantly change the phase diagrams computed in Ising Monte Carlo or any other approach neglecting such contributions. Figure 7 shows the results obtained by using this Monte Carlo method for calculating the composition of surface layers in Cu_3Au, in the disordered state above the critical temperature. Close to the experimental values, the calculations show that along the [100] direction normal to the free surface and in her close neighborhood, chemical order persists above the critical temperature while the in- plane chemical order is entirely lost. A straightforward conclusion is that a two-dimensional order parameter is needed at least for an accurate description of the order-disorder transition in this compound to be made.

6. Real time and Monte Carlo

Unlike Molecular Dynamics, Monte Carlo calculations do not associate explicitly time with the successive configurations of the Markov chain. Even though physical processes such as mixing of species or phase separation, mark naturally the direction of the arrow of time, the 'temporal' path drawn by successive Monte Carlo configurations is not physical i.e. a real system starting from the same initial condition would not follow the same trajectory in the phase space. However, besides artifacts introduced by underperforming random number generators, correlations exist by definition between successive configurations in a Markov chain. These correlations can subsist over a large number of Monte Carlo steps thus reducing the accuracy of the calculations. A better understanding and study of correlations is obtained by adopting a dynamic interpretation relying on the master equation [22,23]:

$$\frac{d\pi_i(t)}{dt} = -\sum_j a_{ij}\pi_i(t) + \sum_j a_{ji}\pi_j(t) \qquad (29)$$

where, $\pi_i(t)$, is the occupation probability of state i and a_{ij} represents the transition probability per unit time from state i to state j. By doing so, 'time' averages of thermodynamical observables and more specifically 'time'-dependent correlation functions can be calculated [24]:

$$\langle X(t)\rangle = \sum_i \pi_i(t)X_i \equiv \sum_i \pi_i(t_0)X_i(t) \qquad (30a)$$

$$\langle X(t_1)Y(t_0)\rangle = \sum_i \pi_i(t_0)X_i(t_1)Y_i(t_0) \qquad (30b)$$

A widely used model of this kind is the kinetic Ising model [23,25] and related extensions examples of which are given by other contributions in this volume.

Discussing the nature of 'time' in Monte Carlo, Choi and Huberman [26] showed that the discreteness of states in Monte Carlo calculations leads to results different from these of continuous dynamics described by the master equation, thus suggesting that better understanding of this discrepancy is needed before Monte Carlo simulations could be regarded as «flexible replacements of natural systems». More recently, Gunn, McCallum and Dawson [27] have introduced a Monte Carlo variant simulating dynamical lattices by including in the trial configurations not only displacements but also particle velocities. Thereby, diffusion coefficient, interfacial tension and shear viscosity, dynamical properties that are still inaccessible with standard implementations of Monte Carlo, can now be calculated.

7. Conclusion

This non-exhaustive overview of Monte Carlo methods aims to show the kind of problems these can help studying and that, in spite of over a half-century methodological developments, the field is still growing whereas several questions are still open. A general conclusion is also that the use of these methods, apparently simple, could lead to incorrect results because sources of errors related with underperforming random number generators are subtle and therefore difficult to detect. Careful validation of the algorithms is therefore required every time a new problem is being studied by Monte Carlo.

8. References

1. Hammersley, J. M., and Handscomb, D. C., (1967) *Monte-Carlo Methods*, Methuen&Co., London.
2. James, F., (1980) Monte Carlo theory and practice, Rep. Prog. Phys., **43**, 1146-1189.
3. Frenkel, D. and Smit, B., (1996), Understanding Molecular Simulation: From Algorithms to Applications, Academic Press, New York.
4. Meyer, M. and Pontikis, V., (eds.), (1991) *Computer Simulation in Materials Science: Interatomic Potentials, Simulations Techniques and Applications*, Series E: Applied Sciences – Vol. 205 Kluwer, Academic Publishers, Dordrecht.
5. Kirchner, H. O., Kubin, L. and Pontikis, V., (eds.), (1995) *Computer Simulation in Materials Science: Nano / Meso / Macroscopic Space & Time Scales*, Series E: Applied Sciences – Vol. 308, Kluwer Academic Publishers, Dordrecht.
6. Binder, K. and Ciccoti, G., (eds.) (1996) *Monte Carlo and Molecular Dynamics of Condensed Matter Systems*, Vol. 49, Società Italiana di Fisica, Bologna.
7. Korn, G. A. and Korn, T. M., (1961) *Mathematical Handbook for Scientists and Engineers: Definitions, Theorems and Formulas for Reference and Review*, Mc Graw-Hill, New York.
8. Metropolis, N., Rosenbluth, A. W., Rosenbluth, M. N., Teller A. H. and Teller, E., (1953) Equation of State Calculations by Fast Computing Machines, J. Chem. Phys., 32, 1087.
9. Valleau, J. P. and Whittington, S. G., (1976) A Guide to Monte Carlo in Statistical Mechanics, in *Highways in Statistical Mechanics*, Part A: Equilibrium Techniques, Berne, B. (ed.), Plenum, New York.
10. Wolff, U., (1989) Collective Monte Carlo Updating for Spin Systems, Phys. Rev. Lett., 62, 361-364.
11. Hull, T. E. and Dobell, A. R., (1962) Random number generators, SIAM Rev. 4, 230.
12. Coldwell, R. L., (1974) , Correlational Defects in the Standard IBM 360 Random Number Generator and the Classical Ideal Gas Correlation Function, J. of Comp. Phys., 14, 223-226.
13. Ferrenberg, A. M., Landau, D. P. and Wong, Y. J., (1992) Monte Carlo Simulations: Hidden Errors from «Good» Random Number Generators, Phys. Rev. Lett., 69, 3382-3384.
14. Vattulainen, I., Ala-Nissila, T. and Kankaala, K., (1994) Physical Tests for Random Numbers in Simulations, Phys. Rev. Lett., 73, 2513-2516.
15. Kirkpatrick S., Gelatt, C. D. Jr. and Vecchi, M. P., (1983) Optimization by Simulated Annealing, Science, 220, 671-680.
16. Salazar, R. and Toral, R., (1997) Simulated Annealing using Hybrid Monte Carlo, J. Stat. Phys., 89, 1047-1060.

392

17. Pangali, C., Rao, M. and Berne, B. J., (1978) On a Novel Monte Carlo Scheme for simulating water and acqueous solutions, Chem. Phys. Lett., 55, 413-417.

18. Rossky, P. J., Doll, J. D. and Friedman, H. L., (1978) Brownian Dynamics as Smart Monte Carlo Method, J. Chem. Phys. 69, 4628-4633.

19. Allen, M. P. and Tildesley, D. J., (1989) *Computer Simulation of Liquids*, Oxford University Press, London.

20. Hayoun, M., Pontikis, V. and Winter, C., (1998) Computer Simulation Study of Surface Segregation in Cu_3Au, Surface Sci., 398, 125.

21. Reichert, H., Eng, P. J., Dosch, H., and Robinson, I. K., (1995) Thermodynamics of Surface Segregation Profiles at Cu3Au(001) Resolved by X-Ray Scattering, Phys. Rev. Lett. 74, 2006.

22. Müller-Krumbhaar, H. and Binder, K., (1973) Dynamic Properties of the Monte Carlo Method in Statistical Mechanics, J. Stat. Phys., 8, 1.

23. Binder, K., Ed., (1979) *Monte Carlo Methods in Statistical Physics*, Springer Verlag, Berlin.

24. Salsburg, Z. W., Jacobson, J. D., Fickett, W. and Wood, W. W., (1959) Application of the Monte Carlo Method to the Lattice Gas Model. I. Two-dimensional Triangular Lattice, J. Chem. Phys., 30, 65.

25. Halperin, B. I., and Hohenberg, P. C., (1977) Theory of Dynamic Critical Phenomena, Rev. Mod. Phys., 49, 435.

26. Choi, M. Y. and Huberman, B. A., (1984) Nature of Time in Monte Carlo Processes, Phys. Rev. B29, 2796-2798.

27. Gunn, J. R., McCallum C. M. and Dawson, K. A., (1993) Dynamical Lattice-model Simulation, Phys. Rev. E47, 3069-3080.

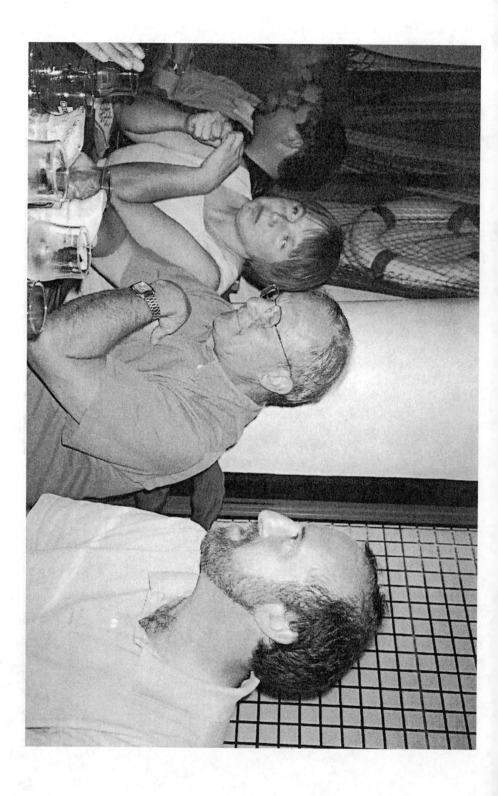

KINETIC MONTE CARLO SIMULATIONS IN CRYSTALLINE ALLOYS: PRINCIPLES AND SELECTED APPLICATIONS

P. BELLON
Department of Materials Science and Engineering
University of Illinois at Urbana-Champaign, Urbana, IL, USA.

1. Introduction

In the past decade, there has been an increased use of kinetic Monte Carlo (KMC) simulations to study diffusion-controlled phenomena, in particular in crystalline alloys. Several factors are contributing to this increased interest, namely the increase in computing power, the need to study microstructures at the atomic or nanometer scale in advanced structural and functional materials to optimise their properties, the development of realistic atomistic energetic models for alloys, and the development of techniques enabling chemical characterisation with sub-nanometer resolution, in particular the three-dimensional atom probe field ion microscopy.

   The main advantage of the KMC method is that it offers a good compromise between the degree of realism that is retained in the description of matter transport at the atomic scale, and the time scales that can be simulated. For short time scales, typically below 1 ns, molecular dynamics (MD) is probably the method of choice to simulate atomistic processes in alloys. While a large amount of important information is obtained by MD simulations, this approach fails when one needs to study the long term evolution of diffusion controlled phenomena, which typically requires to simulate materials evolution over times ranging from 1 s, to thousands of years in the extreme case of nuclear waste storage materials. Continuum methods such as the phase field (PF) method can reach long time scales as well, and even handle larger simulated volumes, but, as they lack atomistic information on diffusion processes, their predictive power is less than that of KMC simulations, in particular for nanoscale microstructural features.

   In this paper, we first briefly review the principles underlying the KMC method, and we then describe its application to two different problems: the coherent precipitation of a second phase during thermal annealing of a supersaturated alloy, and the compositional patterning in alloys under irradiation. We also indicate the key limitations in current KMC simulations, and the solutions that are being proposed and tested to overcome these limitations.

2. Principles of kinetic Monte Carlo simulations

2.1. KINETIC MONTE CARLO AND MASTER EQUATION

We recall here the principles underlying the KMC method. They can be simply stated as follows: Given a system and its phase (configuration) space, a distribution of

A. Finel et al. (eds.), Thermodynamics, Microstructures and Plasticity, 395–409.

configurations at time t=0, and a set of transition rates between configurations, we want to generate temporal trajectories of the system in its phase space. We require that these temporal trajectories are produced with their correct statistical weight, so that both the average evolution of the system and its fluctuations around this average can be properly reproduced from a large set of temporal trajectories.

With such a definition of the KMC method, one recognizes that our goal is equivalent to providing a numerical solution to the corresponding Master Equation for the system and the dynamic processes under consideration. Restricting here to the case of Markovian processes, i.e., processes with no memory effect, this Master Equation yields the temporal evolution of the conditional probability distribution of configurations:

$$\frac{dP_i}{dt} = \sum_j \omega_{ji} P_j - \sum_j \omega_{ij} P_i \qquad (1)$$

where the index "i" labels the configuration, P_i is the time dependent conditional probability to find the system in the configuration "i" at time t, knowing its probability distribution at time t=0 [1]. ω_{ij} is the transition rate from configuration "i" to configuration "j" ($j \neq i$) , i.e., the number of such transitions per unit time for a system that is initially in state "i". The KMC method provides a numerical solution to Eq. (1), and in particular, the whole probability distribution at a given time t can be obtained from an ensemble average of KMC trajectories.

As stressed by several authors, in particular Voter and coworkers [2,3], an important potential limitation of the KMC method is that it relies on a given and fixed list of possible transitions. If an important mechanism has not been included in this list, the simulations may not give a realistic evolution of the system. A classical example of such a problem is the collective migration of adatom clusters on a surface: this phenomenon will not appear by itself in KMC simulations if one assumes that only individual adatoms can migrate [2,2]. Before running any KMC simulations, it is therefore important to determine whether all the relevant processes contributing to transitions between configurations have been properly included.

2.2 RESIDENCE TIME ALGORITHM

While various algorithms can generate temporal trajectories from Eq. (1), the so-called residence time algorithm (RTA) is the reference algorithm for KMC simulations, as it rests on clear physical ground. This algorithm has been derived independently and discussed in several different fields, under different names [4,5,6,7,8,9,10]. It is sufficient for our purpose to recall here the main steps involved in this algorithm. Each KMC step is decomposed into the two following steps:

1/ from the current configuration, calculate all transition probabilities corresponding to all admissible transitions. Assuming that these transitions are independent, the probability for the system to remain in its current state after a time t is given by a Poisson distribution, and the average time that the system spends in its configuration (the residence time) before undergoing a transition is given by:

$$\tau_i = 1 \Big/ \sum_j \omega_{ij} \qquad (2)$$

The clock is then incremented by τ_i. A fluctuating time can also be used, according to the Poisson distribution, but this does not lead to any significant practical difference [9].

2/ the new configuration is chosen among the set of possible ones, taking into account their relative weight. A random number is pulled, *rnd*, homogeneously distributed in the interval $[0, \tau_i^{-1}[$. The kth transition is selected according to:

$$\sum_{j=1}^{k-1} \omega_{ij} \leq rnd < \sum_{j=1}^{k} \omega_{ij} \qquad (3)$$

The advantages of this algorithm are the following. Firstly, the physical time is directly and unambiguously built in the algorithm: as the transition rates are given per unit of physical time, the residence time is in units of physical time. Secondly, each KMC step is successful in the sense that the system always undergoes a transition to a new configuration. Thirdly, this algorithm can be directly applicable to systems where several diffusion mechanisms are operating in parallel, e.g., several vacancies, or vacancies and interstitials, or through thermal and non-thermal processes. Examples in the following sections will illustrate this point.

The main drawback of the RTA is that it can be time consuming to re-calculate all the transition rates at each KMC step. This limitation is less severe when transitions proceed by migration of a few defects, as the set of transitions is then quite small. Finally, another potential limitation of the RTA is that the system may be trapped, for instance by oscillating between two configurations. Solutions can be developed to overcome some trapping situations, using the so-called higher order RTA [9].

It has been shown [8,9] that this RTA algorithm is strictly equivalent to another algorithm, often called "kinetic Metropolis", as it is derived by analogy with the Metropolis algorithm for equilibrium MC simulations. Let us simply indicate here that the main drawback of this kinetic Metropolis algorithm is that is suffers from a high rejection rate (the lower the temperature, the higher the rejection rate). This drawback is so significant that in most cases this algorithm is much less efficient than the RTA, despite the fact that only one transition rate needs to be calculated at each step of the algorithm. It is only at high temperature, and for a very large density of diffusing species, that this other algorithm may be considered [10].

2.3 A SIMPLE DIFFUSION MODEL

We now discuss briefly how to calculate the transition rates. As our goal is to simulate diffusion-controlled evolutions of crystalline alloys, it is well suited to use the harmonic approximation to the transient state theory (TST) [11,12]. In this theory the transition rate from "i" to "j" is:

$$\omega_{ij} = \nu \exp[-(E_{ij}^s - E_i)/kT] \qquad (4)$$

where E_i is the energy of the configuration "i", E_{ij}^S is the energy of the saddle point configuration between "i" and "j", and ν, the so-called attempt frequency, for a harmonic solid is given by:

$$\nu = \prod^{3N-3} w_j^{stable} \bigg/ \prod^{3N-4} w_j^{saddle} \qquad (5)$$

where the w_j are the eigen modes of a system containing N atoms, in the stable and saddle point configuration only. The three modes corresponding to the macroscopic translation of the system not included in the products, and for the saddle point case, the unstable mode is also not included. Note that ν, while having the dimensions of a frequency, does not correspond directly to any single eigen mode of the system. In the following, we will concentrate our discussion on the calculation of the activation energies entering Eq. (4), because of their exponential contribution to w_{ij}. The pre-exponential term is weakly temperature dependent and, in general, does not vary too strongly from configuration to configuration. The interested reader is referred to Bocquet's recent review [13].

A simple diffusion model that has been extensively used [14,15,16,17,18] consists in assuming that diffusion proceeds through the migration of one vacancy, and in using effective pairwise interaction parameters to calculate the energies entering Eq. (4). As the migrating atom, say labeled "X", possesses a quite different atomic environment in the saddle point configuration compared to the stable configuration, its contribution to the saddle point energy of the crystal is written separately, as E_{V-X}^S. It is then simple to show that the activation energy for the exchange between "X" and the vacancy is:

$$E_{V-X}^{act} = \sum_{saddle} \varepsilon_{YZ} - \sum_{initial} \varepsilon_{YZ} = E_{V-X}^S - \sum_{Y \; neighbors \; of \; X} \varepsilon_{XY} \qquad (6)$$

where ε_{XY} is the effective pairwise interaction between "X" and "Y" atoms, here restricted to nearest neighbors for simplicity. Another simplification consists in assuming that E_{V-X}^S, the saddle point binding energy, is a constant, E^S, regardless of the nature of the diffusing atom and of its environment. This ensemble of assumptions lead a simple diffusion model that, in the remaining part of this paper, will be referred to as our reference diffusion model. Despite its simplicity, this generic model can produce a broad range of kinetic evolutions, as we will see in the next sections.

2.4 PARAMETRIZATION OF THE REFERENCE DIFFUSION MODEL

In the case of a binary $A_{1-c}B_c$ substitutional alloy, the reference diffusion model just introduced has four parameters, in addition to temperature, which will be one of our control parameters during the simulations. Out of these four parameters, one parameter, the ordering energy defined as $\varepsilon=\varepsilon_{AA}+\varepsilon_{BB}-2\varepsilon_{AB}$, completely determines the equilibrium phase diagram of the A-B system. Phase separation or ordering will take place at low temperatures when ε is negative or positive, respectively. This parameter can be fitted to reproduce a measured or estimated critical temperature, or alternatively to reproduce a solubility curve in the case of a dilute alloy.

Three parameters control the kinetics of the alloy. The attempt frequency gives simply the time scale, and it can be fitted on the pre-exponential diffusion term of a chemical species in the considered alloy. For instance, for cubic crystals, this pre-exponential term is related to the attempt frequency through $D_0 = z\lambda^2 v/6$, where z is the coordination number of the lattice and λ is the jump distance. For $D_0 = 10^{-5} \text{ m}^2 \text{ s}^{-1}$, a typical value for self-diffusion in copper, one gets $v = 10^{14} \text{ s}^{-1}$. The second kinetic parameter is the saddle point energy E^S. As E^S is a constant in the reference model, it can be factorized out of the jump frequencies, and it thus corresponds to a temperature-dependent re-scaling of the time. The third and last kinetic parameter is the so-called asymmetry energy $a^* = (\varepsilon_{AA}-\varepsilon_{BB})/\varepsilon$. It is only recently that the importance of this parameter has been fully recognized [15,9,17,19,20,21]. By varying this parameter only, alloy systems with the same equilibrium phase diagram but with different kinetics can be compared. As shown and discussed in the first application below, very different kinetic behaviors can be obtained, though the final, equilibrium states will be identical for these various alloys. This asymmetry parameter is related to the difference in cohesive energies between the two elements, or to the difference in the vacancy formation or migration energy between the two elements. Because of the approximations used in the reference model, only one of these three differences can be fitted to a^*. Finally, while this asymmetry parameter could be interpreted as a thermodynamic parameter of the ternary alloy A-B-V, it makes more sense to interpret it as a kinetic parameter for the binary alloy A-B, as the alloy is extremely dilute in vacancy.

3. Coarsening and morphologies of precipitates formed during thermal annealing

In this section we illustrate the use of the reference diffusion model for the case of precipitation in a supersaturated solid solution. KMC simulations are performed with the residence time algorithm. While precipitation has already been studied in detail with KMC simulations, see for instance the contribution by Soisson in this volume [22] or by Athènes and coworkers [17], we emphasize here the role played by the asymmetry parameter on the precipitate morphology and composition, and on coarsening rates (see ref. [20] for more details). This work was initially motivated by recent three-dimensional atom probe field ion microscopy results [23] on the morphology of small Co precipitates in Cu. Unexpected features were reported: some Co precipitates contain up to 20 at.% Cu, exceeding the equilibrium solubility by a factor 200, and most precipitates have rather diffuse interfaces, around 1 nm in thickness, whereas simple KMC simulations predicted atomically sharp interfaces. We have demonstrated that our reference model is rich enough to reproduce these unexpected features, because of the effect of the asymmetry parameter on atomic migration. The main parameters and findings of this study are now given.

We consider a $A_{1-c}B_c$ alloy on an fcc lattice, containing 64^3 or 128^3 sites. One of this site is left vacant, and atom migration proceeds by exchanges between this vacancy and its 12 nearest neighbors. The ordering energy ε is set to -55.3 meV in order to reproduce a moderate positive heat of mixing, typical of the Cu-Co system, with a critical temperature of the miscibility gap, T_c=1573 K. The simulations are performed at T=0.258 T_c. The attempt frequency ν is set to 10^{14} s^{-1}, and E^S is set to reproduce a 0.8 eV vacancy migration energy in pure A. In this parametric study, the asymmetry parameter a* will be varied between -1.5 to +0.5, while a reasonable value for Cu-Co would range between -1 and -2.5. In addition to KMC simulations, mean field models are used to calculate equilibrium and kinetic quantities that are then compared to the results obtained in the simulations.

The first main point is that, by varying a*, one changes dramatically the vacancy distribution in the system. Consider that we have one pure B precipitate, containing L atoms on average, embedded in a pure a matrix. Fig. 1 gives the probabilities to find the vacancy in the matrix, P_v^A, in the precipitate, P_v^B, and at the interface between the precipitate and the matrix, on the precipitate side or on the matrix side, P_v^- and P_v^+, respectively. The consequences are that, in the case of negative a*, the vacancy is mostly spending its time in the matrix and little mobility is available, either at the interface or inside the precipitate, whereas in the case of positive a*, the vacancy is trapped in the precipitate and at the interface.

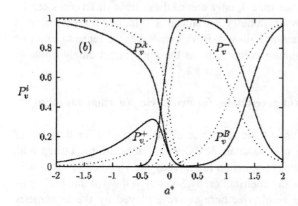

Figure 1: Probabilities to find the vacancy in the four different regions: in the matrix, P_v^A, in the precipitate, P_v^B, and at the interface between the precipitate and the matrix, on the precipitate side or on the matrix side, P_v^- and P_v^+, respectively. Solid lines are for a precipitate containing 20 atoms, the dotted line for 200 atoms [20].

The second main point is that the growth and coarsening rates are strongly affected by a*. Fig. 2 shows the evolution of the average number of atoms in the precipitates, L(t), as a function of the KMC time. By keeping track of the size evolution of all the precipitates (Fig. 3), it is demonstrated that, for zero or positive a*, coarsening takes place by coagulation (stairlike behavior in fig. 3), whereas for negative a*, coarsening proceeds by evaporation-condensation (fountainlike behavior in fig. 3).

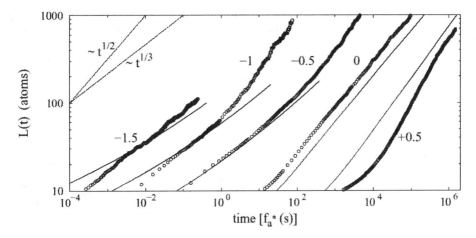

Figure 2: Monte Carlo time evolution of the average cluster volume L(t) (s) at T=0.258Tc with c=0.05, C_V=4.77×10-7, and for a*=-1.5, -1, -0.5, 0, 0.5. Simulation cells contain 128^3 fcc sites except for a*=-1.5, -1 (64^3 fcc sites only). The curves are shifted along the x-axis for the sake of clarity: the Monte Carlo time is scaled by a factor f_{a*}: $f_{-1.5}$=4×10^4, f_{-1}=8×10^2, $f_{-0.5}$=10, f_0=8, and $f_{0.5}$=1. Solid lines represent predictions of the coagulation model. From ref. [20]

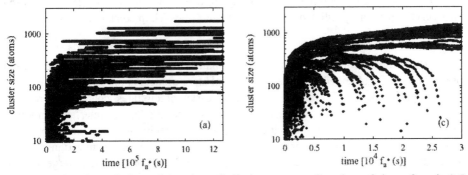

Figure 3: KMC evolution of the size of all clusters as a function of time, for a*=0.5 (left) and a*=-0.5 (right). The time scaling factors are as in Fig.2. From ref. [20].

The third main result is that the solubilities of solute (B species) in the matrix, c_m^B, and of solvent (A species) in the precipitates, c_p^A are also strongly affected by the asymmetry (Fig. 4). For negative or zero asymmetry, after a short transient, the dependence of these solubilities as a function of the precipitate size follows very well the curves calculated from the Gibbs-Thomson equation. In the case of sufficiently negative asymmetry (a*≤-1), however, the solubility of A atoms in the precipitates exceeds largely the Gibbs-Thomson curve. This supersaturation starts to develop from the very first steps of precipitate growth and coarsening. Direct visualizations of the kinetic evolution of the precipitates, coupled with measures of their excess interfacial energy, establish that this supersaturation proceeds by the burrowing of interfacial roughness into the precipitates, as new B atoms are brought by the vacancy migration. This interfacial roughness cannot be annealed fast enough compared to the arrival rate of new B atoms from the matrix to the interface, as the vacancy spends very little time at the interface for these negative asymmetry parameter values (see Fig. 1). Another consequence is that interfaces are rather diffuse.

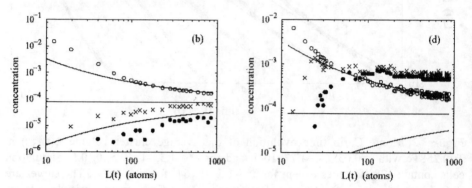

Figure 4: KMC solubilities as a function of average cluster size L(t) for a*=0 (left) and a*=1 (right). Symbols represent the concentration of B atoms in the matrix (o), of isolated A atoms in precipitates (•), and of total A atoms in precipitates (×). The straight solid line indicates the equilibrium solubility for a flat interface at T=0.258 T_c, and the two curves give the predicted solubilities from Gibbs-Thomson's equation. From ref [20].

We can now propose a consistent explanation to the unexpected results on Co precipitates in Cu obtained by three dimensional atom probe. As the vacancy mobility is much larger in the Cu-rich matrix than in the Co-rich precipitates, coarsening proceeds always by the evaporation-condensation mechanism, but the lack of interfacial mobility precludes the formation of sharp interfaces. As the precipitates coarsen their rough interface is buried into the precipitates, leading to large supersaturations. These supersaturations therefore result from a kinetic effect, and are, to a large extent, unaffected by the thermodynamics of the alloy.

It is instructive to analyze why this kinetic behavior has not been identified in previous simulations. On the one hand, previous simulations relying on a vacancy mechanism, as in this work, have ignored the role played by the asymmetry (except for the work by Athènes an coworkers [15,9,19,17]), and this has led all authors to adopt the "natural" choice of setting a*=0 [24,25,26,27,28]. That choice, however, has important consequences: with a*=0, at low enough temperatures, the early stage of coarsening is dominated by the coagulation mechanism and interfacial mobility is large, so that interfaces are sharp, and therefore no solvent supersaturation can build-up in the precipitates. On the other hand, simulations relying on a direct atom-atom exchange mechanism for diffusion lead to coarsening by evaporation-condensation, as a*=0 was also assumed (see ref. [24] and references therein). In contrast with our present diffusion model, however, the direct-exchange mechanism, combined with the choice a*=0, leads to a homogeneously distributed atomic mobility. In particular, any A atom that would have been buried into a B-rich precipitate can easily migrate inside the precipitate and reach the interface. As a consequence, no solvent supersaturation can build up in these simulations.

Finally we want to stress the limitations of the present simple diffusion model: migration barriers are calculated using a very simple broken-bond model with a constant saddle point binding energy. This approach, while well suited for a parametric study and for identifying general trends, cannot be very predictive for a given alloy system. In particular, the present model ignores stress effects on migration barriers. Several authors have proposed to remedy this weakness by calculating migration barriers on a relaxed lattice, using semi-empirical interatomic potentials or ab-initio calculations to calculate the energy of the alloy in stable and saddle point configurations. The main limitation of such relaxed calculations is that they are very CPU intensive, and that, with present computers and algorithms, only very short simulations are possible, typically a few hundred vacancy jumps per atom in a crystal containing 256 atoms [13]. Speed-up by several orders of magnitude is required to obtain kinetic evolutions as in Figs. 2-4. Various approximation schemes are being proposed and tested to increase the speed of calculation of activation barriers, while still retaining as much as possible accurate values [13,22,29].

A last limitation of the present KMC simulations, irrespective of the accuracy of diffusion model, is that, in certain situations, the "metallurgical work" performed per vacancy jump can be very small. Consider for instance the case where vacancies are spending most of their time in the matrix. A very large fraction of the vacancy jumps simply involve the migration of solvent atoms, and therefore do not lead to any advancement of the precipitation reaction. It is this limitation that has prevented us in the present work to study asymmetry parameter values more negative than -1.5. In such situations, it would be probably better to devise another type of simulation technique, where the migration of the vacancy in the matrix is no longer built jump by jump, but rather using fairly large diffusion distances. Such techniques remain to be developed.

4. Compositional patterning under irradiation

We now turn to the case of alloys subjected to a sustained external forcing, in the present case irradiation by energetic particles. We summarize our recent results on the stabilization of compositional patterns under irradiation to illustrate the use and the contribution of KMC simulations in driven alloys.

4.1 ANALYTICAL MODEL

In a seminal paper published in 1984 [30], G. Martin proposed to model phase stability in alloys under irradiation by using a kinetic description that explicitly takes into account the presence of several dynamics in these alloys. Indeed in such alloys atoms can migrate on the one hand because of the thermally activated motion of point defects. and on the other hand because of the forced relocations that take place during nuclear collisions in metallic targets. Note that vacancy and interstitial concentration in an alloy under irradiation are usually in excess of their equilibrium value because of the production of point defect by nuclear collisions. Point defect concentrations can be calculated using rate equation models that include the defect production, the vacancy-interstitial recombination and the defect elimination on sinks [31]. Now for the ballistic jumps, because of the rather large energies involved in these collisions (greater than a few eV) compared to thermal energies, Martin proposed to approximate these forced relocations as purely ballistic, i.e., random. Such a kinetic model belongs to the broad class of systems with competing dynamics, which have been extensively studied as prototypes for non-equilibrium dynamical systems [32,33,34].

Applying this model to the case of alloys with positive heat of mixing in the case of nearest neighbour ballistic exchanges, Martin showed analytically that the *nonequilibrium steady-state* reached under irradiation at a temperature T and at a displacement rate Γ_{dis}, is the *equilibrium* state of the same system but at a higher, *effective*, temperature. This criterion, the effective temperature criterion, has been successfully used, e.g., to account for the extension of solubility under light ion irradiation in various binary alloys, [35] or, in an enriched version, to explain the stabilization of solid solution by ion-beam mixing of Cu-Ag multilayers [36].

As revealed by molecular dynamics simulations (MD), however, the average relocation distance in displacement cascades can vary significantly, from essentially one nearest neighbor distance for 10 keV cascades in Ni to 4 times this distance for 1 keV cascades in Au [37]. Furthermore, the tail of the distribution can extend well beyond this average value. Early kinetic Monte-Carlo simulations [38] as well as mean-field modelling calculations [39] indicated that the presence of medium or long range ballistic relocations could lead to the stabilization of compositional patterning under irradiation. We recently introduced an analytical model that explicitly takes into account this finite range of ballistic jumps [40]. This model predicts that, for appropriate irradiation conditions, an alloy may spontaneously organize into stable patterns at the nanometer scale. We now briefly review this model and its main prediction.

In our model, a kinetic equation is established for the time evolution of the deviation of the composition field with respect to its average value, c(x), in an alloy under irradiation where the ballistic jumps take place at a frequency Γ, with a relocation distribution w_R that has an average distance R:

$$\frac{dc}{dt} = M\nabla^2\left(\frac{\delta F}{\delta c}\right) - \Gamma\left(c - <c>_R\right) \qquad (7)$$

where $<c>_R$ is the average value of c(x) weighted by w_R, M is the thermal mobility, and F{c(x)} is the equilibrium free energy functional of the alloy, and we use a Ginzburg-Landau expression for an alloy that phase separates at low temperature:

$$F\{c(x)\} = \int\left(-Ac(x)^2 + Bc(x)^4 + C|\nabla c(x)|^2\right)dx \qquad (8)$$

The first term in Eq. (7) is the classical diffusional term for a globally conserved order parameter, while the second term represents the contribution of the ballistic mixing. We have shown that, when M is independent of c, this equation can be re-written as:

$$\frac{dc}{dt} = M\nabla^2\left(\frac{\delta E}{\delta c}\right) \qquad (9)$$

where E is an *effective* free energy functional for the alloy under irradiation and expressed as:

$$E = F + \frac{\Gamma}{M}G \qquad (10)$$

The new term G in Eq. (10) represents *effective* interactions that originate from the ballistic jumps. Its exact expression is complicated [40] and it suffices here to say that it is comprised of *effective* repulsive interactions which decay similarly to w_R. We then use this *effective* free energy to build a dynamical phase diagram. For that purpose, we perform a variational minimization of E, using two different class of parametric functions for the composition profile, c(x)=αsin(kx) and c(x)=αtanh[m/k sin(kx)], α, m and k being some coefficients that are varied during the minimization procedure.

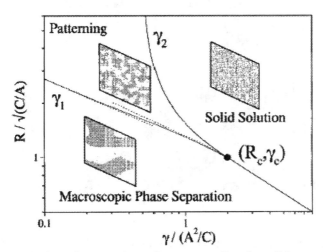

Figure 5: Dynamical steady-state phase diagram as a function of the average relocation distance for the ballistic jumps, R and the reduced irradiation intensity γ. R and γ are expressed in terms of the materials constants A and C (see eq. 8). Solid lines are the transition lines predicted from the minimization of the *effective* free energy functional E for an alloy with C_B=50%, and insets are (111) cuts of three-dimensional atomistic kinetic Monte-Carlo simulations (see also Fig. 6) [40].

The resulting dynamical steady-state phase diagram is shown in Fig. 5. For average relocation distances lower than a critical value, the only possible stable steady-states are macroscopic phase separation and solid solution, at low and high irradiation intensity γ=Γ/M respectively. For relocation distances greater than the critical value of unity in the reduced units used, a new field opens in the phase diagram where the composition field of the alloy self-organises into patterns. These patterns continuously evolve into a solid solution when γ increases, whereas the transition into macroscopic phase separation when γ is decreased is discontinuous and takes place when the decomposition is characterised by a wave vector $k_1=1/(2R)$. As shown in the insets in Fig. 5, and as discussed in the next section, the existence of these three fields has been confirmed by 3D KMC simulations [41].

The above model corresponds to a processing where the composition remains basically unchanged, as for instance during ion beam mixing of thin films, but the main results also hold for films processed by ion implantation, as long as, in the regions of interest, the scale of composition variation is larger than the scale of the patterns.

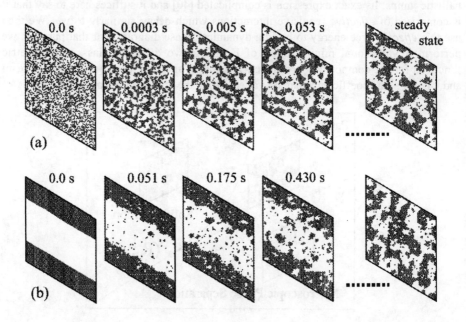

Figure 6: Temporal evolution of (111) cuts starting from different initial conditions: (a) random solid solution and (b) pure A/B bilayer, at an intermediate frequency of atomic relocations $\Gamma = 200$ s^{-1}, with R = 3 nearest neighbour distances (a_{nn}) and at a temperature kT = 0.05 eV. The picture shows (111) cuts of the 3D KMC simulation cell. B atoms are indicated by black circles. For either initial condition the system reaches a steady-state microstructure exhibiting a labyrinthine compositional pattern [42].

4.2 KMC SIMULATIONS

Kinetic Monte-Carlo simulations, based on the reference diffusion model introduced in Sec. 2.3, have been used to test the validity of the results obtained with the analytical model. The parameters used for the simulations is described in detail elsewhere [41,42].The simulation cell size contains 64^3 sites on an fcc lattice. Results given here are for a binary $A_{50}B_{50}$ alloy, with a single vacancy present in the crystal. The thermal parameters are identical to the ones used in the previous example, discussed in Sec. 3.

Ballistic exchanges between sites separated, on average, by a distance R, are introduced at a frequency Γ_b. The relocation site is chosen by performing a random walk with R^2, steps starting from the initial site selected for the ballistic exchange. A residence-time algorithm that includes both thermal and ballistic dynamics is used to build the time evolution of the system. Different initial configurations are used, either a random solid solution or a macroscopic bilayer.

Regardless of the initial configurations, the same steady-state is reached as illustrated on Fig. 6. The simulations confirm that patterning can occur when the relocation distance is greater than a critical distance, R_c (see insets in Fig. 5). Simulations also confirm that the transition from solid solution to patterning is continuous, whereas the transition from patterning to macroscopic phase separation is discontinuous, i.e., of first order. Furthermore, the locations of these transitions, summarized in Fig. 7 in a dynamical phase diagram, are in good agreement with the mean field model. One difference is that, in the patterned state, the model predicts ordered patterns (a stripe phase), whereas disordered patterns are always obtained in the simulations. We believe that the kinetic roughening of interfaces under irradiation [43] is responsible for the formation of undulated interfaces under irradiation. This point deserves further investigation.

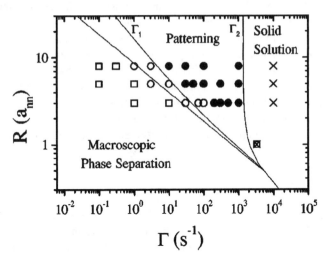

Figure 7: Dynamical phase diagram showing the steady-state regimes as a function of the forcing parameters. Symbols denote the results from KMC simulations, solid lines correspond to the prediction from the continuum model. Squares, crosses and circles denote macroscopic phase separation, solid solution, and patterning, respectively. Full circles are for patterns well contained in the simulation cell, open circles for patterns affected by finite size effects [42].

Recently, we have used molecular dynamics (MD) to determine accurately the spectrum of relocation distances in $Cu_{50}Ag_{50}$ irradiated with various ions: 62 keV He, 270 keV Ne, 500 keV Ar and 1 MeV Kr [44]. The energies are such that the average projected path of the ions in $Cu_{50}Ag_{50}$ is the same in four cases, ≈ 2600 Å. MD reveals that, for the three heavy ions, the relocation distribution can be approximated by a decaying exponential, with a decay length of ≈ 3 Å. In the case of He irradiation, the distribution is much more weighted toward shorter relocation distances, yielding a decay length of 1.4 Å if an exponential fit is used. KMC simulations performed with these distributions reveal that only the three heavy ions can induce compositional patterning. These results are consistent with the prediction of the existence of a critical relocation distance, R_c, for compositional patterns to be stabilized during irradiation. We have confirmed experimentally that 1 MeV Kr irradiation of $Cu_{50}Ag_{50}$ thin films leads to the stabilization of nanocomposites (or compositional patterns), when the irradiation temperature is adjusted so that both irradiation and thermal effects are of similar intensity [45]. Further experimental work is needed to test the prediction that light ion irradiation cannot lead to the stabilization of such patterns, regardless of the irradiation parameters.

5. Conclusion

We have presented the principles underlying kinetic Monte Carlo simulations of crystalline alloys. The advantages in using a residence time algorithm to generate temporal trajectories of alloy system are stressed, namely the explicit incorporation of the physical time scale, and the ability to take into account several defects and mechanisms responsible for atomic migration. We have illustrated the use of KMC simulations on two examples: the morphology and composition of precipitates that are forming and coarsening during the thermal annealing of a supersaturated solid solution, and the stabilization of compositional patterns in alloys under irradiation. In both cases, specific features have been identified in the KMC simulations, owing to the ability in these simulations to retain atomistic information and to reach long time scales. Some limitations of KMC simulations have been discussed, in particular regarding the need to develop predictive simulations for specific alloy systems.

Acknowledgements

We are pleased to acknowledge fruitful collaborations and discussions with Profs. R. S. Averback, F. Haider, and J.-M. Roussel, and Drs. G. Martin, M Athènes, J.-L. Bocquet, R. Enrique, M. Nastar, K. Nordlund and F. Soisson. This material is based in part upon work supported by the U.S. Department of Energy, Division of Materials Sciences under Award No. DEFG02-ER9645439, through the Frederick Seitz Materials Research Laboratory at the University of Illinois at Urbana-Champaign.

References

1. N. G. Van Kampen (1992) *Stochastic Processes in Physics and Chemistry*, NorthHolland, Amsterdam, p. 96
2. A. Voter (1997) *Phys. Rev. Lett.* **78**, 3908.
3. M. D. Sorensen, A. F. Voter (2000) *J. Chem. Phys.* **112**, 9599.
4. W. M. Young, E. Elcock (1966) *Proc. Phys. Soc.* **89**, 735.

5. J.-M. Lanore (1972) Commissariat à l'Energie Atomique, Gif-sur-Yvette, France, Report No CEA-N4565; ibid (1974), *Rad. Effects* **22**, 153.
6. A. B. Börtz, M. H. Kalos, J. L. Lebowitz (1975) *J. Comput. Phys.* **17**, 10.
7. C. Jacoboni, L. Reggiani (1983) *Rev. Mod. Phys.* **55**, 645.
8. M. A. Novotny (1995) *Computers in Phys.* **9**, 46.
9. M. Athènes, P. Bellon and G. Martin (1997) *Philos. Mag* A **76**, 565.
10. E. Adam, L. Billard, F. Lançon (1999) *Phys. Rev. E* **59**, 1212.
11. G. H Vineyard (1957) *J. Phys. Chem Solids* **3**, 121.
12. G. H Vineyard, J. A. Krumhansl (1985) *Phys. Rev. B* **31**, 4929.
13. J.-L. Bocquet (2002) *Defect Diffusion Forum* **203-205**, 81.
14. T. Abinandanan, F. Haider, G. Martin (1998) *Acta Mater.* **46**, 4243.
15. M. Athènes, P. Bellon, F. Haider and G. Martin (1996) *Acta Mater* **44**, 4739.
16. F. Soisson, A. Barbu and G. Martin (1996) *Acta Mater.* **44**, 3789.
17. M. Athènes, P. Bellon and G. Martin (2000) *Acta Mater.* **48**, 2675.
18. D. Le Floc'h, P. Bellon and M. Athènes (2000) *Phys. Rev. B* **62**, 3142.
19. M. Athènes and P. Bellon (1999) *Philos. Mag* A **79**, 2243.
20. J.-M. Roussel and P. Bellon (2001) *Phys Rev* B **63**, 184114.
21. F. Soisson and G. Martin (2000) *Phys. Rev. B* **62**, 203.
22. F. Soisson, this volume.
23. I. Rozdilsky (1998), Ph.D. thesis, Oxord University; I. Rozdilsky, A. Cerezo, G. D. W. Smith, and A. Watson (1998) *Mater. Res. Soc. Symp.* **481**, 521.
24. P. Fratzl and O. Penrose (1994) *Phys. Rev B* **50**, 3477.
25. C. Frontera, E. Vives, T. Castán, A. Planes (1996) *Phys. Rev B* **53**, 2886.
26. P. Fratzl and O. Penrose (1996) *Phys. Rev B* **53**, 2890.
27. P. Fratzl and O. Penrose (1997) *Phys. Rev B* **55**, R6101.
28. T. T. Rautiainen and A. P. Sutton (1999) *Phys. Rev. B* **59**, 13682.
29. Y. Le Bouar and F. Soisson (2001) *Phys. Rev. B* **65**, 094103.
30. G. Martin (1984) *Phys. Rev. B* **30** 1424.
31. R. Sizman (1978) *J. Nucl. Mater.* **69-70**, 386.
32. J. L. Lebowitz, P. Bergmann (1957) *Annals of Phys.* **1**, 1.
33. R. Kubo, K. Matsuo, and K. Kitahara (1973) *J. Stat. Phys.* **9**, 51.
34. G. Martin, P. Bellon (1997) *Solid. Stat. Phys.* **50**, 189.
35. A. Traverse, M.-G. Le Boite, G. Martin (1989) *Europhysics Lett.* **8**, 633.
36. L. C. Wei, R. S. Averback (1997) *J. Appl. Phys.* **81**, 613.
37. R. S. Averback, T. de la Rubia (1997) *Solid State Phys.* **51**, 281.
38. F. Haider (1995) Habilitation thesis, University of Goettingen.
39. V. Vaks and V. Kamyshenko (1993) *Phys. Lett.* A **177**, 269.
40. R. A. Enrique, P. Bellon (2000) *Phys. Rev. Lett.* **84**, 2885.
41. R. A. Enrique, P. Bellon (1999) *Phys. Rev. B* **60**, 14649.
42. R. A. Enrique, P. Bellon (2001) *Phys. Rev. B* **63**, 134111.
43. P. Bellon (1998) *Phys. Rev. Lett.* **81**, 4176.
44. R. A. Enrique, K. Nordlund, R. S. Averabck, and P. Bellon (2002), submitted.
45. R. A. Enrique, P. Bellon (2001) *Appl. Phys. Lett.* **78**, 4178.

ROLE OF ATOMIC-SCALE SIMULATION IN THE MODELING OF SOLIDIFICATION MICROSTRUCTURE

M. ASTA, D. Y. SUN
Northwestern University, Department of Materials Science and Engineering, Evanston, IL, USA

AND

J. J. HOYT
Sandia National Laboratories, Albuquerque, NM, USA

Abstract. A brief overview is given illustrating applications of atomic-scale simulation methods in the hierarchical-multiscale modeling of dendritic solidification in metals and alloys. Molecular-dynamics and Monte-Carlo simulation-based methods are described for calculating the bulk and interfacial, thermodynamic and kinetic properties entering mesoscale models of dendrite growth.

1. Introduction

Over the past decade significant advances have been realized in the development of mesoscale simulation methods for modeling morphological evolution of phase and defect microstructures. For a growing number of problems, phase-field and sharp-interface methods have been developed to the point that a primary factor limiting their application in quantitative, system-specific studies is a lack of experimentally measured values of the materials parameters required as input to the simulation models. Atomic-scale simulations have thus begun to play an expanding role in the modeling of microstructural evolution by providing a means for directly computing the thermodynamic and kinetic properties which govern morphological changes on microscopically large length and time scales. In this paper we provide specific examples describing the application of atomic-scale Molecular-Dynamics (MD) and Monte-Carlo (MC) simulation methods in calculations of bulk and interfacial properties related to dendritic

A. Finel et al. (eds.), Thermodynamics, Microstructures and Plasticity, 411–425.

growth in the solidification of metals and alloys. These examples are taken from recent work devoted to the integration of atomic-scale and phase-field methods as the basis for quantitative modeling of dendritic solidification [1]-[3]. Further examples illustrating integration of atomistic and mesoscale methods in the modeling of microstructure evolution can be found in recent studies of solid-state precipitation [4, 5], grain growth [6] and crystal nucleation from the melt [7].

In dendritic solidification of a pure element, the growth rate is typically governed by transport of latent heat away from the solid-liquid interface as the crystal solidifies. Modeling of dendritic growth thus requires solution of a diffusion equation for the temperature field subject to a boundary condition reflecting conservation of heat at the moving solid-liquid interface (e.g., [8]): $LV_n = c_p D \hat{n} \cdot (\vec{\nabla}T|_S - \vec{\nabla}T|_L)$, where L is latent heat, V_n is the normal interface velocity, and c_p and D denote, respectively, the specific heat and thermal diffusivity. The terms $\hat{n} \cdot \vec{\nabla}T|_S$ and $\hat{n} \cdot \vec{\nabla}T|_L$ give the normal gradients of temperature on the solid (S) and liquid (L) sides of the interface. The interface velocity V_n is governed by both capillary and kinetic factors as follows:

$$T_I = T_M - \frac{T_M}{L} \left[(\gamma(\hat{n}) + \gamma_1''(\hat{n})) \, \kappa_1 + (\gamma(\hat{n}) + \gamma_2''(\hat{n})) \, \kappa_2 \right] - \frac{V_n}{\mu(\hat{n})} \qquad (1)$$

where T_I and T_M denote the interface and bulk melting temperatures, respectively. The second term on the right-hand side of Eq. 1 represents the Gibbs-Thomson capillary correction to the equilibrium melting temperature at a curved interface (e.g., [9]) with local normal $\hat{n}$. In this term, κ_1 and κ_2 denote principal curvatures, $\gamma(\hat{n})$ is the solid-liquid interfacial free energy, while γ_1'' and γ_2'' denote second derivatives of γ with respect to interface orientation along the directions of the two principle curvatures. The final term in Eq. 1 gives the undercooling associated with ad-atom attachment kinetics, where $\mu(\hat{n})$ is the interface mobility, also referred to as the kinetic coefficient. For a flat interface, Eq. 1 reduces to $V_n = \mu(\hat{n})(T_M - T_I)$, with $\mu(\hat{n})$ the constant of proportionality between the interface velocity and undercooling.

From the above discussion it is clear that quantitative models of dendritic growth require precise knowledge of the capillary and kinetic interface parameters $\gamma(\hat{n})$ and $\mu(\hat{n})$. Furthermore, it is well established from both theoretical and numerical calculations (e.g., [1, 10]) that dendrite morphologies and growth kinetics are extremely sensitive to small changes in the crystalline anistropy of these interfacial parameters. Applications of quantitative phase-field models in the simulations of dendrite growth are thus hindered by the fact that solid-liquid interfacial-free-energy anisotropies have been measured for only a few transparent organic materials [11]-[13] and two metallic alloys (Al-Cu, and Al-Si) [14, 15], while no direct measurements of the anisotropy of $\mu(\hat{n})$ are available to date. The very recent quantitative phase-field study of Ni dendrite growth by Bragard *et al.* [1] was thus facilitated by the availability of values of $\gamma(\hat{n})$ and $\mu(\hat{n})$ derived from the atomistic calculations of Hoyt *et al.* [2, 3]. A critical step in this study of Ni dendrite growth was therefore the development of robust atomistic-simulation strategies (to be reviewed in Section 4) for calculating $\gamma(\hat{n})$ and $\mu(\hat{n})$ with sufficiently high statistical precision to resolve the small, yet critically important, crystalline anisotropies in these interfacial properties.

2. Interatomic Potentials and the Pair-Functional Formalism

Atomistic-simulation calculations of the transport and thermodynamic properties relevant to studies of microstructure evolution typically require simulation sizes and time scales well beyond the current reach of the most accurate *ab-initio* molecular-dynamics (AIMD) methods. Classical interatomic-potential models thus represent the only computationally tractable framework for performing the large-scale simulations required in this area of study. In this discussion we focus on a particular class of such interatomic-potential models which have come to be referred to as *pair-functionals* [16]. Over the past fifteen years such potentials have formed the basis for a wide variety of interface and defect simulation studies in metals and alloys (see, for example, [17]). These studies have proven highly successful particularly for metallic systems with closed d shells (where angular contributions to the interatomic forces can be safely neglected). Discussions of the range of applicability of pair-functional models, as well as descriptions of methods that go beyond such schemes, can be found in the books by Phillips [18] and Pettifor [19], the comprehensive reviews by Carlsson [16] and Heine *et al.* [20], as well as papers by Baskes [21], Pettifor *et al.* [22], and Moriarty and co-workers [23].

In a pair-functional model the total energy of a collection of ions with positions $\{\mathbf{R}_i\}$ takes the following form:

$$E(\{\mathbf{R_i}\}) = \sum_i \mathbf{U}\left(\sum_j \mathbf{g}(\mathbf{R_{ij}})\right) + \frac{1}{2}\sum_{i,j}\mathbf{V}(\mathbf{R_{ij}}) \qquad (2)$$

where V is a pair potential depending only on the distance between neighboring atoms ($R_{ij} = |\mathbf{R}_i - \mathbf{R}_j|$), while U gives a contribution to the energy of atom i that depends on its local environment, as characterized by the sum over the radial functions $g(R_{ij})$. The form of Eq. 2 can be motivated by two conceptually different physical pictures. Within a tight-binding model of the electronic structure (e.g., [16, 24]) the first-term on the right-hand side of Eq. 2 represents a second-moment approximation to the cohesive band energy, while the pair potential V gives rise to a repulsive interaction between ions at close distances. An alternative starting point for the derivation of Eq. 2 is based on an "embedding energy" [25]-[28] picture of metallic bonding. In this formalism, which underlies the development of effective-medium theory (EMT) [25, 26], and the embedded-atom method (EAM) [27, 28], the function U is viewed as the gain in cohesive energy associated with embedding a given atom i in the local background electronic density ρ due to the neighboring atoms. In the EAM and EMT models it is assumed that this background density can be derived from a sum of radial functions centered on the neighboring atoms j, $\rho = \sum_j g(R_{ij})$, where $g(R_{ij})$ thus corresponds to the (spherically averaged) atomic density at site i due to an atom at site j.

Insight into the physics embodied in the pair-functional model can be derived by recasting $U(\rho)$ in Eq. 2 as a Taylor-series expansion about a reference value of the local electron density $(\bar{\rho})$ [16, 17] (where for clarity we adopt the embedded-atom picture). Retaining only the term linear in $\rho - \bar{\rho}$ in the expansion of $U(\rho)$, Eq. 2 takes the following form:

$$E(\{\mathbf{R_i}\}) \approx \mathbf{E_0}(\bar{\rho}) + \frac{1}{2}\sum_{\mathbf{i,j}}\left[\mathbf{V(R_{ij})} + \mathbf{2U'}(\bar{\rho})\mathbf{g(R_{ij})}\right] \tag{3}$$

where E_0 is a structure-independent term, and the second contribution on the right-hand side of Eq. 3 is an effective pair potential: $\Phi(R_{ij}) = V(R_{ij}) + 2U'(\bar{\rho})g(R_{ij})$. In both the tight-binding and embedded-atom formulations U is a non-linear convex function of its argument $(U'' > 0)$. Consequently, the derivative of the embedding function $(U'(\bar{\rho}))$, and therefore $\Phi(R_{ij})$, are increasing functions of $\bar{\rho}$. This point is illustrated in Fig. 1 which plots $\Phi(R_{ij})$ for Al-Al and Al-Cu pairs, derived from an EAM potential developed for Al-rich Al-Cu alloys [29]. In deriving the effective pair potentials plotted in Fig. 1, reference densities $\bar{\rho}$ were taken for atoms in fcc crystalline environments; the solid and dashed lines were evaluated with atomic volumes for the lattice corresponding to the solid and liquid phases of bulk Al, respectively. The important point to note is that the higher atomic volume for the liquid phase leads to a lower value of $\bar{\rho}$, and consequently a pair potential with a deeper minimum.

Fig. 1 illustrates a local-environment dependence of the bond strengths derived from a pair-functional model. Specifically, atoms embedded in lower background densities, resulting from expanded local atomic volumes or reduced coordinations, share stronger bonds with their neighbors. This effect is critical to reproduce, for example, the experimentally observed fact that the vacancy formation energy in a metal is typically roughly half the co-

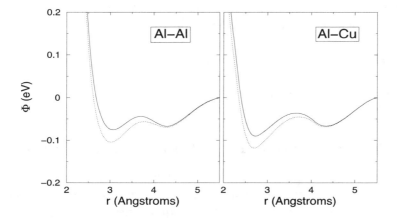

Figure 1. Solid and dashed lines plot effective Al-Al and Al-Cu pair potentials derived from embedded-atom potentials at reference electron densities corresponding to solid and liquid phases of elemental Al, respectively.

hesive energy, whereas the two properties would be equal in a simple pair-potential model [16, 19]. Since the function U in Eq. 2 is non-linear, higher order terms arise in the Taylor-series expansion leading to Eq. 3, and pair-functionals give rise to multi-body interatomic interactions. Furthermore, pair-functional methods are capable of reproducing the non-zero Cauchy pressures characterizing the elastic constants of many metallic systems. Pair functionals thus capture much of the essential physics of bonding in simple metals, at a computational cost comparable to that associated with simpler pair-potential models.

Pair-functional potentials have traditionally been derived by fitting the various functions in Eq. 2 to available room-temperature experimental data for bulk and defect properties in crystalline phases. In recent years improved transferability of the potentials and greater accuracies in calculated finite-temperature properties have been realized through the use of expanded fitting databases supplemented by the results of accurate *ab-initio* calculations (e.g., [30]-[37]). In the force-matching method (FMM) of Ercolessi and Adams [30], for example, the functional forms of U, g and V in Eq. 2 are modeled numerically (with cubic splines) to achieve optimal fits to databases including interatomic forces derived from finite-temperature AIMD trajectories constructed to sample a wide range of atomic environments. Schemes such as the FMM approach offer a promising framework for bridging between accurate first-principles calculations for relatively small systems, and the large-scale simulations required in studies related to microstructural evolution. The potentials used to generate Fig. 1

[29] were derived starting from the Al-Cu FMM potentials developed by Liu *et al.* [36], with modifications made to the Al-Cu pair potential to fit first-principles calculated energies and interatomic force data for Al-rich solid solutions, as well as the measured heats of solution in the liquid phase. Below we show that these potentials provide a realistical model of the solution-thermodynamic properties and solid-liquid phase diagram for dilute Al(Cu) alloys.

3. Bulk Thermodynamic Properties and Phase Diagrams

A first step in the calculation of the interfacial parameters $\gamma(\hat{n})$ and $\mu(\hat{n})$ entering Eq. 1 is a precise knowledge of the equilibrium phase diagrams governing coexistence of bulk solid and liquid phases, as predicted by the interatomic potential model employed in the simulations. Below we discuss MD and MC calculations of elemental melting points, alloy composition-temperature phase diagrams, and associated bulk thermodynamic properties. The discussions in this and the following sections assume a basic familiarity with MD and MC methods, as reviewed in Refs. [38]-[42].

Consider first the calculation of $\Delta g(T) = g_s(T) - g_l(T)$, the difference in molar Gibbs free energies between solid (g_s) and liquid (g_l) elemental phases as a function of bulk undercooling at constant pressure. Such free-energy differences provide the thermodynamic driving force for solidification, and thus enter phenomenological models of crystallization kinetics (e.g., [43]). From the following thermodynamic relation:

$$\left(\frac{\partial (g(T,P)/T)}{\partial T} \right)_P = -\frac{h(T,P)}{T^2} \tag{4}$$

Δg can be derived from an integration of the temperature-dependent enthalpy difference $h_s(T) - h_l(T)$ from the melting point (T_M, where $\Delta g(T) = 0$) to the desired temperature T at constant P. Fig. 2a shows calculated values of $\Delta g(T)$ for elemental Al modeled with the pair-functional potential of Ercolessi and Adams [30]. For these calculations the melting temperature was determined using the so-called coexistence approach [44, 45, 46], where T_M is obtained as the average temperature in an equilibrium MD simulation of coexisting solid and liquid phases in a microcanonical ensemble (fixed particle number, energy and volume).

The enthalpy data required to integrate Eq. 4 was obtained from MC calculations for solid and liquid Al employing 4000-atom simulations in which fluctuations in atomic positions and strain were sampled by attempting random ionic displacements (with a maximum value of 0.1 Å in each of the three Cartesian directions) and changes to the simulation-cell lengths (with maximum magnitude of 0.1 percent), respectively. Each

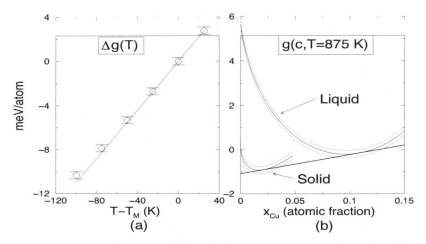

Figure 2. (a) Calculated values of $\Delta g(T)$ for elemental Al are plotted with open circles. The solid line corresponds to the linear relationship described in the text. (b) Thin solid lines correspond to Gibbs free energies calculated for solid and liquid Al-Cu alloys. The thick straight line gives the common-tangent. Dashed lines illustrate estimated statistical uncertainties in the calculated free energies.

such attempted change to the state of the system was accepted according to the Metropolis algorithm for an ensemble with fixed temperature, particle number and stress [47]. From linear fits to the resulting data for $h(T)$, Eq. 4 was integrated to compute the zero-pressure values of $\Delta g(T)$ plotted with open symbols in Fig. 2a. Also plotted with the solid line in Fig. 2a is the often-used linear relation $\Delta g^0(T) = L\Omega(T - T_M)/T_M$, where Ω is the atomic volume. This expression is seen be highly accurate for undercoolings up to approximately 50 K.

Consider now the computation of solidus and liquidus phase boundaries for binary alloys from MC-thermodynamic-integration calculations of composition-dependent Gibbs free energies. The starting point for these calculations is the following thermodynamic relation:

$$\left(\frac{\partial g}{\partial x}\right)_{T,P} = \mu_A(T, P) - \mu_B(T, P) \equiv \Delta\mu(T, P) \tag{5}$$

where x denotes the mole-fraction of solute species (A), while μ_A and μ_B are the chemical potentials for each of the chemical constituents in a binary $A - B$ alloy [48]. The strategy is to integrate Eq. 5 using the functional relationship between $\Delta\mu$ and x obtained from Monte-Carlo simulations at constant T and P. As detailed in Ref. [49], this relationship can be readily derived from Metropolis MC simulations performed within the so-called *semi-grand-canonical* (SGC) ensemble. In such simulations the total number of particles is fixed, while the alloy composition is allowed to fluctuate in a manner consistent with imposed values of $\Delta\mu$, T and P. In a SGC MC simulation for a binary alloy, compositional fluctuations are sampled through "transmutation" steps involving changes in the chemical identity of randomly-chosen particles. MC trial moves are accepted with probabilities governed by the Metropolis algorithm and the equilibrium partition function for a SGC ensemble, as discussed in detail by Frenkel [38, 50].

Fig. 2b plots composition-dependent alloy free energies for solid and liquid Al-Cu alloys derived from MC simulations employing the pair-functional potential represented in Fig. 1. These free energies were computed by integrating Eq. 5 from pure Al ($x = 0$) using the relationship between $\Delta\mu$ and the average composition ($< x >$) obtained in 4000-atom, zero-pressure, semi-grand-canonical MC simulations at 875 K. For pure Al the free energy is taken to be zero in the solid phase, while for the liquid it is taken from the values of Δg shown in Fig. 2a. The thick straight line in Fig. 2b denotes a common-tangent constructed from the calculated free-energy curves, indicating the solidus and liquidus phase-boundaries computed at this temperature. Similar calculations at temperatures of 860 and 900 K led to the calculated composition-temperature phase diagram plotted in Fig. 3a. Open symbols and solid lines represent calculated and measured [51] solidus and liquidus phase boundaries between 860 and 925 K. The agreement between experiment and theory is seen to be quite reasonable, with calculated solidus and liquidus compositions reproducing the eutectic nature of the phase diagram, as well as the high measured value of the equilibrium partition coefficients. A further comparison between our calculations and experiment is given in Fig. 3b, where open symbols and solid lines denote measured [52] and computed activities for liquid Al-Cu alloys at 1373 K. It can be seen that the highly non-ideal solution-thermodynamic properties measured for Al-Cu liquid alloys are well reproduced.

4. Static and Dynamic Properties of Solid-Liquid Interfaces

The calculation of γ and its associated crystalline anisotropy for atomically-rough solid liquid interfaces has been the focus of a number of MD studies in both elemental [3], [53]-[58], and alloy [59] systems. In these calculations use has been made of two distinct approaches. In the thermodynamic-integration technique pioneered by Broughton and Gilmer [53] externally imposed "cleaving potentials" are employed to calculate the reversible work to form a crystal-melt interface from isolated bulk solid and liquid phases [53, 54, 55]. Recently, Hoyt et al. [3] demonstrated an alternative approach for the calculation of γ from an analysis of equilibrium capillary fluctuations in MD simulations for molecularly rough solid-liquid interfaces. This

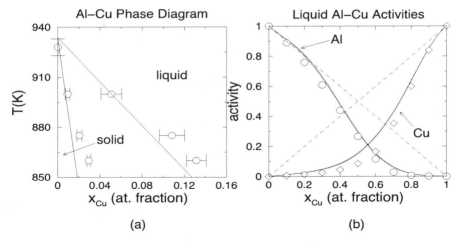

Figure 3. (a) Measured (solid lines) and calculated (open symbols) solidus and liquidus boundaries for Al-Cu alloys. (b) Calculated (solid lines) and measured (open symbols) activities for liquid Al-Cu at 1373 K.

latter approach is based upon the relation between the equilibrium capillary fluctuation spectrum of the solid-liquid interface and its *stiffness*. For a one-dimensional interface the relation takes the form:

$$< |A(k)|^2 > = \frac{k_B T_M}{a(\gamma + \gamma'')k^2} \tag{6}$$

where a is the area of the solid-liquid interface, $A(k)$ is the Fourier transform of the interface height profile and angular brackets denote equilibrium values. The term $\gamma + \gamma''$ corresponds to the interface stiffness, where γ'' is the second-derivative of γ as a function of the angle of the local interface normal relative to its average orientation, as in Eq. 1.

As an example illustrating the application of the capillary-fluctuation approach to calculations of solid-liquid interfacial free energies, Fig. 4a plots on a log-log scale values of $< |A(k)|^2 >$ vs. wavenumber derived from MD simulations for elemental Ni [60] modeled with the EAM potential of Foiles *et al.* [61]. The results in Fig. 4a were derived from microcanonical MD simulations for three different interface orientations using both quasi-two-dimensional and three dimensional cells containing between $5 \times 10^4 - 10^5$ atoms. Sampling of interface fluctuations was performed in simulations lasting on the order of 1 ns. The method used to distinguish between solid and liquid atoms and the procedure for identifying the interface boundary are described in detail in Ref. [3]. For each interface orientation the predicted k^{-2} dependence of $< |A(k)|^2 >$ is observed to hold to within the estimated

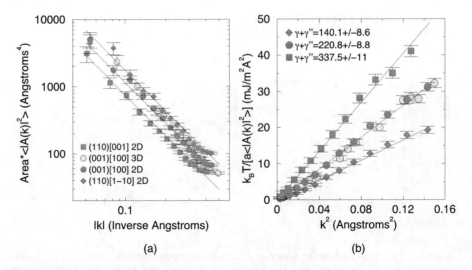

Figure 4. MD-calculated interface-fluctuation spectra for elemental Ni. Open and filled results correspond to results obtained with three-dimensional and quasi-two-dimensional simulation-cell geometries, respectively.

statistical accuracy of the MD data for small and intermediate wavenumbers. At the highest values of k the fluctuation wavelengths approach atomic dimensions and the observed deviations from the predictions of Eq. 6 are thus expected. Stiffness values can be calculated from the MD data through a least-squares fit of $k_B T/[bW < |A(k)|^2 >]$ vs. k^2, as illustrated in Fig. 4b.

From the stiffness values given in Fig. 4b the orientation dependence of the interfacial free energy can be computed using an analytical expansion of $\gamma(\hat{n})$ in terms of cubic harmonics [3, 59] (i.e., linear combinations of spherical harmonics consistent with the cubic symmetry of the crystal). In previous studies for elemental Ni [3] and Al [58] it has been found that the anisotropy of $\gamma(\hat{n})$ can be accurately parametrized using the fourth and six-order terms of such an expansion. The three coefficients in such an expansion can thus be computed from the stiffness results in Fig. 4b to obtain: $\hat{\gamma} = 285 \pm 7 mJ/m^2$, $[\gamma_{100} - \gamma_{110}]/2\hat{\gamma} = 0.014 \pm 0.002$ and $[\gamma_{100} - \gamma_{111}]/2\hat{\gamma} = 0.032 \pm 0.005$, where $\hat{\gamma}$ denotes the orientationally-averaged solid-liquid interfacial free-energy. It is seen that the fluctuation approach outlined above allows interfacial free energies and associated anisotropies to be computed with relatively low statistical uncertainties of 2% and 15%, respectively. The anisotropies of $\gamma(\hat{n})$ are seen to be on the order of a few percent, as has been found in related EAM-based calculations for Al [57, 58], Au and Ag [56], Cu and Pb [62], and a Cu-Ni alloy [59]. These anisotropies are also in the range of values measured recently by Liu, Napolitano and

Trivedi for Al-Cu and Al-Si metallic alloys [14, 15].

Our calculated $\hat{\gamma}$ is in excellent agreement with a value predicted by Laird's [63] recently-proposed generalization of the Turnbull model [64]. In this model Laird assumes that the excess free energy of solid-liquid interfaces in simple fcc metals is dominated by the contribution due to geometrical hard-sphere packing entropy. This assumption, combined with the results of recent MD-calculations of γ for hard-spheres [55], allows one to estimate a value for the Turnbull coefficient $C_T \equiv \gamma \rho^{-2/3} N_A / L$ (where ρ is the atomic density and N_A is Avogadro's number) equal to 0.51. For the Ni system considered here Laird's model predicts $\hat{\gamma} = 278 mJ/m^2$, which is within five percent of our calculated value.

The above discussion demonstrates a method for computing $\gamma(\hat{n})$ from the static capillary-fluctuation spectrum of a solid-liquid interface. As described recently by Hoyt et al. [65], an analysis of fluctuation *dynamics* also provides a method for extracting the kinetic coefficient $\mu(\hat{n})$ defined in Eq. 1. For the relatively small wavelengths probed in MD simulations, the relaxation rate of capillary fluctuations is expected to be limited by atomic attachment kinetics [66]. In other words, if the temperature field relaxes much faster than the height of a capillary fluctuation, we can set $T_I = T_M$ in Eq. 1 to obtain an expression relating the local velocity of the interface (V) to its curvature. An analysis based on the Langevin formalism of Ref. [66] thus yields a value for the interface relaxation time (τ):

$$\tau(k) = \frac{L}{\mu(\gamma + \gamma'')T_M k^2}. \tag{7}$$

Fig. 5a plots values of the autocorrelation function $C(t) = < A(k,t)A^*(k,0) >$ derived from MD simulations for (100) solid-liquid interfaces in Ni [60]. The MD data shows the expected exponential decays $C(t) = < |A(k)|^2 > \exp(-t/\tau)$ for fluctuation wavenumbers ranging from $|k| = 0.085 \mathring{A}^{-1}$ to $0.5 \mathring{A}^{-1}$.

With the values of $\tau(k)$ derived from the exponential fits in Fig. 5a, Eq. 7 yields a value of the kinetic coefficient for a Ni (100) solid-liquid interface of $\mu = 39 \pm 4 cm/s/K$. This value is roughly five times lower than an upper bound provided by the model of Turnbull [67] (widely used in dendrite-growth models) which relates μ to the speed of sound. The magnitude of our calculated kinetic coefficient is, however, consistent with more recent theories for the mobilities of elemental, atomically-rough solid-liquid interfaces [68, 69], which relate the kinetic coefficient to the average thermal velocity: $v_T = (k_B T/m)^{1/2}$, where m is the atomic mass. In these theories, $\mu \propto v_T/T_M$, where the best fit to the MD results of Broughton, Gilmer and Jackson [68] for Lennard-Jones (100) interfaces yields a constant of proportionality ~ 1.2 [69]. For Ni, the resulting estimate of this model

422

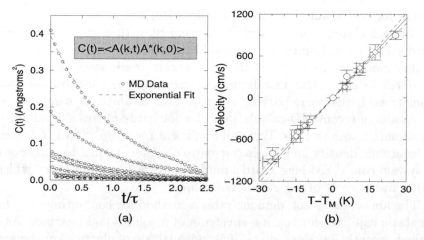

Figure 5. (a) Decay of interface fluctuations in elemental Ni calcualted by MD. (b) Comparison of linear solid-liquid interface growth kinetics derived from equilibrium (lines) and non-equilibrium (open symbols) MD methods, respectively.

for the kinetic coefficient is $\mu \approx 41 cm/s/K$, in very good agreement with the present MD results.

As reviewed recently [65], a number of alternative methods have been proposed for the calculation of μ by MD simulations. In the most commonly-used approach (e.g., [70]) μ is derived through the linear relation between measured growth rates and imposed values of $\Delta T = T - T_M$ in the limit of small undercoolings. In such non-equilibrium MD simulations, use is typically made of Nosé-Hoover [71, 72] and Parinello-Raman [73] dynamics to maintain constant values of the temperature and pressure (stress). Fig. 5b features a comparison between results for Ni(100) obtained from such non-equilibrium MD simulations vs. the equilibrium fluctuation approach described above. The solid line corresponds to the linear growth law derived from the value of μ extracted from capillary-fluctuation kinetics, while the open symbols plot measured interface velocities as a function of imposed undercooling in non-equilibrium MD simulations [60]. It can be seen that to within the estimated statistical uncertainties, each of these methods gives consistent results for the kinetic coefficient μ.

From the stand-point of dendrite-growth modeling, it is important to know not only the magnitude of μ, but also the strength of its crystalline anisotropy. The non-equilibrium MD simulations of Hoyt *et al.* [2, 56, 65] and Celestini and Debierre [74], performed for a variety of fcc metals, have produced the result $\mu_{100} > \mu_{110} > \mu_{111}$. Further, the ratio μ_{100}/μ_{110} is found to be ~ 1.4. This result is consistent with the model of both

Broughton, Gilmer and Jackson [68, 43], where the mobility ratio can be related to the ratio of interplanar spacings $\mu_{100}/\mu_{110} \approx d_{100}/d_{110} = \sqrt{2}$. A similar result is obtained from the theory of Mikheev and Chernov [69]. Both of these theoretical models, however, predict $\mu_{111} > \mu_{110}$, in contrast to the MD results, and further work is required to clarify the source of this discrepancy.

MD calculations and theoretical models for the growth of rough solid-liquid interfaces both predict substantial anisotropy of the kinetic coefficient, with $[\mu_{100} - \mu_{110}]/[\mu_{100} + \mu_{110}] \approx 0.17$. While this anisotropy is comparable to that of the interface stiffness, the kinetic coefficient has been taken as isotropic in most phase-field modeling of dendrite growth. Recently, the anisotropy of $\mu(\hat{n})$ has been shown to have important consequences for dendritic solidication in highly undercooled melts [1]. Specifically, in three-dimensional phase-field simulations for Ni, Bragard et $al.$ [1] examined the sensitivity of calculated dendrite growth rates to changes in the input values of the anisotropies for μ and γ. At an undercooling of 250 K, calculated solidification rates (V) were practically unaffected by removing the anisotropy in γ, while a change in the anisotropy of μ from the calculated value of 0.17 to zero led to roughly a factor-of-two decrease in V accompanied by a change from dendritic to dense-branching growth morphologies. The results of the simulation study of Bragard et $al.$ [1] clearly illustrate the critical role of anisotropic interfacial properties in dendritic solidification. These results also point to the need for further development of atomistic simulation methods to compute $\mu(\hat{n})$ and $\gamma(\hat{n})$ in future quantitative simulation studies of dendritic solidification in metals and alloys.

5. Acknowledgements

This research was supported by the U.S. Department of Energy, Office of Basic Energy Sciences, Materials Science Division, under Contracts No. DE-FG02-01ER45910 and DE-FG02-92ER45471, as well as the DOE Computational Materials Science Network program. Use was also made of resources at the National Energy Research Scientific Computing Center, which is supported by the Office of Science of the Department of Energy under Contract No. DE-AC03-76SF00098. We are grateful to Alain Karma and Axel van de Walle for numerous helpful suggestions.

References

1. Bragard, J., Karma, A., Lee, Y. H. and Plapp, M. (2002) *Int. Science*, **Vol. 10**, pg. 121.
2. Hoyt, J. J., Sadigh, B., Asta, M., and Foiles, S. M. (1999) *Acta Mater.*, **Vol. 47**, pg. 3181.
3. Hoyt, J. J., Asta, M., and Karma, A. (2001) *Phys. Rev. Lett.*, **Vol. 86**, pg. 5530.

424

4. Chen, L. Q., Wolverton, C., Vaithyanathan, V., and Liu, Z. K. (2001) *MRS Bulletin*, **Vol. 26**, pg. 192.
5. Vaithyanathan, V., Wolverton, C., and Chen, L. Q. (2002) *Phys. Rev. Lett.*, **Vol. 88**, art. no. 125503.
6. Upmanyu, M., Kazaryan, A., Holm, E. A., Wang, Y., Patton, B., and Srolovitz, D. J. (2002) *Int. Science*, **Vol. 10**, pg. 201.
7. Gránásy, L., Börzsönyi, T., and Pustzai, T. (2002) *Phys. Rev. Lett.*, **Vol. 88**, art. no. 206105.
8. Davis, S. H. (2001), *Theory of Solidification* (Cambridge University Press, New York).
9. Trivedi, R., in *Lectures on the Theory of Phase Transformations, 2nd Edition*, ed. by H. I. Aaronson (TMS Publishing).
10. Karma, A. and Rappel, W. (1998), *Phys. Rev. E*, **Vol. 57**, pg. 4323.
11. Huang, S.-C. and Glicksman, M. E. (1981) *Acta Metall. Mater.*, **Vol. 29**, pg. 701.
12. Glicksman, M. E. and Singh, N. B., (1989), *J. Cryst. Growth*, **Vol. 98**, pg. 277.
13. Muschol, M., Liu, D., and Cummins, H. A. (1992), *Phys. Rev. A*, **Vol. 46**, pg. 1038.
14. Liu, S., Napolitano, R. E., and Trivedi, R. (2001), *Acta Mater.*, **Vol. 49**, pg. 4271.
15. Napolitano, R. E., Liu, S., and Trivedi, R. (2002), *Interface Science*, **Vol. 10**, pg. 217.
16. Carlsson, A. E. (1990), *Solid State Phys.*, **Vol. 43**, pg. 1.
17. Daw, M. S., Foiles, S. M., and Baskes, M. I. (1993), *Mater. Sci. Rep.*, **Vol. 9**, pg. 251.
18. Phillips, R., (2001) *Crystals, Defects and Microstructures - Modeling Across Scales*, (Cambridge University Press, New York).
19. Pettifor, D. (1995) *Bonding and Structure of Molecules and Solids* (Oxford, New York, 1995).
20. Heine, V., Robertson, I. J., and Payne, M. C. (1991), *Phil. Trans. Roy. Soc. Lond.*, **Vol. A334**, pg. 393.
21. Baskes, M. I. (1992), *Phys. Rev. B*, **Vol. 46**, pg. 2727.
22. Pettifor, D. G., Oleinik, I. I., Nguyen-Manh, D., and Vitek, V. (2002), *Comput. Mater. Sci.*, **Vol. 23**, pg. 33.
23. Moriarty, J. A., Belak, J. F., Rudd, R. E., Soderlind, P., Streitz, F. H., and Yang, L. H. (2002), *J. Phys.-Condens. Matter*, **Vol. 14**, pg. 2825.
24. Finnis, M. W., and Sinclair, J. E. (1984), *Phil. Mag. A*, **Vol. 50**, pg. 45.
25. Nørskov, J. K., and Lang, N. D. (1980), *Phys. Rev. B*, **Vol. 21**, pg. 2131.
26. Jacobsen, K. W., Nørskov, J. K., and Puska, M. J. (1987), *Phys. Rev. B*, **Vol. 35**, pg. 7423.
27. Daw, M. S., and Baskes, M. I. (1983), *Phys. Rev. Lett.*, **Vol. 50**, pg. 1285.
28. Daw, M. S., and Baskes, M. I. (1984), *Phys. Rev. B*, **Vol. 29**, pg. 6443.
29. Asta, M. and Hoyt, J. J., unpublished.
30. Ercolessi, F., and Adams, J. B. (1994), *Europhys. Lett.*, **Vol. 26**, pg. 583.
31. Mishin, Y., Farkas, D., Mehl, M. J., Papaconstantopolous, D. A. (1999), *Phys. Rev. B*, **Vol. 59**, pg. 3393.
32. Lenosky, T. J., Sadigh, B., Alonso, E., Bulatov, V. V., de la Rubia, T. D., Kim, J., Voter, A. F., and Kress, J. D. (2000), *Model. Simul. Mater. Sci. and Engin.*, **Vol. 8**, pg. 825.
33. Belonoshko, A. B., Ahuja, R., Erikkson, O., and Johansson, B. (2000), *Phys. Rev. B*, **Vol. 61**, pg. 3838; Belonoshko, A. B., Ahuja, R., and Johansson, B. (2000), *Phys. Rev. Lett.*, **Vol 84**, pg. 3638.
34. Liu, X. Y., Adams, J. B., Ercolessi, F., and Moriarty, J. A. (1996), *Model. Simul. Mater. Sci. and Engin.*, **Vol. 4**, pg. 293.
35. Liu, S. Y., Ohotnicky, P. P., Adams, J. B., Rohrer, C. L., and Hyland, R. W. (1997), *Surf. Sci.*, **Vol. 373**, pg. 357.
36. Liu, X. Y., Xu, W., Foiles, S. M., and Adams, J. B. (1998), *Appl. Phys. Lett.*, **Vol. 72**, pg. 1578.

425

37. Landa, A., Wynblatt, P., Siegel, D. J., Adams, J. B., Mryasov, O. N., and Liu, X. Y. (2000), *Acta Mater.*, **Vol. 48**, pg. 1753.
38. Frenkel, D., and Smit, B. (1996), *Understanding Molecular Simulation: From Algorithms to Applications*, (Academic Press, New York).
39. Allen, M. P., and Tildesley, D. J. (1987), *Computer Simulation of Liquids*, (Clarendon Press, Oxford).
40. Allen, M. P., and Tildesley, D. J. (1993), editors, *Computer Simulation in Chemical Physics, NATO ASI Series C*, **Vol. 397** (Kluwer Academic Publishers, Boston).
41. Binder, K., and Heerman, D. W. (1997), *Monte Carlo Simulation in Statistical Physics: An Introduction*, Third Edition (Springer, Berlin).
42. Newman, M. E. J. and Barkema, G. T. (1999), *Monte Carlo Methods in Statistical Physics*, (Clarendon Press, Oxford).
43. Jackson, K. A. (2002), *Interface Science*, **Vol. 10**, pg. 159.
44. Morris, J. R., Wang, C. Z, Ho, K. M., and Chan, C. T. (1994), *Phys. Rev. B*, **Vol. 49**, pg. 3109.
45. Alfé, D., Gillan, M. J., and Price, G. D. (2002), *J. Chem. Phys.*, **Vol. 116**, pg. 6170.
46. Morris, J. R., and Song, X. Y. (2002), *J. Chem. Phys.*, **Vol. 116**, pg. 9352.
47. Najafabadi, R., and Yip, S. (1983), *Script Metall.*, **Vol. 17**, pg. 1199.
48. Note that in Section 3 μ denotes a chemical potential which should not be confused with the kinetic coefficient discussed in Sections 1 and 4.
49. Ramalingam, H., Asta, M., van de Walle, A., and Hoyt, J. J., *Interface Science*, **Vol. 10**, pg. 149.
50. Frenkel, D. (1993), *Advanced Monte Carlo Techniques*, in *NATO ASI Series C*, **Vol. 397**, edited by M. P. Allen and D. J. Tildesley (Kluwer Academic Publishers), pg. 93.
51. Murray, J. L. (1985), *International Metals Review*, **Vol. 30**, pg. 220.
52. Hultgren, R. (1973), *Selected Values of the Thermodynamic Properties of Binary Alloys* (ASM International, Metals Park, Ohio).
53. Broughton, J. Q., and Gilmer, G. H. (1986), *J. Chem. Phys.*, **Vol. 84**, pg. 5759.
54. Ravelo, R., and Baskes, M. I. (1996), *MRS Symp. Proc.*, **Vol. 398** (1996).
55. Davidchack, R. L. and Laird, B. B. (2000), *Phys. Rev. Lett.*, **Vol. 85**, pg. 4751.
56. Hoyt, J. J. and Asta, M. (2002), *Phys. Rev. B*, **Vol. 65**, art. no. 214106.
57. Morris, J. R., Lu, Z. Y., Ye, Y. Y., and Ho, K. M. (2002), *Int. Science*, **Vol. 10**, pg. 143.
58. Morris, J. R. (2002), *Phys. Rev. B* (in press).
59. Asta, M., Hoyt, J. J., and Karma, A. (2002), *Phys. Rev. B* (in press).
60. Sun, D. Y., Hoyt, J. J., Asta, M., and Karma, A., unpublished.
61. Foiles, S. M., Baskes, M. I., and Daw, M. S. (1986), *Phys. Rev. B*, **Vol. 33**, pg. 7983.
62. Hoyt, J. J., and Asta, M., unpublished.
63. Laird, B. B. (2001), *J. Chem. Phys.*, **Vol. 115**, pg. 2887.
64. Turnbull, D. (1950), *J. Appl. Phys.*, **Vol. 21**, pg. 1022.
65. Hoyt, J. J., Asta, M., and Karma, A. (2002), *Int. Science*, **Vol. 10**, pg. 181.
66. Karma, A. (1993), *Phys. Rev. Lett.*, **Vol. 70**, pg. 3439; *Phys. Rev. E*, **Vol. 48**, pg. 3441.
67. Turnbull, D. (1981), *Metall. Trans. A*, **Vol. 12**, pg. 695.
68. Broughton, J. Q., Gilmer, G. H., and Jackson, K. A. (1982), *Phys. Rev. Lett.*, **Vol. 49**, pg. 1496.
69. Mikheev, L. V., and Chernov, A. A. (1991), *J. Cryst. Growth*, **Vol. 112**, pg. 591.
70. Tepper, H. L., and Briels, W. J. (2001), *J. Chem. Phys.*, **Vol. 115**, pg. 9434.
71. Nosé, S. (1984), *J. Chem. Phys.*, **Vol. 81**, pg. 811; *Mol. Phys.*, **Vol. 52**, pg.255.
72. Hoover, W. G. (1985), *Phys. Rev. A*, **Vol. 31**, pg. 1695; (1986), *Phys. Rev. A*, **Vol. 34**, pg. 2499.
73. Parinello, M., and Rahman, A., (1980), *Phys. Rev. Lett.*, **Vol. 45**, pg. 1196; (1981), *J. Appl. Phys.*, **Vol. 52**, pg. 7182; (1982), *J. Chem. Phys.*, **Vol. 76**, pg. 2662.
74. Celestini, F., and Debierre, J. M. (2001), *Phys. Rev. E*, **Vol. 65**, art. no. 041605.

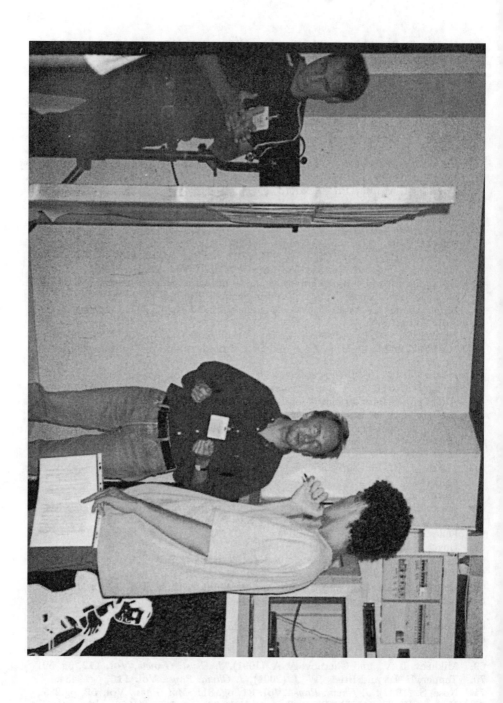

APPLICATIONS OF MONTE CARLO SIMULATIONS TO THE KINETICS OF PHASE TRANSFORMATIONS

F. SOISSON
Service de Recherches de Métallurgie Physique
CEA Saclay, DMN/SRMP, 91191 Gif-sur-Yvette Cedex, France

1. Introduction

The phase transformation kinetics in alloys are now widely studied by atomistic Monte Carlo (MC) simulations. The first MC methods used to study precipitation or ordering kinetics were based of simple algorithms, such as the standard Metropolis algorithm, derived mainly from equilibrium considerations. Recent progresses [1] have been achieved with the development of fully Kinetic Monte Carlo methods (KMC) which take into account both the *equilibrium* properties (i.e. they fulfil the detailed balance principle and drive the system towards its proper equilibrium state) and the *kinetic* properties of an alloy. It has been shown how the details of diffusion mechanisms (vacancy or interstitial jumps), migration barriers, and correlation effects can control the kinetic pathway and the time scale of a phase transformation [2-10].

We present here some recent applications of these methods to systems of industrial interest. We first briefly recall the atomistic diffusion model and MC algorithm we used, paying special attention to the parametrization method used to fit the properties of a given alloy. We present various examples of phase transformations of increasing complexity: simple copper precipitation in iron, precipitation and ordering in nickel based superalloys, precipitation of niobium carbide in steels (where two diffusion mechanisms compete). The Monte Carlo simulations are used to test the main assumptions and the limitations of more simple models and classical theories used in the industry (e. g. the classical theory of incubation and nucleation, the cluster dynamics models, etc...).

2. Atomistic Diffusion Models and Monte Carlo Algorithms

2.1. Atomistic Diffusion Model

The diffusion model and the Residence Time Algorithm (RTA) we use have been explained in previous papers [1,4-9]. We briefly recall that various atoms A, B, C, etc. interact through constant pair interactions on a rigid lattice. Diffusion occurs by thermally activated jumps of vacancies (or interstitials), with jump frequencies such as:

A. Finel et al. (eds.), Thermodynamics, Microstructures and Plasticity, 427–436.

$$\Gamma_{AV} = \nu_A \exp\left(-\frac{\Delta E_{AV}^{act}}{k_b T}\right) \qquad (1)$$

for an A-V exchange. ν_A is an attempt frequency and the activation energy is:

$$\Delta E_{AV}^{act} = \mathcal{E}_{sp}^{tot} - \mathcal{E}_{ini}^{tot} = e_{sp}^A - \sum_i V_{Ai} - \sum_j V_{jV} \qquad (2)$$

It is computed as the difference between the *total* energy of the system when the jumping atom is at the saddle-point (SP) position ($\mathcal{E}_{sp}^{tot}$) and the *total* energy of the system before the jump ($\mathcal{E}_{ini}^{tot}$). In the frame of a simple broken bond model, ΔE_{AV}^{act} can be written as the binding energy e_{sp}^A between the A atom and the system when A is at the SP position, minus the V_{ij} which are broken during the jump.

2.2. Residence Time Algorithm

At each Monte Carlo Step (MCS), one among the n possible jumps is chosen (for a simulation box with one vacancy and no interstitial n is simply the coordination number). The probability to leave a given configuration by a specific jump i is given by $\Gamma_i / \sum_{j=1}^n \Gamma_j$ where j labels all the possible diffusion events. The physical time that corresponds to the MCS is: $t_{MCS} = (\sum_{j=1}^n \Gamma_j)^{-1}$. Then, the KMC method and the RTA algorithm provide the kinetic pathway of the alloy, i.e. the sequence of configurations the system will go through and the time at which it reaches each configuration. This kinetic pathway depends only on the set of jumps frequencies (1-2) and on the point defect concentrations.

2.3. Parametrization

To reproduce the diffusion and thermodynamics properties of a given system, all the parameters involved in the jump frequencies must then be fitted on the proper data (if available). The pair interactions can be chosen in order to reproduce the cohesive energy of pure metals (e. g.: $E_{coh}^i = -zV_{ii}/2$ if pair interactions are restricted to the first-nearest neighbors), the point defect formation energies (in pure i: $E_{for}^v(i) = -zV_{ii}/2 + zV_{iv}$) and the phase diagram (which depend on the mixing energy: $V = V_{AA} + V_{BB} - 2V_{AB}$). Both the attempt frequency and the saddle-point binding energies are fitted [6,7] on the diffusion coefficients using the Le Claire and Lidiard theory of diffusion in alloys [11].

2.4. Absolute Time Scale

Point defect concentrations used in the simulation boxes (typically one vacancy for 10^6 lattice sites) are usually larger than the actual ones: the time given by the RTA must be rescaled. During the simulation the relative point defect concentrations in each phases will tend to be fixed according to their respective formation energies. If no strong vacancy trapping effects occurs the vacancy concentration in the solid solution will be almost constant and the RTA time will rescaled according to: $t = t_{MCS} \times (C_V^{MC}/C_V^{exp})$. Where $C_V^{MC} = N_v/N$ is the vacancy concentration is the simulation box (N_v: number of vacancies, N: number of lattice) and $C_V^{exp} = \exp(-\Delta E_v^{for}/k_b T)$ in the real system [12]. In case of strong trapping in one of the phases, the evolution of the vacancy concentration in the solid solution must be measured during the simulation and the correction of the time scale is no more a constant factor [7].

3. Copper precipitation in α-Fe

Although pure copper has a face centered cubic (FCC) structure, the first step of copper precipitation in the body-centered cubic lattice of the α-iron is coherent (as long as the precipitates radius is smaller than ~ 2 nm). Monte Carlo simulations on rigid lattice have then been applied to this system, with the parametrization method described in the previous section [12]. The evolution of the precipitation microstructure (number and size of precipitates, precipitate distribution, morphology of the precipitate/matrix interface, etc.) have been simulated and successfully compared to electrical resistivity measurements and to 3D atom probe observations [12].

3.1. EAM Potential – Test of the Rigid Lattice Model (RLM)

However, it is often difficult to find all the necessary experimental data: in case of Fe-Cu alloys for example, the V_{CuCu} pair interactions cannot be fitted on the cohesive energy of the BCC copper, which is not experimentally known (in ref. [12] it was then kept as a free parameter to fit the experimental kinetics). The saddle-point binding energies of Fe and Cu *in pure iron* can be fitted on the self-diffusion coefficient of iron and on the impurity diffusion of Cu in iron, but the dependence of e_{sp}^{Fe} and e_{sp}^{Cu} on the local atomic configuration would require the measurement of Fe and Cu tracer diffusion coefficients in solid solutions of various Cu compositions. In such cases, it can be useful to use a more realistic potential such EAM empirical potential or *ab initio* calculations.

Le Bouar et al. have used the iron-copper EAM potential developed by Ludwig and Farkas [14] to parametrize a KMC simulation and to test the rigid lattice approximation. It has been shown that the *relaxed* energies of dilute Fe-Cu configurations computed with the EAM potential (with a conjugate gradient algorithm) can be reproduced on a rigid lattice with first nearest-neighbor, including vacancy-atom interactions [7]. Migration barriers of Fe and Cu atoms in pure iron or in dilute alloys have been computed by the usual static technique [7]. The saddle-point binding energies are then obtained using eq. (2) and the previous first-neighbor interactions.

The SP binding energies of iron and copper atoms are significantly different [7]. Furthermore, e_{sp}^{Fe} strongly depends on the atomic configuration surrounding the SP position (while e_{sp}^{Cu} does not). In the BCC structure, the SP position has six nearest-neighbor sites. Computing more than 30 different migration barriers corresponding to various occupations of this six sites by Fe or Cu atoms, it is found that e_{sp}^{Fe} increases linearly with the number of Cu atoms, but is not sensitive to their respective positions among the six sites. In other words, the SP binding energies of Fe and Cu can be written as sum of pair interactions between the SP atom and its six first neighbors.

This dependence of the SP binding energies does not modify thermodynamic properties of the system (since the solubility limit, the interfacial energies, the vacancy concentrations in various phases do not depend on the SP properties). Moreover, it is has been shown that it only slightly affects the diffusion coefficients of Fe and Cu in iron. Nevertheless, it can strongly affect its precipitation kinetic pathway. Figure 1 displays the evolution of the copper short order parameter α_{Cu}, the number and the size of copper precipitates as measured in KMC simulations, using two sets of parameters derived from the EAM potential of Ludwig and Farkas: these two sets only differs by the value e_{sp}^{Fe}, which depends on the local atomic configuration in set 1, while the dependence is not taken into account in set 2.

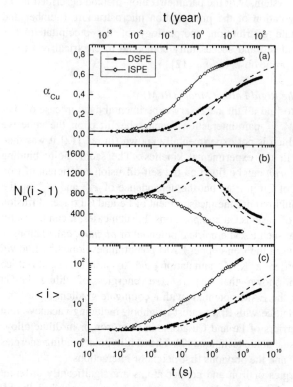

Figure 1 : Precipitation kinetics in a Fe-3at.%Cu alloy at $T = 573$ K. Evolution of (a) the degree of the copper short-range order, (b) the number of supercritical precipitates and (c) the averaged size of supercritical precipitates. Monte Carlo simulations with e_{sp}^{Fe} depending on the local atomic configuration (•) or not (○).

The difference completely modifies the kinetic pathway especially at low temperature: for example, the nucleation regime observed on fig. 1(b) when e_{sp}^{Fe} depends on the local atomic configuration disappears when e_{sp}^{Fe} is constant. This can be indeed related to the migration of small Cu clusters. With the two sets of parameters, the vacancies are strongly trapped in the BCC precipitates (the cohesive energy and the vacancy formation energy are lower in BCC copper than in BCC iron) : this results in a mobility of small Cu clusters, which accelerate the precipitation kinetics [4,5]. Although a modification of e_{sp}^{Fe} does not change the thermodynamic vacancy trapping in both phases (i.e. the time spent by the vacancy in each phase), it changes the mobility of small Cu clusters. When the SP binding energies increase in a given phase, the vacancy spends indeed the same amount of time in this phase, but by a small number of slow jump rather than by a big number of rapid ones. With the second set of parameter, e_{sp}^{Fe} is lower than with the first one in the Cu precipitates and near the Cu/Fe interfaces : the small (< 5 atoms) copper clusters are more mobiles than Cu monomers and the precipitation kinetics is accelerated. Furthermore, because two small copper clusters can meet more rapidly than a Cu monomer can reach a Cu precipitate, the nucleation regime of fig. 1(b) vanishes.

3.2. Limitations of EAM potentials

The possibility to find RLM parameters which fit the EAM potential properties (i.e. its configurational and saddle-point energies) is clearly linked to the small size effect between Fe and Cu atoms. Such an analysis has indeed appeared as impossible in some alloys with strong size effect, such as Ni-Au and Ni-Al alloys [13]. Furthermore it must be noticed that EAM are still empirical potentials and that the previous results could depend on the particular potential used: they must be consider as possible effects typical of what one can expect in this class of phase transformations and alloys. *Ab initio* calculations could give more predictable simulations (see 4.2).

4. Precipitation and ordering

The parameters which control the kinetic pathway during simple unmixing (mainly the asymmetry of cohesion $V_{AA} - V_{BB}$ [4,5], which controls the vacancy trapping and the asymmetry of SP binding energies $e_{sp}^A - e_{sp}^B$ [7]) have similar effects in ordered phases where they control complex diffusion events (e.g. the 6-jump cycles in B2 phases [8]) and the morphology of growing ordered zones and antiphase boundaries [9]. They must be carefully fitted in order to reproduce the kinetics of a phase transformations.

4.1. γ' precipitation in Ni-Cr-Al alloys

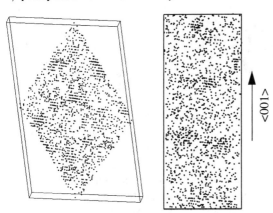

⟨100⟩

Figure 2: Microstructure of a Ni-15at.%Cr-5at.%Al alloy after a thermal ageing of 1h at 600°C. (a) Monte Carlo Simulation and (b) 3D Atom Probe (each dot corresponds to an Al atom)

Figure 2 gives an example of microstructure in a Ni-Cr-Al superalloy where Al-rich (L1$_2$) γ'-precipitates are formed in the Ni rich γ solid solution [13]. The results are compared to a direct 3D atom probe observation. In weakly super-saturated solid solutions γ' precipitates are found to be well ordered, with nearly equilibrium compositions, from the very beginning of the thermal ageing (fig. 2). One observes classical growth and coarsening regimes. Monte Carlo simulations gives similar results: the evolution of the solid solution composition, of the number and size of γ' precipitates are in good agreement with the experimental one. The KMC simulation shows successive ordering stages in the beginning of the phase transformation with a competition between Ni-Cr and Ni-Al types of short-range order [13].

4.2. Ab initio *calculations in Al-Zr alloys*

The precipitation kinetics of ordered $L1_2$ clusters in Al-Zr alloys has been recently performed using KMC simulations and ab initio calculations [16]. Clouet has used the same diffusion model by vacancy jumps and algorithm described above. However, in order to fit the energies of many configurations computed by the FP-LMTO method, triangle and tetrahedron interactions has been include in the model, in addition to the pair interactions. Saddle-point binding energies of Al and Zr atoms have been fitted on experimental diffusion data. (The *ab initio* calculation of the migration barriers is indeed still too time consuming: the simple determination of the an impurity diffusion coefficient for example, requires the computation of typically 5 to 10 migration barriers, some of them with non symmetrical saddle-point configurations [7]). KMC simulations of the nucleation kinetics have been performed and compared with the classical theory of nucleation (see section 6).

5. Vacancy and interstitial diffusion: NbC precipitation in steels

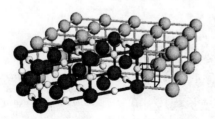

Figure 3: Lattice model for the simulation of NbC precipitation in α-Fe. The Fe atoms are represented in light gray, the Nb atoms in dark gray, the C atoms in white (small spheres). A small cubic cell is represented to show the interstitial sites for C diffusion.

The study of carbide precipitation (such as Nb, Ti or V carbides) in steels is of great industrial interest, because of the strong interaction with grain growth and mechanical properties. In Fe-Nb-C alloys, the kinetic pathway is control by two diffusion mechanisms: vacancy jumps for Fe and Nb atoms, interstitial jumps for C atoms. The diffusion of C atoms is extremely fast compared to that of Nb and Fe. Gendt [17] has developed KMC simulations of NbC precipitation, which takes these features into account. One simple cubic lattice is share by Fe, Nb and C atoms. For the sake of simplicity the NbC and α-Fe are build on the same BCC rigid lattice which fills ¼ of the simple cubic lattice (fig. 3). The NbC carbide is then assumed to be highly constrained in the matrix. The C atoms diffuse on the interstitial sites.

Each C atom can jump towards four interstitial sites: it results in a set of thousands of jump frequencies to be considered. The RTA algorithm has then been modified: the jumps frequencies are sorted and set into segments before the choice of a given jump. Moreover to keep an acceptable ratio between interstitial and vacancy jumps, the C diffusion is slowed down, but in such a way that it is still ~ 10^2 faster than the diffusion of substitutional atoms (while the actual ratio is indeed ~ 10^5 to 10^6): it is assumed that C atoms are still sufficiently rapid to be constantly in thermodynamic equilibrium with respect to the Nb and Fe configuration.

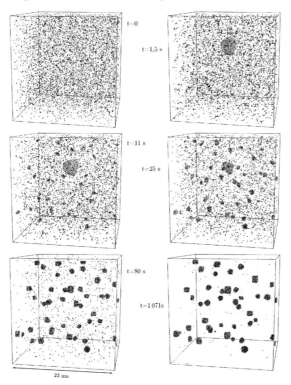

Figure 4: Monte Carlo simulation of precipitation in a Fe-0.5at.%Nb-0.5at.%C during thermal ageing at T = 900 K. The Nb atoms are represented in black, the C atoms in gray.

Depending on the alloy composition and temperature, the simulations exhibits two kinds of kinetic pathways. For low supersaturated solid solutions, one observes a direct precipitation of NbC clusters, with the classical nucleation-growth-coarsening sequence. For higher supersaturation a transient FeC iron carbide precipitation occurs before the formation of stable NbC precipitates (fig. 4). Because of the limitations of a rigid lattice model with constant pair interactions, neither the cementite Fe_3C (which has a complex crystallographic structure) nor any other iron carbide phase of Fe_3C composition, can be stabilized as an equilibrium phase on the lattice used in this study. An FeC carbide fitted to give the same driving force for precipitation is substitute for the cementite. Then in an real alloy, it is expected that a transient precipitation of cementite Fe_3C should occur for such composition and temperature conditions.

The simulation shows a strong vacancy trapping in the transient iron carbides. It can be taken into account by direct rescaling the simulation time according to 2.4, which assumes that the vacancy concentration is at constant equilibrium value in the iron matrix. But the kinetics of vacancy formation (which depends on the vacancy sources – dislocations, grain boundaries – microstructure) has been also simulated by introducing a simple model of vacancy source and sink. Some new events, which correspond to a vacancy formation or annihilation on the sink/source, are introduced in the simulation with probabilities, which tend to maintain the equilibrium concentration.

6. Comparison with Classical Kinetics Theories

The above KMC simulations are still often too time consuming for industrial purposes, and more simple models have been developed, usually based on the classical theories of nucleation, growth and coarsening, or spinodal decomposition. It is then interesting to test such models by comparison with KMC results.

In the case of a binary A-B alloy with a clustering tendency, when a metastable solid solution is quenched in the miscibility gap, the phase separation occurs through thermal fluctuations which eventually lead to nucleation of small B-rich cluster of β phase in the A-rich solid solution α. The free energy change on forming a spherical β cluster of radius R (or with i B atoms) is [18]:

$$\Delta F(R \text{ or } i) = \frac{4}{3}\pi R^3 \Delta f_v + 4\pi R^2 \sigma = iV_\beta \Delta f_v + A(iV_\beta)^{2/3} \qquad (3)$$

where Δf_v is the precipitation driving force and σ is the α/β interfacial free energy. The evolution of the number of β precipitates:

$$\frac{dN(i)}{dt} = -(\alpha_i + \beta_i)N(i) + \beta_{i-1}N(i-1) + \alpha_{i+1}N(i+1) \qquad (4)$$

depends on the emission and condensation rates α_i and β_i for a cluster of size i. β_i is usually be computed from a microscopic model of diffusion (e. g.: $\beta_i = 4\pi R_i D_B C_B / V_\beta$ [6,20]) and α_i from a relation of microscopic reversibility:

$$\alpha_{i+1} = \beta_i \frac{N_{eq}(i)}{N_{eq}(i+1)} \qquad (5)$$

where the equilibrium cluster size distribution is: $N_{eq}(i) = N_0 \exp(-\Delta F(i)/k_b T)$.

Various approximations can be used to get analytical expressions of the nucleation rate ($J_{st} = J_0 \exp[-K\sigma^3/(\ln S)^2]$) and incubation time from Eq. (3-5), or to compute the growth rate of a precipitate in the growth and in the coarsening regimes. To get the evolution of the clusters size distribution it is also possible to perform numerical integration of Eq. (4) (the clusters dynamics method [19]) or to introduce the classical incubation time, nucleation rate, growth rate, coarsening rate in kinetic models, such as the one of Kampmann and Wagner [20] or Maugis *et al* [17].

In simple cases, all the quantities of the classical theories (the precipitate driving force or the solubility limit, the diffusion coefficient of B atoms, the interfacial energy σ, etc.) can be related to the microscopic parameters of the KMC simulations

[6]. However it is usually possible to choose several sets of MC parameters which differ by the diffusion properties of small β clusters (see sect. 3). If small solute clusters are immobile classical models and KMC are usually in good agreement: the incubation times, nucleation rates and generally the whole precipitation kinetics are very close (Figure 5). On the contrary, the migration of small clusters brings about direct coagulation between precipitates. This phenomena, which is not taken into account in classical theories, leads to a acceleration on the first precipitation stages which can reach several orders of magnitude [6]. Similar comparison for the case of precipitation of ordered phases have been recently carried out by Clouet [16].

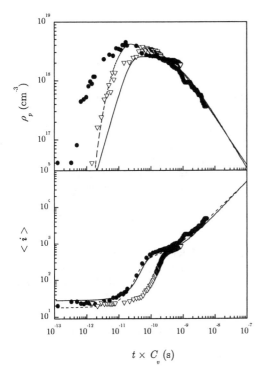

Figure 5: Precipitation kinetics in a binary A-3at.%B alloy at $T = 0.5\ T_c$. Evolution of the density of precipitate and precipitate size computed by the model of Maugis (full line, [17]), the cluster dynamic method (dotted line [19]) and KMC simulations with immobile (▽) or mobile (●) clusters.

7. Conclusion

Monte Carlo simulations based on a simple atomistic diffusion model have been fitted on the thermodynamic and kinetic properties of particular binary and ternary alloys, to study several cases of phase transformation kinetics. These methods are especially useful to study nucleation processes, because they naturally take into account thermal fluctuations. It has been shown that the kinetic pathway depends not only the diffusion mechanism, but also on the details of the model and the microscopic parameters of the jump frequencies and migration barriers. One of the main limitations is due to the rigid lattice approximation.

In some simple cases, it is possible to fit the RLM parameters on the properties of a more realistic EAM potential model or partly on *ab initio* calculations. However,

436

this method may not be straightforwardly used in alloys with strong size effects (in Ni-Au alloys for example, Bocquet has found that neither the stable configuration energies nor the migration barriers could be reproduced with constant pair or even three or four body interactions). Monte Carlo simulations with relaxation of the both the stable and the saddle-point atomic positions will overcome this limitation, but they are still much more time consuming, and have been therefore limited to short times (e.g. to compute tracer diffusion coefficients [21]). The simulation of a phase transformation kinetics in the presence of long-range elastic interactions remains possible, but with a more simple description of diffusion barriers and activated states [22].

References

[1] P. Bellon, this volume.
[2] P. Fratzl an O. Penrose, *Phys. Rev B* **50**, 3477 (1994)
[3] T. T. Rautiainen and A. P. Sutton, *Phys. Rev. B* **59**, 13682 (1999)
[4] M. Athènes, P. Bellon and G. Martin, *Acta Mater* **48**, 2675 (2000)
[5] J.-M. Roussel and P. Bellon, *Phys Rev* B **63**, 184114 (2001)
[6] F. Soisson and G. Martin, *Phys. Rev. B* **62**, 203 (2000).
[7] Y. Le Bouar and F. Soisson, *Phys. Rev. B* **65**, 094103 (2001).
[8] M. Athènes, P. Bellon and G. Martin, *Philos. Mag* A **76**, 565 (1997).
[9] M. Athènes, P. Bellon, F. Haider and G. Martin, *Acta Mater* **44**, 4739 (1996)
[10] E. Vives and A. Planes, *Phys. Rev. B*, 2557 (1993).
[11] J. Philibert, *Atom movements : diffusion and mass transport in solids* (Editions de Physique, Les Ulis, 1991)
[12] F. Soisson, A. Barbu and G. Martin, Acta Mater 44, 3789 (1996).
[13] C. Pareige, F. Soisson, G. Martin and D. Blavette, *Acta Mater* **47**, 1889 (1999).
[14] M. Ludwig, D. Farkas, D. Pedraga and S. Schmauder, *Modell. Simul. Sci. Eng.* **6**, 19 (1998).
[15] J.-L. Bocquet, private communication.
[16] E. Clouet and M. Nastar, proceedings of the *third International Alloy Conference*, Lisbon 2002, in the press.
[17] D. Gendt, P. Maugis, G. Martin, M. Nastar and F. Soisson, *Defects and Diffusion Forum* **194-199**, 1779 (2001).
[18] J. W. Christian, *Transformations in Metals and Alloys* (Pergamon Press, Oxford 1975)
[19] C. Sigly, this volume.
[20] R. Wagner and R. Kampmann, in *Phase Transformations in Materials* (ed. P. Haasen, VCH, Weinhem, 1991), Chap. 4.
[21] J.-L. Bocquet, *Defect and Diffusion Forum* **203-205**, 81 (2002).
[22] A. Finel, in *Phase Transformations and Evolution in Materials* (eds. P. E. A. Turchi, and A. Gonis, TMS, Warrendale, 2000) p. 371.

VI. Experimental investigations of microstructures

EXPERIMENTAL INVESTIGATIONS OF MICROSTRUCTURES

G. KOSTORZ, R. ERNI, H. HEINRICH
ETH Zurich, Institute of Applied Physics
CH-8093 Zurich, Switzerland

Abstract
Materials properties are controlled by microstructural features on various scales ranging from several centimeters down to less than interatomic distances. This paper presents some experimental methods suitable for microstructural investigations in the range of (tenths to tens of) nanometers. Diffraction methods employing X-rays, neutrons or electrons yield average information on the microstructure for a sampling volume which may range from several nm^3 (for electrons) to several cm^3 (for neutrons). The advantages and limitations of the different types of radiation and the corresponding diffraction techniques, especially diffuse scattering and small-angle scattering, are discussed. Current research results on the evolution of microstructures in alloys are presented to highlight the major points. Electron microscopy offers a broad variety of tools for chemical and structural analysis. Some of these are briefly described, and their use is illustrated by results of current research on the decomposition of alloys and on inhomogeneities in thin-film semiconductor heterostructures.

1. Introduction

Apart from real-space analytical and topographical tools (e.g. the tomographic atom probe [1]), methods based on scattering and diffraction of X-rays, electrons and neutrons are widely used to resolve the microstructure of materials on the atomic and mesoscopic scales. There are many introductory textbooks on crystallographic and spectroscopic methods. References [2, 3] address more precisely the points relevant to the present subject; here we concentrate on resolving local structural and chemical features, i.e., deviations from the macroscopically averaged structures (which is the domain of crystallography). The desired resolution may be achieved by using sufficiently narrow beams to probe a given region (as, e.g., convergent-beam electron diffraction) or by analyzing the diffraction phenomena besides Bragg peaks, i.e., diffuse scattering around the incident beam (small-angle scattering), between Bragg peaks or near them. These scat-

A. Finel et al. (eds.), Thermodynamics, Microstructures and Plasticity, 439–464.

tering phenomena will yield, if we restrict the view to static materials properties, local deviations from the macroscopic average, but as an average over the volume sampled by the radiation used. Although X-rays, electrons and neutrons may all have sufficiently small wavelengths to provide subatomic resolution, they differ considerably in intensity, penetration power and strength of interaction with the sample, thus leading to immense variations in sample volume. Choosing the appropriate type of radiation is thus an important prerequisite for a successful microstructural study.

2. Scattering

For simplicity, we treat here primarily distorted crystals with still well-defined Bragg peaks [4]. Many microstructural details, i.e., local deviations from the average structure, lead to measurable scattering intensities in the wide reciprocal space between Bragg peaks. As most of this scattering is relatively weak, kinematic scattering theory is usually sufficient [4].

A perfect crystal has full translational symmetry, with equidistantly spaced scattering centers $\mathbf{R}_n = n_1\mathbf{a}_1 + n_2\mathbf{a}_2 + n_3\mathbf{a}_3$. For a monatomic crystal, the resulting scattering amplitude is given by

$$F(\mathbf{Q}) = b\sum_n \exp(-i\mathbf{Q}\cdot\mathbf{R}_n) \tag{1}$$

where b is the scattering amplitude of an atom and $\mathbf{Q} = \mathbf{k} - \mathbf{k}_0$ is the scattering vector. For quantitative measurements (scattering cross-sections), b is the scattering length. Variations of b with $\mathbf{Q}$ are included in b. In a real crystal, the different sites can be occupied by different atoms, and local displacements from the positions of the average lattice may occur. Taking into account substitutional sites ($\mathbf{R}_n$) and interstitial sites ($\mathbf{S}_j$), the scattering amplitude becomes

$$F(\mathbf{Q}) = \sum_n b_n \exp\{-i\mathbf{Q}\cdot(\mathbf{R}_n + \mathbf{u}_n)\} + \sum_j b_j \exp\{-i\mathbf{Q}\cdot(\mathbf{S}_j + \mathbf{v}_j)\} \tag{2}$$

where $\mathbf{u}_n$ and $\mathbf{v}_j$ are the local displacements.

The scattering intensity, i.e., what we measure, can be obtained from various approximations. The experimental consequences are very simple in principle. Instead of δ-functions for a large perfect crystal, we may get a change in the average lattice constants, in the shape of the Bragg peaks, and also scattering between Bragg peaks (Fig. 1). Some specific cases will be briefly addressed.

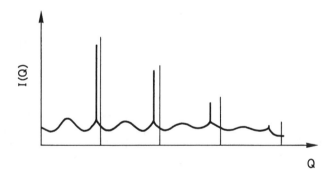

Figure 1. Schematic scattering intensity for a crystal containing defects. The vertical lines indicate the Bragg position of a perfect crystal.

2.1. DISPLACEMENT SCATTERING

If the concentration c of the defects is small, the single-defect approximation may be used. The total scattering intensity is composed of two terms, the Bragg peaks and the diffuse scattering

$$I(\mathbf{Q}) = I_D(\mathbf{Q}) + I_{Bragg}(\mathbf{Q}) \tag{3}$$

The so-called Huang scattering, i.e., the intensity near the Bragg peaks for small s values ($\mathbf{Q}=\mathbf{g}+\mathbf{s}, s \ll g$), turns out to be proportional to the defect concentration, to the square of the product of the reciprocal lattice vector $\mathbf{g}$ and the Fourier transform $\mathbf{u}(\mathbf{Q})$ of the displacement field $\mathbf{u}(\mathbf{r})$ of the defect,

$$I_D(\mathbf{Q}) = cNb_H^2 |\mathbf{g} \cdot \mathbf{u}(\mathbf{Q})|^2 \tag{4}$$

with N = number of atoms in the beam, b_H = scattering length of the host lattice atoms. The displacement field of point defects may be related to the dipole-force tensor $\mathbf{P}$ of the defect,

$$\mathbf{u}(\mathbf{Q}) = \frac{i}{sV_c} \mathbf{S}(\mathbf{s}/s) \cdot \{\mathbf{P} \cdot [\mathbf{s}/s]\} \tag{5}$$

with $P_{jk} = P_{kj} = \sum_n p_{jn} x_{kn}$ where $\mathbf{p}_{jn}$ are the components of the forces $\mathbf{p}_n$ and x_{kn} the components of the position vectors $\mathbf{r}_n$, V_c is the volume of the unit cell, and $\mathbf{S}(\mathbf{s}/s)$ is the tensor of elastic compliances depending on the direction of $\mathbf{s}$.

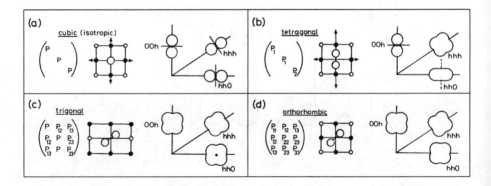

Figure 2. Schematic lines of equal intensity in the $(1\bar{1}0)$ plane of reciprocal space of a cubic crystal for various defect symmetries (for anisotropic defects, see the indicated dipole force tensor, averaged over all crystallographically equivalent orientations of the defect). After Peisl [6].

We see from Eq. (5) that Huang scattering has the same general Q dependence as first-order thermal diffuse scattering (TDS) for $Q \rightarrow g$, especially inversion symmetry around g, but the intensity distribution is different. Fig. 2 shows some examples.

To analyze defects by Huang scattering, the host crystal has to be very perfect (apart from the specific defect). For X-rays, temperatures should be sufficiently low (TDS!). With neutrons, most of the TDS may be separated off experimentally owing to the good energy resolution. The diffuse scattering far away from Bragg peaks is particularly sensitive to displacements in the immediate vicinity of the defects. Especially, the symmetry of the defect environment may be very clearly revealed, which is important for the confirmation of certain models (e.g. for interstitials, small clusters, etc.). In solid solutions, atomic displacements may be studied in the dilute case (see e.g. [3]), but in more concentrated alloys, displacements ("positional" disorder) and short-range order ("compositional" disorder) will always be coupled.

2.2. SHORT-RANGE ORDER SCATTERING

Deviations from random site occupancy (we restrict the discussion to substitutional solid solutions) are generally called short-range order (SRO). For a binary alloy containing species A and B with scattering lengths b_A and b_B (any Q dependence included), and with an atomic fraction c of B atoms, the diffuse scattering intensity without any displacement scattering is

$$I(\mathbf{Q}) = N|c(\mathbf{Q})|^2|b_A - b_B|^2 \qquad (6)$$

where

$$|c(\mathbf{Q})|^2 = c(1-c)\,\alpha(\mathbf{Q}) \qquad (7)$$

and

$$\alpha(\mathbf{Q}) = \sum_n \alpha_n \exp(-i\mathbf{Q}\cdot\mathbf{R}'_n) \qquad (8)$$

Here, $\mathbf{R}'_n$ is a distance vector from an arbitrary reference site 0 to the site labeled n, and the α_n are the Warren-Cowley SRO parameters defined by use of the conditional probability P_n^{ij} of finding an atom of type j (A or B) at the site n if an atom of type i is at the reference position:

$$\alpha_n = \frac{P_n^{BB} - c}{1 - c} = \frac{c - P_n^{AB}}{c} \qquad (9)$$

If A and B atoms are randomly distributed on all available sites, all α_n for $n \neq 0$ will be zero. Then, Eq. (6) yields the monotonic Laue scattering which serves as a reference unit (Laue unit) for each alloy.

As mentioned, SRO and displacement scattering are coupled and must be separated for careful analysis (from experiments using single crystals). This is done by fitting short-range-order Fourier coefficients and displacement coefficients, considering appropriate terms of the expansion of the displacement scattering. For cubic crystals, including displacement terms to second order, this leads already to ten different terms in the diffuse scattering intensity (see, e.g., [3] for details). Very often, linear displacement terms are sufficient, which reduces the number of functions to four.

Fig. 3 shows the SRO scattering obtained from such an analysis for a Ag–13.4 at.% Al crystal aged at 673 K for 184 h and subsequently quenched in iced water [7]. The diffuse scattering was measured at room temperature using Mo K_α radiation, for about 11'000 different positions of the sample placed on an Eulerian cradle. The instrument was calibrated using the scattering from polystyrene. Compton scattering and TDS were calculated and subtracted. Finally, two different evaluation schemes were used to separate SRO and linear displacement scattering, with essentially identical results. Fig. 3 shows that two types of diffuse maxima appear in the SRO scattering; the absolute maximum is around ½ ½ ½, and local maxima are seen at $2k_F$ positions. Particularly flat pieces of the Fermi surface perpendicular to <110>, spanned by the scattering vector $2\mathbf{k}_F$ (where $\mathbf{k}_F$ is the Fermi wave vector for <110> directions), are responsible for these maxima.

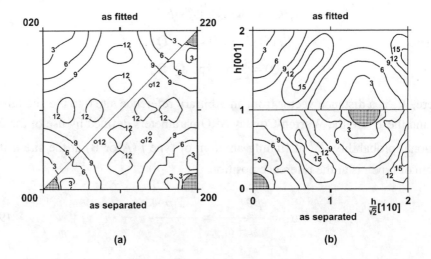

Figure 3. Short-range order diffuse scattering (in 0.1 Laue units) for (a) a (001) plane and (b) a (110) plane of a Ag-13.4 at% Al single crystal. The as-separated data are compared with patterns recalculated from 27 SRO parameters.

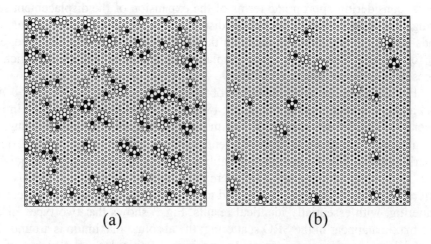

Figure 4. One of the (111) planes of (a) a modeled short-range-ordered crystal (f.c.c. Ag-13.4 at.% Al) and (b) a crystal of the same composition with random site occupation. Open circles are Ag, filled circles are Al atoms. Symbols for atoms belonging to the C 9 configuration (see Fig. 5) are enlarged.

From a given set of SRO parameters, the local atomic arrangement can be simulated on a computer by rearranging many thousands of atoms (e.g. in 32 x 32 x 32 f.c.c. unit cells) until sufficient agreement is obtained with the SRO parameters. Simulating local atomic arrangements including displacements is not yet fully established, but first attempts have been reported (see [8]). Subtle differences relative to random site occupancy may thus be visualized. Fig. 4 shows an example. The C9 "Clapp configuration" highlighted there belongs to the set of 144 nearest-neighbor configurations enumerated by Clapp [9] (see Fig. 5 for a few examples relevant to Ag-Al). The configurations found to be largely enhanced in the Ag-Al crystal in the short-range ordered state are indicated in the caption. These three configurations are all typical structural elements of the A_5B structure, a possible ground-state structure obtained for Ag-Al from *ab-initio* electronic structure calculations [10]. Experimentally, no long-range-ordered phase has been identified for Ag-Al in this range of composition.

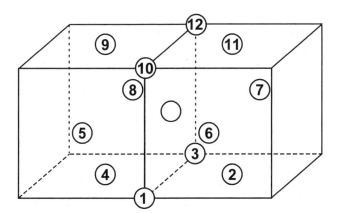

Figure 5. Key to Clapp configurations [9] for f.c.c. crystals. Nearest-neighbor sites of a reference site (here taken by a Ag atom) are labeled by numbers. In short-range-ordered Ag-13.4 at.% Al, three configurations are enhanced; C 4 (5, 7) where sites 5, 7 are occupied by Al, the others by Ag, C 5 (6, 12) and C 9 (1, 7, 9).

If a set of SRO parameters belongs to an equilibrium state, they may be used to calculate basic alloy properties, especially effective pair interactions $V_n = \frac{1}{2}(V_n^{AA} + V_n^{BB}) - V_n^{AB}$. Apart from approximate analytical methods (see Reinhard and Moss [11] for an assessment), exact numerical calculations are now feasible. In the Inverse Monte Carlo method proposed by Gerold and Kern [12], a large number of virtual exchanges of nearest neighbors are introduced to "test" the equilibrium state, and the large set of corresponding nonlinear equations (conditions of detailed balance) is solved to extract the effective pair interactions V_n. In the present example of Ag-Al, a set of at least 14 pair interactions is required to reproduce the measured SRO scattering in a subsequent Monte Carlo simulation. If one assumes that the values of V_n do not depend on temperature and do not vary too much with composition, ground-state energies of hypothetical long-range-ordered phases can be calculated and compared, and Monte Carlo simulations may be performed with such phases to estimate order-disorder temperatures. For Ag_5 Al (A_5B structure), $T_c \cong 135$ K is found [7], while theory yields about 220 K [10], both too low to expect any long-range order under standard experimental conditions.

Apart from this fundamental relevance, SRO and effective pair interactions may be useful in understanding other physical properties, such as electrical resistivity (see Rossiter [13]) and deformation behavior of solid solutions. While pure metals and very dilute solid solutions show finely distributed, "wavy" slip lines after plastic deformation, more concentrated solid solutions deform by "planar" glide, i.e., slip is concentrated in fewer, but more prominent slip steps. This phenomenon, though often attributed to a decrease of stacking-fault energy with increasing concentration, may somewhat more convincingly be related to SRO (see, e.g., [14, 15]). When a unit dislocation with Burgers vector $\frac{1}{2}$ <110> moves on a {111} plane, the number of AA, BB and AB pairs (nearest, next-nearest and further neighbors considered) changes, weighted by the appropriate SRO parameters. This (static) shift changes the configurational energy of the crystal according to the effective pair interactions involved. Further dislocations moving on the same slip plane will find a new, non-equilibrium situation and will change this again. In analogy to the long-range-ordered case, the energy change per unit area resulting from these slip events may be called diffuse antiphase boundary energy [16]. Fig. 6 shows the diffuse antiphase boundary energy γ_d (after a large number of elementary shears, a constant value is reached) calculated from the SRO parameters and the effective pair interactions obtained from diffuse scattering experiments for some Cu alloys (X-rays for Cu-Al [17], neutrons for Cu-Mn [18] and Cu-Zu [19]). Usually, the energy change is larger for the first dislocation than for the following ones. This may lead to the emission of a larger number of dislocations from the same source while the leading one remains at rest until the necessary shear stress is reached at the head of the pile-up.

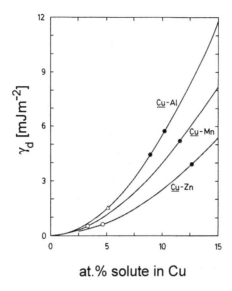

Figure 6. Diffuse antiphase boundary energy γ_d as a function of solute concentration in some Cu-base alloys (continuous lines calculated from results of diffuse scattering). Open circles indicate observations of wavy slip, full circles planar glide, and the triangle mixed behavior. The transition, certainly influenced by a number of other parameters, takes place between 1 and 3 mJ/m^2.

We have restricted the discussion here to solid solution with a tendency towards local ordering. Clustering systems may also be analyzed, but as most of the relevant scattering intensity is more concentrated around the Bragg peaks, single crystals of very high perfection and a good instrumental resolution are required. A recent review [20] gives a more detailed assessment of the whole field.

2.3. SCATTERING FROM NANOSCOPIC OBJECTS (ESP. SMALL-ANGLE SCATTERING)

Inhomogeneities in scattering length density extending over distances larger than an interatomic distance will manifest themselves primarily in scattering around the incident beam (for crystalline substances possibly also around the Bragg peaks, but there, the displacement scattering may be difficult to separate). For such extended objects examined at scattering vector magnitudes less than about π/d_a where d_a is the interatomic distance, the discrete positions of the scattering centers will not be resolved, and instead of individual scattering lengths per site (as in Eq. 2), a scattering-length density $\rho(\mathbf{r})$, locally averaged in each volume element $d^3\mathbf{r}$ around the position vector $\mathbf{r}$, may be used. The scattering amplitude of a sample of volume V is

$$F(\mathbf{Q}) = \int_V \rho(\mathbf{r}) \exp(-i\mathbf{Q} \cdot \mathbf{r}) d^3\mathbf{r} \qquad (10)$$

The resulting small-angle scattering (SAS) intensity may be calculated for any model distribution $\rho(\mathbf{r})$, but the inverse problem has no unique solution.

In many cases, the two-phase model may be used to interpret SAS results. Here, small particles of a homogeneous scattering length density ρ_p are embedded in a homogeneous matrix of scattering-length density ρ_m. Then

$$F(\mathbf{Q}) = \int_V (\rho_p - \rho_m) \exp(-i\mathbf{Q} \cdot \mathbf{r}) d^3\mathbf{r} \qquad (11)$$

If N_p identical uncorrelated particles are contained in the sample volume, the total SAS intensity is

$$I(\mathbf{Q}) = N_p |F_p(\mathbf{Q})|^2 |\rho_p - \rho_m|^2 V_p^2 \qquad (12)$$

where

$$F_p(\mathbf{Q}) = V_p^{-1} \int_{V_p} \exp(-i\mathbf{Q} \cdot \mathbf{r}) d^3\mathbf{r} \qquad (13)$$

is the single-particle scattering amplitude and V_p is the particle volume.

Eq. (12) may be used for (very) dilute solutions of particles. If the particles are anisometric, they must all be equally oriented, or an appropriate average must be taken. In more densely packed arrangements, interparticle interference must be considered (see [21], [3] and references therein for a more extensive coverage).

For sufficiently small Q, the Guinier approximation (see [21]) is very useful, as it provides some measure of the size of scattering objects without intensity calibration. Starting from an expansion of the exponential in Eq. (13) to first order in $\mathbf{Q}$, one obtains a quadratic term in the scattering intensity (the linear term may be eliminated by a proper choice of the particle center). Reinterpreting this as stemming from an expansion of the scattering function, we have

$$|F(\mathbf{Q})|^2 = \exp(-Q^2 R_d^2) \qquad (14)$$

where

$$R_d^2 = V_p^{-1} \int_{V_p} r_d^2 \, q(r_d) dr_d \qquad (15)$$

Here, r_d is the distance within the particle from the center, measured along a direction **d** perpendicular to the incident wave vector $\mathbf{k}_0$ in the $(\mathbf{k}_0, \mathbf{Q})$ plane, and q (r_d) is the cross-sectional area of the particle perpendicular to **d**. The quantity R_d is the average "inertial distance" [21] along **d**. The random orientational average of Eq. (14) is

$$|F(Q)|^2 = \exp\left(-Q^2 R_G^2 / 3\right) \tag{16}$$

with the "radius of gyration" R_G obtained from

$$R_G^2 = V_p^{-1} \int_{V_p} r^2 \mathrm{d}^3 r \tag{17}$$

For spheres of radius R_S, $R_G^2 = 3R_S^2/5$, and the Guinier approximation holds for $QR_G < 1.2$. For other particle shapes, the Q-range may be smaller.

Independent of any interparticle interference (which plays no essential role at larger scattering vectors), the Porod approximation may yield some measure of the particle surface. The orientational average yields

$$|F(Q)|^2 = 2\pi A_p / V_p^2 \, Q^4 \tag{18}$$

and thus the surface-to-volume ratio of a particle (and any ensemble of particles as, e.g., in catalysts). An extension of the Porod analysis for anisotropic scattering shows that in any direction of **Q**, the intensity at sufficiently large Q is controlled by the reciprocal values of the Gaussian curvature of the scattering objects' surfaces perpendicular to **Q**. However, so-called power-law scattering may also be related to other features. For example, fractal dimensions of self-similar objects may be deduced [23], or special size distributions may cause deviations from the exponent 4 in Eq. (18) [24].

If $I(\mathbf{Q})$ can be measured with sufficient accuracy at large enough Q, the integrated intensity $\tilde{I}$ may be determined with some reliability;

$$\tilde{I} = V^{-1} \int I(\mathbf{Q}) \mathrm{d}^3 \mathbf{Q} = (2\pi)^3 \overline{[\rho(\mathbf{r}) - \bar{\rho}]^2} \tag{19}$$

where bars indicate averages over the whole sample and the integration extends over the entire reciprocal space (which implies extrapolations beyond the experimentally accessible **Q** range). As $\tilde{I}$ is insensitive to detailed structural features but only reflects the mean-square fluctuations of the scattering length density, it is sometimes called the "invariant". It is also a special value (for $r=0$) of the Fourier transform of $I(\mathbf{Q})$, i.e. of the correlation function $\gamma(\mathbf{r})$. For the two-phase model,

$$\tilde{I} = C_p \left(1 - C_p\right)\left(\rho_p - \rho_m\right)^2 \tag{20}$$

where $C_p = N_p V_p / V$ is the volume fraction of particles. Eq. (20) illustrates clearly that the "invariant" may vary for a system with some evolution of the two-phase structure.

As an example of the use of SAS, we refer to the long-lasting research on the decomposition kinetics of Ni-rich Ni-Ti (see [25] and references therein). As Ti has a negative scattering length for neutrons, the contrast between Ni and Ti is particularly large, and the small-angle scattering of neutrons upon phase separation can be measured with high precision. Fig. 7 shows the metastable coherent miscibility gap for Ni-Ti. The symbols γ'' and γ' indicate that two stages are observed in the (metastable) phase-separation process, where the γ' regime ($L1_2$ structure) is expected, but the intermediate γ'' stage (showing also $L1_2$-type order) has now been confirmed by a series of experiments [25].

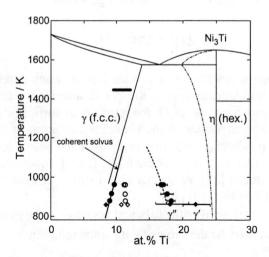

Figure 7. Ni-rich part of the Ni-Ti phase diagram [26] including the approximate *loci* of the metastable (coherent) miscibility gap and the concentration of matrix and precipitates (filled symbols) evaluated from small-angle neutron scattering of alloys (of average concentration as indicated by the open symbols) aged at four different temperatures after quenching from the primary solid solution range (horizontal bar).

Indication and proof of the two-stage behavior came from the analysis of the integrated intensity as shown in Fig. 8. These results were obtained by *in-situ* studies using a new high-temperature cell. It is now clearly established that there is an intermediate plateau in the integrated intensity for a range of concentrations and temperatures and that a second stage follows after some delay.

Details of the phase separation are certainly influenced by the coherency strains building up upon compositional changes (enrichment in Ti increases the lattice constant). The influence of elastic interactions is clearly visible in the small-angle scattering of single crystals.

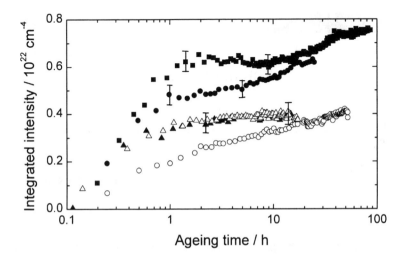

Figure 8. Integrated small-angle neutron scattering intensity as a function of ageing time for Ni-11.3 at.% Ti aged at 870 K (filled squares), 900 K (filled circles) and 950 K (full triangles), for Ni-11.1 at.% Ti at 950 K (open triangles) and for Ni-10.1 at.% Ti at 900 K (open circles).

In Fig. 9, a sequence of scattering patterns is shown for the *in-situ* ageing of a Ni-11.3 at.% Ti single crystal. The strong peaks along <100> are due to inter-particle interference. The particles align preferentially along these elastically soft directions. This reduces the elastic energy. One might think of a possible change

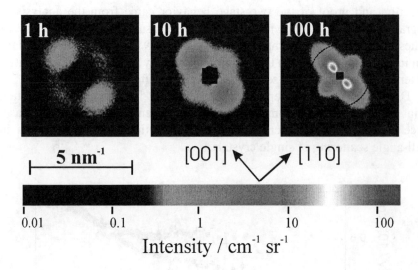

Figure 9. Small-angle neutron scattering patterns of a Ni-11.3 at.% Ti single crystal aged in the neutron beam at 870 K. The inner part of the right pattern was taken at a sample-to-detector distance of 4 m while all the other measurements were taken at 2 m. The wavelength of the incident neutrons was 0.8 nm.

in the internal order of the particles, but large angle X-ray diffraction studies of the evolution of the superstructure reflections ($L1_2$ type) show that the degree of order always takes on the maximum possible value given by the Ti concentration in the particles [27].

At sufficiently high temperatures, *in-situ* SAS with neutrons also signals the appearance of the stable η-phase in Ni-Ti. With single crystals, there are very dramatic effects, as the hexagonal η-phase is semi-coherent with the matrix across the close-packed planes and forms thin, large lamellae. Fig. 10 shows a sequence of SAS patterns taken at 1200 K. The incident beam is again along <110>, and [001] lies in the horizontal direction. Fig. 10(b) shows that even at 1200 K, the metastable coherent phase forms first, as there is higher scattering intensity along [001], similar to Fig. 9. Additional sharp streaks along <111> first appear after about 10 min. In Fig. 10(c), they are well developed and correspond to lamellae of η-phase parallel to the cubic {111} planes. The variation of the onset time for η formation varies strongly with ageing temperature.

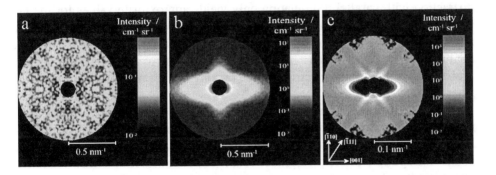

Figure 10. Symmetrized small-angle neutron scattering patterns of a Ni-12.3 at.% Ti single crystal; (a) solution-treated at 1440 K, and subsequently aged *in-situ* at 1200 K for (b) 6 min and (c) 47 min.

Many other applications of SAS in the study of microstructures may be found in the literature. The triennial SAS conference proceedings are usually published in the Journal of Applied Crystallography.

3. Transmission Electron Microscopy (TEM)

Best results on microstructural features are often obtained from a combination of X-ray or neutron diffraction and electron microscopy. Modern transmission electron microscopes offer a variety of methods in addition to "conventional" techniques like electron diffraction and imaging at various resolutions with the direct and/or diffracted beams (see [28, 29] for general introductions).

The introduction of field-emission guns as electron sources allow for small electron probes with diameters down to about 0.2 nm. Scanning such a small electron probe over the sample provides atomic-resolution images using the signals from a bright-field (BF) detector, an annular dark-field (ADF) detector or a high-angle annular dark-field (HAADF) detector in the diffraction plane of the sample. The HAADF signal shows a strong atomic number contrast (Z contrast), which can be used to analyze the composition of individual atomic columns (see, e.g., [30]). The use of an electron energy filter allows electron energy-loss spectra to be obtained with a spatial resolution corresponding to the electron probe size. With a homogeneous illumination of a sample area, energy-filtered images can be obtained (using a GATAN imaging filter (GIF) or a LEO omega filter). Taking these images at element-specific edges in the energy-loss spectrum, one maps elemental distributions with a resolution better than 1 nm (see, e.g., [31]).

A transmission electron microscope can also be used to determine lattice parameters of small particles or the local changes of lattice parameters of a phase, caused by chemical gradients or local stress fields. While selected-area diffraction with parallel illumination only yields approximate values for the lattice parameters ($\Delta a/a \approx 5 \times 10^{-3}$), convergent-beam electron diffraction (CBED) allows lattice parameters to be determined with an accuracy of about 10^{-4}. The quality of CBED patterns is highly improved when using the small electron probe of a field-emission electron source in combination with an electron-energy filter.

The use of some of the "old" and new techniques will now be illustrated by referring to some recent work on some alloys and photovoltaic heterostructures. A full characterization of a multi-phase microstructure comprises the evaluation of the average size of the precipitates (e.g., for Ni-Ti of both the metastable $L1_2$ phase and the platelets of the hexagonal stable $D0_{24}$ phase), the size distribution, the average shape, the spatial distribution (alignment), the chemical compositions, chemical order and the lattice parameters of matrix and precipitates. The mutual influence of different types of precipitate on each other and the influence of grain boundaries are also of interest, e.g. as a function of ageing.

3.1. DIFFRACTION CONTRAST

To complement SAS and other diffraction results of alloys containing precipitates, conventional imaging is very important, as precipitate shape and (by the use of superstructure reflections) internal order can be revealed. Dark-field imaging with a superstructure reflection of long-range ordered precipitates with $L1_2$ structure is a standard technique for Ni-based alloys. An example for Ni-11.3 at.% Ti is shown in Fig. 11. Size, shape and arrangement of the long-range ordered particles can be analyzed. The contrast of the particles depends strongly on the local sample orientation. Sample bending is frequently observed in thin TEM foils. This leads to variations in the deviation parameters s_g. If s_g of the used superstructure reflection increases, the brightness of precipitates in the micrograph decreases. As the image contrast in dark-field imaging depends not only on the long-range order parameter of the precipitates, but also on their thickness and local lattice tilts, some regions in Fig. 11 appear to be gray with only little contrast of the precipitates. Additionally, some parts of a precipitate might be removed during preparation of the thin TEM foil. Thus, not the whole precipitate thickness contributes to the dark-field image, and this leads to a reduced contrast. For thicker regions of TEM samples, overlap in the projected images of precipitates is frequently found. Therefore, the determination of the three-dimensional arrangement of particles, their distance distribution and volume fraction is, if at all possible, not free from large errors.

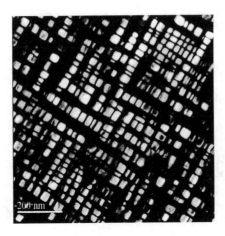

Figure 11. Dark-field micrograph of L12-ordered metastable precipitates imaged along [100] with a 001 superstructure reflection. The Ni-11.3 at.% Ti sample was quenched in brine after ageing for 100 h at 950 K.

3.2. STRAIN FIELDS

Precipitates coherently embedded in a matrix usually induce local strain fields because of different lattice parameters of the two phases. The original method to image the strain-field around precipitates is the two-beam technique, where the crystal is in perfect Bragg condition ($s_g = 0$ nm^{-1}) for one specific reflection **g**. All other reflections are not significantly excited. This leads to micrographs with typical coffee-bean contrast (see, e.g., [32]). In these micrographs, only components of the strain field parallel to **g** lead to a change in image contrast. Therefore, the shape and the size of precipitates can generally not be analyzed by this method. A better approach is the symmetrical bright-field zone-axis (BFZA) method [33]. For exact zone-axis orientation of the sample, no Bragg reflection in the zero-order Laue zone is in exact Bragg condition. Small local sample tilts change this, and one or more Bragg reflections are strongly excited. The intensity of the 000 reflection is reduced by enhanced scattering contributions into these reflections. In BFZA imaging, these local lattice tilts are especially prominent near the particle-matrix interfaces, where the lattice distortions are highest (see Fig. 12). The shapes and the arrangement of precipitates are clearly revealed using BFZA imaging.

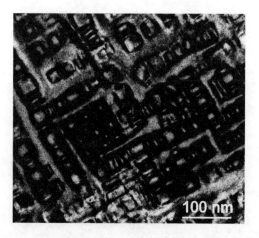

Figure 12. Bright-field zone-axis micrograph of the same sample as in Fig. 11, imaged with the incident beam along [100].

3.3. CBED

CBED is most suitable to determine lattice parameters with a high lateral resolution of a few tens of nanometers. The convergent beam focused on the sample leads to diffraction discs instead of a pattern of diffraction spots in the case of parallel illumination. If for any of the directions of the incoming beam (the central diffraction disk) the Bragg condition is exactly fulfilled for any reflection *hkl* in a higher-order Laue zone (HOLZ), the intensity for this position in the central 000 disk is reduced and transferred to the corresponding position in the respective *hkl* diffraction disk. These exact Bragg conditions lie on cones (Kossel cones) around the *hkl* diffracting planes. The positions of the dark HOLZ lines in the 000 disk are very sensitive to changes of lattice parameters or the electron wavelength. For the evaluation of HOLZ-line patterns, the line positions are determined and distances between intersection points of HOLZ lines are compared with simulated CBED patterns. To match the experimental and the simulated patterns, the lattice parameters used in the simulations are modified and refined. With kinematical scattering simulations, this refinement procedure has been successfully applied to many Ni-base superalloys (see, e.g., [34]). The lattice parameters have been determined for special positions in the γ' particles (center, corners and edges) and in the γ-matrix channels. Tetragonal or orthorhombic distortions have been assumed according to the local symmetries in the particles or the matrix channels between the particles. Therefore, not all six lattice parameters (the three possibly distinct dimensions *a, b, c* of the unit cell and the three angles α, β, γ in the unit cell) are free fitting parameters. This reduction of free parameters is, in this case of large (about 1 μm³) cuboidal precipitates, appropriate and yields reliable data for the strain distribution in Ni-base superalloys.

However, for many materials, assumptions of special symmetries for the lattice distortions can not be made, and all six lattice parameters are to be determined. This turns out to be impossible from just one CBED pattern (see, e.g., [35]), and CBED patterns from at least two significantly different incident-beam directions must be evaluated for each sample volume of interest [36]. This is especially necessary with a heterogeneous stress distribution like in aluminum-based metal matrix composites containing fiber reinforcements. During cooling of the alloy, interfacial reactions and the different thermal expansion coefficients of the ceramic fibers and the metallic alloy induce high stresses directly at the interfaces, especially around small reaction products like spinels at the surfaces of the fibers [37].

Another limitation of the CBED method becomes apparent for materials with small precipitates completely embedded in the thin TEM foil. To obtain a sharp HOLZ-line pattern, the sample should not be too thin. Samples about 100 to 300 nm thick yield acceptable HOLZ patterns, especially, if an electron energy filter can be used to eliminate electrons with energy losses of about 5 or more eV. For samples thinner than 100 nm, the HOLZ lines are usually too broad for an accurate determination of the line positions. Now, for small precipitates completely embedded in a relatively thick TEM sample, the strain field is not homogeneous throughout the sample along the electron-beam direction. Therefore, HOLZ lines are no longer represented by lines of minimal brightness and some side minima (Fig. 13a), but their intensity modulations are irregularly distributed across parts of the 000 disk (Fig. 13b). From CBED patterns of this type (Fig. 13b), lattice parameters for both matrix and precipitates can not be evaluated.

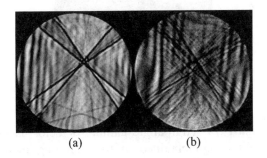

(a)   (b)

Figure 13. (a) Central 000 disk of a HOLZ-line pattern of an homogenized Ni-11.3 at.% Ti sample; [711] pole, 300 kV. (b) The same sample orientation, but for a Ni-11.3 at.% Ti sample aged for 96 h at 870 K. Small precipitates about 20 nm in diameter cause the strong distortions of the HOLZ pattern.

3.4. ATOMIC NUMBER (Z) CONTRAST

For high scattering angles, the contribution of the extended electron-density distribution around the nuclei can be neglected compared with the Rutherford scattering at the nuclei. The use of an annular detector centered about the 000 reflection in the diffraction plane of the sample allows the integrated scattering intensity to be measured for a certain range of scattering angles (see, e.g. [38]). Thermal diffuse scattering causes only an intensity redistribution in reciprocal space but not a significant change of the integrated intensity for large scattering angles (>40 mrad). With a field-emission electron source, the electron probe diameter can be reduced to about 0.2 nm on the sample. In the scanning mode, high-resolution images along low-indexed zone axes can be produced with such a small probe size. A HAADF micrograph of silver-rich Guinier-Preston zones in Al-3 at.% Ag is shown in Fig. 14. The atomic columns in the Ag-rich Guinier-Preston zones are clearly brighter than in the Al-rich matrix, as the scattering intensity for large scattering angles is approximately proportional to Z^2. In contrast to conventional high-resolution TEM with a field-emission source, the image formation process for HAADF micrographs is highly incoherent. The HAADF intensity at the position of an atomic column is the sum of the individual contributions of the atoms in the column independent of atoms in neighboring columns. For binary alloys with components differing significantly in their atomic numbers, a quantitative compositional analysis of the HAADF micrographs is possible if the sample thickness is known. The inset in Fig. 14 shows an example.

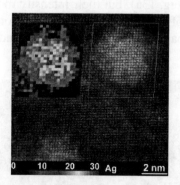

Figure 14. High-angle annular-detector dark-field image obtained in the scanning mode (HAADF-STEM) for an Al-3 at.% Ag alloy aged for 15 h at 380 K. The inset shows the number of silver atoms in and around the Guinier-Preston zone outlined by the frame.

Even for lower magnifications and larger electron probe sizes, the Z-contrast method reveals the distribution of the phases with different densities (Fig. 15). The signal-to-noise ratio of the HAADF signal is typically very low, as only 10^{-5} to 10^{-4} of all electrons are scattered to these high scattering angles. Image distortions by sample drift and scan-coil non-linearities are additional disadvantages of the HAADF-STEM method. The HAADF signal is not sensitive to light elements surrounded by heavier atoms. On the other hand, the HAADF signal is not strongly influenced by orientational differences in the sample. Furthermore, the intensity measured with the HAADF detector increases linearly with the sample thickness. Owing to the incoherent imaging process, contrast reversal and image delocalisation (dynamical scattering effects and thickness contours, as they are apparent for diffraction and strain contrast methods) - well known for the highly coherent imaging in conventional high-resolution TEM - are avoided in the Z-contrast technique.

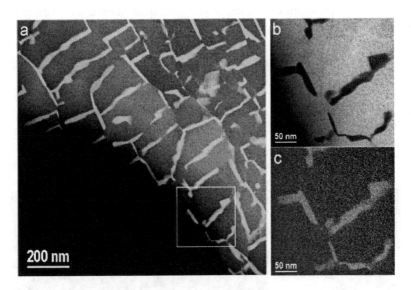

Figure 15. (a) Z-contrast image of Ni-10 at.% Au aged for 24 h at 920 K. (b, c) Energy-filtered Ni (b) and Au (c) maps of the area marked in (a).

3.5. ELECTRON AND X-RAY SPECTROSCOPIES

The methods mentioned so far are based on elastic scattering. The interaction with the sample also leads to energy losses of some of the electrons, which can be detected using an electron spectrometer (see, e.g., [39]). Element-specific edges in the electron energy-loss spectrum can be used to determine the local chemical composition. By scanning a small electron probe across the sample and acquiring energy-loss spectra for each probe position (called spectrum imaging), compositional maps can be obtained (see, e.g., [40]). Spectrum imaging [scanning transmission electron microscopy (STEM) for elemental analysis] usually requires long measuring times (usually a few seconds for each sampling position). The lateral resolution is determined by the size of the electron probe and the step size for the electron probe in the scanning mode. Additionally, the lateral resolution can be strongly reduced or deteriorated by sample drift during the acquisition of the spectrum map.

Similar resolution limits are found for compositional maps from X-ray fluorescence spectra acquired during scanning (see, e.g., [28]). Examples of elemental distribution maps are shown in Fig. 16 for a CdS/CdTe thin-film solar cell heat-treated after $CdCl_2$ deposition. Diffusion of Cl along the grain boundaries and interdiffusion of S and Te are revealed.

While X-ray spectra can be quantitatively evaluated for thin samples and elements with high atomic number $Z>10$, the quantitative evaluation fails for lighter elements like oxygen, and elements with $Z<6$ cannot be detected. In contrast, electron energy-loss spectra are especially useful for lighter elements (except hydrogen). Elements with high Z like Au are usually difficult to quantify because of delayed edges without prominent features. Another problem of electron energy-loss spectroscopy (EELS) arises sometimes from the overlap of edges (the background extrapolation of the pre-edge intensity to extract the edge intensity fails) or from the large energy differences of the edges (large differences in the intensities of the edges).

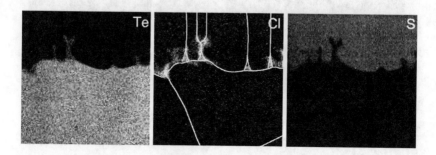

Figure 16. X-ray maps of a CdS/CdTe interface. An area of 1.5 μm x 1.5 μm is shown with 128 x 128 pixels.

Transmission electron microscopes equipped with electron-energy filters for imaging (GIF or Omega filter) allow for qualitative distribution maps without scanning the electron beam over the sample. Instead, the sample area of interest is homogeneously illuminated, and an energy-selecting slit in the energy filter is used to take images with electrons of the corresponding energy losses. With two images taken at (different) energy losses before the loss for a specific edge sets in, the background signal of a third image taken with electrons suffering the specific energy loss can be determined. By this procedure, elemental distribution maps can be obtained. An example for a Ti-distribution map in Ni-11.3 at.% Ti is shown in Fig. 17.

Energy-filtered transmission electron microscopy (EFTEM) can also be used with an electron-energy-selecting slit at energy losses of a few 10 eV corresponding to plasmon losses of the material instead of element-specific edges. The energy losses ΔE for plasmon excitations depend on the density n of the free electrons in the material ($\Delta E \sim n^{1/2}$). Therefore, by using an energy-selecting slit at a plasmon loss of one phase for the formation of an image, the distribution of this phase can be obtained if the differences in the plasmon energies of the different phases are significant (see, e.g., [41]).

Figure 17. Energy-filtered (Ti-L edge) TEM image along [100] of a Ni-11.3 at.% Ti sample aged for 10 h at 950 K, showing the Ti distribution.

4. Conclusion

Scattering methods and transmission electron microscopy are suitable and mutually complementary tools for the investigation of microstructures on the (sub-) nanometer scale. Some recent research results obtained at the authors' laboratory have been presented to illustrate some of the important points. The presentation is neither complete nor well-balanced, and the interested reader is encouraged to consult the rich literature for further insight and information.

5. Acknowledgments

The authors are very grateful to many of their colleagues, especially to B. Schönfeld for contributions on diffuse and small-angle scattering, M. Kompatscher for his work on small-angle scattering and M. Terheggen for Fig. 16. The provision of beam time and local support for neutron scattering experiments at the Institut Laue-Langevin, Grenoble, France, and the Paul Scherrer Institute, Villigen, Switzerland, is gratefully acknowledged. Part of this work has been supported by the Swiss National Science Foundation.

6. References

1. Blavette, D. (2003) This volume.
2. Rühle, M. and Wilkens, M. (1996) Transmission electron microscopy, in R.W. Cahn and P. Haasen (eds.), *Physical Metallurgy* 4th ed., North Holland, Amsterdam, pp. 1033-1113.
3. Kostorz, G. (1996) X-ray and neutron scattering, in R.W. Cahn and P. Haasen (eds.), *Physical Metallurgy* 4th ed., North Holland, Amsterdam, pp. 1115-1199.
4. Krivoglaz, M. (1996) *X-ray and Neutron Diffraction in Nonideal Crystals* and *Diffuse Scattering of X-Rays and Neutrons by Fluctuations*, Springer, Berlin.
5. Dederichs, P.H. (1973) *J. Phys.* **F 3**, 471-496.
6. Peisl, H. (1976) *J. Physique* **37**, C7. 47-53.
7. Yu, S.Y., Schönfeld, B. and Kostorz, G. (1997) *Phys. Rev.* **B 56**, 8535-8541.
8. Schönfeld, B., Kostorz, G., Celino, M. and Rosato,V. (2001) *Europhys. Lett.* **54**, 482-487.
9. Clapp, P.C. (1971) *Phys. Rev.* **B 4**, 255-270.
10. Asta, M.D. and Johnson, D.D. (1997) *Comput. Mater. Sci.* **8,** 64-70.
11. Reinhard, L. and Moss, S.C. (1994) *Ultramicroscopy* **52**, 223-232.
12. Gerold, V. and Kern, J. (1987) *Acta Metall.* **35**, 393-399.

13. Rossiter, P.L. (1987) *The Electrical Resistivity of Metals and Alloys*, Cambridge University Press.
14. Gerold, V. and Karnthaler, H.P. (1989) *Acta Metall.* **37**, 2177-2183.
15. Kostorz, G. (1995) Short-range order, slip coarsening and slip instabilities in alloys, in G. Ananthakrishna, L.P. Kubin and G. Martin (eds.), *Non Linear Phenomena in Materials Science III, Solid State Phenomena,* Trans Tech Publications, Switzerland, pp. 187-194.
16. Büchner, A.R. and Pitsch, W. (1985) *Z. Metallkd.* **76**, 651-656.
17. Schönfeld, B., Roelofs, H. Malik, A., Kostorz, G., Plessing, J. and Neuhäuser, H. (1996) *Acta Mater.* **44**, 335-342.
18. Roelofs, H., Schönfeld, B., Kostorz, G. and Bührer, W. (1995) *Phys. Stat. Sol. (b)* **187**, 31-42.
19. Reinhard, L., Schönfeld, B., Kostorz, G. and Bührer, W. (1990) *Phys. Rev.* **B 41**, 1727-1734.
20. Schönfeld, B. (1999) *Progr. Mater. Sci.* **44**, 435-543.
21. Guinier, A. and Fournier, G. (1955) *Small-Angle Scattering of X-Rays,* Wiley, New York.
22. Ciccariello, S., Schneider, J.-M., Schönfeld, B. and Kostorz, G. (2000) *Europhys. Lett.* **50**, 601-607.
23. Teixeira, J. (1988) *J. Appl. Cryst.* **21**, 781-785.
24. Schmidt, P.W. (1982) *J. Appl. Cryst.* **15**, 567-569.
25. Kompatscher, M., Schönfeld, B., Heinrich, H. and Kostorz, G. (2000) *J. Appl. Cryst.* **33**, 488-491.
26. Massalski, T.B. (ed.) (1990) *Binary Alloy Phase Diagrams, vol. 2,* ASM International, Materials Park, OH, USA.
27. Kompatscher, M., Schönfeld, B., Heinrich, H. and Kostorz, G. (2002) to be published.
28. Williams, D.B. and Carter, C.B. (1996) *Transmission Electron Microscopy,* Plenum Press, New York.
29. Fultz, B. and Howe, J.M. (2001) *Transmission Electron Microscopy and Diffractometry of Materials,* Springer, Berlin.
30. Pennycook, S.J. (1997) Scanning Transmission Electron Microscopy: Z Contrast, in S. Amelinckx, D. Van Dyck, J. Van Landuyt and G. Van Tendeloo, *Handbook of Microscopy, vol. 2,* VCH, Weinheim, Germany, pp. 595-620.
31. Freitag, B. and Mader, W. (1999) *J. Microscopy* **194**, 42-57.
32. Edington, J.W. (1976) *Practical Electron Microscopy in Materials Science,* Van Norstrand Reinhold, New York.
33. Matsumura, S., Toyohara, M. and Tomokiyo, Y. (1990) *Phil. Mag.* **A 62**, 653-670.
34. Fährmann, M., Fratzl, P., Paris, O., Fährmann, E. and Johnson, W.C. (1995) *Acta Metall. Mater.* **43**, 1007-1022.

464

35. Wittmann, R., Parsinger, C. and Gerthsen, D. (1998) *Ultramicroscopy* **70**, 145-159.
36. Heinrich, H. Vananti, A and Kostorz, G. (2000) Unequivocal determination of lattice parameters by CBED, in J. Gemperlová and I. Vávra (eds.), *Proc. 12th European Congress on Electron Microscopy EUREM 12, vol 2*, Czechoslovak Society for Electron Microscopy, Brno, Czechoslovakia, pp. 489-490.
37. Heinrich, H., Vananti, A and Kostorz, G. (2002) *Phil. Mag.* **A 82**, 2269-2285.
38. Nellist, P.D and Pennycook, S.J. (2000) *Adv. Imag. Electr. Phys.*, **113**, 147-203.
39. Egerton, R.F. (1986) *Electron Energy-Loss Spectroscopy in the Electron Microscope*, Plenum, New York.
40. Browning, N.D., Wallis, D.J., Nellist, P.D. and Pennycook, S.J. (1997) *Micron* **28**, 333-348.
41. Csontos, A.A., Tsai, M.M. and Howe, J.M. (1998) *Micron*, **29**, 73-79.

ATOMIC SCALE TOMOGRAPHY OF MICROSTRUCTURES AND PLASTIC PROPERTIES

D. BLAVETTE, E. CADEL, B. DECONIHOUT, F. DANOIX
Groupe de Physique des Matériaux, UMR CNRS 6634
Université de Rouen – site du Madrillet
Avenue de l'Université BP12 76801 St Etienne du Rouvray FRANCE

Abstract Mechanical properties of materials is shown to depend on numerous fine-scale microstructural features that were characterised using Tomographic Atom Probe. Phase composition, short range ordering, solid solution clustering of rhenium, as well as the solute distribution close to interfaces during creep in nickel base superalloys were investigated. The beneficial influence of boron on ductility has been shown to be caused by its segregation in the form of a very thin (1nm) film of boron covering the whole surface of grain boundaries. The mechanisms of segregation of impurities (boron) to line defects (Cottrell atmospheres) in iron-aluminium intermetallics is discussed. The spinodal decomposition that occurs at a nanometer scale in the ferrite of primary coolant pipes of nuclear power plants has been identified and was shown to control the microhardness evolution during service.

1. Introduction

Mechanical properties and plastic behaviour of materials depend on numerous fine-scale microstructural features (precipitates, grain boundary (GB) segregation, chemical order, solute clustering or segregation to line or planar defects). Among the relevant techniques for observation and microanalysis, the Tomographic Atom Probe (TAP) has several unique possibilities [1, 2]. This instrument, developed in the early nineties, is the only 3D quantitative analytical microscope with an atomic resolution. In this paper, some links between plastic properties and microstructure will be highlighted through various illustrations. Concomitant diffusion processes and phase transformation kinetics will be tackled.

Frequently, materials of industrial interest derive their mechanical performance from the presence of a hardening second phase. The phase composition is a key parameter because it controls the lattice misfit and the volume fraction of the hardening phase. An example is that of nickel base superalloys that are implemented in high temperature parts of aircraft engines (disks, single crystal blades..). Superalloys derive their exceptional creep properties from the presence of a high volume fraction of ordered precipitates.

In the following, we shall show that not only precipitates are important. The presence of

A. Finel et al. (eds.), Thermodynamics, Microstructures and Plasticity, 465–482.

solute clusters, or that of short range order in the solid solution, the segregation of "impurities" to grain boundaries (GB) have a drastic effect on mechanical properties. In the subsequent part, impurity interactions with linear defects and the formation of so-called Cottrell atmospheres in boron-doped FeAl intermetallics will be discussed.

A fundamental approach of the physics of phase transformation kinetics that control the microstructure of material under service is not only important for academic reasons but can also be a necessity. Prediction of life times without a clear identification of the decomposition regime (nucleation and growth versus spinodal decomposition) may lead to quite erroneous extrapolations and predictions. Duplex stainless steels implemented in primary coolant pipes of power water reactor (PWR) nuclear plants are a good example. Using TAP, a regime of spinodal decomposition of the ferrite at the nanometer scale was identified and the amplitude of concentration fluctuations was found to control the toughness evolution during service.

2. The tomographic atom probe

The tomographic atom probe (TAP) is based on the field evaporation of surface atoms and the identification of generated ions by time of flight mass spectrometry. The high electric field required to evaporate atoms from the sample surface is created by a high positive voltage ($V_0 \approx 10$ kV) applied to the sample fabricated in the form of a very sharp tip (curvature radius ≈ 50 nm).

The location of individual atoms at the tip surface is calculated from the position of ion impacts on a position-sensitive detector. The initial position of ions at the specimen surface (x,y) is derived using a simple point projection law (close to a stereographic projection). The third dimension is generated by the field evaporation of atomic layers at the tip surface. The depth scale (z) is taken as proportional to the cumulative number of detected ions. More details can be found elsewhere [1].

3. Nickel base superalloy

Nickel base superalloys are key materials in aerospace industry for turbine blades and disks. These engine components are submitted to very high stresses (~200 Mpa) under temperatures in the range 700-1000°C. Thanks to the addition of numerous elements (Al, Cr, Ti, W, Ta...), nickel base superalloys combine high creep performance at high temperatures and excellent corrosion resistance. Modern superalloys contain a high volume fraction (~ 50-70 %) of ordered $Ni_3(Al,Ti)$ type precipitates (called γ' hereafter). The γ' phase has the $L1_2$ ordered structure.

Some features are illustrated in figure 1. These 3D images give the distribution of solute atoms (Al, B, Cr) within a small volume (16 x 16 x 28 nm$^{3)}$) of polycrystalline nickel base superalloy N18 that was analysed using TAP [3]. The γ' particles appear as Al-enriched and Cr-depleted regions while the surrounding matrix is Al-depleted and Cr-enriched. The alternation of Al-enriched planes with Al-depleted planes within the small γ' ordered particle (left hand) is characteristic of concentration waves in the L1$_2$ order of the γ' phase along the <001> direction.

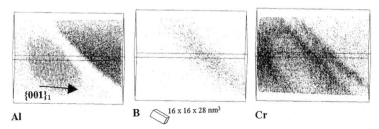

Al   **B** 16 x 16 x 28 nm^3 **Cr**

{001}$_1$

Figure 1. Spatial distribution of Al, B and Cr in the close vicinity of a boron enriched and aluminium-depleted grain boundary (superalloy N18). Note the presence of a small Al enriched γ' precipitate on the left and that of a larger γ' on the right side of GB

The measured phase composition is given in table 1. Samples were subjected to a three step heat treatment: homogenisation for 4 hours at 1165°C followed by a rather slow cooling (50-100°C / minute) and two precipitation treatments (24 hours at 700°C + 4 hours at 800°C). The γ' precipitates are enriched in Al, Ti, Hf whereas the γ solid solution contains a large amount of Cr, Co and to a lesser extent Mo (table 1). Measured concentrations are consistent with the nominal composition (C_n) as given by the manufacturer (ONERA) and for a volume fraction of γ' phase f = 57%, derived from the lever rule:

$$f = (C_n - C_\gamma) / (C_{\gamma'} - C_\gamma) \tag{1}$$

Table 1. Nominal composition (at.%) and composition of both γ and γ' phases. Statistical fluctuations caused by sampling errors are given by the standard deviation σ (ΔC = 2σ) with σ = √(C(1-C)/N) and N the number of ions which were taken into account in the estimate.

at. % ± 2σ	Ni	Cr	Al	Ti	Mo	Co	Hf	B	C	Fe
Nominal composition	54.42	12.3	9.15	5.11	3.77	14.82	0.16	0.083	0.075	0.11
γ	38.28	25.71	1.85	0.25	8.36	25.28	0.01	0.04	0.01	0.20
18400 ions	0.72	0.64	0.20	0.07	0.41	0.64	0.02	0.03	0.02	0.07
γ'	67.71	1.60	12.95	8.65	2.20	6.64	0.21	0.01	0.00	0.03
84784 ions	0.32	0.09	0.23	0.19	0.10	0.17	0.03	0.01	0.00	0.01

Boron is added in small proportion (<100 at. ppm) to such polycrystalline superalloys to improve ductility and mechanical resistance. The beneficial influence of boron is caused by its segregation to the whole surface of GBs which reduces the risk of intergranular rupture (figure 1). The better ductility of boron doped materials is not specific of superalloys but is found in many intermetallics including FeAl, Ni$_3$Al, NiAl...

Because of the limited depth that can be explored with atom probe (100 nm) as compared to the grain size (10 μm), TAP investigation of GBs requires tips to be first examined with TEM. Tips were thus backpolished until a grain boundary (GB) was positioned close to the tip apex. Even though the atomic structure of GBs cannot be determined thoroughly, the grain misorientation, the interface plane index and the type of GB (twin, general, low angle, etc...) can be determined for most of GBs [4]. Most of GBs analysed were found to be general boundaries. All GBs observed in superalloys were found to be enriched in B and Mo. The boron concentration in the GB shown in figure 1 is close to 1.5 at.% (the nominal composition is close to 0.08 at.%).

The apparent thickness of the boron enriched zone at GB shown in figure 1 is a few nanometers. However, ion trajectory aberrations near such interfaces are known to lead to an artificial enlargement in reconstructed images. The actual thickness is less than a nanometer, a distance that is typical of equilibrium interfacial segregation. Non-equilibrium segregation promoted by the elimination of vacancy-boron complexes to defect sinks (GBs) may also occur in some materials. There is thus a risk that the beneficial segregation of boron disappears under service at high temperature.

The relevant parameter to quantify interfacial segregation is the Gibbsian interfacial excess of segregated species i (Γ_i) [5]. In contrast to the local concentration at the GB core, the interfacial excess has the advantage not to depend on the spatial extent of the sampling box used to measure the GB composition. The Gibbsian interfacial excess is the number of solutes i per unit area of interface that are in excess with respect to the surrounding medium. If Υ is the interfacial free energy per unit area and μ_i the chemical potential of species i, then the Gibbsian interfacial excess Γ_i for a given type of GB (5 degrees of freedom related to geometry) can be expressed as the partial derivative of Υ:

$$\Gamma_i = - \partial\Upsilon/\partial\mu_i)_{T,P} \tag{2}$$

One advantage of TAP is that Γ_i can be directly derived from experimental images (Γ_B = 6.4 +/- 0.8 at/nm^2 for boron in figure 1). This value can be compared with the atomic density of <111> close packed planes of nickel (18 at./nm^2) in order to get an expression of segregation in terms of monolayers (here ~0.35 monolayer).

Two models have been proposed to explain the beneficial influence of boron intergranular segregation on mechanical resistance. The first one assumes that boron leads to a loss of order in the vicinity of the GB, which results in an increase of the mobility of dislocations through the interface. This would reduce the piling-up of dislocations at the GB and improve the local plasticity and decrease the risk of crack initiation [6,7]. The second model suggests a reinforcement of GB cohesion originating from the formation of strong interatomic bonds with boron [8]. It is however difficult to determine which mechanism prevails on the only basis of the present observations.

Note the presence of a thin γ film on both sides of the segregated boundary. Chromium is observed to be present at GBs but at a lower extent than in the γ matrix. An important

Al depletion is present all along GBs, in particular those flanked by γ' precipitates on both sides. This local depletion, confined in a narrow zone at the GB (1-2 nm), implies that local long range order between both γ' precipitates, if such a concept is meaningful at this scale in such a small and disturbed region, is very low if compared with that existing in γ'. The existence of such a disordered zone could improve the GB plasticity.

3.1. CONCENTRATION GRADIENTS NEAR INTERFACES

Atom probe techniques has are of great help in the understanding of oriented coarsening during creep tests. According to the sign of misfit and the direction of the stress applied (compression or tension), γ' platelets perpendicular or parallel to the stress axis may develop during creep. The physical mechanisms behind the formation of these « rafts » has been the subject of many speculations, experiments and models [9]. Phase field simulations of free coarsening kinetics were also conducted [10].

Atom-probe investigations of the early stages of creep in superalloy MC2 at 1050°C revealed non-equilibrium concentration gradients near γ-γ' interfaces [11]. Rafts of γ' formed under a compression stress applied along <001> were found parallel to the stress axis (P rafts). Local depletion of Al, Ta, Ti (γ' formers) and a Cr (γ-like element) enrichment were detected in the γ corridor near γ/γ' interfaces that were normal to the stress direction (N interfaces). This suggests a growth process of N interfaces. Opposite gradients were detected at γ-γ' interfaces parallel (P interfaces) to the deformation axis, suggesting a dissolution of particles along P interfaces.

These gradients reveal the nature of diffusion mechanisms of atoms in the internal stress field resulting from both the lattice misfit between phases and the applied stress; γ'-like elements (Al, Ta, Ti) generated by the dissolution of P interfaces diffuse towards N interfaces through γ corridors. The concomitant migration of N interfaces of two neighbouring γ' particles leads finally to the coagulation of γ' precipitates and thus to the formation of rafts parallel to the stress axis. The driving force of the solute migration is the elastic energy gradient between N and P corridors. The local stress field was calculated using finite element modelling and the hydrostatic stress in N corridors was found to be higher, suggesting a smaller stability of these regions in agreement with experiments (N corridors shrink during P rafting) [11].

3.2. DISTRIBUTION OF SOLUTES IN THE γ SOLID SOLUTION

Many additional elements are incorporated in superalloys in order to optimise their mechanical properties at high temperatures. The addition of Re for instance, was found to be beneficial for creep performance of a number of superalloys (CMSX-2, PWA 1480 [12, 13]). In particular, the addition of 1.3 at.% of Re to single crystal PWA1480 was shown to increase drastically the rupture life (580 h at 950°C under 240 MPa as compared to 354 h without Re).

Atom probe analyses showed that Re is a γ-like element: 3.71at.% Re in γ but only

0.76%Re in γ' [14]. Furthermore, tiny Re clusters were found in the matrix [12]. These small clusters may be more efficient obstacles against dislocation motion during creep tests than randomly distributed Re atoms.

More recently, Warren et al. [15] investigated a Re-containing superalloy (RR3000). They found a non-equilibrium Re pile-up at the γ-γ' interfaces of γ' rafts in crept materials. This local enrichment appears as a bow wave ahead of precipitates in concentration profiles. This was related to the rejection of this element from γ' during the particle growth during cooling. Interestingly, the authors suggested that because Re is a slow diffuser, it can reduce coarsening rate.

3.3. SHORT RANGE ORDER IN THE γ SOLID SOLUTION

Combined investigations of the links between deformation mechanisms and ordering in the γ solid solution using TEM, diffuse scattering, and atom probe were conducted by Clément et al. [16]. Short range order of Cr in the solid solution was revealed by atom-probe and confirmed by X ray diffuse scattering. Diffraction patterns short range order of {1½0} special point. This local ordering was found in numerous superalloys as well as in model γ single-phase Ni-Cr-Co alloys. TEM observation of in-situ deformed materials revealed that these ordering effects gave rise to high friction stresses hindering the propagation of dislocations during deformation [16].

4. Cottrell atmospheres in FeAl intermetallics

Cottrell proposed the concept of « atmospheres » in 1949 to explain the behaviour of materials during plastic deformation (yield point, Portevin-LeChatelier effect) [17,18]. Interstitial impurities like carbon or boron for instance, lead also to a deformation of the crystal lattice which may be released in part if these elements segregate along the dislocation lines and form so-called Cottrell atmospheres.

The stress field created by dislocations is here the driving force for solute migration. The total diffusion flux **J** can express as the sum of a Fick term (random diffusion) and a transport term due to the interaction of solute i ($U(r,\theta)$) with the line defect stress field:

$$\mathbf{J} = -D\nabla C - (CD/k_BT)\ \nabla U(r,\theta) \tag{3}$$

Where r and θ are the cylindrical co-ordinates of solute i with respect to the dislocation line (figure 2). D is the diffusion coefficient of solutes and k_B, T the Boltzman constant and temperature respectively.

The elastic energy of interaction of solute i with an edge dislocation U (r,θ) is:

$$U (r,\theta) = \mu b\ \delta V \sin(\theta) / (3\pi r) \tag{4}$$

μ, b, δV are respectively the shear modulus, the modulus of the Burgers vector, and the size effect parameter.

Segregation is expected for U (r,θ) <0. Interstitial elements (as well as over-sized substitutional atoms, δV>0) are therefore expected to segregate in the dilated part of the crystal (θ<0 U <0, equation 4) situated below the extra-half plane (figure 2). Conversely, under-sized substitution elements (δV<0) are likely to migrate towards the compressed regions above the dislocation line (θ>0).

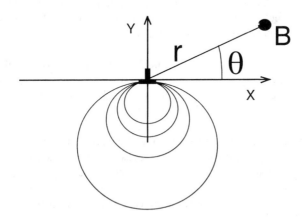

Figure 2. Interaction of a solute B with an edge dislocation. Iso-concentration surfaces are cylinders tangential to the dislocation glide plane (perpendicular to the figure). Interstitial elements (δV>0) segregate below the line defect. The larger the diameter, the smaller the concentration.

Iso-concentration surfaces can be calculated under steady state conditions ($\mathbf{J} = \mathbf{0}$) using equations (3) and (4),

$$C (r,\theta) = C_0 \exp [-U(r,\theta) / k_B T] \qquad (5)$$

Constant concentration surfaces correspond to iso-interaction surfaces that are cylinders tangent to the dislocation line (figure 2). The concentration of solutes C (r) increases for decreasing distances (r) with respect with the dislocation line (provided that U (r,θ) < 0). The lateral extension of a Cottrell atmosphere for a given concentration threshold (2 times the bulk composition for instance) is in the nanometer range. Many refinements of the Cottrell and Bilby model were brought about. The divergence of C (r,θ) when r ~ 0, caused by that of U(r,θ), is obviously meaningless in the framework of this simple approach and correction factors were implemented in order to avoid singularities [19]. Also, the short-ranged repulsive interactions between solute prevents dual site occupancy so that the concentration C (r,θ) is rather given by by Fermi-Dirac statistics instead of Boltzman distribution (equation (5) [20].

Cottrell atmospheres play a major role in mechanical properties of materials. Their presence makes more difficult the movement of dislocations, which can be pinned by solute atmospheres. At the macroscopic scale, they give rise to two major features [17]. The first one appears at the very beginning of plastic deformation : the additional stress needed to move solute-locked dislocations gives rise to a yield point in the deformation curve followed by a sharp decrease related to the subsequent release of dislocations form their atmospheres. The second effect is observed when the material is tested at high temperatures. Atomic mobility of locked impurities may then be sufficient to allow the Cottrell atmospheres to follow the "free" dislocations. The yield stress during deformation thus oscillates due to the cyclic pinning and releasing of moving dislocations. This effect, known as "dynamic strain ageing" or "Portevin-LeChatelier instabilities", leads to serrations in stress-strain curves.

At the nanometer scale, the segregation of interstitial elements to stacking faults (SF) (Suzuki effect [21,22]) may modify the SF energy and modify the dissociation of dislocations. In ordered alloy, a segregation-induced modification of the antiphase boundary energy may also occur and influence plastic properties.

Such Cottrell atmospheres were imaged for the first time using TAP a few years ago [23]. The 3D reconstruction shown in figure 3 reveals the presence of a boron enriched atmosphere in FeAl (40at.%) intermetallics. The atmosphere appears as a rod-shaped cloud, 3 nm in diameter and was found to contain 3 at.% of boron. This concentration, measured in the core of the atmosphere, is equivalent to a thousand times the nominal concentration (0.04at.%B).

These ordered materials (B2 structure) are intrinsically brittle at room temperature but small boron addition improve drastically their mechanical resistance through a reinforcement of grain boundaries [24]. Atom probe experiments revealed that the role of boron in plasticity is in fact much more complex. Boron also segregates to (001) stacking faults. These "high energy" defects are only observed in boron doped intermetallics and it is therefore thought that boron has a stabilising effect [25,26]. When doped with only 100 at. ppm of boron, the maximum yield strength was found to be 200 MPa higher than boron-free FeAl. Moreover, it was observed 200°C higher [28]. This strongly suggests that boron modifies the elementary mechanisms of crystal plasticity.

FeAl samples that were studied were submitted to a two step heat treatment (950°C/1h (quenching) + 400°C/24 h (air cooling)). The annealing at 400°C reduces the concentration of thermal vacancies retained in a large quantities (up to 0.2 % [27]) after quenching from 950°C. Vacancy elimination gives rise to the formation of dislocations, mainly with <100> Burgers vector [28].

The boron-enriched atmosphere shown in figure 3 was found to be parallel to the <100> direction. The thin slice (thickness = 6 nm) cut perpendicular to the boron rod shown in figure 3 is represented in figure 4. The basic stacking sequence of the superlattice (001) planes of this B2 ordered material is exhibited: Al enriched planes alternate with Al depleted (i.e. Fe-rich) planes. Plane spacing is 0.29 nm (the lattice parameter). The boron-enriched atmosphere is Al depleted. The diameter of the depleted region is close to 10 times the lattice parameter (~3 nm). The Burgers circuit drawn in figure 4 confirms that, as expected, this is a perfect a<001> dislocation of B2-ordered FeAl. This confirms that the boron cloud is a Cottrell atmosphere.

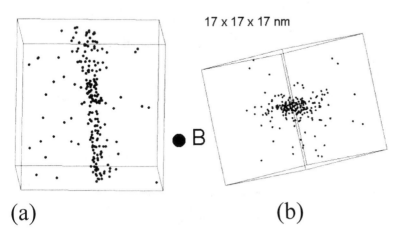

Figure 3. 3D images showing the presence of a Cottrell atmosphere enriched in boron (black dots) in FeAl intermetallics. (a) Side view. The analysis started on the left side and the volume was reconstructed layer by layer along the depth axis (horizontal scale). (b) The same as figure 3a after rotation. The rod-shaped atmosphere is almost perpendicular to the figure.

The boron cloud appears to be slightly shifted with respect to the dislocation line (figure 4). Boron is an intersticial element ($\delta V > 0$), it is therefore expected to segregate below the dislocation line. The approximate position of the dislocation line given in figure 4 is debatable and may be partly at the origin of the observed shift. Artefacts resulting from the reconstruction procedure or the evaporation process of TAP occur and might partly explain the right-hand shift. This is discussed in more details elsewhere [26]. The physical origin of the upward shift of the atmosphere could be the dissociation of the dislocation core. Boron could also segregate along the line terminating the Fe-rich (001) half-plane (not represented) of the dislocation situated above the Al-rich half-plane in figure 4.

Present measurements give a mean boron concentration of 3 at.% in the core of the dislocation ($r_0 \sim 1$ nm). This leads to an estimate of the interaction energy $U(r_0)$ (see equation (5)) of ~ 0.2 eV, a value comparable to the energy of intergranular segregation in metals (0.5 eV for carbon in iron, 0.6 eV for boron in superalloys).

474

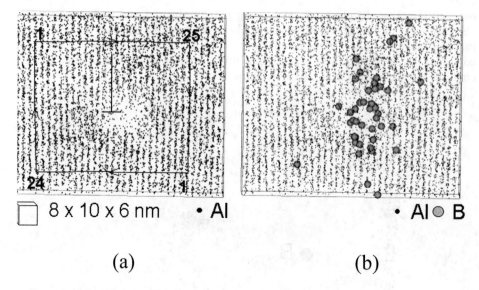

Figure 4. (a) Map of aluminium showing the sequence of Al-rich planes along the <001> direction. Iron atoms are omitted for clarity. The Burgers circuit indicates the presence of an <001> edge dislocation (approximate position shown). (b) Map of both boron and aluminium showing the presence of a boron enriched Cottrell atmosphere along the edge dislocation.

5. The spinodal decomposition of the ferrite phase in primary coolant pipes

Stainless steels are widely used in the nuclear industry, in particular as the main constituent of pipes for the primary cooling system in PWR plants. Their service temperature ranges from 300 to 500°C, where they face a severe problem of embrittlement [29]. After long term service, a loss of impact toughness and ductility is observed, which may lead to the ruin of the component.

The iron-chromium system, basis of the whole stainless steel family, has been known for half a century to be susceptible to embrittlement known for decades as '475°C embrittlement' [30]. It is related to the presence of a large miscibility gap in the Fe-Cr phase diagram and the decomposition of the ferrite into α (Fe-rich) and α' (Cr-rich) phases [31]. The critical temperature is observed at 560°C for 50at%Cr.

Depending on the supersaturation, the system can decompose via two distinct regimes, namely nucleation and growth (NG) and spinodal decomposition (SD). Classical nucleation is generally observed for low supersaturated systems (i.e. low driving force for unmixing). This metastable state is observed when the curvature of the free energy curve F(C) is positive. In the framework of classical theories, nuclei are expected to have abrupt interfaces and a composition (C) very close to the equilibrium state. Such a regime was observed by TAP and supported by Monte-Carlo simulations in model Ni-Cr-Al superalloys [32].

For higher supersaturation, i.e. when quenched well inside the miscibility gap, alloy becomes unstable (negative curvature of the free energy curve F(C) and spinodal decomposition is observed. The limits between NG and SD regimes are defined by the spinodal line $(\partial^2 F/\partial C^2 = 0)$. In a first approach, the Fe-Cr system can be treated as a regular solution. Effective interaction terms (U_{FeCr}) can be adjusted so as to fit the experimental phase diagram. Consequently, a broad assessment of the spinodal line that separates both regimes (NG and SD) in the phase diagram can be calculated.

The concept of spinodal was introduced by Cahn some forty years ago [33]. In contrast to nucleation and growth regime, no more nucleation barrier has to be overcome. Up-hill diffusion is thus observed: small concentration fluctuations initially present in the material gradually increases with ageing time and phase interfaces are diffuse. Theory predicts that large wavelength fluctuations develop at the expense of small ones. In the linear theory [33], the concentration amplitude increases exponentially as a function of time provided that the wavelength λ is larger than a critical value λ_C (or $k_C > k$ with $k = 2\pi/\lambda$):

$$A(k,t) = A(k) \exp (R(k) \cdot t) \tag{6}$$

$$\text{With } R(k) = 2DV_m K_i (k_C^2 - k^2) k^2 / k_B T \tag{7}$$

where D is the diffusivity of solute, V_m the molar volume, T the temperature and K_i a coefficient related to the chemical interfacial energy.

Only Fourier components of the concentration profile that have a wavelength $\lambda > \lambda_C$ can see their amplitude increasing, with:

$$\lambda_C = 2\pi (2K_i V_m / (-(\partial^2 f/\partial C^2)_{C*} - 2K_e) \tag{8}$$

f is the free energy density, C* the nominal atomic fraction of solute and K_e a coefficient related to misfit elastic energy density f_e,

$$f_e = K_e(\delta C)^2 \tag{9}$$

Similarly to nucleation theory, this critical wavelength λ_C results from the competition that occurs between the effective driving force for decomposition (minus the elastic term K_e) and the interfacial energy $(\sim K_i (\nabla C)^2)$ that opposes to phase separation.

An unambiguous identification of the decomposition regime is extremely difficult experimentally as it is necessary to measure the composition of phases in the very early stages of decomposition for which the size of phases is at the nanometer scale. Whereas small angle scattering of X-rays or neutrons yields valuable information on the evolution of fluctuation wavelengths, TAP appears more able to bring about quantitative composition data at this scale. Pionneer work was conducted in the early 80' by Brenner [34]. Spinodal decomposition was also shown to occur in complex CF8 duplex stainless steels (Fe-20Cr-10Ni-1Si-2.5Mo-0.5Mn-0.04C in wt%) [35].

One major contribution of TAP to this topics is that this instrument is able to show directly the topology of α and α' phases at the nanometer scale, and that fluctuation amplitudes can be confronted to theoretical predictions. This is illustrated in figure 5 where the location of Cr atoms in the ferrite of a complex duplex stainless steel is shown. This 3D elemental map clearly shows the typical three-dimensional α/α' interconnected structure (the map of Fe atoms (α phase) is not represented but would give a similar image). Correlation length between Cr enriched regions is close to 5 nm.

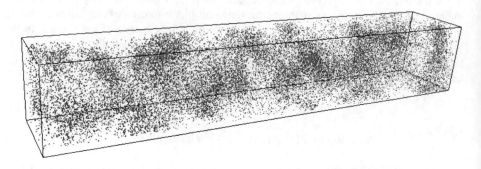

Figure 5: 3D map of Cr atoms within the ferrite phase in duplex stainless steels aged for 30,000 hours at 350°C. Cr- rich regions are related to the α' phase. The represented volume is 8x8x80 nm³.

In order to quantify the phase composition and to follow the decomposition kinetics, one-dimensional concentration profiles through such a microstructure can be drawn. A typical evolution of concentration profiles as a function of ageing time at 350°C is shown in figure 6. The related amplitude distribution of fluctuations, called hereafter frequency distribution, is represented for every ageing time. In this representation the frequency of occurrence is plotted as a function of Cr composition.

As expected for spinodal decomposition, the amplitude of Cr fluctuations increases gradually. In the unaged conditions, no phase separation is detected. Only statistical fluctuations related to sampling errors are observed. This is proved by the corresponding concentration frequency distribution that is identical to the binomial distribution, as expected for a random solid solution [36]. Frequency distributions clearly indicate a departure from the random distribution (binomial) and significant concentration fluctuations appear and develop as ageing proceeds. The concentration profile is not purely sinusoidal but is composed of various Fourier components. Qualitatively, one can see that larger wavelengths develop with increasing ageing time. Correlation length or wavelength follow a power law $\lambda \sim t^a$ with a ~ 0.16.

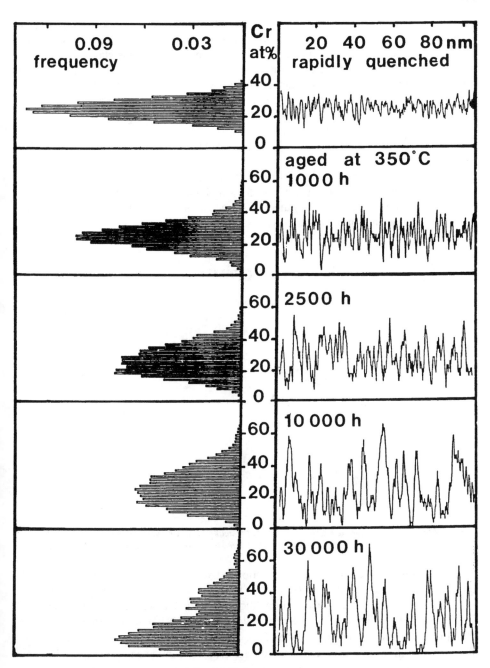

figure 6 The kinetics of spinodal decomposition of the ferrite in duplex stainless steels. Concentration profiles and related frequency distributions after ageing at 350°C..

One effective way to characterise the amplitude of fluctuations is to assess the statistical distance between the observed frequency distribution O (i) and a reference distribution corresponding to pure statistical fluctuations (binomial B (i)). This statistical distance was chosen as the sum of the differences between the observed frequency distribution and the binomial distribution (random solution) over the whole range of concentration (i) [37]. This scalar parameter, also called the "variation V", is an assessment of the effective amplitude of Cr fluctuations observed in profiles such as shown in figure 6.

$$V = \Sigma_i \, |\, O(i) - B(i)\,| \tag{10}$$

The linear Cahn-Hilliard theory [33] predicts an exponential evolution of amplitudes (equations (6) and (7)) that is valid only for small ageing times. In contrast, the non-linear Langer-Baron-Miller (LBM) theory [38] was found to account experimental data for longer ageing times more accurately.

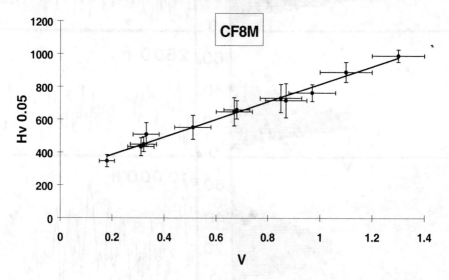

Figure 7 *Evolution of the ferrite microhardness with the amplitude of Cr fluctuations (~ V) that develop in the ferrite phase of primary coolant pipes. Data related to specimens aged at 300, 350 and 400°C.*

As shown figure 7, there is a clear correlation between the evolution of the "variation" and the microhardness of the ferrite. The microhardness increases almost proportionally with the amplitude of Cr concentration fluctuations in the ferrite. Although the size difference between Fe atoms (0.126nm) and Cr atoms (0.127 nm) is small (less than 1%), the α/α' spinodal decomposition of the ferrite induces a stress field (misfit) which hinders dislocations to propagate. The lattice misfit between the emerging Fe-rich α and Cr-rich α' domains, and the resulting elastic strain, increases with ageing time and can be regarded as the main cause of hardening. The spinodal decomposition of the ferrite was shown to be the main source of embrittlement of this alloy family [39].

In addition to spinodal decomposition, Mo-bearing duplex stainless steel grades may exhibit an extensive precipitation of small Ni and Si enriched intermetallic G-phase particles. The G phase was shown to incorporate most of the solute elements present in the steel [40]. As shown on the 3D reconstruction figure 8, G-phase particles are located at α-α' interfaces. No particle has been observed in the inner part of α or α' domains. The nucleation of the G-phase could here be favoured by the opposite diffusion fluxes of Ni and Si across the α/α' interface, these elements being known to partition to α (Ni) and α'(Si) during the decomposition of the ferrite. This strongly suggests that the α-α' spinodal decomposition drives the precipitation of the G-phase.

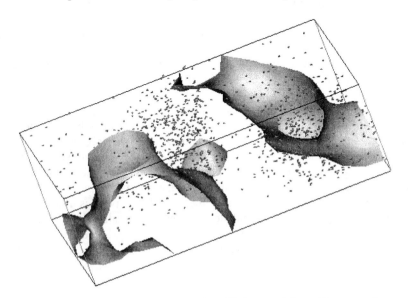

Figure 8 Respective positions of α-α' network and G phase precipitates. The grey surface are iso-concentration surfaces and corresponds to α/α' interfaces at a given threshold (30 at% of Cr). Dots represent the position of Ni atoms. Represented volume is $8x8x25nm^3$.

Confirmation of this correlation between both phase transformations is provided by the comparison of the kinetics of both the spinodal decomposition and the precipitation of G phase. The temporal evolution of the diameter of G-phase particles ϕ during ageing at 350°C was shown to follow a t^a law with a = 0.15. This temporal exponent is significantly different from that expected for pure diffusional growth (a=1/2) or for classical coarsening (a = 1/3). Instead, the exponent appears to be close to that observed for the correlation length ($\sim\lambda$) of a spinodally decomposed solid solution (a = 0.15-0.18) [41].

6.   Conclusion

The various examples show that the links between the microstructure, the microchemistry and plastic properties are often complex. An element such as boron,

480

added in very small proportion (400 ppm) can have a multiple role and can interact with a large variety of crystal defects such as antiphase boundaries, stacking faults, dislocations or grain boundaries and it is not obvious to ascertain a hierarchy between their respective influences.

The emergence of the 3D atom probe in the early nineties has provided a fresh boost to material science. TAP has been the first analytical microscope showing Cottrell atmospheres in 3D at the atomic-scale. Experiments revealed unanticipated effects related to the interactions between dislocations and solute atoms and this should motivate new simulations or modelling.

The combination of high resolution electron microscope with TAP is a promising approach since both instruments are very complementary. Because of its higher lateral resolution, the former is better for structural characterisation while the latter gives the chemical nature of imaged atoms in an unambiguous way.

TAP is also a very attractive tool to study the early stages of phase separation. It is the only instrument providing 3D images that can be confronted directly to Monte-Carlo simulations of kinetics. Both provide images for similar size objects (a million of atoms). This dual approach was applied to the study of the transformation pathways during the ordering and phase separation kinetics in model systems notably in model superalloys Ni-Cr-Al [42]. This was shown to be very fruitful as TAP can validate the simulation parameters and more generally the underlying physics implemented in atomistic simulations. Mean field simulation of phase separation kinetics is also an attractive approach that can be used together with TAP. This was recently applied to the Ni-Al-V system where two ordered phases ($L1_2$ and DO_{22}) occur concurrently [43].

References

1   Blavette D., Bostel A., Sarrau J.M., Deconihout B. and Menand A. (1993) An atom-probe for three dimensional tomography, *Nature* **363** 432-435
2   Blavette D. and Menand A. (1994) New developments in atom-probe techniques and potential applications to material science *invited article for MRS bulletin*, **19**, 21-26
3   Cadel E., Lemarchand D., Chambreland S. and D. Blavette (2002) Atom probe tomography investigation of the microstructure of superalloys N18 *Acta Mater.* **50** 957-966
4   Lemarchand D., Cadel E., Chambreland S. and Blavette D. (2002) Investigation of grain boundary struture-segregation relationship on a N18 nickel-based superalloy *Phil. Mag. A* **82** 1651-1669
5   Krakauer B.W, Seidman D.N. (1993) Absolute atomic-scale measurements of the Gibbsian interfacial excess of solute at internal interfaces Phys. Rev. B 48 6724-6727
6   Schulson E.M., Weihs T.P., Baker I. and Horton J.A. (1986) Grain boundary accommodation of slip in Ni_3Al containing boron *Acta Met.,* **34**, 1395-1399
7   Lee T.C., Robertson I.M. and Birnbaum H.K. (1992) Interaction of dislocations with grain boundaries in Ni_3Al *Acta Met. Mater.,* **40**, 2569-2579
8   Messmer R.P. and Briant C.L. (1982) The role of chemical bonding in grain boundary embrittlement *Acta Met.,* **30**, 457-467
9   Louchet F., Hazotte A. (1997) A model for low stress cross-diffusional creep and directional coasening of superalloys, Scripta Mater. 37, 589-597
10 Wang Y., Banerjee D., Su C.C., Katchaturyan (1998) Field kinetic model and computer simulation of

precipitation of L1$_2$ ordered intermetallics from FCC solid solution, Acta Mater. **46**, 2983-3001

11 Blavette D., Letellier L., Racine A. and Hazotte A. (1996) Concentration gradients near heterophase boundaries in crept single crystal nickel base superalloys, *MMM* **7**, 185-193

12 Blavette D., Caron P. and Khan T. (1986) An atom-probe investigation of the role of rhenium additions in improving creep resistance of nickel-base superalloys, *Scr. Met.* **20**, 1395-1400

13 Blavette D. (1990) Etude de là structure fine de superalliages à base nickel par microscopie ionique et microanalyse à la sonde atomique, *Rev. Sci. et Tech. de la Déf.*, 61-69

14 Blavette D., P. Caron and T. Khan (1988) An atom-probe study of some fine-scale microstructural features in Ni-based single crystal superalloys, Seven Spring, Pen./USA, "Superalloys 88", *The Met. Soc.* 305-314

15 Warren P.J., Cerezo A. and Smith G.D.W. (1998) An atom probe study of the distribution of Re in a nickel base superalloy, *Mater. Sci. Eng. A* **250**, 88-92

16 Clément N., Coujou A., Calvayrac Y., Guillet F., Blavette D. and Duval S. (1996) Local order and associated deformation mechanisms of the γ phase of nickel base superalloys, *Microscopy Microstructures Microanalysis* **1**, 65-84

17 Cottrell A.H. and Bilby B.A. (1949) Dislocation theory of yielding and strain ageing of iron *Proc. Phys. Soc. Lond.*, **A62** , 49

18 Cottrell A.H. (1953) Dislocations and plastic flow in crystals *(Oxford Clarendon)* 134

19 Nandedkar A.S., Johnson R.A. (1982) Carbon atmosphere near an edge dislocation in α-iron Acat Metall. **30**, 2055-2059

20 Wolfer W.G., Baskes M.I. (1985) Interstitial solute trapping by edge dislocations, Acata Metall. **33**, 2005-2011

21 Suzuki H. (1952) Chemical interactions of solute atoms with dislocations *Sci. Repts Tohoku Univ.*, **A4**, 455-463

22 Suzuki H. (1962) Segregation to solute atoms to stacking faults *Jour. Phys. Soc. jap.*, **17**, 322-325

23 D.Blavette, E. Cadel, A. Fraczkiewicz, A. Menand, (1999), Three-dimensional atomic-scale imaging of impurity segregationto line-defects, SCIENCE **17** 2317-2319

24 Crimp M. A. and Vedula K. (1986) Effect of boron on the tensile properties of B2 FeAl, *Mat. Sci. Eng.*, **78**, 193-200

25 Pang L., Chisholm M.F. and Kumar K.S. (1998) {001} faults in B2 Fe-40 at.% Al-0.7 at.% C-0.5 at.% B *Phil. Mag. Lett.*, **78**, 349-455

26 Cadel E., Launois S., Fraczkiewicz A. and D. Blavette (2000) Investigation of boron-enriched cottrell atmospheres in feal on an atomic-scale by a three-dimensional, APFIM, *Phil Mag Lett.* **vol. 80**, 725-736

27 Paris D. and Lesbats P. (1978) Vacancies in Fe-Al alloys, *J. Nucl. Mat.* **69-70**, 628–632

28 Morris M.A. and Morris D.G. (1998) Quenching and ageing effects on defects and their structures in FeAl alloys, and the influence on hardening and softening, *Scripta Metall* **38**, 509-516

29 Auger P., Danoix F., Menand A., Bonnet S., Bourgoin J. and Guttmann M. (1990) Atom-probe and transmission electron microscopy study of aging of cast duplex stainless steels, *Materials Science and Technology* **6**, 301-313

30 Newell H.D. (1949) Properties and characteristics of 27% chromium-iron, *Metals Progs.* **49**, 977-1006

31 Williams R.O. and Praxton H.W. (1957) The nature of the ageing of binary iron-chromium alloys around 500°C, *J. Iron Steel Inst.* **185**, 358-374

32 Schmuck C., Caron P., Hauet A., Blavette D., Ordering and precipitation in low supersaturated NiCrAl model alloy : an atomic scale investigation, (1997) Phil. Mag. A. **76**, 527-542

33 Cahn J.W. (1968) Spinodal decomposition, *Trans. AIME* **242**, 166-180

34 Brenner S.S., Miller M.K. and Soffa W.A. (1982) Spinodal decomposition of Fe-32at%Cr at 470°C, *Scripta Met.* **16**, 831 836

35 Miller M.K., Bentley J., Brenner S.S. and Spitznagel J.A. (1984) Long term thermal aging of type CF8 stainless steel, *Journal de Physique* **45-C9**, 385-390

36 Blavette D., Grancher G. and Bostel A. (1988), Statistical analysis of atom-probe data (I) : derivation of some fine scale features from frequency distributions for finely dispersed systems, *Journal de Physique* **49-C6**, 433-438

37 Auger P., Menand A. and Blavette D. (1988) Statistical analysis of atom-probe data (II) : theoretical frequency distributions for periodic fluctuations and some applications, *Journal de Physique* **49-C6**, 439-444

38 Langer J.S., Bar-On M. and Miller H.D. (1975) New computational method in the theory of spinodal decomposition *Physical Review A* **11**, 1417-1729

39 Park K.H., LaSalle J.C., Schwartz L.H. and Kato M. (1986) Mechanical properties of spinodally decomposed Fe-30wt%Cr: yield stength and embrittlement, *Acta Met.* **34**, 1853-1865

482

40 Danoix F., Auger P., Chambreland S. and Blavette D. (1994) A 3D study of G-phase precipitation in spinodally decomposed α-ferrite by tomographic atom probe analysis, *Microsc. Microanal. Microstruct.* **5**, 121-132

41 Danoix F., Auger P. and Blavette D. (1992) An atom-probe investigation of some correlated phase transformation in chromium, nickel and molybdenum containing supersaturated ferrites, *Surface Science* **266** 364-369

42 Pareige C., Soisson F., Martin G., Blavette D., Ordering and phase separation in nicral alloys: monte carlo simulation and 3d atom probe study (1999) Acta met mater. **47-6** 1889-99

43 Zapolsky H., Pareige C., Marteau L., Blavette D., Chen L.Q., Atom Probe analyses and numerical calculation of ternary phase diagram in Ni-Al-V(2001), Calphad, **25**, 125-134

ANALYSIS OF DISPLACEMENT AND STRAIN AT THE ATOMIC LEVEL BY HIGH-RESOLUTION ELECTRON MICROSCOPY

Application to dislocations

M.J. HYTCH
Centre National de Recherche Scientifique
Centre d'Etudes de Chimie Métallurgique-CNRS, 15 rue Georges Urbain, 94407 Vitry-sur-Seine, France

1. Introduction

High-resolution electron microscopy (HREM) differs from conventional electron microscopy in that images are formed of the atomic lattice. Lattice fringes are created by the interference of diffracted beams with the transmitted beam. In general, the number of lattice fringes is limited to two or three and practically all the information contained in an HREM image can be obtained by analysing these few components to the image intensity. It is not our aim to explain how the images are formed; it is sufficient for the present purposes to know that the lattice fringes seen in the image are closely related to the atomic planes in the specimen.

A typical high-resolution image, in this case of a metal, with its decomposition into different lattice fringes, is shown in Figure 1a. The image contrast is dominated by the fringes corresponding to the {111} atomic planes; the other periodicities are only weakly present, as can be seen in the Fourier transform of the image. The crossing of the different fringes produces the dot-like contrast corresponding to the atomic columns viewed in projection. The principle behind strain mapping in HREM is the measurement of the lattice fringe positions from the image and relating this to a displacement field in the specimen.

The measurement is schematically presented in Figure 1b. An undistorted reference lattice is chosen and the displacements u_n calculated with respect to it. This is a very important step as strain in elastic theory is defined with respect to the undeformed initial state. There is no way of knowing this from electron microscopy. The nearest that can be obtained is to use a region of crystal which is undistorted, far from defects and interfaces. A second point is that the local chemistry is not always known. If the composition varies the lattice spacing will vary irrespectively of any strain present. These considerations are independent of the way the displacements are measured, and must be born in mind throughout the ensuing description.

Another important point is that only the displacements in the plane of observation will be measured. It will generally be assumed that the atomic columns be unbent in the viewing direction and move as one. This is often a good approximation in high-resolution microscopy because the sample is extremely thin and allows relaxation in the projection direction. However, this also means that the measured displacements can differ from the displacement field in the original bulk sample [1].

The technique of measurement is derived from a method used in optical interferometry [2] though developed independently for high-resolution electron microscopy [3,4].

A. Finel et al. (eds.), *Thermodynamics, Microstructures and Plasticity*, 485–494.

486

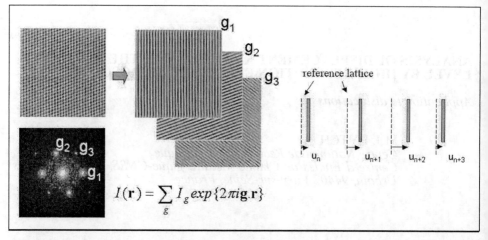

Figure 1. Basis of high-resolution image analysis (a) image of Al taken at [110] orientation, Fourier transform of the image, decomposition of the image in terms of principal lattice fringes, description as a Fourier sum; (b) measurement of displacements, u_n, of atomic planes with respect to a chose reference lattice.

2. Phase and the Local Fourier components

Displacements can be measured directly from high-resolution images by calculating the local Fourier components of the lattice fringes [3]. In this formulation, the intensity in an image, $I(\mathbf{r})$, is written as

$$I(\mathbf{r}) = \sum_g I_g(\mathbf{r}) \exp\{2\pi i \mathbf{g} . \mathbf{r}\} \qquad (1)$$

where $\mathbf{g}$ are the reciprocal lattice vectors describing the undistorted lattice. In this representation, the Fourier components $I_g(\mathbf{r})$ become a function of position in the image. When considered in Fourier space, this information concerning the variations in the lattice is contained in the diffuse intensity around the sharp reciprocal lattice spots. By placing a mask around the periodicity of interest in Fourier space, and taking the inverse Fourier transform, the local Fourier components can be obtained. They have an amplitude and a phase:

$$I_g(\mathbf{r}) = A_g(\mathbf{r}) \exp\{iP_g(\mathbf{r})\} \qquad (2)$$

such that the amplitude, $A_g(\mathbf{r})$, describes the local contrast of the fringes and the phase, $P_g(\mathbf{r})$, their position. The phase is related to the displacement field $\mathbf{u}(\mathbf{r})$ by the following expression:

$$P_g(\mathbf{r}) = -2\pi \mathbf{g} . \mathbf{u}(\mathbf{r}) \qquad (3)$$

This can most readily be understood by considering a set of perfect lattice fringes, of intensity $B_g(\mathbf{r})$:

$$B_g(\mathbf{r}) = 2A_g \cos\{2\pi \mathbf{g} . \mathbf{r} + P_g\} \qquad (4)$$

In this case, the amplitude and phase are fixed quantities and equal the amplitude and phase of the Fourier component corresponding to **g**. Imagine now a displacement field where $\mathbf{r} \to \mathbf{r} - \mathbf{u}$ then:

$$B_g(\mathbf{r}) \to 2A_g \cos\{2\pi\mathbf{g}.\mathbf{r} - 2\pi\mathbf{g}.\mathbf{u} + P_g\} \tag{5}$$

so that an extra phase term is introduced, as given by Eq. (3). The image $B_g(\mathbf{r})$ is the "Bragg filtered" image produced traditionally by placing a mask around a particular spot in the Fourier transform.

The phase gives the component of the displacement field in a direction perpendicular to the lattice fringes. The two-dimensional displacement field can therefore be determined by measuring two phase images, $P_{g1}(\mathbf{r})$ and $P_{g2}(\mathbf{r})$, providing the two sets of lattice fringes are non-parallel. The resulting expression takes on a very simple form when written in terms of the real-space lattice [4]:

$$\mathbf{u}(\mathbf{r}) = -\frac{1}{2\pi}\left[P_{g1}(\mathbf{r})\mathbf{a}_1 + P_{g2}(\mathbf{r})\mathbf{a}_2\right] \tag{6}$$

where the basis vectors for the lattice in real-space, $\mathbf{a}_1$ and $\mathbf{a}_2$, are related to the reciprocal-space lattice such that $\mathbf{g}_i.\mathbf{a}_j = \delta_{ij}$.

Having calculated the displacement field, the local deformation follows automatically. If the lattice distortion **e** is written as a matrix:

$$\mathbf{e} = \begin{pmatrix} e_{xx} & e_{xy} \\ e_{yx} & e_{yy} \end{pmatrix} = \begin{pmatrix} \dfrac{\partial u_x}{\partial x} & \dfrac{\partial u_x}{\partial y} \\ \dfrac{\partial u_y}{\partial x} & \dfrac{\partial u_y}{\partial y} \end{pmatrix} \tag{7}$$

The matrix **e** can be separated into two terms, a symmetric term ε and an antisymmetric term ω, defined as follows:

$$\varepsilon = \tfrac{1}{2}\left(\mathbf{e} + \mathbf{e}^T\right) \tag{8}$$

and

$$\omega = \tfrac{1}{2}\left(\mathbf{e} - \mathbf{e}^T\right) \tag{9}$$

where T denotes the transpose. The strain is given by ε and the rigid lattice rotation by ω for small distortions, thus completing our description.

2.1. ANTIPHASE BOUNDARIES IN Cu₃Au

As an illustration of the phase image technique, Figure 2 shows a high-resolution image of a Cu_3Au sample [5]. In its long-range ordered form, the Au atoms position themselves on alternate Cu planes. It should be noted that as ever in high-resolution work, the image needs to be interpreted with care: the interference fringes correspond to the ordered superlattice, and show only every other atomic plane, i.e. the {200} atomic plane periodicity is missing. This was achieved at the microscope by eliminating all but the {100} Bragg spots with an objective aperture. The image of the superlattice reveals the translation domains, or antiphase domains in this case. These occur when the Au atoms order themselves on an adjacent Cu plane. At each antiphase boundary, the position of the Au rich plane translates by exactly half a {100} lattice spacing (which corresponds to the spacing between atomic columns).

The original image can be filtered to show only the (100) lattice fringes for example (Figure 2c) by masking the Fourier transform (Figure 2b) where these lattice fringe displacements can be seen clearly. This Bragg filtered image is not a necessary step in the calculation of the phase image but is an aid to interpretation.

Figure 2d shows the corresponding phase image. The reference lattice has been chosen in the centre of the image so the phase is zero here, and corresponds to perfectly periodic fringes. The phase changes by π on crossing the antiphase boundary which corresponds to exactly half a lattice spacing as can be deduced from Eq. (3). If bending of the planes occurred at the antiphase boundaries, the phase would be modulated – which is not the case here. Instead it changes abruptly from one value to another indicating the rigid shift of the whole lattice.

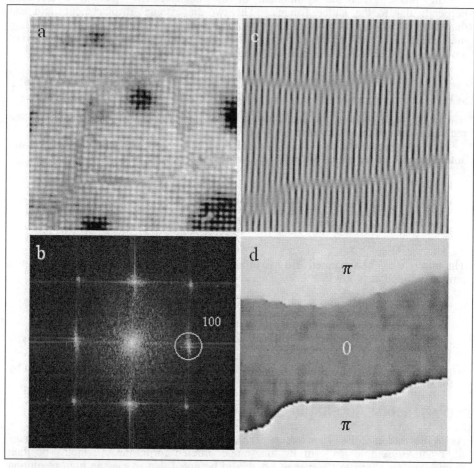

Figure 2. Phase analysis of translation domains (a) high-resolution image of long-range ordered Cu₃Au in [001] orientation taken on a JEOL 4000FX, an objective aperture limits resolution to the {110} spots; (b) Fourier transform of image (amplitude only), mask for phase analysis indicated; (c) lattice fringes (100) obtained by Bragg filtering; (d) phase of (100) lattice fringes, phase change of π indicates a displacement of half a lattice spacing.

The complete two-dimensional displacement field can be determined by calculating the phase image for the (010) lattice fringes and the result is shown in Figure 3. In this case, Eq. (6) for the displacement field takes on a very simple form:

$$\mathbf{u}(\mathbf{r}) = \begin{pmatrix} u_x \\ u_y \end{pmatrix} = -\frac{a}{2\pi} \begin{pmatrix} P_{100} \\ P_{010} \end{pmatrix} \qquad (10)$$

where 'a' is the lattice parameter and the x- and y-axes correspond to [100] and [010] respectively. In such a way, even complicated rigid body displacements and arrangements of translations domains can be analysed.

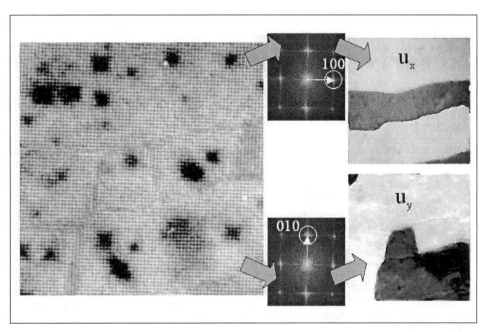

Figure 3. Determination of 2-dimensional displacement field: high-resolution image is processed to produce phase images for the (100) and (010) lattice fringes which directly give the displacements along x and y axis.

3. Phase analysis of dislocations

The analysis of dislocations requires a special introduction both because of their importance but also by the particular considerations brought into play by the discontinuous displacement fields they generate. For the analysis by high-resolution electron microscopy we assume that the dislocation is seen end-on assuring that in projection, the displacements are identical along the viewing direction. If this were not the case, the atomic columns would be inclined or bent and the image would be blurred, especially in the core region. This implies particular constraints in terms of the specimen preparation. More importantly, the major part of the displacements produced by screw components will be invisible as they lie along the dislocation line, the viewing direction.

A dislocation is defined by its Burgers vector **b** and line ξ. The most apt description for our purposes is in terms of the displacement field **u**:

$$\mathbf{b} = \oint_L \nabla \mathbf{u} . \mathbf{dl}$$

(11)

where L is a loop containing the dislocation core. Using the expression for $\mathbf{u}$ in terms of the phase, Eq. (6), this produces:

$$\mathbf{b} = -\frac{1}{2\pi} \left[\mathbf{a}_1 \oint_L \nabla P_{g1} . \mathbf{dl} + \mathbf{a}_2 \oint_L \nabla P_{g2} . \mathbf{dl} \right].$$

(12)

The phase will therefore be discontinuous at the dislocation core, just as the displacement field. This can be understood by taking the simplest case of an edge dislocation in an idealised cubic crystal with $\mathbf{b} = [100]$ and $\xi = [001]$. With $\mathbf{g}_1 = [100]^*$ and $\mathbf{g}_2 = [010]^*$ and hence $\mathbf{a}_1 = [100]$ and $\mathbf{a}_2 = [010]$ we see that:

$$\mathbf{b} = -\frac{1}{2\pi} \mathbf{a}_1 \oint_L \nabla P_{100} . \mathbf{dl}$$

(13)

and since $\mathbf{b} = \mathbf{a}_1$,

$$\oint_L \nabla P_{100} . \mathbf{dl} = -2\pi$$

(14)

The phase will therefore have a discontinuity of 2π on going *clockwise* around the dislocation core. This is natural as a phase of 2π represents one lattice period, indicating the extra half plane inserted by the dislocation. Generalising, if the clockwise phase discontinuities for P_{g1} and P_{g2} are $2n_1\pi$ and $2n_2\pi$ respectively, the Burgers vector will be:

$$\mathbf{b} = n_1 \mathbf{a}_1 + n_2 \mathbf{a}_2$$

(15)

where n_1 and n_2 will typically be integers. The Burgers vector can therefore be determined by observing the discontinuities in the phase. A interesting way of measuring these discontinuities, and hence the Burgers vectors, has been developed by Kret et al. [6].

3.1. LOMER DISLOCATION IN SILICON

Figure 4 shows a high-resolution image of a Lomer dislocation in silicon taken on a JEOL 200CX operating at 200 kV (point resolution 0.22 nm) [7]. The crystal is oriented along a $[1\,1\,0]$ zone axis so that the dislocation is seen end-on with the core at the centre of the image. For these imaging conditions, each black spot corresponds to a pair of atomic columns as in this orientation they are too close to separate – the so-called silicon dumbbells.

In order to determine the displacement field around the dislocation, phase images were calculated for the (111) and $(11\bar{1})$ lattice fringes, these having the strongest image contrast, and hence giving results with the best signal-to-noise ratio. The results are shown in Figure 5. The basis vectors for the calculation are therefore:

$$\mathbf{g}_1 = [111]^* \quad \mathbf{g}_2 = [11\bar{1}]^* \quad \text{and} \quad \mathbf{a}_1 = \tfrac{1}{4}[11\bar{2}] \quad \mathbf{a}_2 = \tfrac{1}{4}[112]$$

(16)

There is a phase discontinuity of 2π for both sets of lattice fringes and hence:

$$\mathbf{b} = \mathbf{a}_1 + \mathbf{a}_2 = \tfrac{1}{2}[110]$$

(17)

the well known pure edge Lomer dislocation.

Taking the x-axis parallel to [220] and the y-axis parallel to [002], the displacement

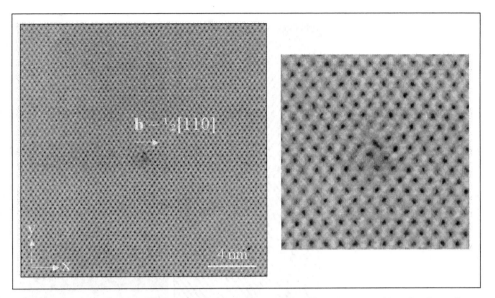

Figure 4. HREM image with zoom of a Lomer dislocation in [1 $\overline{1}$ 0] Silicon (courtesy of Jean-Luc Putaux).

field **u**=(u$_x$,u$_y$) calculated from Eq. (6) is shown in Figure 6. The double discontinuity is due to the fact the largest lattice spacing in the x-direction corresponds to the (220) lattice planes for which n=2 (b=2d$_{220}$). Also shown on Figure 6 is the displacement field calculated using anisotropic inelastic theory [8]. Qualitatively it can be seen that the match is excellent. In order to analyse quantitatively the results and due to the complexity of the anisotropic equations, it is useful to consider the result as given by isotropic elastic theory:

$$u_x = \frac{b}{2\pi}\left(\theta + \frac{\sin 2\theta}{4(1-\nu)}\right) \tag{18}$$

$$u_y = -\frac{b}{2\pi}\left(\frac{1-2\nu}{4(1-\nu)}\ln r^2 + \frac{\cos 2\theta}{4(1-\nu)}\right) \tag{19}$$

where r and θ are the polar coordinates centred on the core position, and ν the Poisson's constant. The first term in u$_x$ describes the discontinuity in the displacement field and increases linearly in θ and is the main contribution seen in the calculated and experimental displacement fields. The displacement in u$_y$ has again two terms, this time one depending uniquely on r and the other in θ. The results of isotropic and anisotropic theory are in fact very close. We shall consider now the second term sinusoidal variations in u$_x$ and u$_y$.

These secondary sinusoidal terms have been calculated from the experimental displacements by subtracting the theoretical first terms, giving the results shown on Figure 6b. In order to carry out a quantitative comparison with anistropic theory, circular line profiles have been taken 7.5 nm from the dislocation core (Figure 6c). It can be seen that isotropic elastic theory would not have been sufficient to describe the experimental behaviour as the two curves have a different amplitude. The fit, however, with anisotropic theory is excellent: average deviation 0.03Å. The amplitudes are predicted with 0.01Å accuracy which, to our knowledge, is the highest precision in displacement measurements so far obtained at this scale.

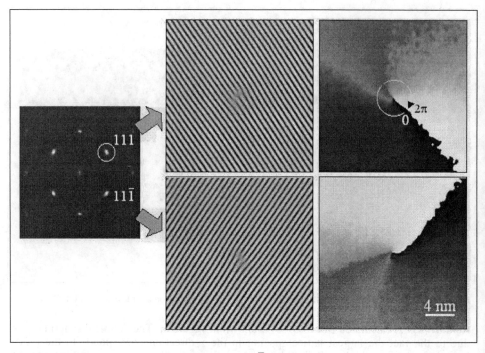

Figure 5. Phase images calculated for the (111) and $(11\bar{1})$ lattice fringes. Colour scale is from 0 radians (black) to 2π radians (white). Phase discontinuity of 2π at dislocation core position. Lattice fringe image has been zoomed by a factor of two with respect to original image for display purposes.

4. Conclusions

Measurements of displacements can be carried out at the nanometre scale to very high precision using high-resolution electron microscopy coupled with image analysis. The measurement of strain follows directly from the derivatives of the displacement field. However, it is usually best to compare the displacement field directly with theoretical models to avoid noise amplification when taking gradients. The problem of image artefacts has not been considered here, but it has been shown elsewhere that these are limited when studying centrosymmetric crystals (as here), when the displacement fields are slowly varying and where rotations are concerned [9].

Acknowledgements

Part of this work was carried out in the framework of the European Research Group "Quantification and measurement in transmission electron microscopy" involving France, the United Kingdom, Germany and Switzerland.

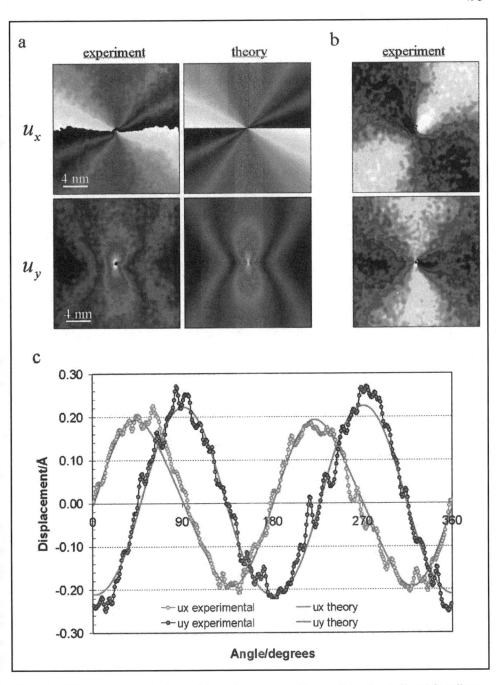

Figure 6. (a) Displacement field calculated from phase images of Lomer dislocation (left) and from linear anistropic elastic theory (right). Colour ranges for u_x =0 nm (black) to 0.192 nm (white), and for u_y −0.271 nm to 0 nm (lattice spacings d_{002}=0.271 nm, d_{220}=0.192 nm); (b) sinusoidal contributions to experimental displacement field for u_x and u_y. Colour range −0.3Å (black) to 0.3Å (white); (c) circular line scan around dislocation core for experimental results and anisotropic elastic theory. Root mean squared departure from theory = 0.03 Å.

494

References

1. Treacy, M.M.J, Gibson, J.M. and Howie, A. (1985) On elastic relaxation and long wavelength microstructures in spinodally decomposed $In_xGa_{1-x}As_yP_{1-y}$ epitaxial layers. *Philos. Mag. A* 51, 389–417.
2. Takeda, M., Ina, H. and Kobayashi, S. (1982) Fourier-transform method of fringe pattern analysis for computer-based topography and interferometry. *J. Opt. Soc. Am.* 72, 156–160.
3. Hÿtch, M. J. (1997) Analysis of variations in structure from high resolution electron microscope images by combining real space and Fourier space information. *Microsc. Microanal. Microstruct.* 8, 41–57.
4. Hÿtch, M.J., Snoeck, E. and Kilaas, R. (1998) Quantitative measurement of displacement and strain fields from HREM micrographs. *Ultramicroscopy* 74, 131–146.
5. Hÿtch, M.J. and Potez, L. (1997) Geometric phase analysis of high-resolution electron microscope images of antiphase domains: example Cu_3Au. *Philos. Mag. A* 76, 1119–1138.
6. Kret, S., Dluzewski, P., Dluzewski, P. and Sobczak, E. (2000) Measurement of dislocation core distribution by digital processing of high-resolution transmission electron microscopy micrographs : a new technique for studying defects. *J. Phys. C* 12, 10313–10318.
7. Hÿtch, M.J., Putaux, J.-L. and Pénisson, J.-M. (2002) Measurement of displacement fields around dislocations to 0.01Å by high-resolution electron microscopy, *in preparation.*
8. Hirth, J. P. and Lothe J. (1992) *Theory of dislocations,* Krieger, Malabar Florida.
9. Hÿtch, M.J. and Plamann, T. (2001) Imaging conditions for reliable measurement of displacement and strain in high-resolution electron microscopy. *Ultramicroscopy* 87, 199–212.

DISLOCATION ORGANIZATION UNDER STRESS : TiAL

P. VEYSSIÈRE
LEM, CNRS-ONERA
BP 72, 92322-Châtillon cedex. FRANCE

1. Introduction

The present contribution is intended to provide examples of dislocation microstructures organized under stress. The cases investigated are restricted to mechanisms occurring in the γ phase of the Ti-Al system, either in the single-phased γ alloy or in the two-phase $\gamma + \alpha_2$ system. γ-TiAl and α_2-Ti$_3$Al have the L1$_0$ face centered tetragonal ordered structure and the D0$_{19}$ hexagonal ordered structure, respectively. In the case of microstructures self-organized in single slip, we will nevertheless show that the relevant dislocation micro-mechanisms are essentially independent of material structure.

The L1$_0$ structure offers a variety of potential slip directions (shortest unit translations) amongst which 1/2<110], <011], 1/2<112] and <001] have been experimentally identified (*Figure 1*). The <*hkl*] notation stands for any crystal direction generated by permutation between the first two indices (i.e. $\pm h$ and $\pm k$) with the third ascribed to $\pm l$. Dislocations with 1/2<110], <011] and 1/2<112] Burgers vectors are created at low and intermediate temperatures and slip takes place preferentially on octahedral planes [1]. Properties of <001] dislocations, operative at high temperatures [2], are poorly documented.

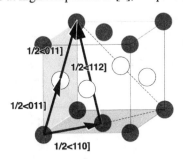

Figure 1. The L1$_0$ unit cell.

The yield stress of γ-TiAl peaks at about 600-800°C regardless of the nature of the operating slip system, but this question is beyond the scope of the present contribution. It is worth keeping in mind though that below approximately 400°C, <011]{111} has the lowest critical resolved shear stress of all possible slip systems, that for 1/2<110]{111} is of the same order of magnitude as, though perceivably larger than $\tau_{<011]\{111\}}$ [1]. On the other hand, it is not clear whether 1/2<112]{111} is activated below the temperature of the flow stress peak (section 2, see however [3, 4]).

1/2<110], <011], and 1/2<112] dislocations have very different core structures of

497

A. Finel et al. (eds.), Thermodynamics, Microstructures and Plasticity, 497–506.
© 2003 *Kluwer Academic Publishers. Printed in the Netherlands.*

which we give a deliberately simplified account. 1/2<110] dislocations — also commonly referred to as *ordinary* dislocations — exhibit a very modest splitting, if any, which makes them similar to dislocations in Ni or Al in this particular respect. Yet, ordinary dislocations show a pronounced tendency, unexplained so far, towards being aligned along the screw direction. This tendency increases dramatically with temperature again for reasons that remain to be elucidated. By contrast, <011] dislocations are dissociated but not into two 1/2<011] partials bordering an APB as is found in a number of intermetallic alloys such as L1$_2$-Ni$_3$Al. At room temperature, <011] dislocations actually split into a 1/6<112] Shockley partial and a 1/6<154] partial bordering a low energy intrinsic stacking fault (ISF) [5]. The latter partial is simply the combination of a 1/6<121] and a 1/2<011] partial dislocation (depending on parameters such as temperature and composition, threefold dissociation of <011] dislocations into two Shockley partials (i.e. 1/6<112] and 1/6<121]) and a 1/2<011] partial has been reported [6]). <011] dislocations form screw locks upon cross-slip of a 1/2<011] partial and creation of an APB strip inclined to the primary slip plane, that is, a configuration somewhat similar to the Kear-Wilsdorf lock in L1$_2$ alloys. 1/2<112] dislocations dissociate into two or three partials with collinear Burgers vectors [3].

What makes γ-TiAl rather unique within face-centered cubic based structures is the potential of <011] dislocations for *decomposition* into two *perfect* dislocations [6] (the term *dissociation* is specifically aimed at the production of *partial* dislocations). Based on the b^2 Frank criterion, the corresponding reaction

$$<011] \leftrightarrow 1/2<110] + 1/2<112] \qquad (1)$$

is energetically indifferent with the double-headed arrow standing for the fact that the reaction is in principle reversible (*Figure 2*(a)). Dislocation decomposition is at the origin of a number of unusual microstructural properties including coplanar slip in several directions and the production of specific debris.

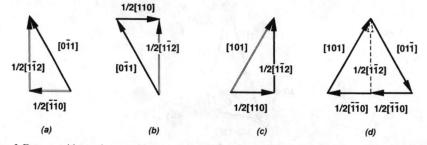

Figure 2. Decomposition and composition reactions within a given {111} habit plane. In (*a*) to (*c*), double-lined arrows represent reaction end-products. (*d*) Reaction between the three Burgers vectors along coplanar close-packed directions in the L$_0$ structure (compare with reactions between 1/2<110> Burgers vectors in fcc crystals).

2. Multiple, coplanar slip directions

Under load orientations selected to activate single slip, it is found that none of the above 1/2<110], 1/2<112] and <011] slip directions can be activated alone during deformation at low and intermediate temperatures [7]. The coexistence of multiple coplanar slip directions is consistent with the operation of decomposition reactions schematized in *Figure 2*.

Suppose load orientation is such that the primary slip direction is [0$\bar{1}$1] and that there is a finite driving force for decomposition of [0$\bar{1}$1] dislocations. Then fragments of these will split apart into coplanar segments with 1/2[$\bar{1}$$\bar{1}$0] and 1/2[1$\bar{1}$2] Burgers vectors (*Figure 2*(a)). An ordinary dislocation that revolves around obstacles generates portions with Burgers vectors of the two signs, i.e. ±1/2[110], that move in opposite directions and that impact other dislocations from time to time. 1/2[110] dislocations may for instance (i) annihilate with 1/2[$\bar{1}$$\bar{1}$0] dislocations, (ii) create dislocations with 1/2[1$\bar{1}$2] Burgers vector by composition with [0$\bar{1}$1] dislocations (*Figure 2*(b)) or else (iii) merge with 1/2[1$\quad$$\bar{1}$2] dislocations to form [101] dislocations (*Figure 2*(c)). Hence, even starting from the single operation of one of the <011] Burgers vectors of a {111} plane, all four possible Burgers vectors of this plane are engendered in almost unpredictable proportions by means of a set of reactions summarized in *Figure 2*(d). It is noted that a similar behavior is to occur in a number of ordered alloys, one of the simplest is the decomposition of <111> dislocations into <100> and <011> dislocations in the B2 structure.

Considering that reaction (1) is energetically indifferent, one may wonder what is the driving force for decomposition. An answer is provided in *Figure 3* that shows that this takes place at cusped configurations where a gain in energy is obtained by pulling a 1/2<110] portion (O) from the parent <011] dislocation (S) so as to reduce the length of the former. For the reaction to proceed, what counts is not self-energy but total energy. It is noted that at low temperatures, meandering lines of 1/2<112] dislocations, which are then relatively sessile, materialize the parent <011] dislocation as it undergoes decomposition.

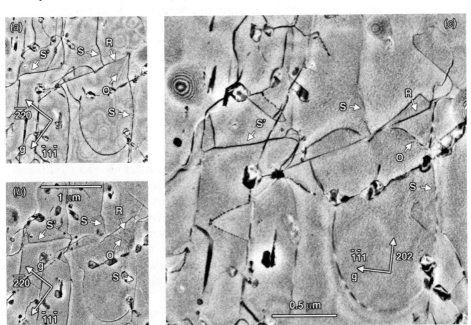

Figure 3. An example of a [101] dislocation (S) forming a hairpin configuration (enlarged in (c)) at which it decomposes into a 1/2[110] and a 1/2[1$\bar{1}$2] dislocation (O and R, respectively). (a) **g** = $\bar{1}$11, ordinaries are invisible. (b) **g** = $\bar{2}$$\bar{2}$0, ordinaries are visible while 1/2[1$\bar{1}$2] dislocations are out of contrast. (c) **g** = $\bar{1}$$\bar{1}$1, [101] dislocations are out of contrast (some residual contrast arises from their dissociation into dissimilar partials). At S', the presence of a dislocation with [011] Burgers vector should be noticed (Courtesy Dr. Y.-L. Chiu).

3. Correspondence between populations and mobility of dislocations

We do not know for sure if 1/2<112] dislocations are poorly mobile at low temperature, but there is consistent indication that the relative proportion of these dislocations does not significantly decrease under load orientations that preclude 1/2<112]{111} slip systems. This raises the more general issue, in post mortem experiments, of the relationship between the density of dislocations of a given type and the mobility of these. It is usually argued that the most numerous dislocations are those with the poorest mobility, that is, those that can hardly be eliminated at free surfaces or by mutual annihilation. In systems such as bcc, $L1_2$ or covalent crystals, a profusion of screw dislocations manifests the tendency of these to form locks. This reasoning hardly applies to several coexisting slip directions in crystals where dislocations undergo decomposition. In this case, it is indeed quite natural to postulate that the dislocation family that appears in the largest density is that which provides most of the strain. The situation in TiAl is therefore rather complex:

- under load orientations that promote the operation of one <011] slip direction, the above sequence of reactions takes place and the deformation microstructure fills up with dislocations of all four coplanar Burgers vectors. Some, such as 1/2<112], are even seen in noticeable proportions even when their Schmid factor is insignificant.
- the range of load orientations favoring one 1/2<110] slip direction is such that the shear stress resolved on the most stressed <011]{111} slip system is close to its CRSS so that in practice, those slip systems are locally operative depending on internal stresses.

It is worth keeping in mind that the presence of members of a given slip system does not prove the operation of that system and therefore is not sufficient to derive a CRSS.

4. Intralamellar boundaries in γ + α₂ TiAl

After adequate annealing, a grain of Ti-rich TiAl is comprised of a succession of lamellae of α_2-Ti_3Al and of each of the six variants of γ-TiAl (see [8]). The close-packed planes of the face-centered tetragonal phase are parallel to the basal plane of the α_2 phase, with the close-packed directions mutually aligned. Alignment is, however, imperfect (i) within the γ variants because of tetragonality and (ii) between α_2 and γ platelets because of differences in lattice parameters and of tetragonality. Lattice mismatches induce elastic shears that are relaxed by interlamellar boundaries [9]. Relaxation is, nevertheless, not complete leaving residual shear stresses parallel to the interfaces [10]. The latter are in turn relaxed by planar intralamellar dislocation networks, also parallel to the interfaces (*Figure 4*). Such boundaries are encountered after deformation at room temperature, and therefore are created entirely by glide. It is again the decomposition reaction that renders such an unusual situation possible [11].

In brief, intralamellar boundaries occur in grains where deformation is dominated by <011] slip but they consist essentially of rectangular cross-grids of screw 1/2<112] and 1/2<110] dislocations (*Figure 4*) and this raises the questions as to why and how they form.

- The driving force arises evidently from the above-mentioned residual shears. A rectangular cross-grid of screws is capable of relaxing elastic shears (pure shear for a

given sign of the pair of Burgers vectors, twist as one of the two signs is changed [12]).
- There is solid experimental indication that intralamellar rectangular networks build up upon intersection of the primary <011]{111} slip system (parallel to the interfaces) by that <110]{111} slip system which can cross-slip into the primary slip plane. Incorporation of ordinary dislocations into arrays of <011] dislocations results in junctions with 1/2<112] Burgers vector that lengthen gradually at the expense of the segments of <011] dislocations.

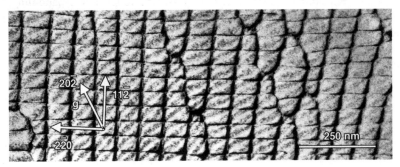

Figure 4. Plane view of a planar intralamellar boundary observed in lamellar TiAl strained at room temperature. The network consists of a rectangular cross-grid of screw dislocations perturbed in places upon incorporation of alien dislocations generated during deformation. In the unperturbed zones, the network is roughly symmetrical (i.e. $b_O/d_O = b_R/d_R$ where d_O and d_R are the distances between the meshes with Burgers vectors b_O and b_R, respectively).

In Ti-rich TiAl, planar rectangular networks result from the conjunction of several factors (i) the stability of a lamellar microstructure (ii) the elastic shears that result from the coexistence of α_2 and γ, (ii) the existence of two orthogonal Burgers vectors in the appropriate plane and (iii) the decomposition reaction undergone by primary <011] dislocations in this plane.

5. Self-organisation of 1/2<110] dislocations in γ-TiAl

The early models of plastic processes were largely founded on the concept of dislocation pile-ups for it was not thought that dislocations could engender obstacles to their own motion. A surprising outcome of the pioneering TEM investigations of metals (late 50's) is that after deformation under single slip conditions at low temperature, dislocations form patterns such as bundles or walls together with a profusion of prismatic loops [13]. This led researchers to contemplate the role of cross-slip whose importance in plastic properties has become increasingly clear (e.g. slip band thickening, work hardening, recovery).

Several decades after their discovery, the physical origin of dislocation bundles under single slip conditions remains largely unclear. Phenomenologically, one can successfully regard crystals as black boxes and then account for gradients of dislocation densities, that is of bundles, by means of adequately parametrized rate equations with no knowledge whatsoever of micro-mechanisms. This approach is indeed largely sufficient for modeling purposes. It is nevertheless useful, if not desirable, to investigate what in the deformation microstructure triggers congregation of dislocations and of deformation debris. A precious indication was provided by Kratochvil *et coll.* who, to the author's knowledge, were the first to contemplate prismatic loops as active components of the microstructure instead of as

502

dead debris [14-16]. Indeed, a prismatic loop is glissile on the prism surface defined by its line and its Burgers vector. However, since the effect of an external stress on a portion of prismatic loop is always counterbalanced by the response of a segment of that loop with equal length and opposite sign, a prismatic loop should remain essentially immobile under an applied load. This is no longer true in the stress gradient due in particular to a moving dislocation where loops are swept when the approach distance is small enough (see [17]).

TEM observations in γ-TiAl have confirmed the instrumental role played by interactions between ordinary dislocations and prismatic loops in self-patterning, but they have also revealed a formation process for the loops that imposes considering the interactions in question under a different perspective.

5.1. NON-DIFFUSIVE GENERATION OF PRISMATIC LOOPS

It has been known for years, that the formation of loops by glide in an obstacle-free crystal involves either mutual annihilation by cross-slip of two portions of dislocations with opposite signs (*Figure 5*) or repeated double cross-slip events on a given line (*Figure 6*). Surprisingly, certain essential implications of these maneuvers have remained largely ignored.

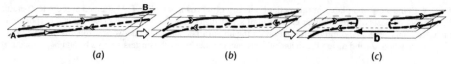

(a)   *(b)*   *(c)*

Figure 5. Two dislocation portions forming a dipole (a) may reorient locally in the screw direction (b) at which stage they can cross-slip to mutually annihilate forming two hairpin-like segments (c). Unless hampered by lattice friction, these segments flip in the edge orientation with their extremities aligned in the screw direction.

Cross-slip annihilation systematically generates two jogs and therefore two hairpin-like half loops. A second cross-slip annihilation is required in order to form a fully closed loop, yet this loop is neighbored by a hairpin at both extremities, and therefore no isolated loops can be formed by glide. Another important property is that the two jogs formed by cross-slip annihilation are coplanar which implies that the nearest-neighbor extremities of loops originating from the same event are aligned in the screw direction. As this process repeats itself, loops become organized as strings with loop extremities exhibiting a two by two, strict correspondence along the screw direction.

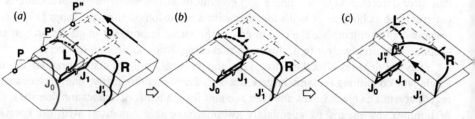

Figure 6. The successive stages of the formation of prismatic loops by double cross-slip. (a) The thick grey line represents the dislocation after a first double cross-slip event has taken place from plane P to plane P' forming a pair of jogs such as J_1 in the cross-slip plane. In black is sketched the same dislocation after it has moved forward under stress generating a hairpin segment. Its portion L has kept slipping in P' while R has double cross-slipped from P to P" forming a second pair of jogs J_1 and J_1'. (b) Those jogs are pulled sideward as R expands laterally under stress until, for instance, J_1 intersects L, closing a prismatic loop. (c) The resulting jog J_1" moves laterally in a direction dictated by line tension. It is noted that jogs J_1' and J_1" are aligned with J_1 in the screw direction.

Double cross-slip is an alternative means to engender prismatic loops (*Figure 6*). Again, no individual loops can be formed. Instead the effect results in loop strings whose extremities are all associated two by two in a neighborly way in the screw direction.

Configurations involving strings are profuse in γ-TiAl deformed by ordinary dislocations in single slip up to 400°C (*Figure 7*). The strings commonly observed comprise an average of 5 to 10 loops but some containing up to 18 members have been encountered [18].

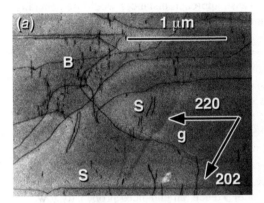 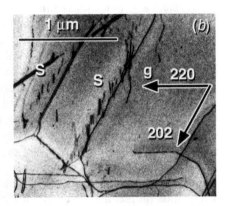

Figure 7. Examples of loop strings encountered in γ-TiAl strained to 2% at room temperature. (a) The central region marked S exhibits several paired loops forming two strings, while the bottom feature consists of three individual loops. At B one can see the congregation of a significant number of loop onto which a dislocation seems to be anchored. (b) Two organized sets of loop walls containing paired loops. A dislocation appears to be stopped by the central wall. The horizontal segments at the bottom are ordinaries locked in the screw direction (Courtesy Dr. F. Grégori).

The two mechanisms generate dissimilar strings, and that enables one to discriminate between the two formation processes. The strings formed by cross-slip annihilation are comprised of loops all of the same sign (either vacancy or interstitial, see loop sign indicated by arrows in *Figure 5*(c)) and whose upper and lower branches are coplanar. In the double cross-slip maneuver, loop sign alternates (see signs in *Figure 6*(c)) and the lower branch of a given loop is coplanar with the upper branch of its neighbor. Experimental evidence indicates that all the members of a string have the same sign, thus supporting the fact that loop strings build up by mutual annihilation of independent dislocations. It should be realized that string formation is in fact a totally general maneuver that operates in any crystal whose dislocations have some potential for cross-slip.

Why is experimental evidence of loop strings so recent and why is it restricted so far to γ-TiAl and to similarly "exotic" alloys [19]? There are actually several reasons for this.

- Observing strings requires that several stringent experimental requirements be fulfilled. In effect, two technical prerequisites are that (i) the sample be deformed in single slip, which is obvious if ones wants to exclude intersection processes, and that (ii) thin foils be sliced rigorously parallel to the slip plane otherwise strings are hard to spot for their extension is rapidly limited by the foil free surfaces.
- One should not ignore in addition the vicissitudes of TEM investigations especially when pinpointing tiny dipolar features — especially those resulting from annihilation processes — when these are not specifically looked for. Over more than four decades of

investigation of deformation microstructure in single crystals, researchers have almost invariably concentrated on bundles, that is, on debris whose dislocation density is already far too large for strings to be identified. To the author's knowledge, there is only one documented case of loop alignments in Cu and Cu alloys [20] whose physical implications have, however, not been adequately addressed.

- Consider deformation under single slip conditions with slip band thickening proceeding by double cross-slip. Frequent strings formation requires rather short free-flight distances λ_{CS}, in the cross-slip plane otherwise approach distances would be generally too large resulting in infrequent dipolar annihilation.

Beside their role in microstructural organization (§ 5.2), loop strings may help solve issues on dislocation behavior that would be otherwise difficult to infer from experiments [21]. For instance, with regard to the fact that cross-slip activity is expected to increase with temperature in a thermally-activated manner, should λ_{CS} increase with temperature? Observations of slip band thickening suggest that this is indeed the case. However, as cross-slip frequency is increased, the dislocation free-flight time in the cross-slip plane is reduced either, and λ_{CS} diminishes accordingly, unless dislocation velocity is itself thermally-activated. Alternatively, it is worth considering that as long as loop strings are observed, one should admit that λ_{CS} still scales with the (stress-dependent) cross-"section" for dipolar annihilation (in fact not a section but a distance). Hence, an enhanced propensity towards slip band thickening would not result from an increased λ_{CS} but from the multiplication of short excursions from the primary slip plane. Still work remains to be done before every effect entering the process is identified and adequately interpreted, but the above example illustrates how careful microstructural investigations may help set the stage for reasonably sound computer simulation of dislocation dynamics in single slip conditions.

5.2. SELF-PATTERNING

Dislocations immobilized at loop clusters, as is exemplified B and S in *Figure 7*(a) and (b) respectively, illustrate the well-known property [15, 16, 22] that loops are liable to significantly hamper dislocation motion. This may result either from the formation of helix turns whenever a dislocation intersects a loop or from elastic interactions in the absence of contact. Hence, the main question in analyzing self-patterning is that of the constructive micro-mechanisms that ensue in bundles. As mentioned earlier, a physical explanation was provided by the drifting and sweeping processes of individual loops [14-16] but it is easy to figure out numerically from theoretical analyses provided in the literature [17] that these processes fail to account for the transport of strings. What TEM suggests instead is that, as soon as the first string is created within a slip band, the microstructure bifurcates from a more or less homogeneous dislocation distribution towards the formation of a wall [23].

As the first cross-slip annihilation event takes place forming the first string, the latter opposes an elastic field (attractive or repulsive) against any dislocation approaching within a sufficiently small cross-section, immobilizing this dislocation in its close vicinity (slight rearrangements of the loops within the string cannot be excluded). Then deformation proceeds until another dislocation with opposite sign impacts the above complex and annihilates by cross-slip with the dislocation at rest. This results in two strings comprised of pairs of loops arranged more or less in registry (S in *Figure 7*) for the meandering shape adopted by dislocations at impact is governed by elastic interactions with the string. Since the width and height of the loops are determined by the planes of incidence, there is no reason for the second string to be located at the same height as the initial string, which implies in turn that the cross-section of this aggregate for capture of new dislocations is increased. As the process repeats itself, new strings are incorporated until the loop aggregate is so thick that an impacting dislocation is trapped with no possibility of being annihilated from the other side by a mobile dislocation with opposite sign. As this stage, complicated bundles comprised of segmented dislocations embedded in groups of loops start to be observed. Dislocations are segmented as a result of intersections with loops and formation of helix turns; loops are in general rather narrow and, for a fraction of them, hard to visualize because of destructive dipolar contrast.

It should be kept in mind though that whereas bundle nucleation and building-up at loop strings is satisfactorily consistent with experimental observations and theoretical estimates, as well as with non-diffusive mechanisms for the formation of prismatic loops, the question of the inverse relationship between the mean distance between bundles and the applied stress requires further work within the string-nucleated maneuver.

6. Conclusions

The above three experimental examples, all taken from experimental observations in γ-TiAl, illustrate several circumstances of collective dislocation response to stress. The crystal investigated exhibits many more degrees of freedom than is normally encountered in fcc, bcc or hcp crystals. The question of a relationship between the density of certain dislocations relative to others and their mobility enables one to evaluate the limits of an extrapolation from simple to more complex systems. The formation of planar, rectangular networks by glide in the γ phase of lamellar TiAl provides a rare example of a complex stress-relaxing feature, analyzed theoretically a long time ago [12], whose existence is made possible by the decomposition reaction. The maneuver proposed for concerted loop formation is again a fair illustration of how studies of a complex structure may contribute to clarify and, to some degree, to stimulate certain situations. The consequences of loop string formation on self-patterning illustrate the fact, rather common in plasticity, that explanations offered for a phenomenon may dramatically differ depending on the scale at which investigations are conducted, none of them being fundamentally incorrect.

7. References

1. Inui, H., Matsumoro, M., Wu, D.-W. and Yamaguchi, M. (1997) Temperature dependence of yield stress, deformation mode and deformation structure in single crystals of TiAl (Ti-56 at.% Al), Phil. Mag. A 75, 395-423.
2. Hug, G. (1988) Etude par microscopie électronique en faisceau faible de la dissociation des dislocations dans TiAl : relation avec le comportement plastique., Paris-Sud (Orsay).
3. Jiao, S., Bird, N., Hirsch, P. B. and Taylor, G. (1999) Yield stress anomalies in single crystals of Ti-54.5 at.% Al. II 1/2<112]{111} slip, Phil. Mag. A 79, 609-625.
4. Jiao, S., Bird, N., Hirsch, P. B. and Taylor, G. (1998) Yield stress anomalies in single crystals of Ti-54 at.% Al. I Overview and <011] superdislocation slip, Phil. Mag. A 78,

506

777-802.

5. Grégori, F. and Veyssière, P. (2000) Properties of <011]{111} slip in Al-rich g-TiAl. I Dissociation, locking and decomposition of <011] dislocations at room temperature., Phil. Mag. A 80, 2913-2932.

6. Hug, G., Loiseau, A. and Veyssière, P. (1988) Weak-beam observation of a slip transition in TiAl., Phil. Mag. A 57, 499-523.

7. Grégori, F. (1999) Plasticité de l'alliage g-TiAl : rôle des dislocations ordinaires et superdislocations dans l'anomalie de limite élastique, University of Paris VI.

8. Kishida, K., Inui, H. and Yamaguchi, M. (1998) Deformation of lamellar structure in TiAl-Ti3Al two-phase alloys, Phil. Mag. A 78, 1-28.

9. Kad, B. K. and Hazzledine, P. M. (1992) Shear boundaries in lamellar TiAl, Phil. Mag. Lett. 66, 133-139.

10. Saada, G. and Couret, A. (2001) Relaxation of coherency stresses by dislocation networks in lamellar g-TiAl, Phil. Mag. A 81, 2109-2120.

11. Couret, A., Calderon Benavides, H. A. and Veyssière, P. (2002) Intralamellar dislocation networks formed by glide in g-TiAl I. The mechanism of formation, Phil. Mag. A submitted.

12. Matthew, J. W. (1974) Misfit dislocations in screw orientation, Phil. Mag. 29, 797-802.

13. Kuhlmann-Wilsdorf, D. and Wilsdorf, H. G. F. (1963) On the origin of dislocation tangles and long prismatic loops in deformed crystals, in G. Thomas and J. Washburn (eds): Electron Microscopy and the Strength of Crystals, Interscience Publishers, New York, pp. 575-604.

14. Kratochvìl, J. (1988) Plastic properties and internal stability of deformed metals, Czech. J. Sci. B 38, 421-424.

15. Kratochvìl, J. and Libovicky, S. (1986) Dipole drift mechanism of early stages of dislocation pattern formation in deformed metal single crystals, Scripta Met. 20, 1325-1630.

16. Kratochvìl, J. and Saxlovà, M. (1992) Sweeping mechanism of dislocation pattern formation, Scripta Metall. Mater. 26, 113-116.

17. Kubin, L. P. and Kratochvìl, J. (2000) Elastic model for the sweeping of dipolar loops, Phil. Mag. A 80, 201-218.

18. Grégori, F. and Veyssière, P. (2002) Properties of <110]{111} slip in Al-rich g-TiAl deformed at room temperature. I Transmission electron microscope analysis of deformation debris, Phil. Mag. A 82, 553-566.

19. Shi, X., Saada, G. and Veyssière, P. (1995) The annihilation of superdislocations in Ni3Al during deformation at room temperature, Phil. Mag. Lett. 71, 1-9.

20. Essmann, U. and Mughrabi, H. (1979) Annihilation of dislocations during tensile and cyclic deformation and limits of dislocation densities, Phil. Mag. A 40, 731-756.

21. Veyssière, P. and Grégori, F. (2002) Properties of <110]{111} slip in Al-rich g-TiAl deformed at room temperature. II The formation and the role of strings of prismatic loops during deformation, Phil. Mag. A 82, 567-577.

22. Kroupa, F. (1966) The force between a dislocation dipole and a non parallel dislocation, Acta Met. 14, 60-61.

23. Veyssière, P. and Grégori, F. (2002) Properties of <110]{111} slip in Al-rich g-TiAl deformed at room temperature. III The role of loop strings in dislocation pinning and in dislocation patterning, Phil. Mag. A 82, 579-590.

ANANTHAKRISHNA G.

Materials Research Centre and Centre for Condensed
Matter Theory
Indian Institute of Science
Bangalore-560012
INDIA
garani@mrc.iisc.ernet.in

ASTA Mark

Dpt of Materials Science and Eng
Northwestern University
IL60091 EVANSTON
USA
m-asta@northwestern.edu

BELLON Pacal

Dpt of Materials Science and Eng.
302C MSEB, MC-246
1304 W. Green St.
IL 61801 URBANA
USA
bellon@ux1.cso.uiuc.edu

BLAVETTE Didier

Université de Rouen
Groupe de métallurgie Physique
76821 Mont St Aignan Cedex
FRANCE
didier.blavette@univ-rouen.fr

BRECHET Yves

LTPCM
Domaine Universitaire de Grenoble
BP 75
38042 ST Martin d'Heres
FRANCE
yves.brechet@ltpcm.inpg.fr

BULATOV Vasily

University of California
Lawrence Livermore National Laboratory
Livermore CA 94550
USA
bulatov1@llnl.gov

DEVINCRE Benoit

ONERA-LEM
29 ave. de la Division Leclerc
92322 Châtillon sous Bagneux Cedex
FRANCE
devincre@zig.onera.fr

EMBURY J. David

Materials Science and Engineering
Mac Master University
L85427 HAMILTON, Ontario
CANADA
emburyd@mcmail.cis.mcmaster.ca

508

ESTRIN Juri

Institut für Werkstoffkunde und Werkstofftechnick
Technische Universitat
D-38678 Clausthal-Zellerfeld
ALLEMAGNE
juri.estrin@tu-clausthal.de

GROMA Istvan

Dpt of General Physics
Lorand Eotvos University
Faculty of Science
H1518, Budapest POB 32
HUNGARY
groma@metal.elte.hu

HAATAJA Mikko

Princeton Materials Institute
Dpt Mechanical and Aerospace Engrg
Bowen Hall, 70 Prospect Ave,
Princeton, NJ 08544-5211
USA
mhaataja@princeton.edu

HAHNER Peter

Mechanical Performance Characterization
Institute for Energy
DG-Joint Research Centre
European Commission
NL-1755 ZG Petten
NETHERLANDS
hahner@jrc.nl

HUTCHINSON Christopher

ENSEEG-LTPCM
BP75
38042 Saint Martin d'Hères
FRANCE
chris.hutchinson@ltpcm.inpg.fr

HŸTCH Martin

CECM-CNRS
15 rue Georges Urbain
94407 Vitry sur Seine Cedex
FRANCE
martin.hytch@glvt-cnrs.fr

INDEN Gerhard

Max Planck Institut of Iron research
Max Planck Str. 1,
D-40237 DUSSELDORF
ALLEMAGNE
inden@mpie.de

JACQUES Pascal

UCL-PCIM
2 Place Ste Barbe
B1348 Louvain La Neuve
BELGIQUE
jacques@pcim.ucl.ac.be

KARMA Alain

Northeastern University
360 Huntington Avenue
MA 02115 BOSTON
USA
a.karma@neu.edu

KHACHATURYAN Armen

Dpt of Ceramic and Materials Engineering
Rutgers, The State University of New Jersey
NJ 08854-8065 PISCATAWAY
USA
khach@jove.rutgers.edu

KOSTORZ Gernot

ETH Zurich
Institut für Angewandte Physik
CH-8093 ZURICH
SUISSE
kostorz@iap.phys.ethz.ch

MUGHRABI Haël

Institut fur Werkstoffwissenschaften
Lehrstuhl 1, Martensstr.5
91058 Erlangen
GERMANY
mughrabi@ww.univ-erlangen.de

NAIMARK Oleg

Laboratory of Physical Foundation od Strength
Institute of Continuous Media Mechanics of the
Russian
Academy of Sciences
1 Acad. Korolev Str.
614013 Perm
RUSSIA
naimark@icmm.ru

OLEMSKOÏ Alexander

Physical Electronics Department
Sumy State University
2, Rimskii-Korsakov
Sumy 40007
UKRAINE
olemskoi@ssu.sumy.ua

PONTIKIS Vassilis

CNRS-CECEM
15, rue Georges Urbain
94407 VITRY SUR SEINE
FRANCE
Vassilis.Pontikis@glvt-cnrs.fr

RODNEY David

ENSPG-GPM2
38042 Saint Martin d'Hères
FRANCE
david.rodney@gpm2.inpg.fr

ROLLETT Tony

Carnegie Mellon University
Materials Science Department
5000 Forbes Avenue
Pittsburg, PA 15213
USA
rollett@andrew.cmu.edu

SIGLI Christophe

Péchiney CRV
725 rue Aristide Bergès
38341 VOREPPE
FRANCE
christophe.sigli@pechiney.com

SOISSON Frédéric

CEA Saclay
DEN/DMN/SRMP
91191 Gif sur Yvette
FRANCE
soisson@ortolan.cea.fr

TOKAR Vasily

Institute of magnetism
National Academy of Science
3142 KIEV
UKRAINE
tokar@im.imag.kiev.ua

VAN DER GIESSEN Erik

Materials science and Engineering
Dpt of applied physics Nyenborgh 4
49747 AG GRONINGEN
NETHERLANDS
giessen@phys.rug.nl

VAN SWYGENHOVEN Helena

Paul Scherrer Institute
Nanostrukturierte Materialen,
CH 5232, Villigen
SWITZERLAND
helena.vs@psi.ch

VEYSSIERE Patrick

ONERA-LEM
29 ave. de la Division Leclerc
92322 Châtillon sous Bagneux Cedex
FRANCE
patrickv@onera.fr

VOORHEES Peter

Materials Science and Eng. Dpt
mcCormick School of Eng. and applied Science
IL 60208-3108 EVARISTON
USA
p-voorhees@northwestern.edu

Partners/Sponsors

NATO Scientific Affairs Division

France

CEA : Direction de l'Energie Nucléaire, Direction de la Recherche Technologique
DGA
INPG
IRSID
ONERA

1-MONTH